D0161523

FLUID MECHANICS

FLUID MECHANICS

Irfan A. Khan

Youngstown State University

HOLT, RINEHART AND WINSTON

New York Chicago San Francisco Philadelphia
Montreal Toronto London Sydney
Tokyo Mexico City Rio De Janeiro Madrid

To my parents

Publisher Ted Buchholz
Acquisitions Editor John J. Beck
Senior Project Manager Chuck Wahrhaftig
Production Manager Paul Nardi
Design Supervisor Bob Kopelman

Library of Congress Cataloging-in-Publication Data

Khan, Irfan A.
Fluid mechanics.

Includes index.
1. Fluid mechanics. I. Title.
TA357.K48 1987 620.1′06 86-14250

ISBN 0-03-071473-7

Address correspondence to:
383 Madison Avenue
New York, N.Y. 10017

PRINTED IN THE UNITED STATES OF AMERICA
Published simultaneously in Canada

7 8 9 090 9 8 7 6 5 4 3 2 1

Holt, Rinehart and Winston
The Dryden Press
Saunders College Publishing

PREFACE

This textbook has been written for one and two term courses in Introductory Fluid Mechanics. The suggested prerequisites are introductory courses in rigid body statics, mathematics (through integral calculus), and the computer language, BASIC. Some exposure to thermodynamics may be helpful. Both English and SI units are used in examples and problems with an overall emphasis on SI units.

The primary purpose of this book is to introduce students to the basic concepts and laws of Fluid Mechanics without the confusion of excessive mathematical jargon. To achieve this objective, physical explanations of fluid phenomena have been included wherever possible in addition to mathematical formulations. For example, equations of Continuity, Momentum and Energy have been derived using physical as well as control volume equations. The author believes the two approaches compliment each other and help the student develop a better understanding of the subject. A generous number of solved examples are a part of each chapter to facilitate the understanding and application of the basic concepts and laws of fluid mechanics.

In some chapters, the author has departed from the conventional arrangement of subject matter found in most textbooks. The chapter on fluid kinematics, for instance, contains many definitions and topics conventionally introduced elsewhere. Concepts of stream function, potential flow, and flow net are included in that chapter because they are closely related to the concept of streamline. Introducing them elsewhere would break continuity of coverage. Similarly, lubrication of bearings is introduced with viscosity. The pressure and flow measuring devices such as Bourdon gauge, manometers, venturimeters, and weirs are introduced as soon as their underlying theory is covered, rather than lumping them together later on in the book as is done in many fluid mechanics textbooks. The author believes this format will be more useful to readers. A most thorough treatment of the Momentum, Euler, and Bernoulli equations has been presented. It is the author's hope that students will find these chapters challenging and interesting.

The problems at the end of each chapter begin with multiple choice solutions. This helps the student clear up any confusion regarding important concepts and fine points presented in the chapter. Thereafter, problems requiring worked out solutions are organized in order of increasing difficulty. The number of problems are kept to a reasonable limit while exposing the student to a variety of concepts. Care has been taken not to include many similar problems so as to avoid duplication.

Throughout the book, wherever necessary, comments have been added to remind students of the pitfalls in problem solving. These comments are based on many years of experience in the classroom with the most common mistakes made by students in solving fluid mechanics problems.

Computer programs written in BASIC have been included throughout the book to aid in problem solving. It is not the author's intention to replace the basic understanding of problem solving procedures by blindly following the computer program. Recently, an increasing demand has been made by the Accreditation Board for Engineering and Technology (ABET) to include computers in almost every phase of engineering. The inclusion of computer usage in this text meets such demands. Listings of all the programs for IBM, TRS 80, and Apple computers are also available on floppy disks from the author. A manual containing detailed solutions of all the problems in this book accompanies the text.

The author would like to thank the reviewers whose constructive suggestions have improved the final manuscript, and wishes to acknowledge, with great appreciation, the helpful recommendations provided by the following professors in the thermo-fluid sciences: Leonard B. Baldwin, Jr., University of Wyoming; Allen R. Barbin, Auburn University; Morris E. Childs, University of Washington; Chao-Lin Chiu, University of Pittsburgh; L. S. Fletcher, Texas A & M University; T. H. Okiishi, Iowa State University; James R. Pfafflin, United States Coast Guard Academy; Albin A. Szewczk, University of Notre Dame; and Chih Wu, United States Naval Academy. The author welcomes any constructive recommendations that may enlighten this book for future editions, as he believes that there is always room for improvement.

Finally, the author would like to thank his editor, John J. Beck, for his experienced and professional guidance as he prepared the manuscript, and his senior project manager, Chuck Wahrhaftig, who exchanged his typed offering for a beautiful book. The author is grateful to his administration who provided support and encouragement. Most of all, he thanks his family for their understanding and patience during the long hours that go into preparing a textbook.

I. A. K.
Youngstown, Ohio

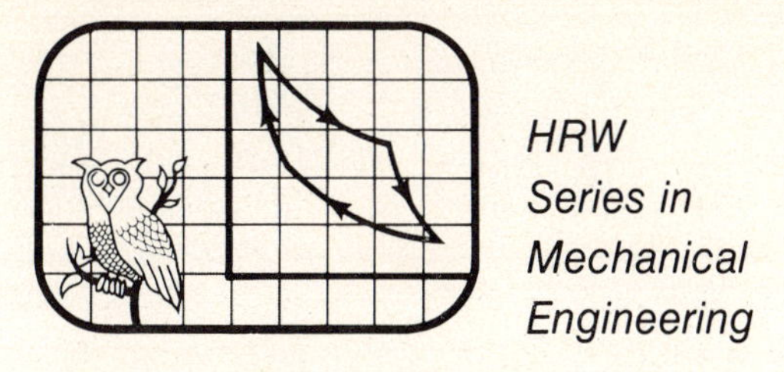

HRW
Series in
Mechanical
Engineering

L. S. Fletcher, Series Editor

A. W. Al-Khafaji and J. R. Tooley COMPUTERIZED NUMERICAL ANALYSIS
A. W. Al-Khafaji and J. R. Tooley NUMERICAL METHODS IN ENGINEERING PRACTICE
I. A. Khan FLUID MECHANICS
F. L. Stasa APPLIED FINITE ELEMENT ANALYSIS FOR ENGINEERS
B. J. Torby ADVANCED DYNAMICS FOR ENGINEERS

CONTENTS

FLUID MECHANICS

Chapter One

FLUID PROPERTIES

In this chapter, we will define a fluid, study its scope and application in engineering. The fundamental properties of a fluid such as density, viscosity, surface tension, etc. will be defined. The knowledge of these properties is essential to the understanding of the fluid behavior and the subject matter presented in the subsequent chapters.

1.1 INTRODUCTION

Fluid mechanics refers to the behavior of fluids when subjected to various forces. The term "fluid" is applied to certain substances, which, whether at rest, or under conditions of static equilibrium, cannot resist a shear force—no matter how small that force is. Thus, a fluid deforms continuously under the influence of a shear force—i.e., the fluid particles continuously change their position relative to one another when subjected to a shear force.

On the other hand, solids may resist a shear force when at rest and if they deform, they do not continue to deform indefinitely. The behavior of a solid and a fluid under the influence of a shear force is shown in Fig. 1.1. In the case of a solid, the deformation is small and θ, a measure of angular deformation, is not a continuous function of time; whereas in the case of a fluid, the deformation is large and θ is a continuous function of time. A more detailed discussion of the behavior of a fluid under the influence of a shear force will be presented in Section 1.5.

In everyday life, the distinction between a solid and a fluid is usually clear; however, some fluids may appear to be solids at first sight. For example, a lump of thick tar may look like a solid. When placed on the ground, it does not spread quickly as water does. However, it does start to deform as soon as it is placed on the ground although the deformation may not be readily discernible. Given sufficient time, perhaps a few days, it will spread just like any other fluid. As stated above, a solid although plastic and soft in appearance will resist a shear force when it is first applied and until it exceeds a certain minimum. Also, within the elastic limit, a solid will resume its original shape after the shear force is removed, but a fluid retains its new shape after removal of the shear force.

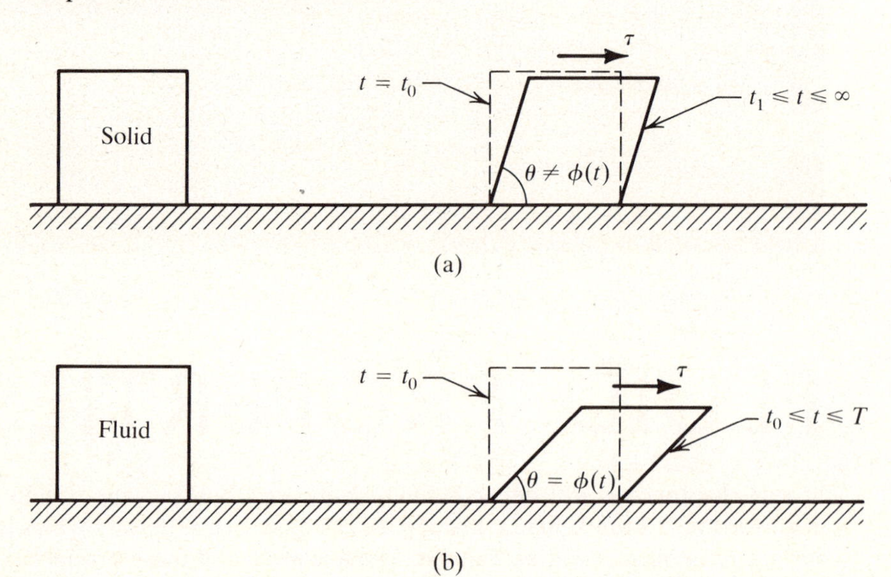

FIGURE 1.1 Effect of shear stress on a solid and a fluid.

Fluids may be divided into three classes: *liquids*, *vapors* and *gases*. Liquids are highly resistant to compression while vapors and gases are easily compressed. Also, liquids are not greatly affected by changes in temperature, while vapors and gases are.

Molecular Structure of Fluids

All substances in nature are made up of molecules separated by empty spaces. The differences between solids, liquids and gases stem from differences in their molecular structures. In solids, the molecules are much closer together than in a liquid; in a liquid the molecules are much closer together than in a gas.

The force of attraction between the molecles is inversely proportional to the square of the distance between them. Since the molecules of a solid are located close to one another, the forces of attraction between them are large and, therefore, they offer a great resistance to any external force which tries to move them relative to one another. If the external force is large enough to overcome the forces of attraction between them, the molecules may change positions but the strong forces between them still remain and they bring the solid back to its original shape after the external force is removed. If the external force is too large, some of the molecules may be permanently dislodged rendering them incapable of returning back to their original positions after the force is removed. In this case, the solid is said to have been deformed beyond its elastic limit.

In a liquid, the forces of attraction between molecules are only large enough to hold them together to give the liquid a definite shape. When an external force is applied to a liquid, its molecules get rearranged continuously until the force is removed. Also, the molecules do not come back to their original positions after the force is removed.

In a gas, the force of attraction between the molecules is negligible because of the large distance between molecules. Therefore, the molecules tend to travel freely away from each other. For this reason, when a gas is enclosed in a container, it will expand until it completely fills the container.

In a solid, individual molecules move only slightly and they do not move readily relative to one another. In a liquid, there is greater movement of the molecules. They move in curved wavy paths around each other. Because the molecules can move past one another, a liquid is not a rigid substance; however, the force of attraction between the molecules is large enough to keep the liquid in a definite volume.

Scope of Fluid Mechanics

The understanding and knowledge of the basic fluid properties and the laws of fluid mechanics are essential for the analysis and design of a system in which fluid is a working medium. Such systems are used in aircraft, rockets, automobiles, trains, ships, submarines, hydraulic structures (dams, bridges, culverts, etc.), hydraulic machines (water pumps, turbines), pneumatic machines (fans, blowers, compressors) and many others.

Aerodynamic considerations are extremely important for designing fuel-efficient automobiles, aircraft, ships, rockets and missiles which move through fluids. It is important to study the effect of wind on tall buildings so that they can be designed for minimum vibrations.

Lubrication is highly important for efficient performance and long life of machines containing friction parts.

The list of applications of fluid mechanics may be extended considerably leading to the conclusion that the subject has great practical value to an engineer.

Fluid as a Continuum

All fluids are made up of molecules which are separated from each other by spaces. The properties of a fluid cannot be defined in these spaces due to the nonexistence of mass. Therefore, serious difficulties may arise if a fluid is to be analyzed molecule by molecule. To overcome these difficulties, a fluid is regarded as a continuum—i.e., a hypothetically continuous substance.

A continuum assumption is basic to all branches of physics and is valid when studying a fluid under normal conditions. However, the model breaks down whenever the mean free path of the molecules approaches the smallest characteristic dimension of the problem. The mean free path of air molecules at standard temperature and pressure is approximately 6.3×10^{-5} mm. Under rare conditions in outer space, the assumption of a continuum may not be valid due to large distances between molecules.

Due to the continuum assumption, fluid properties such as density, temperature, velocity, etc., are considered to have definite value at each point in the fluid, and are defined with respect to the representative elementary volume (REV). Let us now develop the concept of property at a point using the continuum principle.

Let $\forall(x)$ denote the volume of a spatial domain centered at a point P whose position vector is x, and E be the amount of some extensive property of the fluid contained in $\forall$. The concentration ρ_E of E over $\forall$ at time t is defined by

$$\rho_E(x, t) = \frac{E}{\forall(x)} \tag{1.1}$$

Clearly, this is not the property at point P. To obtain the property at point P, $\forall(x)$ must be reduced to some value $V_{\min}$, which is exceedingly small but, nevertheless, large enough to contain a considerable number of molecules. If the ratio $E/\forall(x)$ is plotted against $\forall(x)$, (Fig. 1.2), notice that as $\forall(x)$ is reduced below a certain value, the ratio $E/\forall(x)$ becomes very noisy. This happens because $\forall(x)$ has become so small that the property E changes erratically within it. $\forall_{\min}$ is the smallest volume which can be used to define the density of property E within $\forall(x)$. $\forall_{\min}$ is called the *representative elementary volume* (REV). Mathematically, the density ρ_E of property E at point P is defined as:

$$\rho_E(x, t) = \lim_{\forall(x) \to \forall_{\min}} \frac{E}{\forall(x)} \tag{1.2}$$

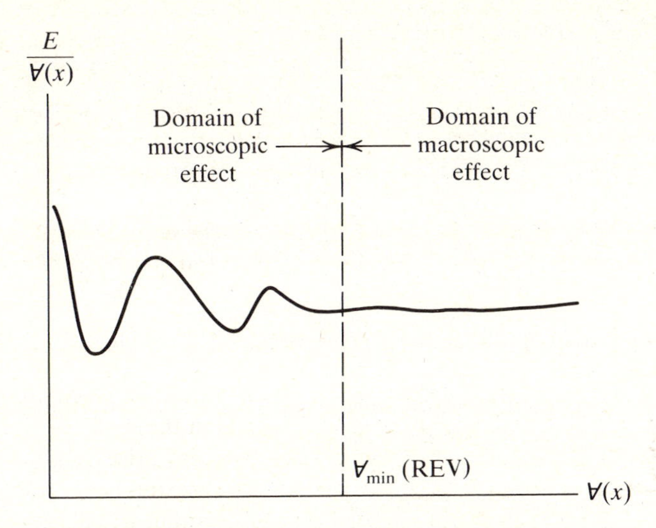

FIGURE 1.2 Property of a fluid as a function of its volume.

For all practical purposes, the mean density $\bar{\rho}_E$ of property E for a fluid of volume $\forall$ may be determined using the relationship:

$$\bar{\rho}_E = \frac{E}{\forall} \tag{1.3}$$

Thus, continuum fluid mechanics involves situations for which even point size control volumes contain enough fluid molecules to make property specification meaningful.

1.2 UNITS OF MEASUREMENT

It is extremely important that consistent units be used in all derivations and problems. The units are consistent when a unit force applied to a unit mass produces a unit acceleration.

At present, two systems of units are commonly used: The international system of units or Système International (SI), and the U.S. Customary System of Units (FPS). In the SI system (Table 1.1), length is specified in meters (m), time in seconds (s), mass in kilograms (kg) and force in Newtons (N). One N is the force required to produce an acceleration of 1 m/s^2 when applied to a mass of 1 kg. In the FPS system (Table 1.1), length is specified in feet (ft), time in seconds (s), mass in slugs and force in pounds (lb). One lb is

Table 1.1 SYSTEMS OF UNITS

System	Length	Time	Mass	Force
SI units	meter (m)	second (s)	kilogram (kg)	Newton (N)
FPS units	foot (ft)	second (s)	slug (slug)	pound (lb)

Table 1.2 PREFIXES USED IN SI SYSTEM

Multiple	SI Prefix	Abbreviation	Multiple	SI Prefix	Abbreviation
10^9	giga	G	10^{-3}	milli	m
10^6	mega	M	10^{-6}	micro	μ
10^3	kilo	k	10^{-9}	nano	n
10^{-2}	centi	c	10^{-12}	pico	p

the force required to produce an acceleration of 1 ft/s^2 when applied to a mass of 1 slug.

The SI system is used predominantly in this text since, eventually, it will become the worldwide standard for measurements.

In the SI system, prefixes are often used to describe a very small or a very large quantity. Some of the prefixes used are given in Table 1.2.

The FPS unit for temperature is the degree Fahrenheit (°F), while the SI unit for temperature is the degree Celsius (°C). The absolute temperature in the FPS system is measured in degrees Rankine (°R), while the absolute temperature in SI is measured in degrees Kelvin (K). Also:

$$\text{K} = 273^\circ + {}^\circ\text{C} \tag{1.4}$$

$$^\circ\text{R} = 450^\circ + {}^\circ\text{F} \tag{1.5}$$

Conversion of Units

Sometimes, it may be necessary to convert from one system to the other. The following equations provide a set of direct conversion factors for the basic quantities:

$$\frac{4.448 \text{ N}}{1 \text{ lb}} = 1 \tag{1.6}$$

$$\frac{0.3048 \text{ m}}{1 \text{ ft}} = 1 \tag{1.7}$$

$$\frac{14.594 \text{ kg}}{1 \text{ slug}} = 1 \tag{1.8}$$

Rules for Use of Units

1. A symbol is always written as a singular and never with a plural "s" to avoid confusion with the unit of second (s).
2. Symbols are always written in lower case letters except those which are named after an individual such as Newton (N). Also G for giga and M for mega are capitalized.
3. Quantities defined by several units are separated by a dot from each other to avoid confusion with the prefix notation. For example, Newton-second/meter^2 will be written as $\text{N}\cdot\text{s/m}^2$ or $\text{N}\cdot\text{s}\cdot\text{m}^{-2}$.

1.3 DENSITY, SPECIFIC WEIGHT, AND SPECIFIC GRAVITY

Some of the fundamental properties of a fluid such as density, specific weight, and specific gravity are now defined.

Density (ρ)

The density of a fluid is defined as its mass per unit volume. The density at a point (ρ_0) in a fluid may be determined using Eq. 1.2, or

$$\rho_0 = \lim_{\Delta \forall \to \Delta \forall_{min}} \frac{\Delta m}{\Delta \forall} \tag{1.9}$$

where Δm is the mass of an elemental volume $\Delta \forall$, and $\Delta \forall_{min}$ is the representative elementary volume.

Average or mean density (ρ) is normally used in practice which is simply obtained by dividing the given mass of a fluid by its volume, or

$$\rho = \frac{\Delta m}{\Delta \forall}$$

The density of a gas is calculated using the perfect gas law, which is

$$p = \rho R T \tag{1.10}$$

where p = absolute pressure; ρ = density; R = gas constant; and T = absolute temperature. The gas constant R is independent of temperature and pressure and may be determined from the relation

$$R = \frac{8312}{M} \text{ m} \cdot \text{N/kg} \cdot \text{K} \qquad \text{(SI units)} \tag{1.11}$$

$$= \frac{49{,}709}{M} \text{ ft} \cdot \text{lb/slug} \cdot {}^\circ\text{R} \qquad \text{(FPS units)} \tag{1.12}$$

where M = molecular weight of the gas. The values of R for various gases are given in Table A-3 of Appendix A.

The density of a fluid depends on its molecular structure—i.e., size and weight of the molecules and the mechanisms which bind the molecules together. The binding mechanisms usually vary with temperature and pressure; therefore, the density of a fluid may change with pressure and temperature. Generally, the density of a liquid varies considerably with temperature but varies only slightly with pressure, while the density of a gas varies considerably with both pressure and temperature. Water has a maximum density of 1000 kg/m^3 at 4°C. Below and above 4°C, the density of water decreases. Because of its special molecular structure, water is one of the few substances which expands when it freezes. The density of water at various temperatures is given in Tables A-1 and A-2 of Appendix A.

The reciprocal of density is called the *specific volume* ($\forall_s$).

Specific Weight (γ)

The specific weight of a fluid is its weight per unit volume, or

$$\gamma = \frac{\Delta W}{\Delta \forall} \tag{1.13}$$

where ΔW is the weight of the elemental volume $\Delta \forall$.

Weight is the product of mass and local acceleration due to gravity, or

$$\Delta W = \Delta m \cdot g \tag{1.14}$$

Substituting in Eq. (1.13), we obtain

$$\gamma = \frac{\Delta W}{\Delta \forall} = \frac{\Delta m \cdot g}{\Delta \forall} = \rho g$$

or

$$\gamma = \rho g \tag{1.15}$$

The specific weight of water at various temperatures is given in Tables A-1 and A-2 of Appendix A.

Specific Gravity (S)

The specific gravity of a fluid is a ratio of its density to that of water at 4°C, or

$$S = \frac{\rho_f}{\rho_w} \tag{1.16}$$

where

ρ_f = density of fluid at 4°C

ρ_w = density of water at 4°C

The specific gravity of water is 1. A specific gravity of less than one indicates a fluid lighter than water while a specific gravity of greater than one indicates a fluid heavier than water. The specific gravities of various fluids are given in Table A-4 of Appendix A.

Comments

1. Unless otherwise specified, γ will be used to denote the specific weight of water throughout this text. The specific weight of any other fluid will be written as γS, where S is the specific gravity of the fluid.
2. The term *specific gravity* is usually used for solids and liquids. It is seldom used for gases. Sometimes, the term *relative density* is used for a gas when its density is compared with that of air or hydrogen.

EXAMPLE 1.1

Determine the density of air at 20°C and 50 psia.

Solution Molecular weight of air at 20°C is equal to 29.0.
Hence:

$$R = \frac{8312}{29} = 287 \text{ m} \cdot \text{N/kg} \cdot \text{K}$$

From Eq. (1.10)

$$\rho = \frac{p}{RT}$$

where

$$p = 50 \times 144 = 7200 \text{ psfa}$$

$$= 7200 \times 47.88 = 344{,}700 \text{ N/m}^2$$

$$R = 287 \text{ m} \cdot \text{N/kg} \cdot \text{K}$$

$$T = 273 + 20 = 293 \text{ K}$$

Hence:

$$\rho = \frac{344{,}700}{(287)(293)} = 4.1 \text{ kg/m}^3$$

1.4 BULK MODULUS OF ELASTICITY

The *bulk modulus of elasticity* (β) of a fluid is a measure of its elasticity and is defined as the ratio of stress to the volumetric strain.

$$\beta = -\frac{dp}{\frac{d\forall}{\forall}} \tag{1.17}$$

where β = bulk modulus of elasticity; dp = change in pressure (final pressure − initial pressure); $d\forall$ = change in volume (final volume − initial volume); and $\forall$ = initial volume. The minus sign indicates that an increase in pressure results in a decrease in volume and vice versa. The bulk modulus of elasticity can also be expressed in terms of density or specific volme of a fluid. The density of a fluid of mass m and volume $\forall$ is given by:

$$\rho = \frac{m}{\forall}$$

or

$$\rho\forall = m = \text{constant}$$

Differentiating, we obtain:

$$\rho d\forall + \forall d\rho = 0$$

$$\frac{d\forall}{\forall} = -\frac{d\rho}{\rho}$$

Substituting in Eq. (1.17), we obtain:

$$\beta = \frac{dp}{\dfrac{d\rho}{\rho}} \tag{1.18}$$

The specific volume of a fluid of mass m and volume $\forall$ is given by:

$$\forall_s = \frac{\forall}{m}$$

or

$$\frac{\forall}{\forall_s} = m = \text{constant}$$

Differentiating, we get:

$$\forall_s d\forall - \forall d\forall_s = 0$$

or

$$\frac{d\forall}{\forall} = \frac{d\forall}{\forall_s}$$

Substituting in Eq. (1.17), we get

$$\beta = -\frac{dp}{\dfrac{d\forall_s}{\forall_s}} \tag{1.19}$$

The bulk modulus of elasticity varies with the variation in temperature and pressure. In the case of a liquid, an increase in pressure causes an increase in the bulk modulus of elasticity. In other words, as a liquid is compressed, its resistance to further compression increases. The β of water, for example, roughly doubles as the pressure is raised from 1 atm to 3500 atm. There is also a decrease of β with increase in temperature. The values of β for water at various temperatures and pressures are given in Table A-5 of Appendix A.

The bulk modulus of elasticity of a gas depends upon its mode of compression. For isothermal compression, the gas temperature remains constant. Therefore, the pressure density relationship is given by:

$$p = \rho C$$

or

$$p\forall_s = C$$

where C is a constant

Differentiating, we obtain

$$dp(\forall_s) + (p)d\forall_s = 0$$

$$\frac{d\forall_s}{\forall_s} = -\frac{dp}{p} \tag{1.20}$$

Substituting in Eq. (1.19), we obtain:

$$\beta = p \tag{1.21}$$

Hence, for isothermal compression of a gas, the bulk modulus of elasticity is equal to the absolute pressure to which the gas is subjected.

For isentropic compression, there is no heat transfer and there is no frictional loss, and the pressure density relationship is given by

$$\frac{p}{\rho^{\kappa}} = \text{constant} \tag{1.22}$$

where κ is the ratio of the specific heats of the gas, or $\kappa = c_p/c_v$, where, c_p = specific heat at constant pressure and c_v = specific heat at constant specific volume. The values of c_p and κ for various gases are given in Table A-3 of Appendix A. In terms of specific volume, Eq. (1.22) becomes

$$p\forall_s^{\kappa} = C$$

Differentiating:

$$p\cdot(\kappa\forall_s^{\kappa-1}\cdot d\forall_s) + \forall_s^{\kappa}\cdot dp = 0$$

or

$$\frac{d\forall_s}{\forall_s} = -\frac{dp}{\kappa p} \tag{1.23}$$

Substituting in Eq. (1.19), we obtain:

$$\beta = \kappa p \tag{1.24}$$

Hence, the bulk modulus of elasticity of a gas under isentropic compression is equal to the absolute pressure multiplied by the constant κ.

Equation (1.24) has a very important application in calculating the speed of sound in gases. A sound wave travels in a gas medium with negligible loss of heat and negligible friction; hence, the isentropic equation applies.

The sonic velocity c, in any fluid is given by

$$c = \sqrt{\beta/\rho} \tag{1.25}$$

Since β for an isentropic process is equal to κp and from the perfect gas law $\rho = p/RT$, by substituting in Eq. (1.25) we obtain

$$c = \sqrt{\kappa RT} \tag{1.26}$$

The bulk modulus of elasticity β_m of a mixture of N different fluids may be calculated using

$$\frac{1}{\beta_m} = \frac{\eta_1}{\beta_1} + \frac{\eta_2}{\beta_2} + \cdots + \frac{\eta_n}{\beta_n} + \cdots + \frac{\eta_N}{\beta_N} \tag{1.27}$$

where η_n is the proportion of the nth fluid by volume and β_n is the bulk modulus of elasticity of the nth fluid.

Also, the speed of sound c_m in a mixture of fluids may be calculated by

$$c_m = \sqrt{\beta_m/\rho_m} \tag{1.28}$$

where ρ_m is the density of the mixture of fluids. For N number of fluids

$$\rho_m = \eta_1\rho_1 + \eta_2\rho_2 + \cdots + \eta_n\rho_n + \cdots + \eta_N\rho_N \tag{1.29}$$

where η_n is as defined earlier and ρ_n is the density of the nth fluid.

In the case of a mixture of a liquid and a gas, Eq. (1.28) is valid at frequencies below gas bubble resonance.

Comment

For the solution of practical problems involving finite changes in pressure and volume, the differential form of Eq. (1.17) may be changed to a finite-difference form. Then:

$$\beta = -\frac{\Delta p}{(\Delta\forall/\forall)}$$

EXAMPLE 1.2

Find the percent reduction in volume of a cubic meter of water when the pressure is increased by 1 MN/m^2.

Solution From Eq. (1.17), for a finite increment

$$\Delta\forall = -\frac{\Delta p \cdot \forall}{\beta}$$

where

$$\Delta p = 10^6\ N/m^2$$

$$\forall = 1\ m^3$$

$$\beta = 2.2 \times 10^9\ N/m^2 \qquad \text{(from Table A-5)}$$

Hence:

$$\Delta\forall = -\frac{10^6 \times 1}{2.2 \times 10^9} = -4.54 \times 10^{-4}\ m^3$$

Therefore, the reduction in volume is equal to 4.54×10^{-4} m^3. This is equal to $4.54 \times 10^{-4}/1 \times 100$ or 0.045% change in volume. For this reason it is safe to consider water as incompressible for practical purposes.

EXAMPLE 1.3

A 10 m^3 volume of water is under 103.4 N/m^2 pressure initially. The pressure is increased tenfold, and the temperature increases from 20°C to 50°C. What is the change in volume?

Solution From Table A-5:

$$\beta \text{ at } 20°\text{C and } 103.4 \text{ N/m}^2 = 2.28 \times 10^9 \text{ N/m}^2$$

$$\beta \text{ at } 50°\text{C and } 1034 \text{ N/m}^2 = 2.94 \times 10^9 \text{ N/m}^2$$

Using linear interpolation we may assume that the average value of the bulk modulus of elasticity is

$$\beta_{av} = \frac{2.28 \times 10^9 + 2.94 \times 10^9}{2}$$

$$= 2.61 \times 10^9 \text{ N/m}^2$$

Also

$$\Delta p = 1034 - 103.4 = 930.6 \text{ N/m}^2$$

$$\forall = 10 \text{ m}^3$$

Now

$$\beta = -\frac{\Delta p}{\dfrac{\Delta \forall}{\forall}}$$

or

$$\Delta \forall = -\frac{\Delta p \forall}{\beta}$$

$$= -\frac{930.6 \times 10}{2.61 \times 10^9}$$

$$= -3.56 \times 10^{-6} \text{ m}^3$$

Hence the volume reduction is equal to 3.56×10^{-6} m^3. The negative sign indicates that the volume is reduced.

EXAMPLE 1.4

Determine the bulk modulus for isothermal air at a pressure of 100 psi at sea level. If the pressure is increased by 2 psi, what would be the change in volume of 1 ft^3 of air?

Solution For isothermal air, from Eq. (1.21),

$$\beta = p$$

Assuming the atmospheric pressure at sea level is 14.6 psi, the absolute pressure is equal to:

$$p = 100 + 14.6 = 114.6 \text{ psia}$$

Hence

$$\beta = 114.6 \text{ psi}$$

Also

$$\beta = -\frac{\Delta p}{\frac{\Delta \forall}{\forall}}$$

$$\Delta \forall = -\frac{\Delta p \cdot \forall}{\beta}$$

When the pressure is increased by 2 psi, the total pressure on the air becomes $100 + 2 = 102$ psi. Then

$$p = 102 + 14.6 = 116.6 \text{ psia}$$

$$\text{the new} \quad \beta = 116.6 \text{ psi}$$

$$\text{average} \quad \beta = (114.6 + 116.6)/2 = 115.6 \text{ psi}$$

Also

$$\Delta p = 102 - 100 = 2 \text{ psi}$$

Hence

$$\Delta \forall = -\frac{(2)(1)}{115.6} = -0.0173 \text{ ft}^3$$

Therefore, an increase in pressure by 2 psi, will result in a volume reduction of 0.0173 ft^3.

EXAMPLE 1.5

What is sonic velocity in air at 68°F?

Solution From Eq. (1.26)

$$c = \sqrt{\kappa RT}$$

$$\kappa = 1.4 \text{ (from Table A-3)}$$

$$R = 1716 \text{ ft} \cdot \text{lb/slug} \cdot {}^\circ\text{R (from Table A-3)}$$

$$T = 68 + 460 = 528 \ {}^\circ\text{R}$$

Therefore:

$$c = \sqrt{(1.4)(1716)(528)} = 1126 \text{ ft/s}.$$

1.5 VISCOSITY

As stated earlier, when a shearing force is applied to a fluid at rest it causes the fluid to deform. The deformation takes place in the form of one layer of fluid sliding over an adjacent one with a different velocity. All fluids in nature are known to offer some resistance to such a movement of one layer over another. This resistance is attributed to a property of the fluid called *viscosity*. Therefore, viscosity is the property of a fluid which is a measure of its internal resistance to relative motion between adjacent layers. Under particular conditions some fluids, like honey and glycerine (so-called "thick fluids"), offer greater resistance to flow than other fluids like water and gasoline (so-called "thin fluids"). It may, however, be pointed out that viscosity is not related to density. Gases, as well as liquids, have viscosity although the viscosity of a gas is less evident in everyday life.

The phenomenon of viscosity can be further investigated by the following experiment.

Consider two plates, one lying on top of another, as shown in Fig. 1.3. The lower plate is fixed while the upper plate can be moved by applying a tangential force. Let the area of the upper plate be A.

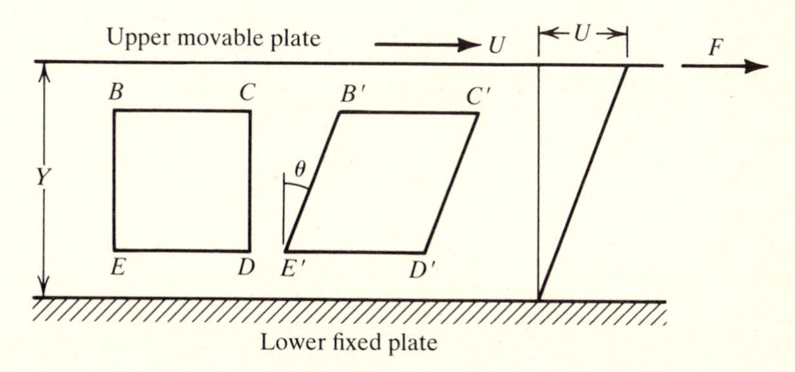

FIGURE 1.3 Deformation of fluid under the influence of a shear stress.

Let the space Y between the plates be filled with a fluid. Let a force F be applied at the upper plate to move it with a velocity U. As a result, all particles of the fluid, except those in contact with the stationary plate, move in the direction of motion of the upper plate. Various layers of the fluid move with various velocities causing a velocity gradient across the flow. For a sufficiently small portion of the fluid it may be assumed that fluid layers move in straight lines. An elemental volume, $BCDE$ (Fig. 1.3) of the fluid assumes the position $B'C'D'E'$ as it moves along. The angular deformation of the element is given by the angle θ.

Newton (1642–1727) observed the following relationships among the various variables:

1. Keeping the area A and the distance Y constant, the velocity U attained by the plate is directly proportional to the applied force F:

$$F \propto U \tag{1.30}$$

2. Keeping the velocity U and the distance Y constant, the force required to move the plate with a velocity U, is directly proportional to the area of the plate:

$$F \propto A \tag{1.31}$$

3. Keeping the velocity U and the area A constant, the force required is inversely proportional to the distance between the plates:

$$F \propto \frac{1}{Y} \tag{1.32}$$

Combining Eqs. (1.30), (1.31) and (1.32), we obtain:

$$F \propto \frac{UA}{Y}$$

or

$$\frac{F}{A} \propto \frac{U}{Y}$$

But F/A is the shear stress being exerted on the fluid through the plate. Denoting it by τ, we obtain:

$$\tau \propto \frac{U}{Y} \tag{1.33}$$

$$\tau = \mu \frac{U}{Y} \tag{1.34}$$

where μ is a constant and is a measure of the internal fluid resistance to relative motion between layers. This constant is called the *viscosity* of the fluid. Viscosity is also referred to as *absolute viscosity*, *dynamic viscosity* or *coefficient of viscosity*.

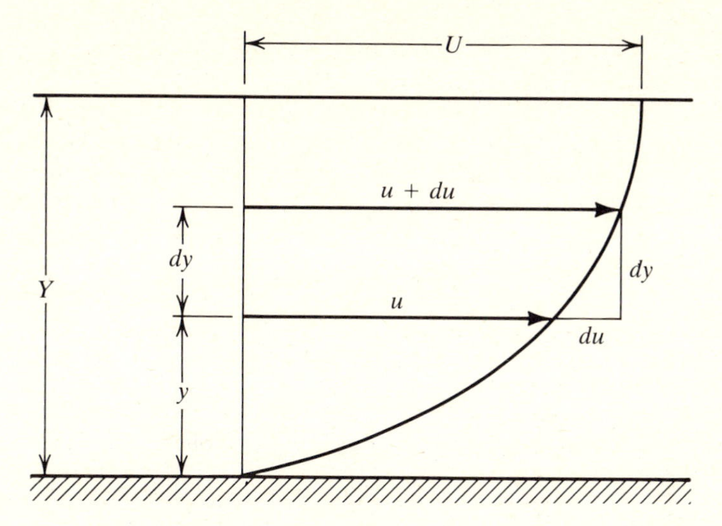

FIGURE 1.4 A fluid with a nonlinear velocity variation between the plates.

In Fig. 1.3 a linear velocity variation across the plate is shown. In actual practice, such a linear variation is uncommon. A nonlinear variation, such as shown in Fig. 1.4, is more common. Since the slope of a nonlinear velocity gradient varies from point to point across the plate, it is convenient to write Eq. (1.34) in the differential form. Hence:

$$\tau = \mu \frac{du}{dy} \tag{1.35}$$

Eq. (1.35) is known as *Newton's Law of Viscosity.*

From Eq. (1.35) it follows that the shear stress is zero where the velocity gradient du/dy is zero and the shear stress is infinite where the velocity gradient is infinite. It is physically impossible to have an infinite shear stress and, hence, an infinite velocity gradient. The velocity gradient can become infinite only if there is an abrupt change in the velocity. Therefore, to prevent shear stress from becoming infinite, the velocity variation across the flow must be continuous without any abrupt change. Hence, a fluid layer near a stationary boundary should not move since such a movement would constitute an abrupt change in the velocity gradient. Thus, a condition of "no slip" must always be satisfied at a stationary boundary.

The Causes of Viscosity

There are two major causes of viscosity in fluids: molecular attraction and transfer of molecular momentum. As stated earlier, the molecules of a liquid attract each other and this mutual attraction tends to obstruct the sliding of one layer over another giving rise to viscous forces. *Molecular attraction is the major cause of viscosity in liquids.* Since the molecular attraction decreases with the increase in temperature, the viscosity of a liquid also decreases with

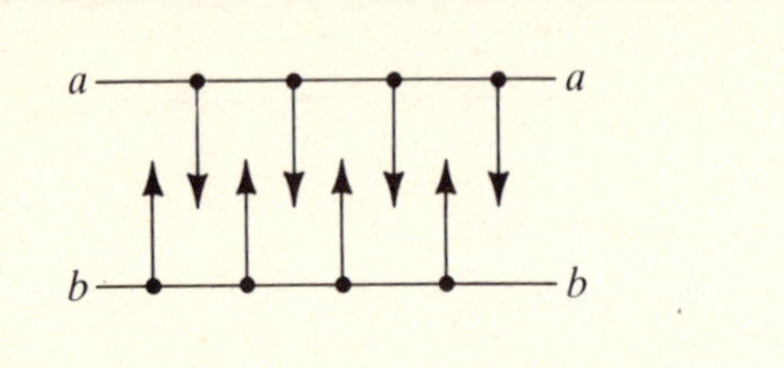

FIGURE 1.5 Transfer of momentum between two fluid layers.

an increase in temperature. Within normal range, pressure has no effect on molecular attraction, and, hence, on viscosity.

Although molecular attraction in gases is negligible, gases do possess viscosity. *Transfer of molecular momentum is the major cause of viscosity in gases.* The phenomenon of transfer of molecular momentum is now explained. Suppose that in a straight and parallel flow of a gas, a layer *aa* is moving faster than another layer *bb* (Fig. 1.5). Since the molecules of a gas move freely, some faster-moving molecules from the layer *aa* enter into the layer *bb* and collide with its slower-moving molecules. This increases the momentum of the layer *bb* resulting in its acceleration. Similarly, slower-moving molecules of layer *bb* when entering into the layer *aa*, decrease the momentum of the layer *aa*, and cause it to decelerate. This acceleration and deceleration give rise to viscous forces in gases. Since the molecular movement increases with the increase in temperature, the viscosity of a gas increases with the increase in temperature.

Although the phenomenon of transfer of molecular momentum does occur in liquids, its effect on viscosity is negligible as compared to that of molecular attraction.

Non-Newtonian Fluids

For most fluids, viscosity is independent of the velocity gradient; hence, the relationship between τ and du/dy is linear with zero intercept and slope equal to μ. Such a fluid is called *a Newtonian fluid* (Fig. 1.6). If the viscosity of a fluid is not independent of the velocity gradient the relationship between τ and du/dy is nonlinear and the fluid is called *non-Newtonian.* A non-Newtonian fluid is a *pseudo-plastic* or a *dilatant* depending on the manner in which τ changes with du/dy (Fig. 1.6). Gelatine, clay, milk, blood and liquid cement are examples of pseudo-plastics. Aqueous solutions of sugar and rice starch are examples of dilatants.

For some non-Newtonian fluids, the viscosity changes with time for the duration of a constant shearing stress. Liquids which exhibit increase in viscosity for the duration of stress are called *rheopectic*: those which exhibit decrease in viscosity for the duration of stress are termed *thixotropic.*

A hypothetical fluid of zero viscosity is termed *an ideal fluid.* Such a fluid may be used for mathematical simplicity to study fluid flow in which viscosity effects are small and may be neglected.

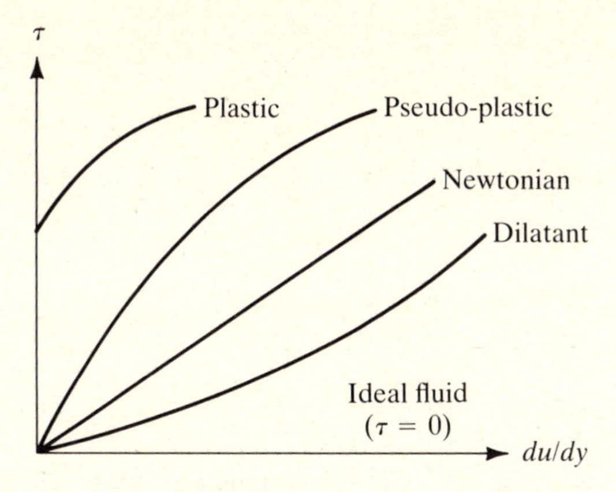

FIGURE 1.6 Relationship between τ and du/dy for various types of fluids and a plastic.

Units of Viscosity

The units of viscosity may be determined using Eq. (1.34), from which

$$\mu = \frac{\tau Y}{U}$$

Using SI units:

$$\mu = \frac{(\mathrm{N/m^2})(\mathrm{m})}{(\mathrm{m/s})}$$

$$= \frac{\mathrm{N \cdot s}}{\mathrm{m^2}}$$

Hence the SI unit of viscosity is $\mathrm{N \cdot s/m^2}$. The corresponding unit in the FPS system is $\mathrm{lb \cdot s/ft^2}$. Another common unit of viscosity is the c.g.s. unit, called *the poise* (P). It is equal to one-tenth of the SI unit.

Kinematic Viscosity

The kinematic viscosity (ν) of a fluid is the ratio of its viscosity to its density:

$$\nu = \frac{\mu}{\rho} \tag{1.36}$$

The SI unit of kinematic viscosity is $\mathrm{m^2/s}$, while the FPS unit is $\mathrm{ft^2/s}$. The c.g.s. unit is $\mathrm{cm^2/s}$ and is called a *stoke* (st).

EXAMPLE 1.6

A piston 50 mm in diameter and 20 cm in length moves coaxially in a cylinder 52 mm in diameter. The space between the piston and the cylinder is filled with oil of viscosity 0.09 $\mathrm{N \cdot s/m^2}$. Assuming a linear variation in velocity, determine the force required to move the piston at 1 m/s.

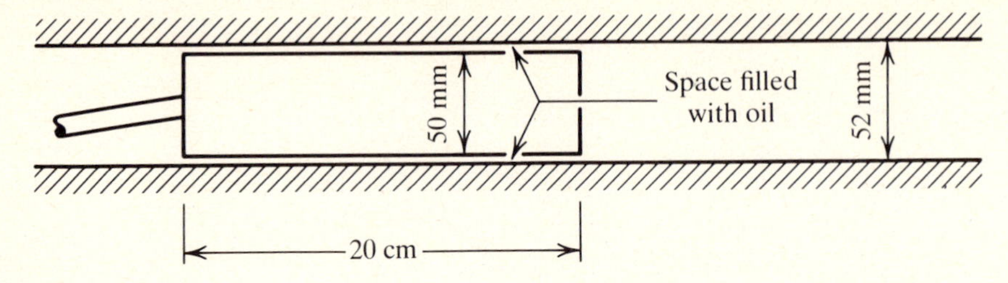

FIGURE 1.7 Example 1.6.

Solution In this problem, the walls of the cylinder behave like a stationary plate, while the surface of the piston behaves like a moving plate (Fig. 1.7). From Eq. (1.34)

$$\tau = \mu \frac{U}{Y}$$

where

$$\mu = 0.09 \frac{\mathrm{N \cdot s}}{\mathrm{m}^2} \qquad \text{(given)}$$

$$U = 1 \text{ m/s} \qquad \text{(given)}$$

$$Y = 1 \text{ mm} = 0.001 \text{ m} \qquad \text{(clearance between the piston and the cylinder)}$$

Hence

$$\tau = \frac{(0.09)(1)}{0.001} = 90 \text{ N/m}^2$$

This shear stress acts on the surface area of the piston which is in contact with the oil.

$$A = \pi \mathrm{DL} = (\pi)(0.05)(0.2) = 0.01\pi \text{ m}^2$$

Hence

$$F = \tau A = (90)(0.01)(\pi) = 2.83 \text{ N}$$

EXAMPLE 1.7

Water at 20°C flows between two walls. The velocity distribution is given by $u = 10(0.01y - y^2)$ m/s, where y is measured in meters from the lower wall. Determine (a) the distance between the walls, (b) the shear stress at the wall, (c) the shear stress at a distance of 20 μm from the wall, (d) the location of zero shear stress, and (e) the location of maximum velocity.

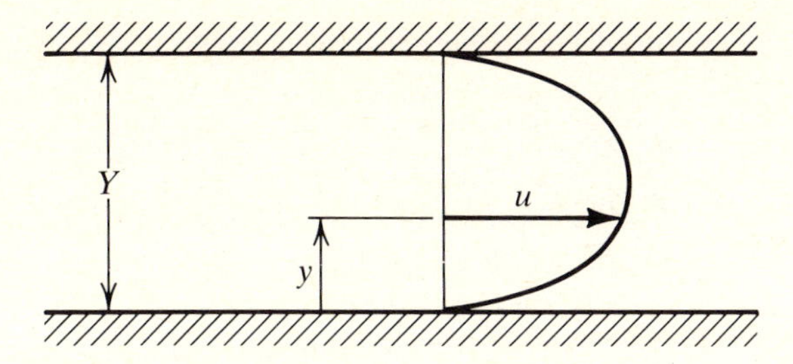

FIGURE 1.8 Example 1.7.

Solution The walls are stationary; hence, the molecules of water attached to the walls do not move and the velocity distribution would be as shown in Fig. 1.8. The velocity distribution is given by

$$u = 10(0.01y - y^2)$$

(a) Two boundary conditions must be satisfied. They are

$$\text{when } y = 0;\ u = 0 \qquad \text{(condition at the lower plate)}$$

and

$$\text{when } y = Y;\ u = 0 \qquad \text{(condition at the upper plate)}$$

Using the second boundary condition, we obtain:

$$0 = 10(0.01Y - Y^2)$$

From which

$$Y = 0.01 \text{ m} = 1 \text{ cm}$$

Hence, the distance between the walls is 1 cm.

(b) The shear stress at the wall is calculated using Newton's Law of Viscosity—i.e.:

$$\tau = \mu \frac{du}{dy}$$

$$\mu = 1.005 \times 10^{-3} \quad \text{N} \cdot \text{s/m}^2 \qquad \text{(viscosity of water at 20°C from Table A-1)}$$

Differentiating the given equation with respect to y, we obtain

$$\frac{du}{dy} = 10(0.01 - 2y)$$

Hence

$$\tau = (1.005 \times 10^{-3}) \times 10(0.01 - 2y) \tag{1}$$

This equation can be used to determine the shear stress at any distance y from the lower wall. At the wall, $y = 0$, hence:

$$\tau = (1.005 \times 10^{-3}) \times 10(0.01) = 1.005 \times 10^{-4}\ \mathrm{N/m^2}$$

Therefore, the shear stress at the wall is equal to 1.005×10^{-4} N/m^2.

(c) Shear stress at a distance of 20 μm from the wall may be found by substituting $y = 20 \times 10^{-6}$ m in Eq. (1). Hence:

$$\tau = (1.005 \times 10^{-3}) \times 10(0.01 - 2 \times 20 \times 10^{-6})$$
$$= 9.98 \times 10^{-5}\ \mathrm{N/m^2}$$

(d) The distance y at which shear stress is zero, may be obtained by letting $\tau = 0$ in Eq. (1) and then solving for y—i.e.:

$$0 = (1.005 \times 10^{-3}) \times 10(0.01 - 2y)$$
$$0.01 - 2y = 0$$
$$y = 0.005\ \mathrm{m}$$

(e) The location of maximum velocity may be found by letting du/dy equal zero and then solving for y. Hence:

$$10(0.01 - 2y) = 0$$

or

$$y = 0.005\ \mathrm{m}$$

The above calculations indicate that the zero shear stress and the maximum velocity occur at the same location which is halfway between the plates. This also is in conformity with the fact that, in a flowing fluid, the velocity is maximum where shear stress is zero.

Viscous Resistance of Oiled Bearings

A shaft revolving in an oiled bearing is separated from the bearing by a very thin film of oil (Fig. 1.9). The oil offers resistance due to its viscosity and some power is lost in overcoming this resistance. Assuming a linear distribution of the velocity gradient within the oil, the shear stress may be approximated by Eq. (1.34). That is, $\tau = \mu U/t$, where U is the tangential velocity of the shaft and t is the thickness of the oil film. The force exerted on the shaft due to the oil resistance is then given by

$$F = 2\pi R_1 L\tau$$

where R_1 and L are as shown in the figure. Substituting for τ we obtain

$$F = 2\pi R_1 L\mu U/t \qquad (1.37)$$

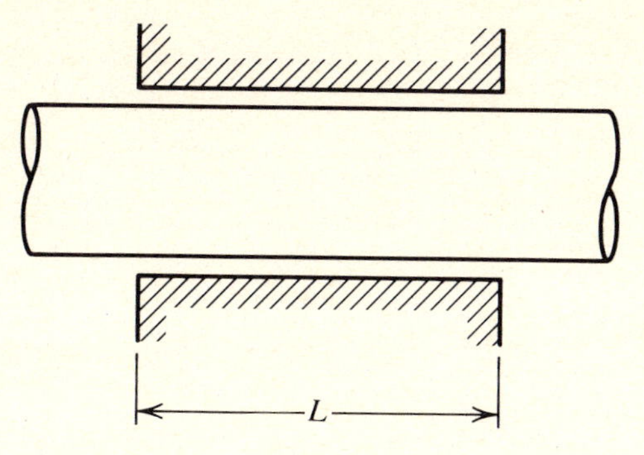

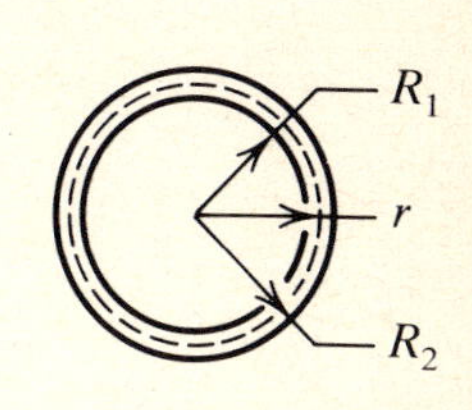

FIGURE 1.9 A shaft revolving in an oiled bearing.

The torque T required to rotate the shaft is given by

$$T = FR_1$$

In the derivation described above, it was assumed that the radial velocity of the fluid varies linearly across the space between the shaft and the bearing. This is not strictly true. A more precise expression for the force may be developed as follows:

We note here that the torque applied to the shaft is transmitted to the stationary bearing through concentric layers of the oil. At any distance r from the center, where $R_1 \leqslant r \leqslant R_2$ we have

$$F = 2\pi r L \tau$$

From Newton's Law of Viscosity:

$$\tau = \mu \frac{du}{dy}$$

If y is measured from the shaft, then

$$y = R_2 - r; \frac{dy}{dr} = -1 \quad \text{or} \quad dy = -dr.$$

Hence:

$$\tau = -\mu \frac{du}{dr}$$

and

$$F = -2\pi r L \mu \frac{du}{dr}$$

$$F \frac{dr}{r} = -2\pi L \mu du$$

The total force on the shaft due to oil can be obtained by integrating the above equation between suitable limits. If the tangential velocity of the shaft is U at $r = R_1$, then

$$\int_{R_1}^{r} F\frac{dr}{r} = -2\pi L\mu \int_{U}^{u} du$$

or

$$F \ln\left(\frac{r}{R_1}\right) = -2\pi L\mu(u - U)$$

At the bearing, $u = 0$ and $r = R_2$. Hence:

$$F = \frac{2\pi L\mu U}{\ln(R_2/R_1)} \tag{1.38}$$

The above expression gives the force which must be applied on the shaft to overcome the oil resistance.

EXAMPLE 1.8

A shaft 10 cm in diameter runs in a bearing of length 20 cm. The two surfaces are separated by a film of oil 0.002 cm in thickness. The oil is SAE 30. Determine the torque necessary to rotate the shaft at 30 rpm assuming (a) linear distribution of velocity, and (b) nonlinear distribution of velocity.

Solution From Table A-4, for SAE 30 oil $\mu = 0.44\ \text{N}\cdot\text{s/m}^2$. Also:

$$R_1 = 0.05\ \text{m}$$

$$R_2 = 0.05 + 0.00002 = 0.05002\ \text{m}$$

$$L = 20\ \text{cm} = 0.2\ \text{m}$$

$$N = 30\ \text{rpm}$$

$$t = 0.002\ \text{cm} = 0.00002\ \text{m}$$

$$U = \pi \times 0.1 \times 30/60 = 0.157\ \text{m/s}$$

(a) From Eq. (1.37):

$$F = 2\pi R_1 L\mu \frac{U}{t}$$

$$= \frac{2\pi \times 0.05 \times 0.2 \times 0.44 \times 0.157}{0.00002}$$

$$= 217\ \text{N}$$

$$T = FR_1 = 217 \times 0.05 = 10.85\ \text{m}\cdot\text{N}$$

Hence the torque required $= 10.85\ \text{m}\cdot\text{N}$

(b) Assuming nonlinear velocity distribution, from Eq. (1.38)

$$F = \frac{2\pi L\mu U}{\ln(R_2/R_1)}$$

$$= \frac{2\pi \times 0.2 \times 0.44 \times 0.157}{\ln(0.05002/0.05)}$$

$$= 217\ \text{N}$$

$$T = FR_1 = 217 \times 0.05$$

$$= 10.85\ \text{m}\cdot\text{N}$$

Notice that for this problem the answers assuming linear or nonlinear velocity distribution are identical.

Viscous Resistance of Collar Bearings

A collar bearing is shown in Fig. 1.10. The viscous resistance of a collar bearing can be obtained by assuming the face of the collar to be separated from the bearing surface by a thin film of oil of uniform thickness. Consider a thin ring of the bearing surface at a distance r and thickness dr, and let v be the tangential velocity at this radius.

From Newton's Law of Viscosity

$$\tau = \mu \frac{v}{t}$$

where t = thickness of the oil film.

If ω = speed of the shaft in radians per second,

$$v = r\omega$$

and

$$\tau = \mu \frac{r\omega}{t} \tag{1.39}$$

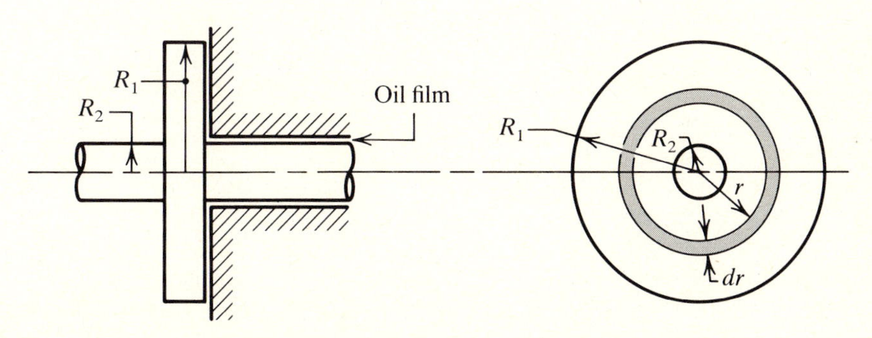

FIGURE 1.10 A shaft revolving in an oiled collar bearing.

The viscous force on a thin ring = δF

$$\delta F = \tau 2\pi r \cdot dr$$

Substituting from Eq. (1.39).

$$\delta F = \frac{2\pi\mu\omega r^2\,dr}{t}$$

The moment of tangential force on the ring = δFr

$$\delta Fr = \frac{2\pi\mu\omega}{t} r^3\,dr$$

The total torque = T

$$T = \frac{2\pi\mu\omega}{t}\int_{R_2}^{R_1} r^3\,dr$$

or

$$T = \frac{\pi\mu\omega}{2t}(R_1^4 - R_2^4) \tag{1.40}$$

Also $\omega = 2\pi N/60$, where N = shaft rpm. Hence:

$$T = \frac{\pi^2\mu N}{60t}(R_1^4 - R_2^4) \tag{1.41}$$

The power absorbed in viscosity

$$= \frac{T \times 2\pi N}{60} \text{ watts in the SI system} \tag{1.42}$$

and

$$= \frac{T \times 2\pi N}{550 \times 60} \text{ HP in the FPS system.} \tag{1.43}$$

EXAMPLE 1.9

The thrust of a shaft is taken by a collar bearing provided with a forced lubrication system which maintains a film of oil of uniform thickness between the surface of the collar and the surface of the bearing. The external and the internal diameters of the collar are 15 cm and 10 cm, respectively. If the thickness of the film of oil which separates the surfaces is 0.025 cm and the viscosity of oil is 0.44 $N \cdot s/m^2$, determine how many watts of energy will be lost in overcoming friction when the shaft rotates at 300 rpm.

Solution The power lost in overcoming friction can be calculated using Eq. (1.42). First calculate the torque using Eq. (1.41).

$$T = \frac{\pi^2 \mu N}{60t}(R_1^4 - R_2^4)$$

Here:

$$\mu = 0.44\ \text{N}\cdot\text{s/m}^2$$
$$N = 300\ \text{rpm}$$
$$t = 0.025\ \text{cm} = 0.00025\ \text{m}$$
$$R_1 = 15\ \text{cm} = 0.15\ \text{m}$$
$$R_2 = 10\ \text{cm} = 0.1\ \text{m}$$

Therefore:

$$T = \frac{\pi^2 \times 0.44 \times 300}{60 \times 0.00025}(0.15^4 - 0.1^4)$$
$$= 35.28\ \text{N}\cdot\text{m}$$

Power lost $$= \frac{T \times 2\pi N}{60}\ \text{watts}$$
$$= \frac{35.28 \times 2\pi \times 300}{60} = 1108\ \text{watts}$$

1.6 SURFACE TENSION AND CAPILLARITY

It has been observed that the surface of contact between two immiscible fluids, like water and air, exhibits the characteristics of a stretched skin. The reason for this stretching effect is now explained.

Consider a vessel containing water as shown in Fig. 1.11. The surface of the water is in contact with the air. A molecule A of water located beneath the surface is acted upon by equal forces in all directions. But a molecule B at the

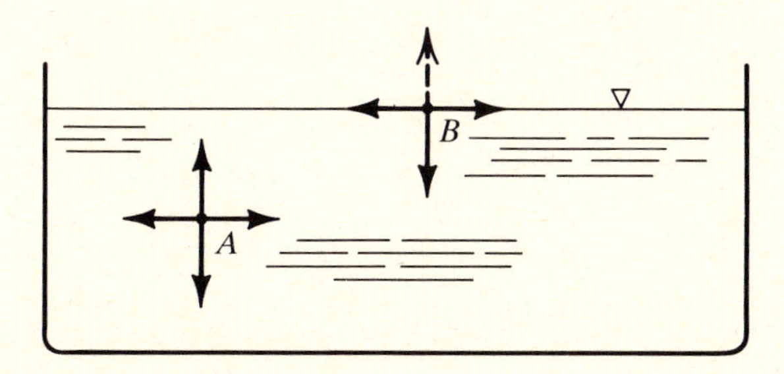

FIGURE 1.11 A vessel containing water.

surface is acted upon by unbalanced forces which have the tendency to drown the molecule. This causes the liquid surface to seek a minimum possible area by developing a force at the surface whose effect is to pull the surface molecules upwards. This force is called *the surface tension.* Normally, bodies having specific gravity greater than one must sink in water; but sometimes it is possible for the surface tension forces to keep such objects afloat. This can be shown by carefully dropping a paper clip on a calm water surface; it will float. Surface tension (σ) is measured in units of force per unit length. The magnitude of the surface tension force depends on the type of liquid and the substance it is in contact with. For example, surface tension for a water-air surface is 0.073 N/m at room temperature. Surface tension decreases with increase in temperature. The values of σ for water at various temperatures are given in Tables A-1 and A-2. The values of σ for various liquids are given in Table A-4.

Small amounts of electrolytes added to the water increase its surface tension. On the other hand, organic substances like soaps, alcohol, or acids decrease the surface tension.

EXAMPLE 1.10

Find the excess pressure inside a water droplet of 1 cm in diameter at 20°C.

Solution The upper half of a water droplet with pressure and surface tension forces is shown in Fig. 1.12. Let

$$\Delta p = \text{excess pressure}$$

$$d = \text{diameter of the bubble}$$

$$\sigma = \text{surface tension force}$$

The force due to the excess pressure

$$= \Delta p \cdot \frac{\pi d^2}{4}$$

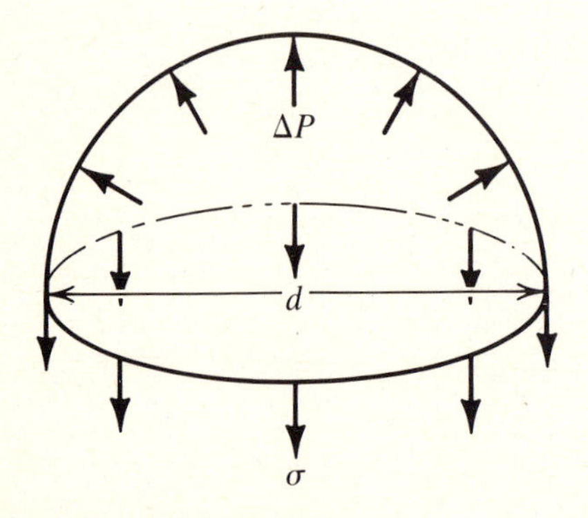

FIGURE 1.12 Example 1.10.

The force due to the surface tension

$$= \sigma \cdot \pi d$$

Equating the two forces, we obtain

$$\Delta p \cdot \frac{\pi d^2}{4} = \sigma \pi d$$

or

$$\Delta p = \frac{4\sigma}{d}$$

In this problem:

$$d = 0.01 \text{ m}$$

$$\sigma = 0.073 \text{ N/m} \qquad \text{(from Table A-1)}$$

Hence:

$$\Delta p = \frac{4 \times 0.073}{0.01} = 29.2 \text{ N}$$

Capillarity

If a glass tube of small diameter is lowered into a vessel full of water, it is seen that the water rises in the tube. On the other hand, if the same tube is lowered into a vessel full of mercury, the mercury level drops inside the tube (see Fig. 1.13). This rise or drop of a liquid in a tube of a solid is known as *capillarity*. This is caused by the combined action of surface tension and adhesion. Adhesion is an important contributor to capillarity. Adhesion between fluids and solids is expressed by the contact angle, θ, at the edge of the contacting surfaces. Those fluids which tend to repel solids have a contact angle larger

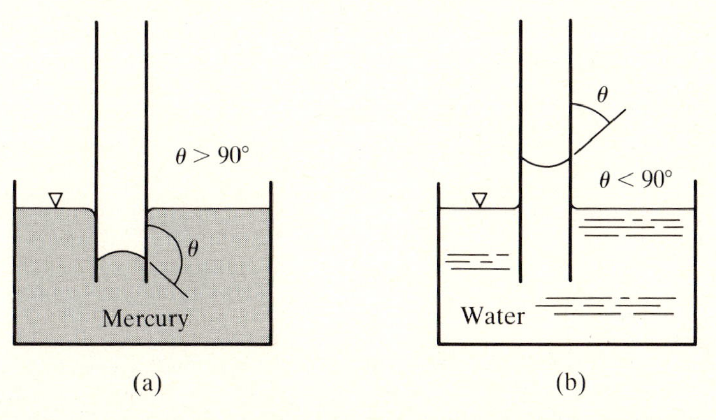

FIGURE 1.13 Capillarity in a small tube immersed in mercury and water.

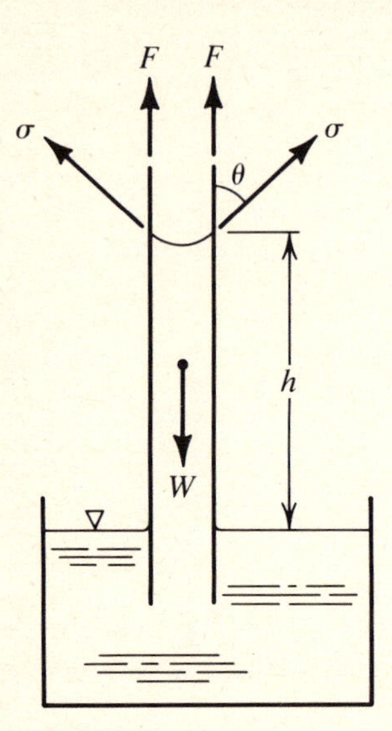

FIGURE 1.14 Forces in capillarity.

than 90 degrees and, therefore, have a tendency to drop out of the tube. On the other hand, those fluids which tend to cling to solids have a contact angle smaller than 90 degrees and, therefore, have a tendency to rise in the tube. The height to which a fluid such as water would rise in a tube can be calculated as described below.

Let a glass tube of diameter d be lowered into water as shown in Fig. 1.14. Let us assume that water rises to a height h in the tube. The surface tension (σ) acts upwards at an angle θ as shown in the diagram. Equilibrium is attained when enough water has risen in the tube so that its weight (W) becomes equal to the surface tension force (F). The vertical component of surface tension is equal to $\sigma \cos \theta$ and it acts along the circumference of the tube. Hence the total force F acting upwards is given by:

$$F = \sigma \cos(\theta) \pi d$$

The weight of the water in the tube is given by

$$W = \gamma \frac{\pi d^2}{4} h$$

At equilibrium:

$$W = F$$

$$\gamma \frac{\pi d^2}{4} h = \sigma \cos(\theta) \pi d$$

$$h = \frac{4\sigma \cos(\theta)}{\gamma d} \tag{1.44}$$

The angle θ for most liquid-solid interfaces is small; therefore, $\cos\theta \simeq 1$. Then:

$$h = \frac{4\sigma}{\gamma d} \tag{1.45}$$

For liquids other than water, Eq. (1.45) may be written as

$$h = \frac{4\sigma}{\gamma S d} \tag{1.46}$$

where S is the specific gravity of the liquid.

EXAMPLE 1.11

A glass tube, 2 mm in diameter, is lowered into a vessel full of alcohol at 20°C. Determine the height to which the alcohol will rise in the tube.

Solution From Eq. (1.46)

$$h = \frac{4\sigma}{\gamma S d}$$

$\sigma = 0.022$ N/m (from Table A-4)

$\gamma = 9789$ N/m^3 (from Table A-1)

$S = 0.79$ (from Table A-4)

$d = 2$ mm $= 0.002$ m

Hence

$$h = \frac{(4)(0.022)}{(9789)(0.79)(0.002)}$$

$$= 0.0057 \text{ m}$$

$$= 5.7 \text{ mm}$$

Therefore, the alcohol will rise to a height of 5.7 mm in the glass tube.

1.7 VAPOR PRESSURE

The molecules of a liquid are in constant motion. Some of these molecules, by virtue of their momentum, escape from the liquid and enter space as vapors. These vapors are in the form of gas and exert a partial pressure in space just like any other gas. This partial pressure due to the vapors is known as *the vapor pressure*. At a given time, not only are vapors escaping from the liquid to space; some are returning from space to the liquid. If, at any stage, the rate at which vapors are leaving the liquid becomes equal to the rate at which they are returning to the liquid, an equilibrium is reached. The vapor pressure at this stage is called the *saturation vapor pressure* (C_v). At this stage, the space above the liquid is fully saturated at that specific temperature and cannot accommodate more vapor. Since the movement of the molecules and, hence, their ability to escape from the liquid increases with an increase in temperature, so does the vapor pressure. The values of saturation vapor pressure for water at various temperatures are listed in Tables A-1 and A-2. The values of saturation vapor pressure for other liquids are given in Table A-4.

If, at any time, the pressure above the liquid surface becomes less than the saturation vapor pressure, the equilibrium is disturbed and a vigorous transfer of vapors from the liquid to space occurs. This phenomenon is known as *boiling*. For pure water, the boiling occurs at about 100°C at sea level because at this temperature the saturation vapor pressure of water is equal to the sea level atmospheric pressure. Boiling can occur at a lower temperature if the pressure above the liquid is reduced below sea level atmospheric pressure. For example, the saturation vapor pressure of water at 68°F is 0.34 lbs/in^2. Therefore, water can be made to boil at 68°F if the pressure on its surface is reduced to 0.34 lbs/in^2. Similarly, the boiling temperature can be raised by increasing the pressure on the surface of the liquid. This principle is used in pressure cookers where a tight lid and a weight controlled valve are used to increase the pressure on the water surface so that it boils at a higher temperature. This helps the food to cook faster.

Similar to boiling, the dissolved gases in a liquid may come out in the form of bubbles if the pressure of the liquid is sufficiently lowered. These bubbles, if subsequently transferred to a region of high pressure, may collapse producing noise and high impact forces. This phenomenon, known as *cavitation*, may occur in pumps and turbines where the pressure may decrease below the saturation vapor pressure on the suction side. A more detailed discusssion on cavitation is presented in Chapter 12.

PROBLEMS

p.1.1 Select the correct statement:

a. A fluid cannot resist a shear force when at rest or in motion.
b. A fluid can resist a small amount of shear stress when at rest.
c. A fluid cannot resist a shear force when at rest, but can develop shear stress when in motion.
d. A fluid can resist a shear force when at rest, but cannot resist a shear force when in motion.
e. None of these statements.

p.1.2 Select the correct statement:

a. The molecules of a solid are spaced closer together as compared to those of liquids and gases.
b. The molecules of a liquid are spaced farther apart as compared to those of a gas.

c. The molecules of gas have great attraction for each other.
d. The molecules of a solid can be rearranged easily.
e. None of these statements.

p.1.3 Using the continuum assumption the density ρ_E of property E at a point P in a fluid is defined as

a. $\rho_E = \lim_{V_x \to \text{REV}} \frac{E}{V_x}$.

b. $\rho_E = \lim_{E \to \text{REV}} \frac{V_x}{E}$.

c. $\rho_E = \lim_{\text{REV} \to V_x} \frac{E}{V_x}$.

d. $\rho_E = \lim_{E \to \text{REV}} \frac{E}{V_x}$.

e. none of these answers.

p.1.4 The perfect gas law is

a. $p\rho = RT$.
b. $p = \rho RT$.
c. $\rho = pRT$.
d. $p = \rho R/T$.
e. none of these.

p.1.5 The molecular weight of air is 29. The gas constant in $\text{m} \cdot \text{N/kg} \cdot \text{K}$ is

a. 28700.
b. 287.
c. 1480.
d. 1722.
e. none of these answers.

p.1.6 The density of water in kg/m^3 at 30°C is

a. 1000.
b. 1.94.
c. 995.7.
d. 9.957.
e. none of these answers.

p.1.7 One cubic meter of a fluid weighs 1050 N at a place where $g = 9.7\ \text{m/s}^2$. At another place where $g = 9.81\ \text{m/s}^2$, one cubic meter of the same fluid will weigh

a. 1061.9 N.
b. 1038.2 N.

c. 1060 N.
d. 9810 N.
e. none of these answers.

p.1.8 The units for the bulk modulus of elasticity in the SI system are

a. N/m^2.
b. N/m^3.
c. N/m.
d. $N \cdot m/s^2$.
e. none of these.

p.1.9 The bulk modulus of elasticity of a gas under isothermal conditions is

a. independent of pressure.
b. equal to the absolute pressure.
c. equal to κp.
d. equal to the gauge pressure.
e. none of these answers.

p.1.10 The percent reduction in volume of 10 m^3 of water when subjected to pressure of 120 kPa is very nearly

a. 0.0455.
b. 0.0545.
c. 5.45.
d. 54.5.
e. none of these answers.

p.1.11 The viscosity of a liquid

a. is independent of temperature.
b. increases with increase in temperature.
c. decreases with increase in temperature.
d. is mainly derived from the transfer of molecular momentum.
e. none of these.

p.1.12 The units of viscosity in the SI system are

a. $N \cdot m/s^2$.
b. $N \cdot s/m^2$.
c. m^2/s.
d. m/s^2.
e. none of these.

p.1.13 The units of kinematic viscosity in the SI system are

a. $N \cdot m/s^2$.
b. $N \cdot s/m^2$.
c. m^2/s.
d. m/s^2.
e. none of these.

p.1.14 A real fluid

a. has zero viscosity.
b. has some viscosity.
c. satisfies $p = \rho RT$.
d. can resist some shear force.
e. is not described by any of the above.

p.1.15 The kinematic viscosity of water at 70°F is 1.059×10^{-5} ft^2/s. The equivalent kinematic viscosity in the SI system in m^2/s is

a. 9.84×10^{-7}.
b. 9.84×10^{7}.
c. 1.059×10^{-3}.
d. 9.84×10^{-5}.
e. none of these.

p.1.16 Surface tension

a. is independent of temperature.
b. increases with increase in temperature.
c. decreases with increase in temperature.
d. is not described by any of these answers.

p.1.17 Water would boil at 80°C if the pressure on its surface in kN/m^2 is

a. 4.86.
b. 46.45.
c. 101.3.
d. 14.7.
e. none of these answers.

p.1.18 Determine the density of air at −5°C and 98.0 kPa absolute.

p.1.19 An oil has a specific gravity of 0.8. Determine its specific weight and density at 20°C.

p.1.20 While undergoing a certain process, the pressure on a gas is doubled while its specific volume is decreased by two-thirds. If the initial temperature is 130°F, what is the final temperature?

p.1.21 Determine the percent reduction in volume of 10 m^3 of water at 20°C, if the pressure is increased by 3 MN/m^2.

p.1.22 What are the isothermal and isentropic bulk moduli of elasticity of air under a pressure of 0.5 MPa? The barometric pressure is 103 kPa.

p.1.23 Determine the velocity of sound in an ocean if the water temperature is 5°C.

p.1.24 The tank shown in Fig. 1.15 contains olive oil and water under atmospheric pressure. A pressure of 150 psi is then applied on the piston. What will be the total downward movement of the oil surface? Assume that there is no change in the volume of the tank itself.

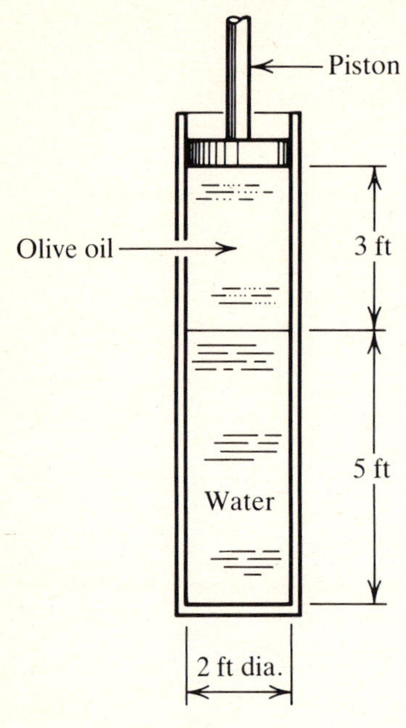

FIGURE 1.15 Problem 1.24.

p.1.25 A steel spherical tank has an outside diameter of 2 ft and wall thickness of 1/4 in. It is full of water at atmospheric pressure. What volume of water can be forced into the tank before yielding takes place? The yield stress and the modules of elasticity for the steel are 50,000 psi and 30×10^6 psi, respectively.

p.1.26 Determine the kinematic viscosity of castor oil at 20°C.

p.1.27 A substance has the following relationship between deformation and shear stress:

du/dy, rad/s	0	0.5	1.0	2.0
τ, N/m^2	2	3	4	6

Classify the substance.

p.1.28 A piston 50 mm in diameter moves inside a cylinder of 52 mm internal diameter and 10 cm long. The clearance between the piston and the cylinder is filled with SAE 30 oil at 20°C. Determine the force required to move the piston at 1 m/s. Neglect the end effects and the weight of the cylinder.

p.1.29 A body weighing 2.5 N and 5 cm $\times$ 5 cm $\times$ 5 cm in size and lubricated with SAE 10 oil at 20°C slides down a surface making an angle of 45° with the horizontal. Determine the speed of the body if the thickness of the lubricating film is 1 mm.

p.1.30 The velocity distribution for the flow of glycerine at 68°F between two fixed plates is given by $u = 50\, y(0.1 - y)$ ft/sec, where y is measured in feet from the lower wall. Determine the spacing between the walls and the shear stress at the walls.

p.1.31 The weight shown in Fig. 1.16 falls at a constant velocity of 5 cm/s. Determine the viscosity of the oil.

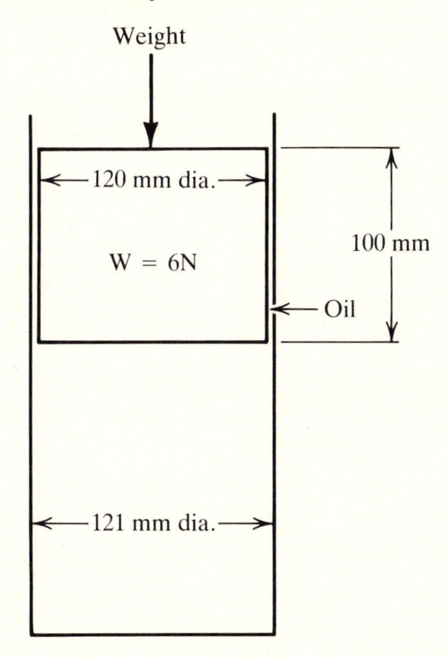

FIGURE 1.16 Problems 1.31, 1.32.

p.1.32 In Fig. 1.16, the viscosity of the oil is 0.75 $N \cdot s/m^2$. Determine the velocity with which the weight would fall through the cylinder.

p.1.33 Calculate the power lost in friction if the shaft in Fig. 1.17 is being rotated at 300 rpm.

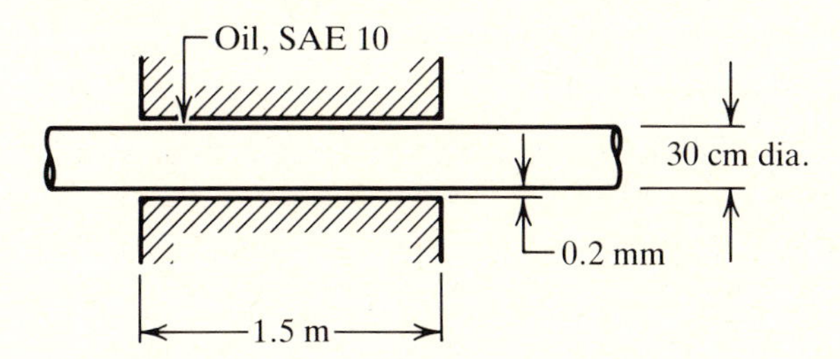

FIGURE 1.17 Problem 1.33.

p.1.34 The viscosity of oil (1) is three times the viscosity of oil (2) in Fig. 1.18. When the plate is pulled with a velocity of 35 cm/s, a shear stress of 30 N/m^2 is exerted on the plate. Calculate the viscosities of the oil.

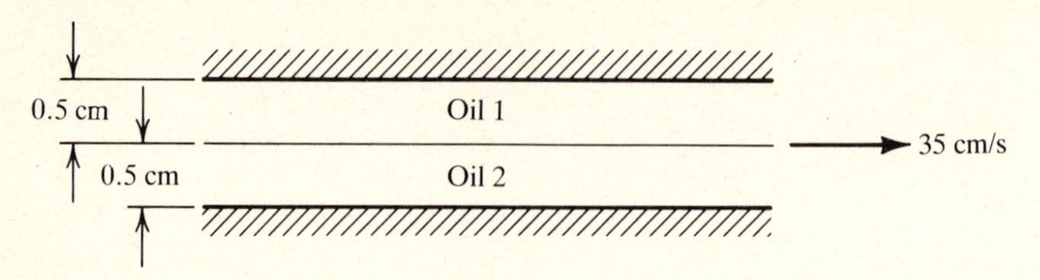

FIGURE 1.18 Problem 1.34.

p.1.35 Determine the pressure within a droplet of water 0.1 mm in diameter if the outside pressure is 101 kN/m^2.

p.1.36 Determine the height to which kerosene oil at 100°F will rise due to capillarity in a glass tube 1/8 in. in diameter.

p.1.37 A liquid at 10°C rises to a height of 15 mm in a 0.5 mm glass tube. The angle of contact is 42°. Determine the surface tension of the liquid if its density is 1200 kg/m^3. (Hint: use Eq. (1.44).)

p.1.38 A glass tube is inserted into mercury as shown in Fig. 1.19. Compute the distance d if the surface tension σ for mercury and air is 0.514 N/m and the angle θ is 40°.

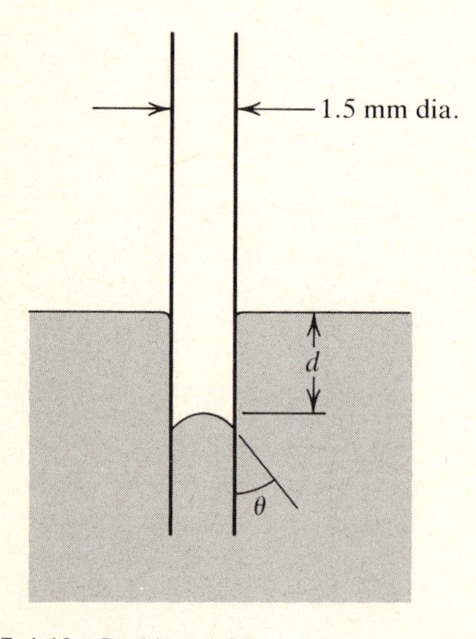

FIGURE 1.19 Problem 1.38.

p.1.39 A narrow tank shown in Fig. 1.20 is completely filled with water at 20°C. If the pressure gauge measures a pressure of 2900 Pa, what is the radius of curvature of the water surface?

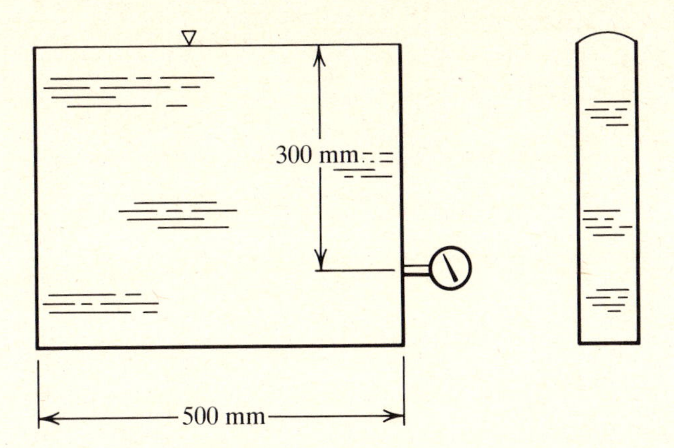

FIGURE 1.20 Problem 1.39.

p.1.40 What force will be required to lift a thin wire platinum ring 20 mm in diameter from a water surface at 20°C? Neglect the weight of the ring.

p.1.41 Determine the pressure at which cavitation would develop at the inlet of a pump if the water temperature is 15°C.

p.1.42 To what value must the absolute pressure over ethyl alcohol, gasoline and mercury by reduced to make them boil at 20°C in separate containers?

Chapter Two

FLUID STATICS

In Chapter One, we learned the basic properties of a fluid. In this chapter, we will study how the pressure varies in a fluid at rest. It will help us determine the forces exerted by a static fluid on the surfaces and the bodies which surround it, or are immersed in it. We will also study two special cases of fluid equilibrium, the linear acceleration and the forced vortex. In these cases, although the fluid as a whole is in motion, there are no shear forces within the fluid. Hence, the laws of fluid statics are applicable.

2.1 INTRODUCTION

Fluid statics is that branch of fluid mechanics which deals with fluids at rest. It can be concluded from the definition of a fluid that a fluid at rest cannot have any shear force acting on it, otherwise it will not be at rest. Therefore, all forces due to the pressure of a fluid must be normal to the surfaces on which they act.

2.2 PRESSURE

The average pressure $\bar{p}$ over a plane area δA in a fluid is the ratio of the normal force δF to the area, or

$$\bar{p} = \frac{\delta F}{\delta A} \tag{2.1}$$

The pressure p at a point is the limit of Eq. (2.1) as the area is indefinitely reduced to $\delta A'$ subject to the restriction of a continuum, or

$$p = \lim_{\delta A \to \delta A'} \frac{\delta F}{\delta A} \tag{2.2}$$

Since force and area are vectors, it is evident from Eq. (2.1) that pressure is a scalar. Just like temperature and density, pressure is the property of a fluid at a point. The units of pressure in the SI and FPS systems are N/m^2 and lbs/ft^2, respectively. The unit N/m^2 is called a *pascal* with the abbreviation Pa. A pressure of 100 kPa is called a *bar*.

Due to molecular collision, any part of a fluid experiences pressure due to the adjoining fluid or the solid boundaries containing it. Therefore, if we divide a fluid into two parts by an imaginary plane passing through it, the plane will experience pressure. At a point in a fluid with no relative motion between the adjacent layers, the magnitude of the pressure is independent of the orientation of the plane passing through that point.

To prove this, consider a wedge-shaped body of unit width with plane faces surrounding the point P, the point under consideration (Fig. 2.1). Since there is no relative motion, there are no shear forces and the only forces are the normal surface forces and gravity. The equation of motion in the x-direction is given by

$$\sum F_x = ma_x$$

or

$$p_x \delta y - p_s \delta s \sin\theta = \rho \frac{\delta x \delta y}{2} a_x$$

Since $\delta s \sin\theta = \delta y$; we can write:

$$(p_x - p_s)\delta y = \rho \frac{\delta x \delta y}{2} a_x$$

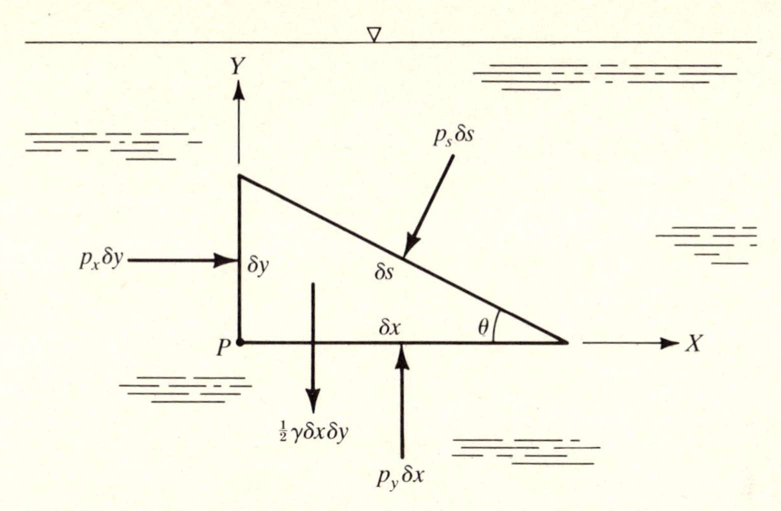

FIGURE 2.1 A wedge-shaped fluid element with pressure forces acting on it.

or

$$p_x - p_s = \rho \frac{\delta x}{2} a_x \tag{2.3}$$

At the limit, when the wedge is reduced to the point P, $\delta x \to 0$ and the right-hand side (RHS) of Eq. 2.3 tends to zero. Therefore, at point P

$$p_x = p_s \tag{2.4}$$

The equation of motion in the y-direction is given by

$$p_y\delta x - p_s\delta s \cos\theta - \gamma \frac{\delta x\delta y}{2} = \rho \frac{\delta x\delta y}{2} a_y$$

Since $\delta s \cos a = \delta x$, we can also write

$$(p_y - p_s)\delta x = \frac{\delta x\delta y}{2}(\gamma + \rho)a_y$$

or

$$(p_y - p_s) = \frac{\delta_y}{2}(\gamma + \rho)a_y \tag{2.5}$$

At the limit, when the wedge is reduced to the point P, $\delta y \to 0$ and the RHS of Eq. (2.5) tends to zero. Therefore, at point P

$$p_y = p_s \tag{2.6}$$

Combining Eqs. (2.4) and (2.6), we obtain

$$p_x = p_y = p_s \tag{2.7}$$

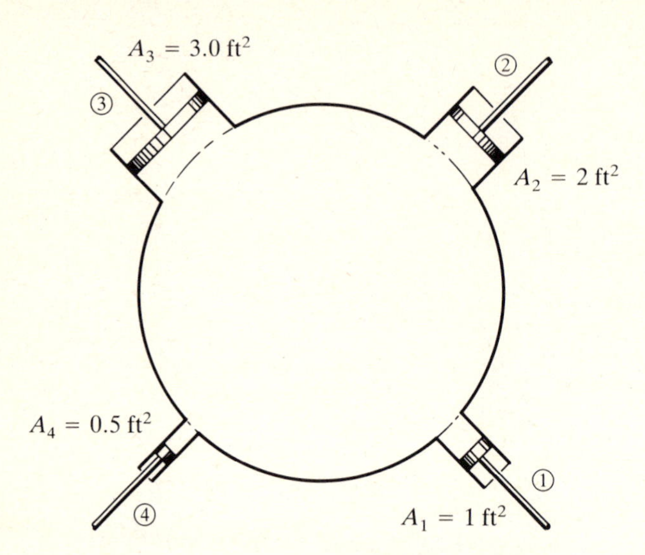

FIGURE 2.2 A vessel with four openings filled with a fluid.

This result is valid for any value of the angle θ and any orientation of the wedge. This means that at any point within a static fluid, the pressure is the same in all directions. This result is also known as *Pascal's Law* after the French mathematician Blaise Pascal (1623–62).

If there is a relative motion between the adjacent fluid layers, shear forces are present and Eq. (2.7) is not valid. However, if the shear forces are small as compared to the normal forces, Pascal's Law may still be approximately true.

To further elaborate Pascal's Law, consider a vessel as shown in Fig. 2.2. The vessel has a unit thickness normal to the plane of the paper and is equipped with four circular openings of different diameters. Each opening is fitted with a piston which can move freely through the mouth of the opening. The areas of the four openings are shown in Fig. 2.2.

Let us assume that the system is in equilibrium and under pressure. Now, if we increase the pressure at piston 1 by 1 lb/ft^2, then, according to Pascal's Law, this pressure change will be transmitted equally in all directions. To keep the system in equilibrium, the same pressure (equal to 1 lb/ft^2) must be applied on pistons 2, 3, and 4. Hence, forces of 2 lbs, 3 lbs, and 0.5 lb will be needed on pistons 2, 3, and 4, respectively, to keep the system in equilibrium.

Pascal's Law has wide application in the development of hydraulic controls for heavy earthmoving equipment, aircraft, hydraulic presses, and hydraulic hoists.

EXAMPLE 2.1

A hydraulic press has a ram of 10 cm diameter and a piston of 1 cm diameter. How much force should be applied on the piston to lift a weight of 1000 N?

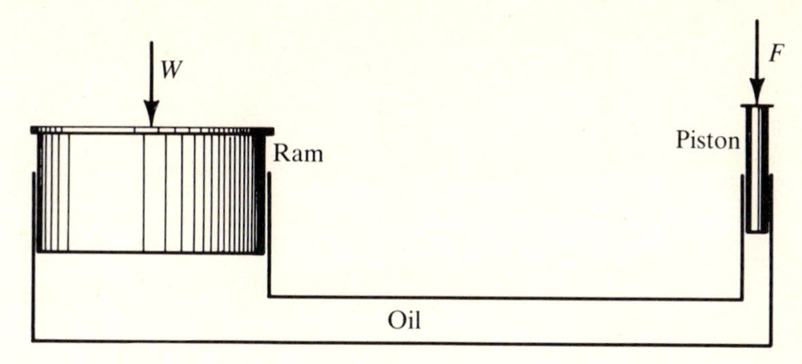

FIGURE 2.3 Example 2.1.

Solution A hydraulic press is shown diagrammatically in Fig. 2.3

$$\text{Area of piston} = a = \frac{\pi(1)^2}{4} = \frac{\pi}{4}\ \text{cm}^2$$

$$\text{Area of ram} = A = \frac{\pi(10)^2}{4} = 25\pi\ \text{cm}^2$$

The important point to remember is that it is the *pressure* and *not the force* which is transmitted equally in all directions.

The weight to be lifted is 1000 N. To lift this much weight, the pressure to be supplied by the oil is

$$p = \frac{W}{A} = \frac{1000}{25\pi} = \frac{40}{\pi}\ \text{N/cm}^2$$

To produce this much pressure, some force F must be applied on the piston, which is

$$F = pa = \frac{40}{\pi} \times \frac{\pi}{4} = 10\ \text{N}$$

Hence, by making use of Pascal's Law, a weight of 1000 N can be lifted by applying a force of only 10 N.

Pressure Measurement

Pressure cannot be measured directly. All pressure measurements are made in terms of the difference of pressure between the point in question and a point of reference. The point of reference is usually the local atmospheric pressure or a complete vacuum. Pressure measured with reference to the local atmospheric pressure is called the *gauge pressure* and the pressure measured with reference to the complete vacuum is called the *absolute pressure*. The

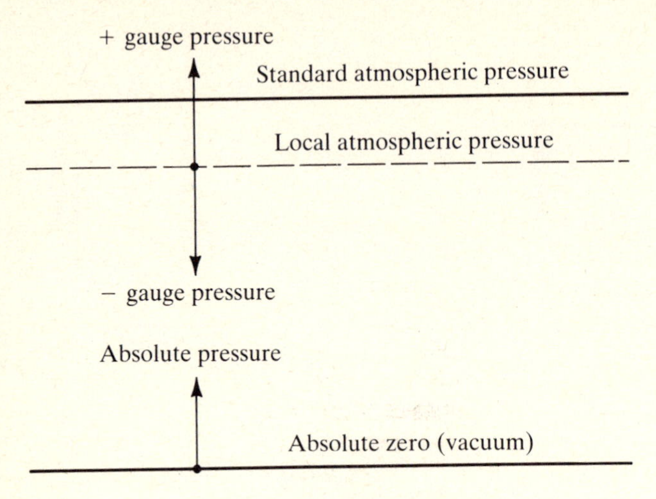

FIGURE 2.4 Various scales of pressure measurements.

relationship between the gauge pressure p, absolute pressure p_{ab} and local atmospheric pressure p_{atm} is given by:

$$p_{ab} = p + p_{atm} \tag{2.8}$$

Gauge pressure may be negative, but absolute pressure is never negative. *Standard atmospheric pressure* is the mean pressure at sea level. It is taken to be equal to 14.7 psi or 101.3 kPa. Fig. 2.4 illustrates the relationship between gauge and absolute pressure.

A typical device used for measuring gauge pressure is the Bourdon gauge invented by Eugene Bourdon (1808–84). The gauge consists of a curved tube of elliptical cross-section (Fig. 2.5). One end of the tube is closed and is free to

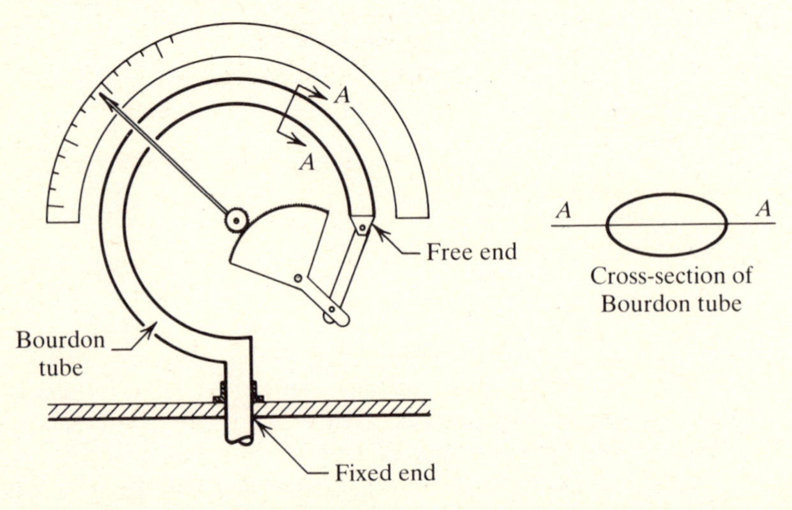

FIGURE 2.5 Mechanism of a Bourdon gauge.

move and the other end is rigidly fixed to the frame. The fluid enters the tube through the fixed end. When the pressure inside the tube, due to the fluid in it, exceeds the pressure outside the tube, which is the local atmospheric pressure, the cross-section tends to become circular. This causes the free end of the tube to move. The movement is transmitted to a dial through a suitable mechanism. The range of pressure which can be measured depends on the stiffness of the tube. A Bourdon gauge should not be subjected to a pressure larger than its maximum limit because it might strain the tube beyond its elastic limit and invalidate the calibration.

For measuring very high pressures, piezoelectric gauges may be used. These gauges utilize a quartz crystal or other material which produces a small but measurable electrical potential difference across itself when subjected to fluid pressure.

Local atmospheric pressure is measured by a mercury barometer or by an aneroid barometer. These devices measure pressure between the atmosphere and an evacuated medium.

2.3 PRESSURE VARIATION IN A STATIC FLUID

Consider a reservoir as shown in Fig. 2.6a containing fluid at rest. Let A be a point within the fluid, and let us investigate how pressure varies along x, y and z directions from the point A. Erect a small elemental volume of sides δx, δy and δz with A at the center as shown in Fig 2.6b.

Let the origin of the coordinate system be located at the point A. Assume that the pressure at point A is equal to p and that the pressure increases in the $+x$, $+y$ and $+z$ directions and that it decreases in the $-x$, $-y$ and $-z$ directions. The pressure at each of the six faces of the elemental volume may be obtained using a Taylor series expansion of the pressure about the

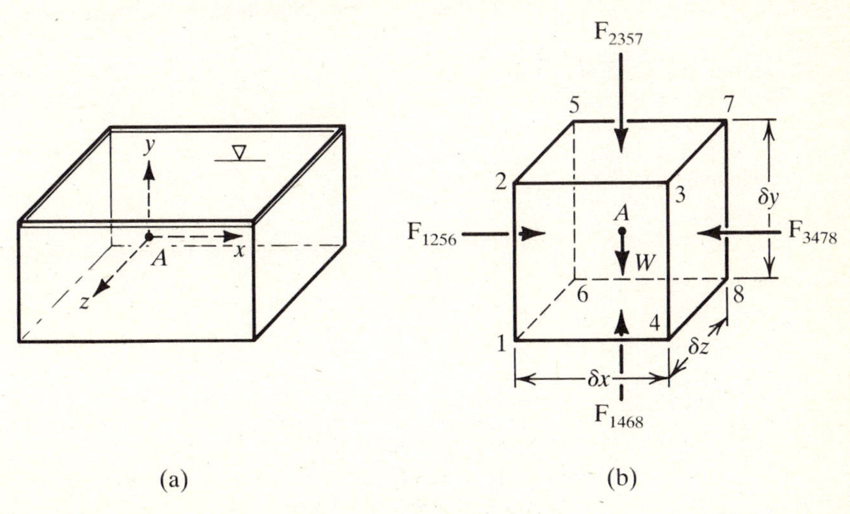

FIGURE 2.6 Elemental volume of a fluid at rest in a tank.

point A. Using only the first two terms and ignoring the higher power terms in the expansion, the pressure at face 1256 is

$$p_{1256} = p - \frac{\partial p}{\partial x}\frac{\delta x}{2}$$

Similarly

$$p_{3478} = p + \frac{\partial p}{\partial x}\frac{\delta x}{2}$$

Now the forces acting on these two faces can be calculated by multiplying the pressure by the area. Or:

$$F_{1256} = \left(p - \frac{\partial p}{\partial x}\frac{\delta x}{2}\right)\delta y \delta z \tag{2.9}$$

$$F_{3478} = \left(p + \frac{\partial p}{\partial x}\frac{\delta x}{2}\right)\delta y \delta z \tag{2.10}$$

These are the only two forces acting in the x-direction. Hence:

$$\Sigma F_x = \left(p - \frac{\partial p}{\partial x}\frac{\delta x}{2}\right)\delta y \delta z - \left(p + \frac{\partial p}{\partial x}\frac{\delta x}{2}\right)\delta y \delta z$$

$$= p\delta y \delta z - \frac{\partial p}{\partial x}\frac{\delta x \delta y \delta z}{2} - p\delta y \delta z - \frac{\partial p}{\partial z}\frac{\delta x \delta y \delta z}{2}$$

or

$$\Sigma F_x = -\frac{\partial p}{\partial x}\delta x \delta y \delta z \tag{2.11}$$

Since the fluid is at rest

$$\Sigma F_x = 0$$

Hence:

$$-\frac{\partial p}{\partial x}\cdot \delta x \delta y \delta z = 0$$

$$\frac{\partial p}{\partial x}(\delta x \delta y \delta z) = 0 \tag{2.12}$$

For Eq. (2.12) to be zero, either $\partial p/\partial x = 0$ or $(\delta x \delta y \delta z) = 0$. But $(\delta x \delta y \delta z)$ is an elemental volume and cannot be zero. Hence:

$$\frac{\partial p}{\partial x} = 0 \tag{2.13}$$

Equation (2.13) implies that the pressure is constant in the x-direction. Equations similar to Eqs. (2.9) and (2.10) may now be written for faces (1234) and (5678), and it can be shown that

$$\frac{\partial p}{\partial z} = 0 \tag{2.14}$$

which implies that the pressure is constant in the z-direction. Equations (2.13) and (2.14) when combined together suggest that the pressure is the same at every point in the x–z plane.

Again, equations similar to Eqs. (2.9) and (2.10) may be written in the y-direction. But in addition to these two forces, there is also a force W, equal to the weight of the fluid in the elemental volume, as shown in Fig. 2.6b.

$$W = \rho g(\delta x \delta y \delta z)$$

Now

$$\Sigma F_y = F_{1468} - F_{2357} - W$$

$$= \left(p - \frac{\partial p}{\partial y}\frac{\delta y}{2}\right)\delta x \delta z - \left(p + \frac{\partial p}{\partial y}\frac{\delta y}{2}\right)\delta x \delta z - \rho g(\delta x \delta y \delta z)$$

Simplifying

$$\Sigma F_y = -\frac{\partial p}{\partial y}(\delta x \delta y \delta z) - \rho g(\delta x \delta y \delta z) \tag{2.15}$$

Since the fluid is at rest,

$$\Sigma F_y = 0$$

Hence:

$$-\frac{\partial p}{\partial y}(\delta x \delta y \delta z) - \rho g(\delta x \delta y \delta z) = 0$$

or

$$\frac{\partial p}{\partial y} + \rho g = 0$$

Since pressure is a function of y only, we may change partial to total derivatives. Hence:

$$\frac{dp}{dy} + \rho g = 0$$

$$\frac{dp}{dy} = -\rho g \tag{2.16}$$

This is the differential equation of the variation of pressure in a static fluid. The equation is valid for a compressible, as well as an incompressible fluid.

The results in the form of Eqs. (2.13), (2.14), and (2.16) show that, in a static fluid, the pressure does not vary from point to point in a horizontal plane, but it does vary from point to point in a vertical plane.

Pressure Variation in an Incompressible Fluid

If a fluid is incompressible, its density ρ may be assumed to remain constant throughout the fluid. Then integrating Eq. (2.16) with respect to y, and replacing ρg with γ, we obtain

$$p = -\gamma y + C$$

where C is a constant of integration. If we change the convention such that the y-axis is positive downwards, then

$$p = \gamma y + C \tag{2.17}$$

Placing the origin at the surface and letting $p = p_0$ at $y = 0$, we obtain $C = p_0$. Then

$$p = p_0 + \gamma y$$

As a matter of convention, y is replaced by h which signifies depth below the liquid surface. Then the above equation becomes:

$$p = p_0 + \gamma h \tag{2.18}$$

If the pressure on the surface of the liquid is atmospheric, $p_0 = 0$ and the above equation simplifies to

$$p = \gamma h \tag{2.19}$$

Comments

1. The equation $p = \gamma h$ assumes that γ is constant. Therefore, it is applicable only to incompressible fluids.
2. The equation $p = \gamma h$ or $p = p_0 + \gamma h$ are valid only if the fluid is at rest or all shear stresses are negligibly small.

Pressure Variation in a Compressible Fluid

If a fluid is compressible, the density ρ in Eq. (2.16) is not constant, but varies with the pressure. Unless the manner of this variation is known, Eq. (2.16) cannot be integrated.

For gases, the relationship between density and pressure may be expressed using $p/\rho^n =$ constant, where n depends on the process involved. For example, $n = 1$ for an isothermal process and $n = k = (c_p/c_v)$ for an isentropic process.

For an *isothermal* process:

$$\frac{p}{\rho} = \frac{p_0}{\rho_0} \quad \text{or} \quad \rho = \frac{p}{(p_0/\rho_0)}$$

Substituting in Eq. (2.16), we obtain

$$\frac{dp}{dy} = -\frac{gp}{(p_0/\rho_0)}$$

or

$$\frac{dp}{p} = -\frac{g}{(p_0/\rho_0)} dy$$

Integrating with respect to y, we obtain

$$\ln(p) = -\frac{gy}{(p_0/\rho_0)} + C$$

where C is a constant of integration. The value of C may be obtained by using the known initial conditions—i.e., when $y = 0$, $p = p_0$. This yields $C = \ln(p_0)$. Also for a perfect gas, $p_0/\rho_0 = RT$. Substituting for C and p_0/ρ_0, we obtain

$$\ln\left(\frac{p}{p_0}\right) = -\frac{gy}{RT}$$

$$\frac{p}{p_0} = \exp\left(-\frac{gy}{RT}\right) \tag{2.20}$$

For an *isentropic* process:

$$\frac{p}{\rho^n} = \frac{p_0}{\rho_0^n}$$

$$\rho = \left[p\bigg/\left(\frac{p_0}{\rho_0^n}\right)\right]^{1/n}$$

$$\rho = \left(\frac{p}{K}\right)^{1/n} \quad \text{where } K = \left(\frac{p_0}{\rho_0^n}\right)$$

Substituting in Eq. (2.16), we obtain

$$\frac{dp}{dy} = -g\left(\frac{p}{K}\right)^{1/n}$$

$$\frac{dp}{p^{1/n}} = \left(\frac{-g}{K^{1/n}}\right) dy$$

Integrating with respect to y, we obtain

$$\frac{n}{n-1} p^{(n-1)/n} = -\frac{g}{K^{1/n}} y + C$$

Where C is a constant of integration. The value of C may be evaluated by using the known initial conditions—i.e., when $y = 0$, $p = p_0$. This gives $C = n/(n-1)p^{(n-1)/n}$. Substituting for C and putting $K = (p_0/\rho_0^n)$, we obtain

$$\frac{n}{n-1}\left[p^{(n-1)/n} - p_0^{(n-1)/n}\right] = -\frac{g\rho_0 y}{p_0^{1/n}}$$

$$\frac{p}{p_0} = \left[1 - \frac{g\rho_0 y}{p_0}\left(\frac{n-1}{n}\right)\right]^{n/(n-1)}$$

For a perfect gas, $p_0 = \rho_0 R T_0$. Therefore:

$$\frac{p}{p_0} = \left[1 - \frac{gy}{RT_0}\left(\frac{n-1}{n}\right)\right]^{n/(n-1)} \tag{2.21}$$

The above equation may be used to describe the variation of pressure with altitude in the atmosphere; hence, it is of interest in aeronautics and meteorology.

In the atmosphere, n varies with altitude and may be stated as

$$n = \frac{1}{\left(1 + \dfrac{R}{g}\dfrac{dT}{dy}\right)}$$

where dT/dy is the rate of change of temperature with altitude, known as the *lapse rate.*

The U.S. Standard Atmosphere has been defined as follows:

1. At sea level, 0 ft, $p = 14.969$ psia (101.325 KPa) and $T = 59.00°$F (15°C).
2. From altitude 0 to 36,151.6 ft (11,019 m), the atmosphere is polytropic with a lapse rate $dT/dy = -3.56°$F/1000 ft ($-6.49°$C/1000 m) ($n = 1.2345$)—that is, the temperature drops linearly from 59°F (15°C) at sea level to $-69.7°$F ($-56.5°$C) at 36,151.6 ft (11,019 m).

$$\frac{p}{\rho^n} = C$$

applies.
3. From altitude 36,151.6 ft (11,019 m) to 65,824 ft (20,063.1 m) the atmosphere is isothermal—that is, the temperature is constant at $-69.7°$F ($-56.5°$C).

$$\frac{p}{\rho} = C$$

applies.
4. From altitude 65,824 ft (20,063.1 m) to 105,500 ft (32,156.4 m) the atmosphere is polytropic with a lapse rate equal to 0.544°F/1000 ft (0.992°C/1000 m) ($n = 0.97177$)—that is, the temperature increases linearly from $-69.7°$F ($-56.5°$C) at 65,824 ft to $-48.1°$F ($-44.5°$C) at 105,500 ft (32,156.4 m).

EXAMPLE 2.2

Find the pressure in kN/m^2 at a point in the ocean which is 1 km below the surface.

Solution γ of sea water = 10,070 N/m^3. Using Eq. (2.19)

$$p = \gamma h$$
$$= (10{,}070)1 = 10{,}070 \text{ kN/m}^2$$

EXAMPLE 2.3

Find the pressure at the bottom of the tank shown in Fig. 2.7.
The air is under a pressure of 150 kN/m^2 gauge. The specific gravity of oil is 0.85.

Solution Let points 1, 2, and 3 be located at the bottom of the air, oil and water respectively, as shown in Fig. 2.7, and let

p_1 = pressure at point 1

p_2 = pressure at point 2

p_3 = pressure at point 3

Then $p_1 = 150 \text{ kN/m}^2$ (equal to the air pressure). From Eq. (2.18)

$$p_2 = p_1 + \gamma_{oil} h$$
$$= p_1 + \gamma \times S_{oil} \times h$$
$$= (150) + (9.810)(0.85)(0.8)$$
$$= 156.7 \text{ kN/m}^2$$

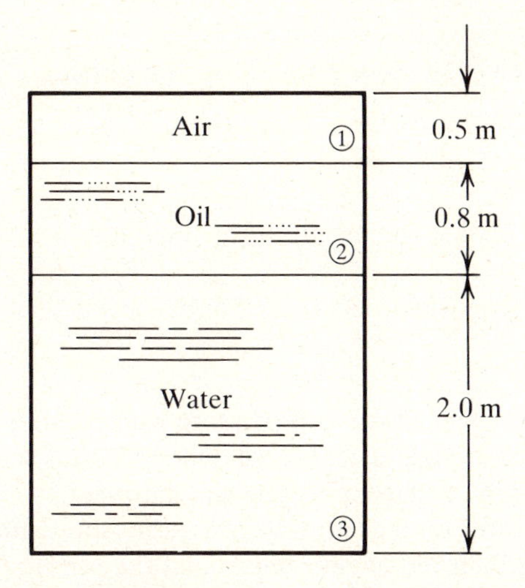

FIGURE 2.7 Example 2.3.

Now

$$
\begin{aligned}
p_3 &= p_2 + \gamma \cdot h \\
&= 156.7 + (9.810)(2.0) \\
&= 176.3\ \mathrm{kN/m^2}
\end{aligned}
$$

Hence, the pressure at the bottom of the tank is 176.3 $\mathrm{kN/m^2}$.

EXAMPLE 2.4

At the top of a mountain, the temperature is $-5°\mathrm{C}$ and a mercury barometer reads 570 mm. At the foot of the same mountain, the barometer reads 750 mm. Assuming a polytropic atmosphere and $R = 287\ \mathrm{J/kg \cdot K}$, calculate the height of the mountain.

Solution For this problem

$$
\begin{aligned}
p_0 &= 570\ \mathrm{mm} \\
T_0 &= -5 + 273 = 268\ \mathrm{K} \\
p &= 750\ \mathrm{mm} \\
n &= 1.2345 \quad \text{(for a polytropic atmosphere)} \\
R &= 287\ \mathrm{J/kg \cdot K}
\end{aligned}
$$

$$\frac{n-1}{n} = \frac{1.2345 - 1}{1.2345} = 0.19$$

$$\frac{n}{n-1} = 5.263$$

Using Eq. (2.21) and substituting the known values, we obtain

$$\frac{750}{570} = \left[1 - \frac{9.81y}{287 \times 268}(0.19)\right]^{5.263}$$

From which, $y = 2209$ m. Hence, the height of the mountain is 2209 m.

Pressure Prism

A prismatic diagram showing the variation of pressure with depth is called a *pressure prism*. Consider a rectangular tank containing water as shown in Fig. 2.8a. Let the pressure at the surface of water be atmospheric.

The pressures at various points on the face *CDFE* may be represented by a pressure prism shown in Fig. 2.8b. The figure shows that the pressure at the

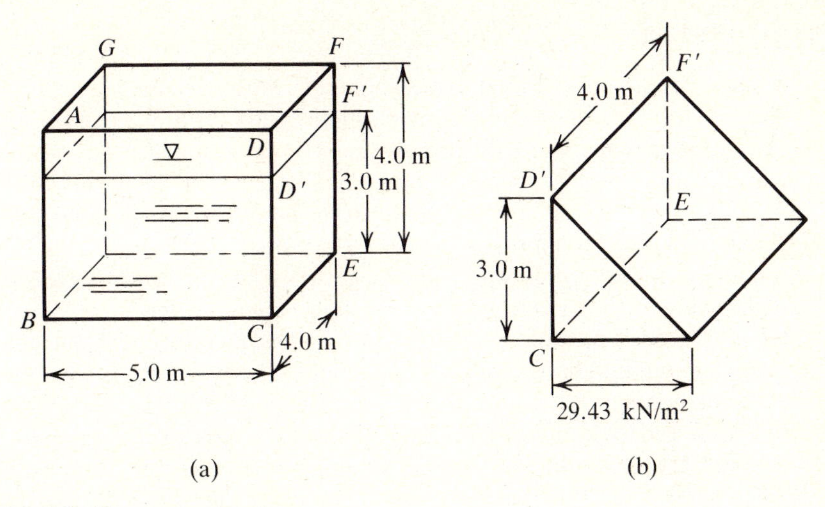

FIGURE 2.8 The pressure prism.

line of intersection (line D'F') of the water and the face *CDFE* is zero. The pressure then increases linearly according to the relationship $p = \gamma h$, until the bottom of the box, where the pressure is maximum and equal to $(9.81 \times 3) = 29.43$ kN/m^2. A pressure prism may be used to determine the total force acting on the face *CDFE*. This would be equal to the volume of the pressure prism. Hence, the force acting on face *CDFE* is $(3.0 \times 29.43)/2 \times 4.0 = 176.6$ kN. The point of application of the force, called the *center of pressure* (C_p), is located at the center of gravity of the pressure prism. The center of pressure is located at a distance $= 2/3 \times 3.0 = 2.0$ m below the water surface.

Pressure prisms, similar to the one described above, may also be drawn for plane areas submerged in a fluid.

Static Head

A pressure may be represented in terms of an equivalent static head of a liquid column using the equation $p = \gamma h$. For example, a pressure of 101.3 kN/m^2 is equivalent to a height of

$$h = \frac{p}{\gamma} = \frac{101{,}300}{9810} = 10.33 \text{ meters of water}$$

$$= \frac{101{,}300}{13.55 \times 9810} = 76.2 \text{ cm of mercury}$$

In other words, if we take a tube and fill it to a height of 10.33 meters with water, or to a height of 76.2 cm with mercury, then the pressure at the bottom of the tube will be 101.3 kN/m^2. This concept of an equivalent static head is widely used in the measurement of pressures using manometers, which are described in the next Section.

2.4 MANOMETERS

A *manometer* is a device used to measure the pressure of a fluid in a pipe. In its simplest form, it consists of a clear plastic or glass tube, one end of which is inserted in the pipe where the pressure is to be measured and the other end is left open to the atmosphere, as shown in Fig. 2.9. The height to which the fluid rises in the tube gives the pressure at the point of insertion in terms of the height of the fluid column. For example, the pressure at the point A in Fig. 2.9 is equal to a height h of water. The height may be converted to pressure by multiplying by the specific weight of the fluid. A manometer can only measure the difference of pressure between its two ends. The pressure at one end can only be found if the pressure at the other end is known. That is why one end of the manometer is left open to the atmosphere. In this way, the pressure at the open end is known to be zero. If the two ends of a manometer are connected to two different points of unknown pressures, then only the difference of pressure between the two points can be determined. The individual pressures at the two points cannot be determined in this way.

A manometer may be in the form of a U-tube, an inverted U-tube or a combination of U-tube and inverted U-tube as shown in Fig. 2.10. The U-portion may contain manometer fluids other than the pipe fluid. The shape of the manometer and the type of manometer fluids are carefully selected so as to keep the size of the manometer manageable for measuring a given range of pressure.

The unknown pressure at a point can be determined by moving along the length of the manometer from the point of unknown pressure to the point of known pressure while adding the column heights when moving down and subtracting them when moving up. For example, in the case shown in Fig. 2.10, to find the pressure at point A, start from A and move towards the open end considering the following items:

1. The pressure is the same within the same continuous fluid at the same level.
2. All heights must be expressed in terms of equivalent water heights. This is done by multiplying the fluid heights by the respective specific gravities.
3. Add the height when moving down and subtract it when moving up.

The manometer equation for Fig. 2.10 may then be written as

$$h_A - h_1S_1 + h_2S_2 - h_3S_3 - h_4S_4 = 0 \tag{2.22}$$

If all the heights and all the specific gravities are known, h_A may be determined from the above equation. If the open end of the above manometer were connected to another pipe B, we could only determine the difference of pressure between the two pipes. We could not determine the actual pressures in the pipes A and B.

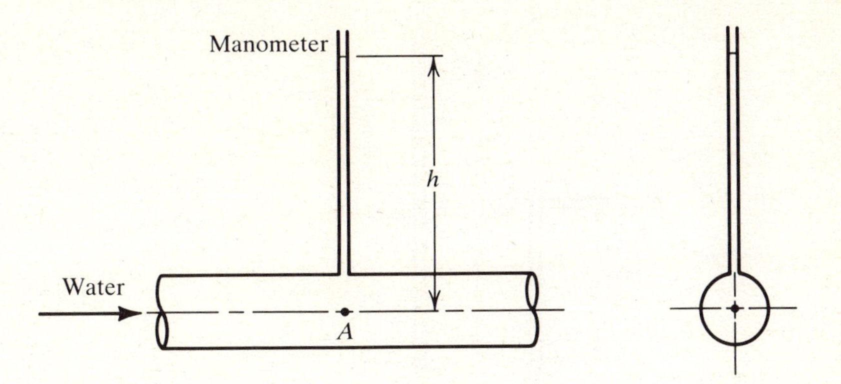

FIGURE 2.9 A simple manometer.

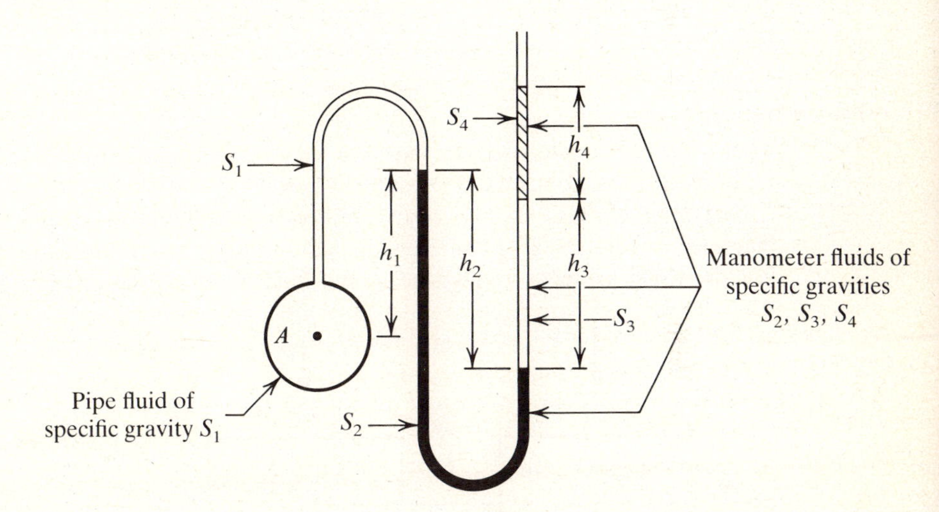

FIGURE 2.10 A manometer with two manometer fluids of different specific gravities.

EXAMPLE 2.5

Determine the pressure in the pipe in kPa for the manometer of Fig. 2.11.

Solution

$$h_A + (500)(1) - (950)(1.5) = 0$$

$$h_A = 925 \text{ mm of water}$$

$$= 0.925 \text{ m of water}$$

$$\text{Pressure at } A = (9810)(0.925) = 9074 \text{ Pa}$$

$$= 9.074 \text{ kPa}$$

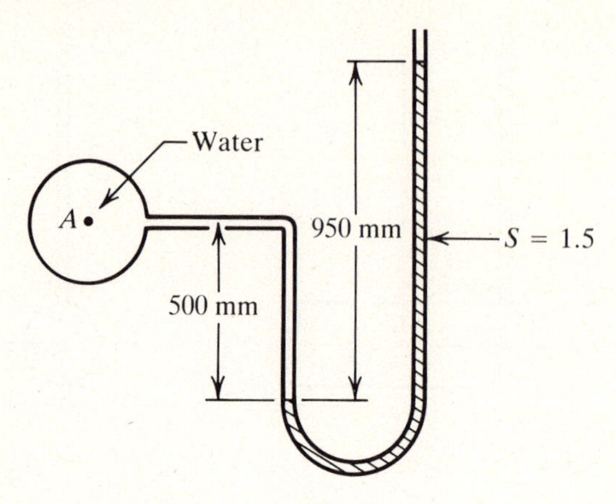

FIGURE 2.11 Example 2.5.

Micromanometers

Micromanometers are used to measure *small* differences of pressure. Two types of micromanometers are described here.

One type consists of two reservoirs connected together with a small diameter U-tube (Fig. 2.12). The device employs two immiscible gauge liquids which are also immiscible in the fluid to be measured. The gauge

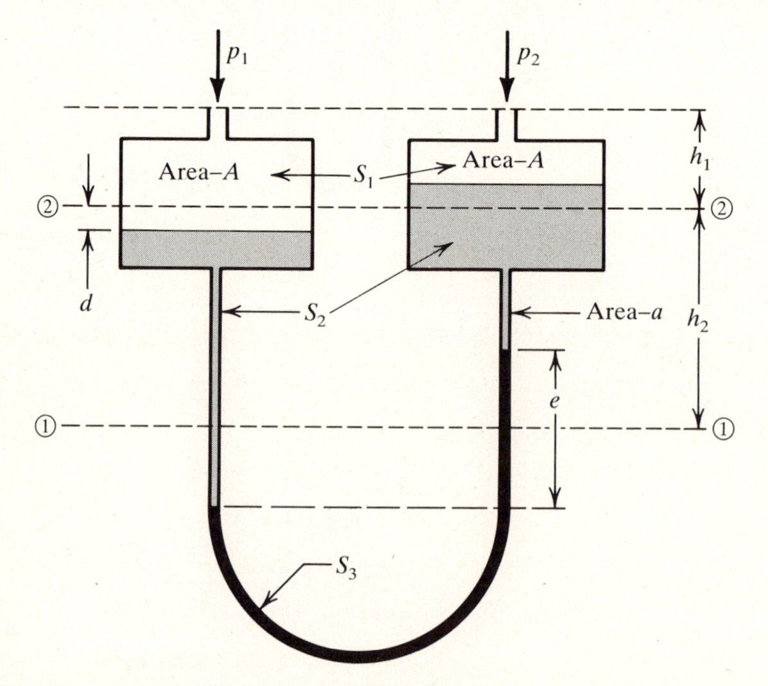

FIGURE 2.12 Micromanometer with two reservoirs.

liquid of the higher specific gravity fills the lower portion of the tube up to the level 1–1. The lighter gauge liquid occupies the space between the marks 1–1 and 2–2. When a pressure difference $(p_1 - p_2)$, $p_1 > p_2$ is applied to the left reservoir, the liquids assume a position as shown. Writing the manometer equation, we obtain

$$\frac{p_1}{\gamma} + (h_1 + d)S_1 + \left(h_2 - d + \frac{e}{2}\right)S_2 - eS_3 - \left(h_2 - \frac{e}{2} + d\right)S_2 - (h_1 - d)S_1 = \frac{p_2}{\gamma}$$

where e is the difference in elevation between the two ends of the heavier gauge liquid as shown in Fig. 2.12.

$$\left(\frac{p_1}{\gamma} - \frac{p_2}{\gamma}\right) = -2dS_1 + 2dS_2 - eS_2 + eS_3 \tag{2.23}$$

Let the cross-sectional area of the reservoir and the U-tube be A and a, respectively. Then, for the displacement d in the reservoir:

$$dA = \frac{ae}{2}$$

or

$$d = \left(\frac{a}{A}\right)\frac{e}{2}$$

Substituting in Eq. (2.23), we obtain

$$(p_1 - p_2) = \gamma e\left[S_3 - S_2\left(1 - \frac{a}{A}\right) - S_1\frac{a}{A}\right] \tag{2.24}$$

From Eq. (2.24), it can be seen that the pressure difference is directly proportional to e for a gauge.

The inclined tube manometer is frequently used to measure small differences in gas pressures. It consists of a reservoir and an inclined tube of small diameter containing an oil of specific gravity S (Fig. 2.13). Let the cross-sectional areas of the reservoir and the tube be A and a respectively, and the angle of inclination of the tube be θ.

In theory, when a small pressure difference is applied at the reservoir, oil in the reservoir goes down by a small amount d but the oil in the inclined tube moves through a large distance e. Applying the manometer theory to the final oil elevations, we obtain

$$\frac{p_1}{\gamma} - (d + e\sin\theta)S = \frac{p_2}{\gamma} \tag{2.25}$$

Also

$$dA = ae$$

$$d = \left(\frac{a}{A}\right)e$$

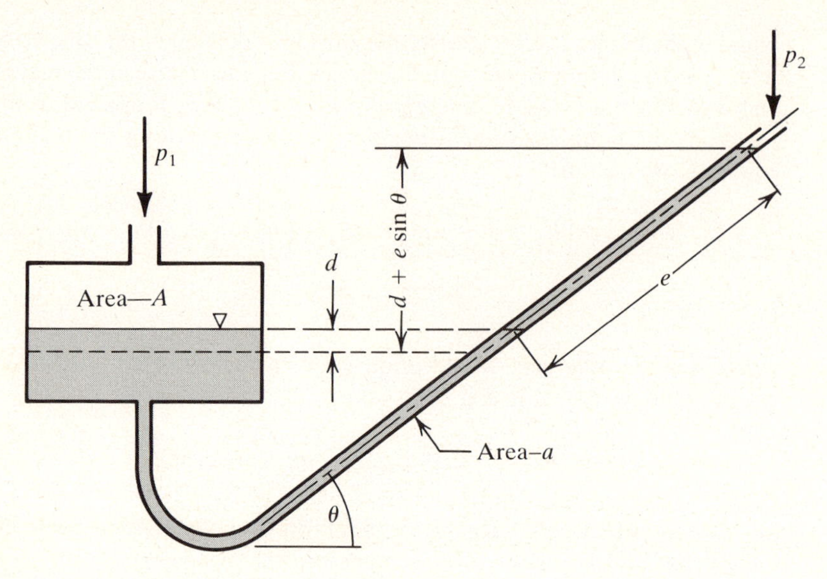

FIGURE 2.13 Inclined tube micromanometer.

Substituting in Eq. (2.25), we obtain

$$\frac{p_1}{\gamma} - \left[\frac{a}{A}e + e \sin \theta\right]S = \frac{p_2}{\gamma}$$

$$(p_1 - p_2) = \gamma Se\left[\frac{a}{A} + \sin \theta\right] \tag{2.26}$$

In Eq. (2.26) the change d in the reservoir elevation has been considered. Failure to consider this change results in an error given by:

$$\text{Error}(\%) = \frac{a/A}{(\sin \theta + a/A)} \times 100$$

Usually, a micromanometer's scale is adjusted and written in such a way that it automatically incorporates the change in the reservoir oil level d, the angle of inclination θ, and the specific weight of the oil γ, such that the pressure difference may be read directly in millimeters or inches of water.

2.5 FORCES ON IMMERSED PLANE AREAS

It was shown in Section 2.3 that every particle below the surface of a fluid is under pressure which is given by the relationship $p = \gamma h$. This pressure causes a force to act on any surface which comes into contact with the fluid. This force, caused by the fluid at rest on surfaces in contact with it, is called the *fluidstatic force*. If the fluid is water, the force is known as the *hydrostatic force*. The point of application of the fluidstatic force is called the *center of pressure*. In this section, we will derive simple formulas which can be used to

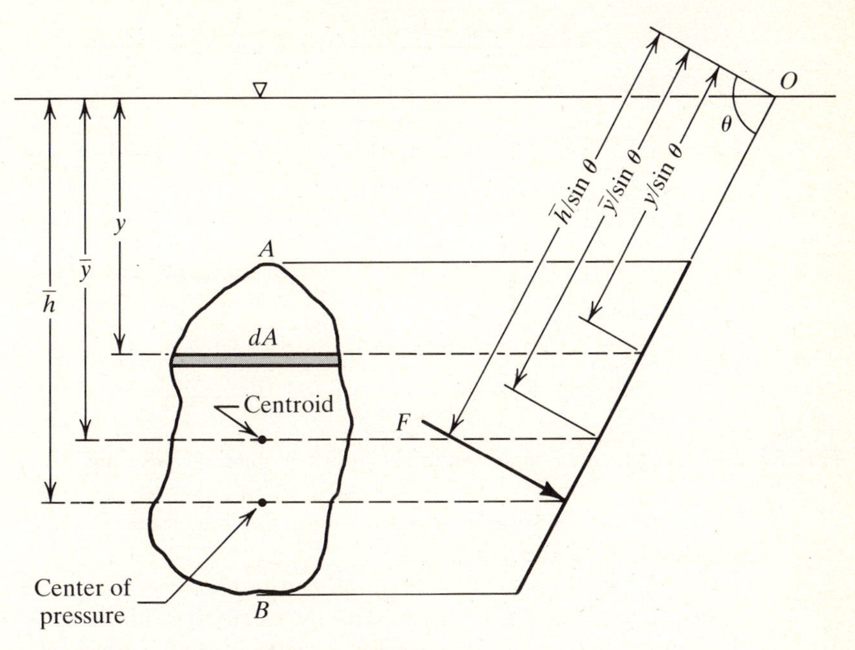

FIGURE 2.14 A plane area immersed in a fluid.

determine the fluidstatic force and its point of application on immersed plane areas.

Consider a plane area AB making an angle θ with the fluid surface as shown in Fig. 2.14. Let dA be a small area on the surface at a distance y vertically below the fluid surface. Considering the fluid to be incompressible, the pressure on the small area is

$$dp = \gamma y$$

The force on the small area is

$$dF = dpdA = \gamma y dA \tag{2.27}$$

The total force on the entire area AB is

$$F = \int_A \gamma y dA = \gamma \int_A y dA$$

$$F = \gamma A \bar{y} \tag{2.28}$$

where A is the total area of the surface and $\bar{y}$ is the vertical depth of the centroid of the area below the fluid surface.

$$\bar{y} = \frac{1}{A} \int_A y dA$$

It is seen from Eq. (2.28) that the fluidstatic force acting on a plane area is equal to its area multiplied by the pressure acting on its centroid. Since shear

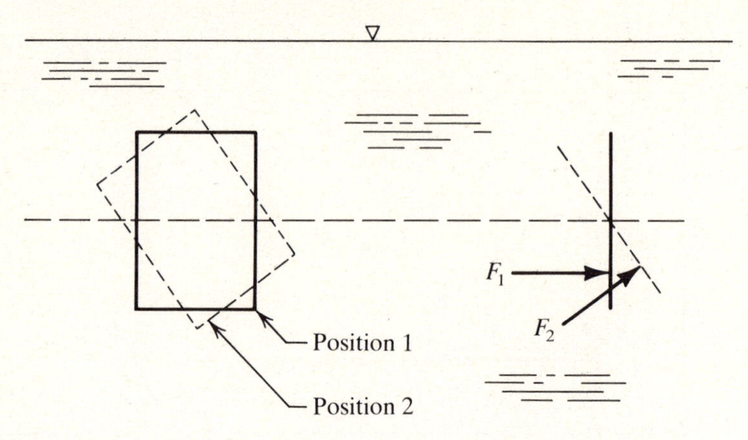

FIGURE 2.15 A plane area immersed in a fluid in two different positions.

forces are not present in a static fluid, the fluidstatic force always acts perpendicular to the plane area.

Since θ does not appear in the expression for the force, the magnitude of the fluidstatic force does not change with the orientation of the area as long as the centroid remains at the same depth below the fluid surface. However, the direction of the force will change with the change in the orientation of the area in such a way that it always remains perpendicular to the area (see Fig. 2.15).

The fluidstatic force acts through the center of pressure which is not the same point as the centroid of the area. In Fig. 2.14, let $\bar{h}$ be the depth of the center of pressure vertically below the fluid surface. Then taking moments about O:

$$F \frac{\bar{h}}{\sin\theta} = \int_A dF \frac{y}{\sin\theta}$$

Substituting for dF from Eq. (2.27) and also multiplying the numerator and the denominator of the right-hand side by $\sin\theta$:

$$\frac{F\bar{h}}{\sin\theta} = \int_A \gamma dA \left(\frac{y}{\sin\theta}\right)^2 \sin\theta$$

$$\frac{F\bar{h}}{\sin\theta} = \gamma I_o \sin\theta \tag{2.29}$$

where $I_o = \int_A dA(y/\sin\theta)^2$ = moment of inertia of the plane area about an axis passing through O.

Substituting for F from Eq. (2.28) and simplifying, we obtain

$$\bar{h} = \frac{I_o \sin^2\theta}{A\bar{y}} \tag{2.30}$$

Also

$$I_o = I_G + A\left(\frac{\bar{y}}{\sin\theta}\right)^2$$

where I_G is the moment of inertia of the plane area about an axis passing through the centroid and parallel to the fluid surface. Substituting for I_o in Eq. (2.30) and simplifying

$$\bar{h} = \bar{y} + \frac{I_G \sin^2 \theta}{A\bar{y}} \tag{2.31}$$

In Eq. (2.31), the term $I_G \sin^2 \theta / A\bar{y}$ is always positive. Therefore, the center of pressure is always located below the centroid. The only exception is when $\theta = 0$. In this case, the center of pressure coincides with the centroid.

The area, centroid and the moment of inertia about a centroidal axis for certain common plane surfaces are given in Fig. 2.16.

Imaginary Free Surface

Equations (2.28) and (2.31) are based on the assumption that the pressure on the surface of the fluid is atmospheric. If the pressure on the surface is greater than the atmospheric pressure, additional force due to this pressure must also act on the immersed area. This additional force may be calculated in one of two ways, as follows:

1. Let p' be the gauge pressure on the surface. Then additional force F' due to this pressure on an immersed plane area A would be given by:

$$F' = p'A$$

The total force on the plane area will be:

$$TF = F + F'$$

The additional force F' acts on the centroid of the area and, therefore, will shift the position of the center of pressure. The new location of the center of pressure, $\bar{H}$, is given by:

$$\bar{H} = \frac{F\bar{h} + F'\bar{y}}{F + F'}$$

2. Another method of dealing with the excess pressure is to convert the pressure into an equivalent height of the fluid column and then create an imaginary fluid surface at this level. The force and the location of the center of pressure may now be calculated as usual using Eqs. (2.28) and (2.31). Both of the above procedures are explained by Example 2.8.

Comments

1. The force F always acts perpendicular to the plane area and is independent of the orientation of the area.
2. The center of pressure is always below the centroid of the plane area unless the area is horizontal. In this case, $\bar{h} = \bar{y}$.
3. While calculating I_G care must be exercised so that it is always about an axis which passes through the centroid of the area and is parallel to the fluid surface.

Shape	Area	Centroid	Moment of Inertia
Rectangle	bh	$\bar{x} = \frac{1}{2}b$ $\bar{y} = \frac{1}{2}h$	$I_{xx} = \frac{1}{12}bh^3$
Triangle	$\frac{1}{2}bh$	$\bar{x} = \frac{b+c}{3}$ $\bar{y} = \frac{h}{3}$	$I_{xx} = \frac{1}{36}bh^3$
Circle	$\frac{1}{4}\pi d^2$	$\bar{x} = \frac{1}{2}d$ $\bar{y} = \frac{1}{2}d$	$I_{xx} = \frac{1}{64}\pi d^4$
Trapezoid	$\frac{h(a+b)}{2}$	$\bar{y} = \frac{h(2a+b)}{3(a+b)}$	$I_{xx} = \frac{h^3(a^2+4ab+b^2)}{36(a+b)}$

Semicircle 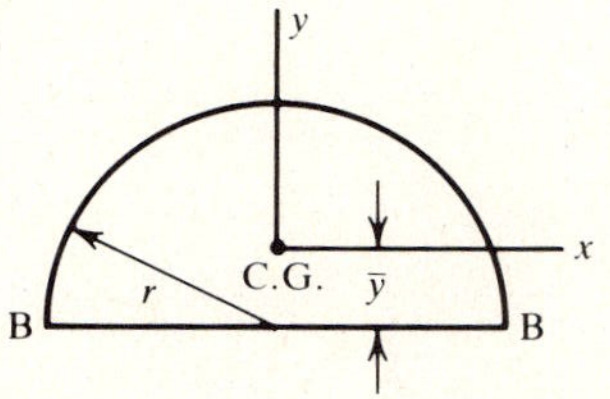	$\frac{1}{2}\pi r^2$	$\bar{y} = \dfrac{4r}{3\pi}$	$I_{xx} = \dfrac{(9\pi^2 - 64)r^4}{72\pi}$
Ellipse 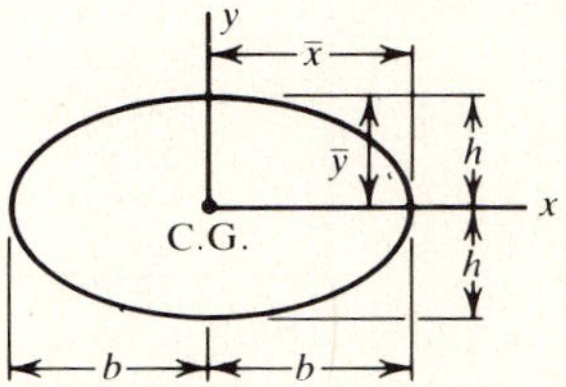	πbh	$\bar{x} = b$ $\bar{y} = h$	$I_{xx} = \frac{\pi}{4}bh^3$
Semiellipse 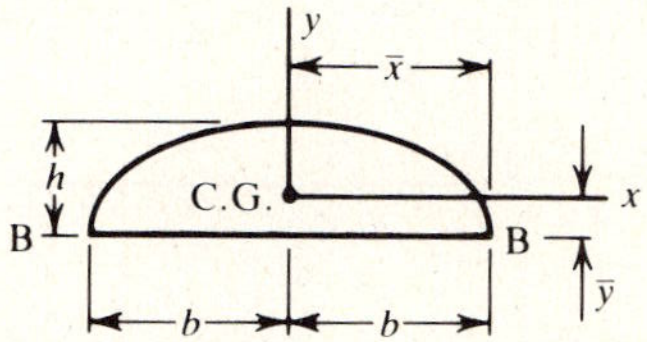	$\frac{\pi}{2}bh$	$\bar{x} = b$ $\bar{y} = \dfrac{4h}{3\pi}$	$I_{xx} = \dfrac{(9\pi^2 - 64)}{72\pi}bh^3$

4. Equations (2.28) and (2.31) are based on the assumption that the pressure on the surface of the fluid is atmospheric. If the pressure is not atmospheric, adjustment must be made in the force and its location as explained.
5. Since Eqs. (2.28) and (2.31) were derived using a constant value of density, they are valid only for a single homogeneous fluid.

EXAMPLE 2.6

A circular gate 1 m in diameter is located as shown in Fig. 2.17. Find the magnitude of the force F_1 required to open the gate. The fluid is water of specific weight 9810 N/m^3.

Solution The hydrostatic force is given by:

$$F = \gamma A \bar{y}$$

where

$$A = \frac{\pi(1)^2}{4} = 0.78 \text{ m}^2$$

$$\bar{y} = 5 + 0.5 = 5.5 \text{ m}$$

Hence:

$$F = 9810 \times 0.78 \times 5.5 = 42{,}080 \text{ N}$$

The depth of the center of pressure is given by

$$\bar{h} = \bar{y} + \frac{I_G \sin^2 \theta}{A\bar{y}}$$

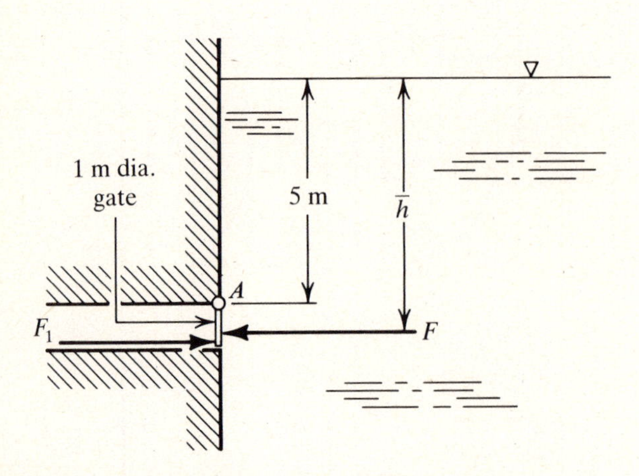

FIGURE 2.17 Example 2.6.

where

$$I_G = \frac{\pi(1)^4}{64} = 0.049 \text{ m}^4$$

$$\theta = 90^\circ$$

Hence:

$$\bar{h} = 5.5 + \frac{0.049 \times 1}{0.78 \times 5.5}$$

$$= 5.51 \text{ m}$$

Taking moments about the hinge A,

$$F_1 \times 1 = 42{,}080 \times (5.51 - 5)$$

$$F_1 = 21{,}460 \text{ N}$$

EXAMPLE 2.7

Find (a) the total force, and (b) the position of the center of pressure on one side of an immersed rectangular plate, 2.0 m long and 1.0 m wide, when the plane of the plate makes an angle of 60° with the surface of the water and the 1.0 m edge of the plate is parallel to, and at a depth of 75 cm below, the surface of the water.

Solution The immersed rectangular plate is shown in Fig. 2.18.

The formulas to be used are:

$$F = \gamma A\bar{y}$$

$$\bar{h} = \bar{y} + \frac{I_G \sin^2 \theta}{A\bar{y}}$$

(a) To find the force F:

$$A = 1 \times 2 = 2.0 \text{ m}^2$$

$$\bar{y} = 0.75 + 1.0 \sin 60^\circ = 1.616 \text{ m}$$

$$F = 9810 \times 2.0 \times 1.616 = 31.71 \text{ kN}$$

(b) To find the depth of the center of pressure:

$$I_G = \frac{1 \times (2)^3}{12} = 0.667 \text{ m}^4$$

$$\bar{h} = 1.616 + \frac{0.667 \times (\sin 60)^2}{2 \times 1.616}$$

$$= 1.616 + 0.155 = 1.771 \text{ m}$$

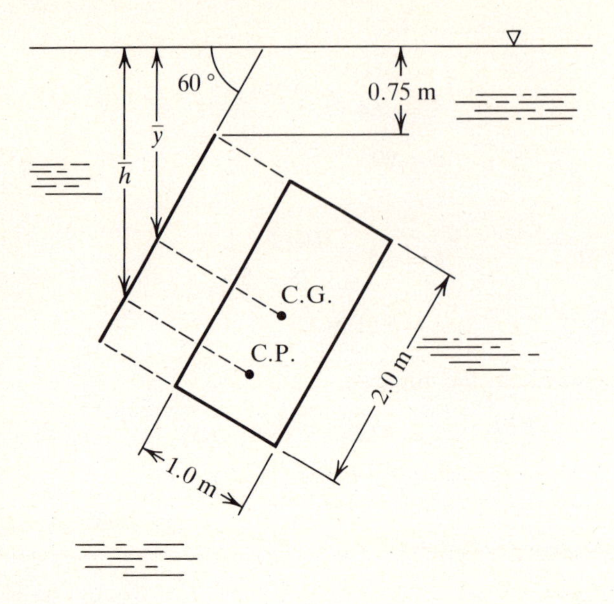

FIGURE 2.18 Example 2.7.

EXAMPLE 2.8

Determine the hydrostatic force and its point of location on the circular gate shown in Fig. 2.19

Solution This problem can be solved using two different approaches described in Section 2.5.

Using method (1):

$$A = \frac{\pi(6)^2}{4} = 28.27 \text{ ft}^2$$

$$\bar{y} = 10 + 2 + 3 \sin 30° = 13.5 \text{ ft}$$

$$F = 62.4 \times 28.27 \times 13.5 = 23{,}814 \text{ lbs}$$

$$F' = p'A = 600 \times 28.27 = 16{,}962 \text{ lbs}$$

Hence:

$$TF = 23{,}814 + 16{,}962 = 40{,}776 \text{ lbs}$$

$$I_G = \frac{\pi(6)^4}{64} = 63.617 \text{ ft}^4$$

$$\bar{h} = 13.5 + \frac{63.617 \times (\sin 30°)^2}{28.27 \times 13.5} = 13.54 \text{ ft.}$$

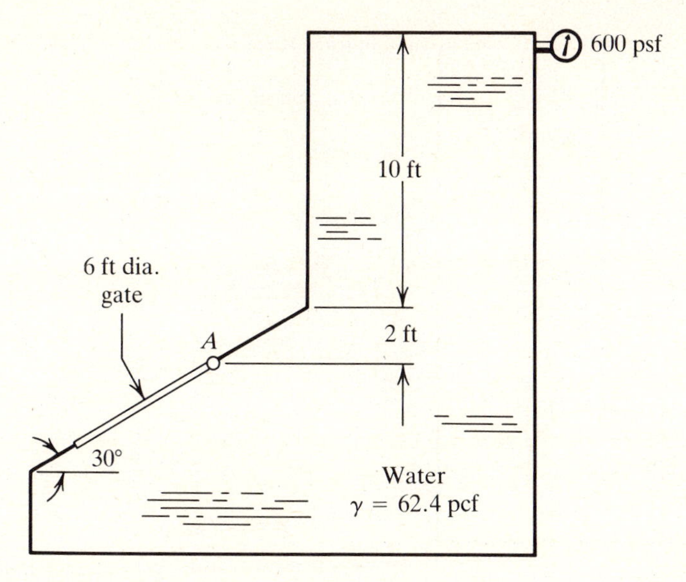

FIGURE 2.19 Example 2.8.

Therefore:

$$\bar{H} = \frac{(23{,}814 \times 13.54) + (16{,}962 \times 13.5)}{40{,}776}$$

$$= 13.52 \text{ ft}$$

Using method (2):

$$\text{equivalent fluid height} = \frac{p'}{\gamma} = \frac{600}{62.4} = 9.62 \text{ ft.}$$

Thus the imaginary fluid surface would be at a distance of 9.62 ft above the point where pressure is measured as shown in Fig. 2.20. Now

$$A = 28.27 \text{ ft}^2$$

$$\bar{y} = 9.62 + 13.5 = 23.12 \text{ ft}$$

Hence:

$$TF = 62.4 \times 28.27 \times 23.12 = 40{,}776 \text{ lbs}$$

$$\bar{h} = 23.12 + \frac{63.617(\sin 30)^2}{28.27 \times 23.115} = 23.14 \text{ ft}$$

Therefore:

$$\bar{H} = 23.14 - 9.62 = 13.52 \text{ ft}$$

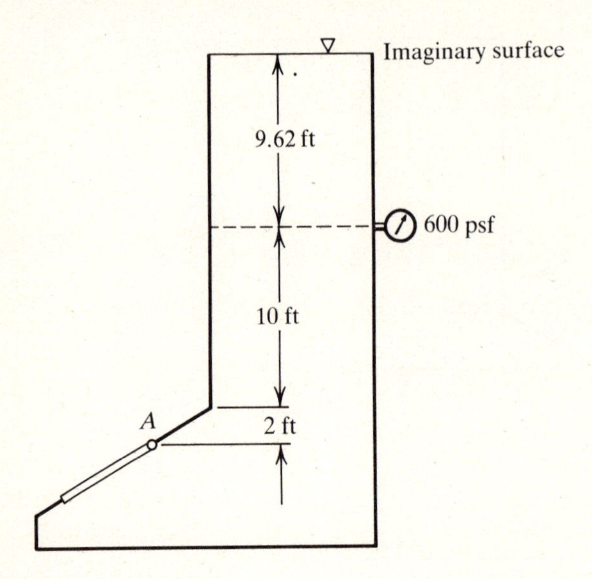

FIGURE 2.20 Example 2.8.

Design of a Lock Gate

The plan of a pair of lock gates AB and BC is shown in Fig. 2.21a. The gates are held in contact by the water pressure. Each gate is hinged at top and bottom. The forces acting on gate AB are: resultant hydrostatic force F, which acts through the middle of line AB; reaction T of gate BC; and the reaction R of the two hinges. Since gate AB is a three-force body, all the forces must pass through a common point D. In triangle ABD, angle DBA = angle DAB = angle θ. Hence $R = T$. Also:

$$F = (R + T)\sin\theta = 2R\sin\theta$$

$$R = \frac{F}{2\sin\theta} \tag{2.32}$$

Consider the hydrostatic forces on gate AB. The view of gate AB normal to itself is shown in Fig. 2.21b. The following forces act on gate of width w:

$$F_1 = \frac{1}{2}\gamma h_1^2 w \quad \text{at} \quad \frac{h_1}{3} \quad \text{from } F$$

$$F_2 = \frac{1}{2}\gamma h_2^2 w \quad \text{at} \quad \frac{h_2}{3} \quad \text{from } F$$

$R_T \sin\theta$ = component of the reaction of the top bearing normal to AB

$R_B \sin\theta$ = component of the reaction of the bottom bearing normal to AB.

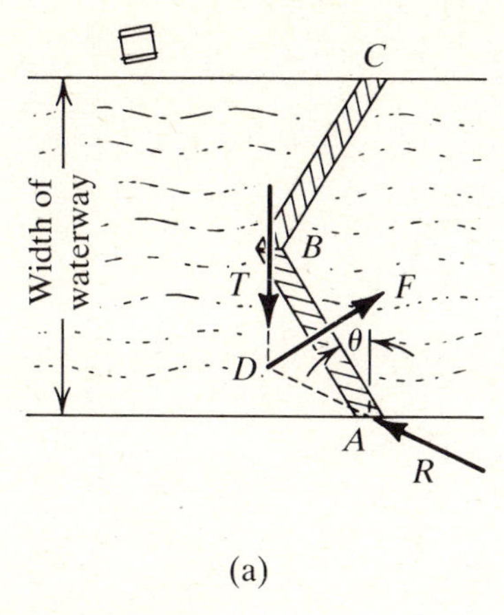

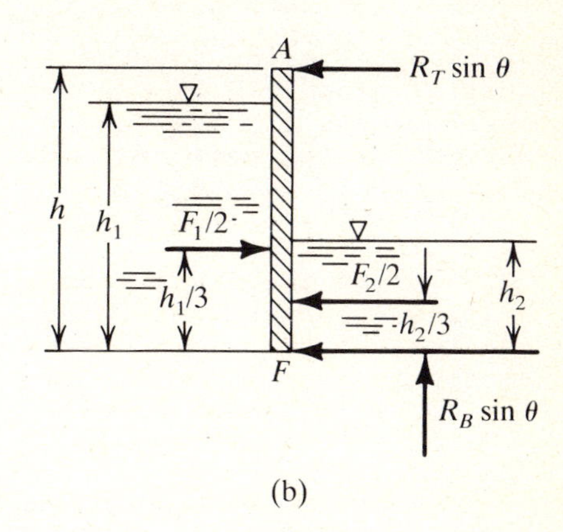

FIGURE 2.21 Lock gate.

It is assumed that only half the water pressure is taken by the hinges while the remaining half is taken by the reaction of the gate *BC*.

Taking moments about *F*:

$$R_T \sin \theta(h) = \frac{F_1}{2}\left(\frac{h_1}{3}\right) - \frac{F_2}{2}\left(\frac{h_2}{3}\right) \tag{2.33}$$

Summing the forces horizontally:

$$\frac{F_1}{2} - \frac{F_2}{2} = R_B \sin \theta + R_T \sin \theta \tag{2.34}$$

Equations (2.33) and (2.34) may be used to calculate the thrust on the hinge bearings. The bearings must be able to withstand this thrust with a reasonable factor of safety, say 1.5. A typical lock gate design problem would involve determining the angle θ, the width w of the gate, and the bearing thrust given the depth of water on both sides of the gate, the height of the gate and the width of the waterway.

A computer program is listed in Appendix B, which can be used to design a lock gate.

EXAMPLE 2.9

Design a lock gate for a waterway of width 30 ft. The gate is to be 20 ft high, with water elevation of 17 ft and 6 ft on its two sides. Use a factor of safety of 1.5. The available bearings have a maximum allowable thrust of 100,000 lbs, 120,000 lbs and 150,000 lbs.

```
RUN
lock gate design program
title? example 2.9
system of units (si/fps) ? fps
height of the gate (ft)? 20
fluid depth on the higher side (ft)? 17
fluid depth on the lower side (ft)? 6
width of the waterway (ft)? 30
sp. wt. of the fluid (lbs/cu.ft)? 62.4
minimum gate angle (degrees) ? 10
maximum gate angle (degrees) ? 45
angle to be incremented by (degrees) ? 5
allowable maximum thrust on each bearing (lbs)? 100000

example 2.9
abbreviations used in the table of printout
 ttb = thrust at the top bearing
 tbb = thrust at the bottom bearing
 safety factor = maximum thrust/allowable thrust
angle        width        ttb          tbb          safety
(deg)         (ft)        (lbs)        (lbs)        factor
  10       15.2314      116970.1     229219.4     .4362634
  15       15.52914     80012.24     156795.3     .6377744
  20       15.96267     62238.48     121965.1     .8199068
  25       16.55068     52224.29     102340.9     .9771267
  30       17.32052     46195.12     90525.87     1.104657
  35       18.31164     42573.67     83429.12     1.198622
  40       19.58114     40623.34     79607.16     1.256168
  45       21.21324     40006.2      78397.81     1.275546
*****
do you wish to try another bearing (y/n) ? y
allowable maximum thrust on each bearing (lbs)? 120000

example 2.9
abbreviations used in the table of printout
 ttb = thrust at the top bearing
 tbb = thrust at the bottom bearing
 safety factor = maximum thrust/allowable thrust
angle        width        ttb          tbb          safety
(deg)         (ft)        (lbs)        (lbs)        factor
  10       15.2314      116970.1     229219.4     .523516
  15       15.52914     80012.24     156795.3     .7653293
  20       15.96267     62238.48     121965.1     .9838881
  25       16.55068     52224.29     102340.9     1.172552
  30       17.32052     46195.12     90525.87     1.325588
  35       18.31164     42573.67     83429.12     1.438347
  40       19.58114     40623.34     79607.16     1.507402
  45       21.21324     40006.2      78397.81     1.530655
*****
do you wish to try another bearing (y/n) ? y
allowable maximum thrust on each bearing (lbs)? 150000
```

```
example 2.9
abbreviations used in the table of printout
 ttb = thrust at the top bearing
 tbb = thrust at the bottom bearing
 safety factor = maximum thrust/allowable thrust
angle        width         ttb           tbb           safety
(deg)         (ft)         (lbs)         (lbs)         factor
  10       15.2314       116970.1      229219.4       .6543951
  15       15.52914      80012.24      156795.3       .9566615
  20       15.96267      62238.48      121965.1      1.22986
  25       16.55068      52224.29      102340.9      1.46569
  30       17.32052      46195.12      90525.87      1.656985
  35       18.31164      42573.67      83429.12      1.797934
  40       19.58114      40623.34      79607.16      1.884253
  45       21.21324      40006.2       78397.81      1.913319
*****
do you wish to try another bearing (y/n) ? n
*****
do you wish to solve a new problem (y/n) ? y
title? example 2.9
system of units (si/fps) ? fps
height of the gate (ft)? 20
fluid depth on the higher side (ft)? 17
fluid depth on the lower side (ft)? 6
width of the waterway (ft)? 30
sp. wt. of the fluid (lbs/cu.ft)? 62.4
minimum gate angle (degrees) ? 25
maximum gate angle (degrees) ? 30
angle to be incremented by (degrees) ? 1
allowable maximum thrust on each bearing (lbs)? 150000

example 2.9
abbreviations used in the table of printout
 ttb = thrust at the top bearing
 tbb = thrust at the bottom bearing
 safety factor = maximum thrust/allowable thrust
angle        width         ttb           tbb           safety
(deg)         (ft)         (lbs)         (lbs)         factor
  25       16.55068      52224.29      102340.9      1.46569
  26       16.68904      50768.51      99488.08      1.507718
  27       16.8349       49450.3       96904.84      1.54791
  28       16.98856      48256.12      94564.69      1.586216
  29       17.15032      47174.38      92444.85      1.622589
  30       17.32052      46195.12      90525.87      1.656985
*****
do you wish to try another bearing (y/n) ? n
*****
do you wish to solve a new problem (y/n) ? n
Break in 650
Ok
```

FIGURE 2.22 Solution to example 2.9.

Solution The computer program listed in Appendix B is used to solve this problem. The results are shown in Fig. 2.22. Two solutions can be identified:

1. $\theta = 40°$, $w = 19.58$ ft with a thrust bearing of 120,000 lbs; safety factor = 1.507
2. $\theta = 26°$, $w = 16.69$ ft with a thrust bearing of 150,000 lbs; safety factor = 1.507

It is noticed that with a smaller bearing, larger angle and wider gates are required, while with a large bearing, smaller angle and narrower gates are required. Both meet the design criteria. The selection should be guided by economic considerations.

2.6 FORCES ON IMMERSED CURVED AREAS

The formulae developed in the previous section are applicable to plane areas only. They cannot be used if the area is curved. In this section, we will develop formulae to determine fluidstatic force and its point of application on a curved area.

Consider the curved surface AB shown in Fig. 2.23. The direction of the forces shown is due to the reaction of the surface. The fluidstatic forces will have directions opposite to the ones shown here. The fluidstatic forces δF acting on individual infinitesimal areas δA do not act parallel to each other due to the curvature of the surface. Hence, simple summation of these forces cannot yield the resultant force. However, these forces can be resolved into their components which can then be summed up vectorially using a convenient coordinate system.

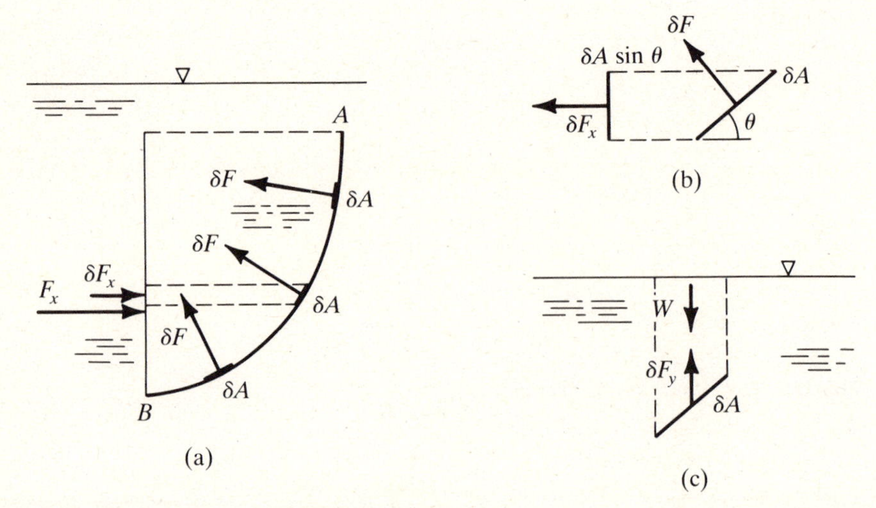

FIGURE 2.23 A curved area immersed in water.

Consider an elemental area δA shown exaggerated in Fig. 2.23b. The fluidstatic force acting on the area is $\delta F = p\delta A$, where p is the pressure at the centroid of the elemental area. The x-component of this force is given by:

$$\delta F_x = p\delta A \sin\theta = p\delta A_x$$

where δA_x is the projection of δA on a vertical plane. The x-component of the total force acting on the curved area is then given by:

$$F_x = \int p\delta A_x = \int \gamma y \delta A_x$$

$$= \gamma \int y \delta A_x$$

Or

$$F_x = \gamma A_x \bar{y} \tag{2.35}$$

where A_x is the area of the projection of the curved surface on a vertical plane. Since this formula is identical to the one for a plane area, the location of F_x can be found by using Eq. (2.31).

The y-component of the small force δF acting on the elemental area δA can be found by considering Fig. 2.23c. For equilibrium in the y-direction, δF_y must be equal in magnitude and opposite in direction to W, which is equal to the weight of the fluid overlying the elemental area. The y-component of the total force acting on the entire curved area will then be equal to the sum of all such elemental fluid weights. Hence:

$$F_y = \gamma \forall \tag{2.36}$$

where $\forall$ is the total volume of the fluid overlying the curved surface. The vertical component acts through the center of gravity of the overlying volume of the fluid.

The x-component and the y-component can then be combined vectorially to find the resultant force acting on the curved surface.

EXAMPLE 2.10

Determine the vertical and the horizontal components of the hydrostatic force acting on the curved gate shown in Fig. 2.24. The radius of the gate is 4 ft and its width is 1 ft.

Solution In this example, the underside of the curved gate is subjected to the hydrostatic pressure; therefore, the problem will be solved using an imaginary water surface as shown in Fig. 2.25. For the gate to be in equilibrium, the magnitude of the components calculated using the imaginary water will be equal in magnitude, but opposite in direction, to the actual components.

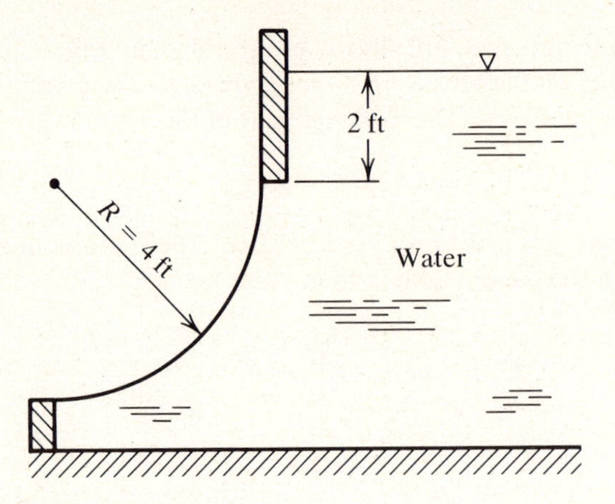

FIGURE 2.24 Example 2.10.

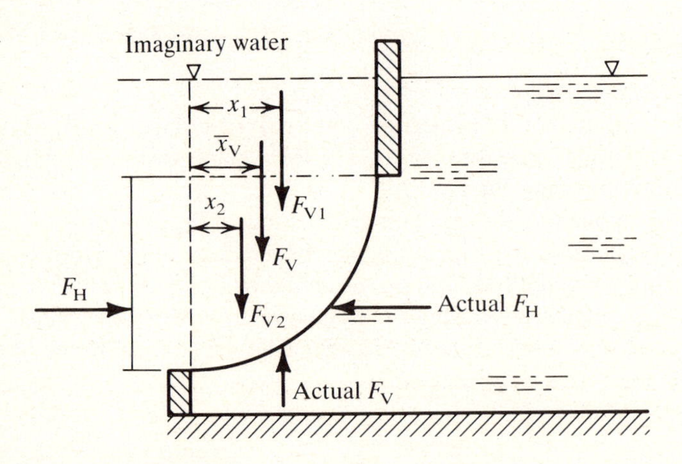

FIGURE 2.25 Example 2.10.

The horizontal component F_H is given by:

$$F_H = \gamma A \bar{y}$$
$$= 62.4(4)(4)$$
$$= 998.4 \text{ lbs}$$

Now, find the location of F_H.

$$I_G = \frac{1 \times 4^3}{12} = 5.33 \text{ ft}^4$$

$$\bar{h}_H = \bar{y} + \frac{I_G}{A\bar{y}}$$

$$= 4 + \frac{5.33}{(4)(4)}$$

$$\bar{h}_H = 4.33 \text{ ft}$$

The vertical component F_V can be divided into two subcomponents; F_{V1} due to the rectangular portion of the fluid and F_{V2} due to the quarter circle portion of the fluid. Then:

$$F_{V1} = 4 \times 2 \times 62.4$$
$$= 499.2 \text{ lbs}$$
$$F_{V_2} = \frac{\pi 4^2}{4} \times 62.4$$
$$= 784.1 \text{ lbs}$$
$$F_V = F_{V_1} + F_{V_2} = 499.2 + 784.1 = 1283.3 \text{ lbs}$$

Hence, the vertical component is equal to 1283.3 lbs. Now, find the location of F_V.

$$\bar{x}_V = \frac{F_{V_1}(x_1) + F_{V_2}(x_2)}{F_V}$$
$$= \frac{499.2(2) + 784.1(1.7)}{1283.3}$$
$$\bar{x}_V = 1.82 \text{ ft}$$

EXAMPLE 2.11

Determine the vertical and horizontal components of the force acting on the curved gate shown in Fig. 2.26. The gate is 1 ft wide.

Solution The water is under pressure. First create an imaginary water surface of height h' equivalent to the pressure of 50 psi.

$$h' = \frac{p}{\gamma} = \frac{50 \times 144}{62.4} = 115.4 \text{ ft}$$

The imaginary problem is shown in Fig. 2.27.
The horizontal component is given by:

$$F_H = \gamma A \bar{y}$$

where

$$A = 4 \times 1 = 4.0 \text{ ft}^2$$
$$\bar{y} = 115.4 + 2.5 + 2 = 119.9 \text{ ft.}$$

Hence:

$$F_H = 62.4 \times 4 \times 119.9$$
$$= 29{,}900 \text{ lbs.}$$

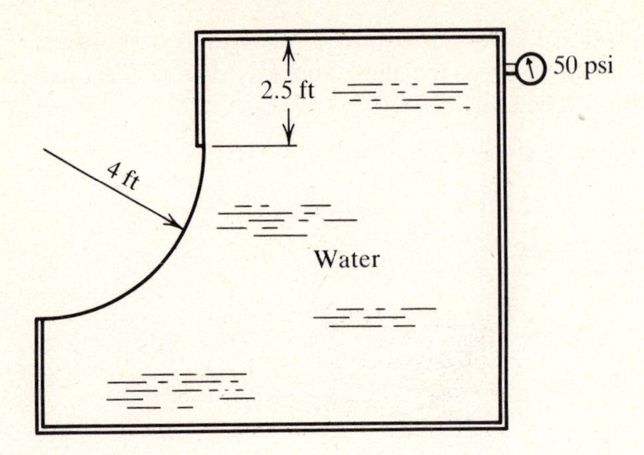

FIGURE 2.26 Example 2.11.

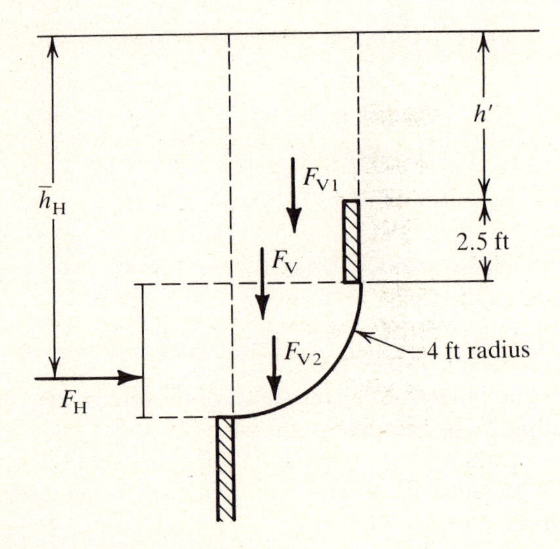

FIGURE 2.27 Example 2.11.

The location of the horizontal component is given by:

$$\bar{h}_H = \bar{y} + \frac{I_G}{A\bar{y}}$$

where

$$I_G = 5.33 \text{ ft}^4$$

Hence:

$$\bar{h}_H = 119.9 + \frac{5.33}{4 \times 119.9}$$

$$= 119.9 \text{ ft}$$

The vertical component is given by:

$$F_{V_1} = 62.4(4 \times 117.9) = 29{,}400 \text{ lbs}$$

$$F_{V_2} = 62.4\left[\frac{\pi(4)^2}{4}\right] = 784 \text{ lbs}$$

Hence:

$$F_V = 29{,}400 + 784 = 30{,}200 \text{ lbs}$$

The location of the vertical component is given by:

$$\bar{x}_V = \frac{29{,}400 \times 2 + 784 \times 1.7}{30{,}200}$$

$$= 1.99 \text{ ft from the left wall.}$$

Design of a Sluice Radial Gate

A sluice radial gate may be used to impound water above the crest of an overflow structure. A typical sluice radial gate is shown in Fig. 2.28.

At point O, a pair of bearings is provided to facilitate the raising and lowering of the gate. The bearings must also withstand the resultant thrust (R) due to water. The resultant thrust passes through the point O.

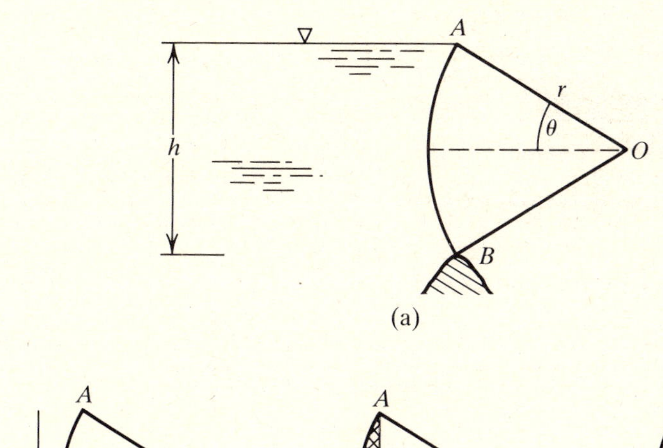

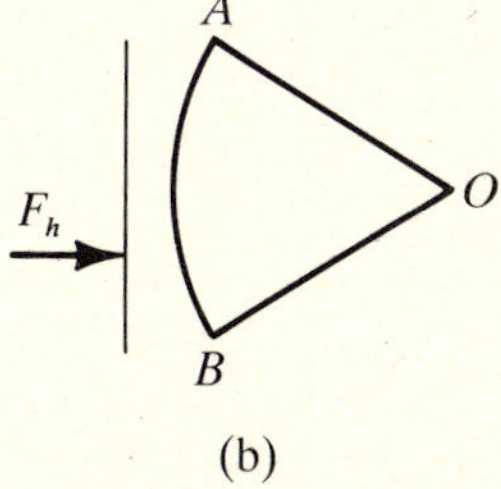

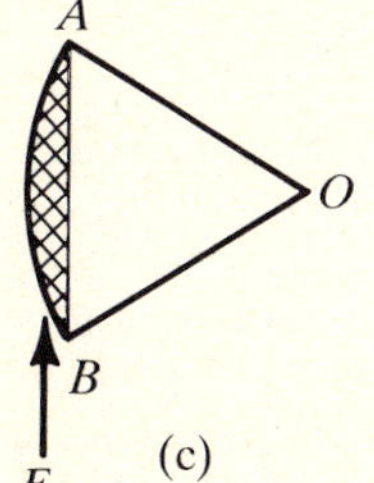

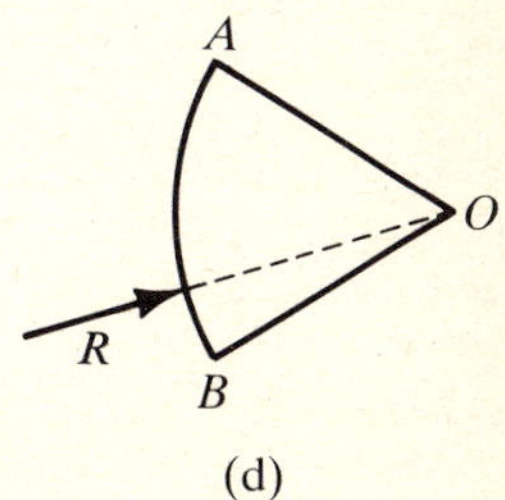

FIGURE 2.28 Sluice radial gate.

Let

r = radius of the gate

w = width of the gate

h = height of water to be impounded

θ = half of the angle subtended by the arc AB at O.

Then, the horizontal component of R is given by (Fig. 2.28b),

$$F_h = \frac{1}{2}\gamma w h^2$$

The vertical component of R is given by the weight of the fluid in the shaded area of Fig. 2.28c

$$F_v = \gamma r^2 w\left(\theta - \frac{\sin 2\theta}{2}\right)$$

And the resultant R is given by

$$R = \sqrt{F_h^2 + F_v^2}$$

The bearings at O must be able to withstand this thrust with a factor of safety of 1.5 to 2. A typical design problem would involve finding the angle θ and the radius of the gate, given the height of a fluid to be impounded and the ratings of various available bearings.

A computer program in BASIC is listed in Appendix B which can be used to design a radial sluice gate.

EXAMPLE 2.12

Design a 6 ft wide radial sluice gate to retain a maximum head of 12 ft of water of specific weight 62.4 lbs/ft^3. Three sizes of thrust bearings are available which can withstand 20,000 lbs, 25,000 lbs, and 30,000 lbs of thrust. Use a safety factor of 2.

Solution The computer program listed in Appendix B was used to solve this problem. The results are shown in Fig. 2.29. It is found that a radial gate having $\theta = 65°$, $r = 6.62$ ft with a pair of thrust bearings of 30,000 lbs each would meet the design criteria with a safety factor of 2.02.

```
RUN
radial sluice gate design program

title ? Example 2.12
system of units (si/fps) ? fps
maximum depth of water to be impounded(ft)? 12
width of the gate (ft)? 6
sp. wt. of the fluid (lbs/cu.ft)? 62.4
minimum angle theta (degrees) ? 30
maximum angle theta (degrees) ? 75
```

```
angle to be incremented by (degrees) ? 5
allowable maximum thrust on each bearing (lbs)? 25000

     angle          radius           thrust           safety
     (deg)           (ft)            (lbs)            factor
      30          11.99997         13697.82         1.825108
      35          10.46066         13784.48         1.813634
      40          9.334324         13889.82         1.799879
      45          8.485266         14016.58         1.783602
      50          7.832431         14168.2          1.764515
      55          7.324636         14349.02         1.742279
      60          6.928193         14564.51         1.716502
      65          6.62026          14821.54         1.686734
      70          6.385061         15128.88         1.652469
      75          6.211652         15497.67         1.613146
*****
do you wish to try another bearing (y/n) ? y
allowable maximum thrust on each bearing (lbs)? 30000

     angle          radius           thrust           safety
     (deg)           (ft)            (lbs)            factor
      30          11.99997         13697.82         2.19013
      35          10.46066         13784.48         2.17636
      40          9.334324         13889.82         2.159855
      45          8.485266         14016.58         2.140322
      50          7.832431         14168.2          2.117418
      55          7.324636         14349.02         2.090735
      60          6.928193         14564.51         2.059802
      65          6.62026          14821.54         2.024081
      70          6.385061         15128.88         1.982963
      75          6.211652         15497.67         1.935775
*****
do you wish to try another bearing (y/n) ? y
allowable maximum thrust on each bearing (lbs)? 35000

     angle          radius           thrust           safety
     (deg)           (ft)            (lbs)            factor
      30          11.99997         13697.82         2.555152
      35          10.46066         13784.48         2.539087
      40          9.334324         13889.82         2.519831
      45          8.485266         14016.58         2.497043
      50          7.832431         14168.2          2.470321
      55          7.324636         14349.02         2.439191
      60          6.928193         14564.51         2.403103
      65          6.62026          14821.54         2.361428
      70          6.385061         15128.88         2.313456
      75          6.211652         15497.67         2.258404
*****
do you wish to try another bearing (y/n) ? n
*****
do you wish to solve a new problem (y/n) ? n
Break in 510
Ok
```

FIGURE 2.29 Solution to example 2.12.

2.7 BUOYANT FORCES

Since the pressure in a static fluid increases with the depth, a body immersed in it experiences a net upward force. This force is known as the *buoyant force.* Let a body $ADBC$ be immersed in a fluid as shown in Fig. 2.30. The fluidstatic forces F_{DAC} acting on face DAC and F_{CBD} acting on face CBD are equal in magnitude and opposite in direction and, therefore, cancel each other out. Similarly, a horizontal force in any direction balances out with a horizontal force in the opposite direction. Hence, there is no net force acting on the body in the horizontal direction. Let

F_{ACB} = force acting on the face ACB of the body. It acts vertically upwards.

F_{ADB} = force acting on the face ADB of the body. It acts vertically downwards.

$F_{ACB} = \gamma$(volume of water in $EACBF$)

$F_{ADB} = \gamma$(volume of water in $EADBF$)

Since volume $EACBF$ is larger than the volume $EADBF$, the net force will be acting upwards.
Let

F_B = net force acting vertically upwards

Then

$$F_B = F_{ACB} - F_{ADB} = \gamma(\text{volume of fluid in } EACBF - \text{volume of fluid in } EADBF)$$

$$= \gamma(\text{volume of fluid in } ADBC)$$

or

$$F_B = \gamma \forall \tag{2.37}$$

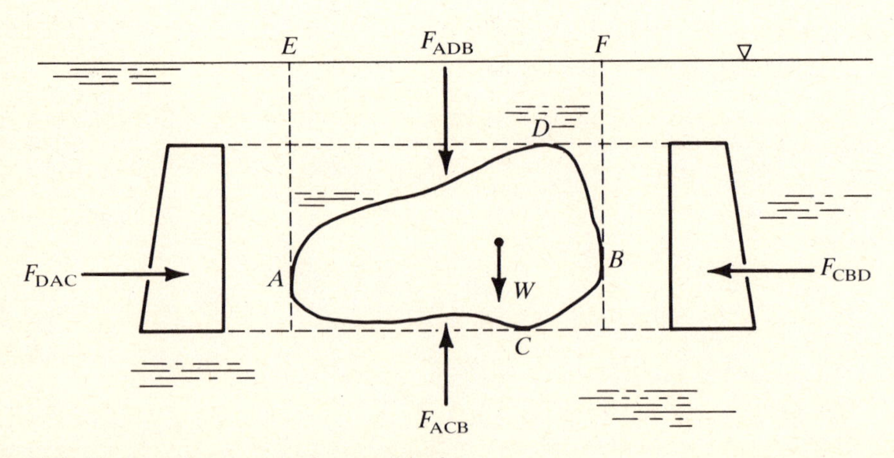

FIGURE 2.30 Buoyant force on a submerged body.

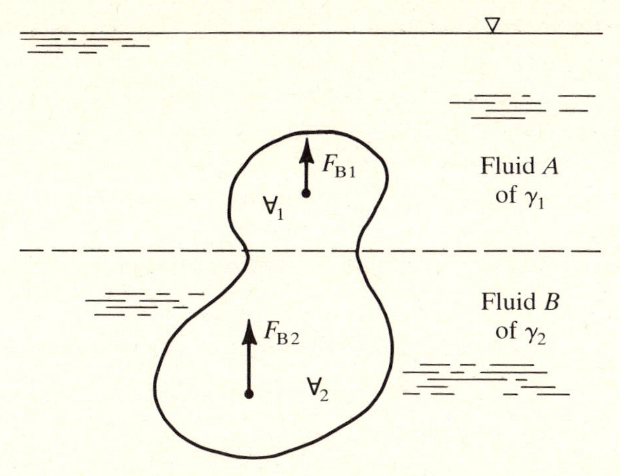

FIGURE 2.31 Buoyant force on a body submerged in two immiscible fluids.

where

F_B = buoyant force,
γ = specific weight of the fluid,
$\forall$ = volume of the body = volume of fluid displaced by the body

From the above equation, the magnitude of the buoyant force is equal to the weight of the fluid displaced by the body and it acts through the center of gravity of the displaced fluid. The point through which the buoyant force acts is known as *the center of buoyancy.*

When a body is immersed in a fluid unsupported, two forces (neglecting the viscous forces) act on the body. One is the weight of the body (W) which acts vertically downwards and the other is the buoyant force which acts vertically upwards. If $W > F_B$, the body will sink; if $W = F_B$, the body will float and if $W < F_B$, the body will rise as in the case of a balloon in the air.

If a body is partially immersed in a fluid, it can be proved by similar considerations that the buoyant force is equal to the weight of the fluid displaced and that it acts through the center of gravity of the displaced volume.

If a body is immersed in two immiscible fluids as shown in Fig. 2.31, the total buoyant force will be the sum of the buoyant forces due to the fluids A and B, or

$$F_B = \gamma_1 \forall_1 + \gamma_2 \forall_2$$

The center of buoyancy in this case may not be located at the center of gravity of the displaced volume because the centers of buoyancy of the volumes $\forall_1$ and $\forall_2$ are not on the same vertical line.

Comments

1. The *center of buoyancy* should be clearly distinguished from the *center of gravity* of the immersed body. In the case of a homogeneous fluid, the center of buoyancy is located at the center of gravity of the displaced

volume of fluid which may or may not coincide with the center of gravity of the body.

2. For everyday calculations, the effect of buoyancy of the atmospheric air on the weight of the objects is mostly ignored, although in very accurate work a correction may be necessary.

EXAMPLE 2.13

The barge shown in Fig. 2.32 has a specific gravity of 0.9. It carries an additional weight of 150 kN. Determine the depth to which water would rise. The water is of specific gravity 1.09.

Solution

$$\text{Weight of barge} = \gamma \text{ of barge} \times \text{volume of barge}$$

$$= (9810 \times 0.9) \times (3 \times 4 \times 15)$$

$$= 1589 \text{ kN}$$

$$\text{Additional weight} = 150 \text{ kN}$$

$$\text{Total weight} = W = 1589 + 150$$

$$= 1739 \text{ kN}$$

Since the barge is floating

$$F_B = W$$

$$= 1739 \text{ kN}$$

Also:

$$F_B = \gamma \forall \qquad \text{(from Eq. 2.37)}$$

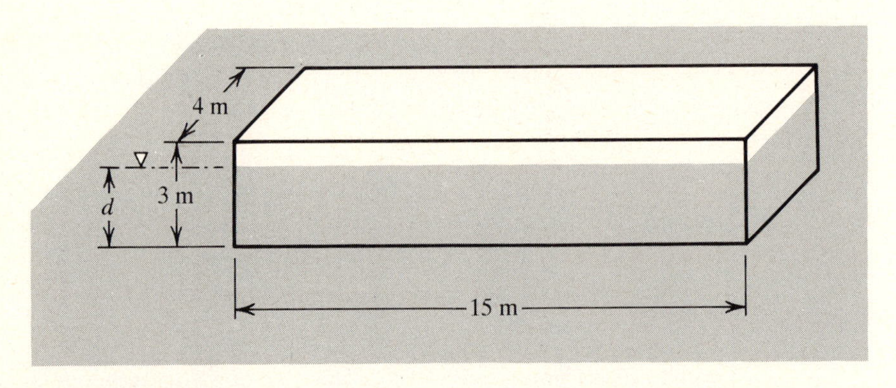

FIGURE 2.32 Example 2.13.

Hence

$$\forall = \text{volume submerged}$$

$$= \frac{F_B}{\gamma}$$

$$= \frac{1739}{(9.81)(1.09)} = 162.6 \text{ m}^3$$

Therefore,

$$\text{Depth submerged} = d = \frac{162.6}{15 \times 4}$$

$$= 2.71 \text{ m}$$

EXAMPLE 2.14

A 10 cm diameter solid cylinder of height 9 cm weighing 0.4 kg is immersed in a liquid of specific gravity 0.8 contained in a tall upright metal cylinder of 14 cm diameter. Before the solid cylinder was introduced the liquid was 8 cm deep. At what level will the solid cylinder float?

Solution Assume that when the solid cylinder is dropped, it displaces a certain volume (A) before coming to equilibrium. The equilibrium position of the cylinder is shown in Fig. 2.33. Let

$$CC = \text{original liquid level}$$

$$DD = \text{liquid level after immersion}$$

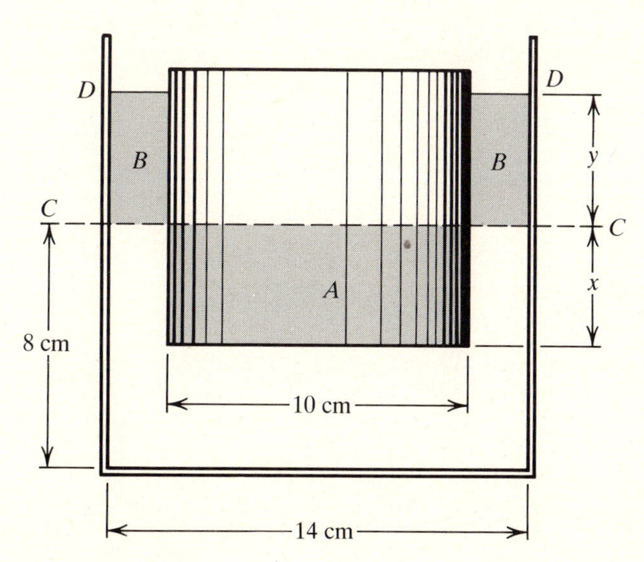

FIGURE 2.33 Example 2.14.

Now

$$\text{Vol. } A = \text{Vol. } B$$

$$\frac{\pi(10)^2}{4}x = \frac{\pi}{4}(14^2 - 10^2)y$$

$$100x = 96y$$

$$x = 0.96y \text{ cm}$$

Since the solid cylinder is floating:

$$F_B = W = 0.4 \text{ kg} = 0.4 \times 9.81 = 3.924 \text{ N}$$

Also

$$F_B = \text{weight of the volume of liquid displaced}$$

$$= 9810 \times 0.8 \times \frac{\pi}{4}(10)^2(x + y) \times 10^{-6} = 0.616(x + y) \text{ N}$$

Hence:

$$0.616(x + y) = 3.924$$

$$x + y = 6.37 \text{ cm}$$

Substituting $x = 0.96y$, we get

$$0.96y + y = 6.37$$

$$y = 3.25 \text{ cm}$$

and, $x = 0.96 \times 3.25 = 3.12$ cm. Therefore, the bottom of the solid cylinder will be $(8.0 - 3.12) = 4.88$ cm above the bottom.

EXAMPLE 2.15

A small metal pan of length 100 cm, width 20 cm, and depth 4 cm is floating in water. When a uniform vertical load of 1.5 N/m is applied as shown in Fig. 2.34, the pan assumes the given configuration. Find the weight of the pan and the magnitude of the righting moment.

Solution Since the pan is in equilibrium, the buoyant force must be equal to the weight of the pan plus the added weight of 1.5 N/m.

$$F_B = \gamma \forall$$

$$= 9810\left[\frac{4 \times 20 \times 100}{(100)^3 \times 2}\right]$$

$$= 39.2 \text{ N}$$

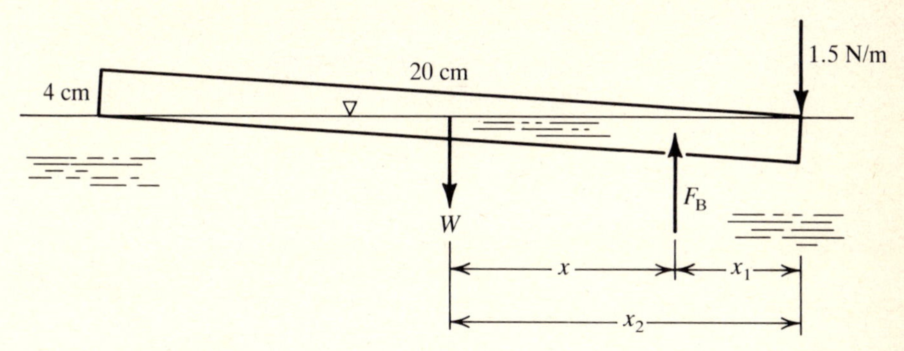

FIGURE 2.34 Example 2.15.

Let the weight of the pan be W. Then total weight is

$$W + (1.5)1 = (W + 1.5)\ \mathrm{N}$$

At equilibrium, this total weight must be equal to the buoyant force. Therefore,

$$W + 1.5 = 39.2$$

or

$$W = 39.2 - 1.5 = 37.7\ N$$

Hence, the weight of the pan is 37.7 N. Righting moment $= Wx$, where x is as shown in Fig. 2.34.

$$x = (x_2 - x_1)\cos\theta,$$

where x_1, x_2 are as shown in Fig. 2.34

$$x_1 = \frac{20}{3} = 6.67\ \mathrm{cm}$$

$$x_2 = \frac{20}{2} = 10\ \mathrm{cm}$$

$$\theta = \tan^{-1}\left(\frac{4}{20}\right) = 11.31^\circ$$

Therefore,

$$x = (10 - 6.67)\cos 11.31^\circ$$
$$= 3.27\ \mathrm{cm}$$

$$\text{Righting moment} = 37.7 \times 3.27$$
$$= 123.3\ \mathrm{N \cdot cm}$$

Stability of a Floating Body

Let $ABCD$ (Fig. 2.35a) be a body floating in a liquid. Let it be heeled through a small angle θ and acquire a new position $A'B'C'D'$ (Fig. 2.35b).

Also let

B = position of the center of buoyancy before heeling

B_1 = position of the center of buoyancy after heeling

G = position of the center of gravity

W = weight of the body

M = the point where a vertical line passing through B_1 meets a line drawn through B and G.

Since the body is floating, the magnitude of the buoyant force is equal to the weight of the body, or

$$F_B = W$$

This creates a couple which has a tendency to bring the body back to its original position. This couple is called *the righting couple.*

Let S = the length of a perpendicular drawn from G to MB_1

Then:

$$\text{Moment of the couple} = W \times S$$

$$= W \times MG \sin \theta \tag{2.38}$$

$$= W \times MG \tan \theta \text{ (as } \theta \text{ is small)} \tag{2.39}$$

$$= W \times MG\,\theta \tag{2.40}$$

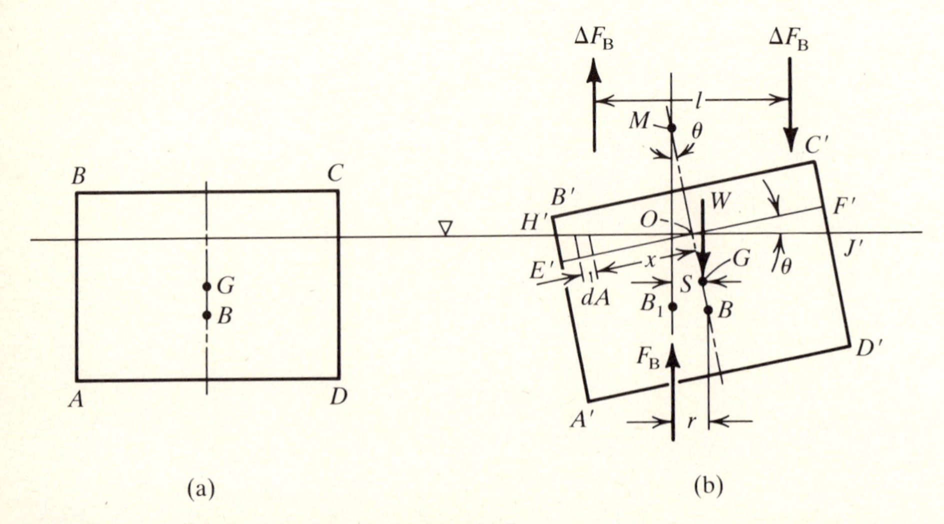

FIGURE 2.35 Stability of a prismatic floating body.

The point M is called the *meta center.*

The distance MG is known as the *metacentric height.*

If M is *above* G, the equilibrium is *stable.*

If M is *below* G, the equilibrium is *unstable.*

If M *coincides* with G, the equilibrium is *neutral.*

A formula is now derived to determine the metacentric height for a very small angle of rotation θ. Notice from Fig. 2.35b that as a result of rotation, a small wedge $OH'E'$ of the body is submerged in the liquid causing an upward force ΔF_B, and an identical wedge $OF'J'$ of the body comes out of the liquid causing a downward force ΔF_B. The horizontal shift in the center of buoyancy r is determined by the change in buoyant forces caused by the wedges $OH'E'$ and $OF'J'$, or

$$\Delta F_B l = Wr$$

and

$$\Delta F_B l = \gamma \forall r \tag{2.41}$$

where $\forall$ is the total volume of the liquid displaced.

The term $\Delta F_B l$ can be evaluated by considering an elemental volume of the wedge having a base area equal to dA (Fig. 2.35b) at a distance x from the point O. Then volume of the elemental wedge is $x\theta dA$. The buoyant force due to this elemental wedge is $\gamma x\theta dA$ and its moment about O is $\gamma\theta x^2 dA$. Then:

$$\Delta F_B l = \int_A \gamma\theta x^2 dA$$

$$= \gamma\theta \int_A x^2 dA$$

or

$$\Delta F_B l = \gamma\theta I_o \tag{2.42}$$

where I_o is the moment of inertia of the area at the waterline about the longitudinal axis passing through O.

Substituting for $\Delta F_B l$ in Eq. (2.41):

$$\gamma\theta I_o = \gamma \forall r$$

or

$$\theta I_o = \forall r \tag{2.43}$$

Also

$$r = MB \times \sin\theta$$

$$= MB \times \theta \text{ (since } \theta \text{ is small)}$$

Hence

$$\theta I_o = \forall MB\theta$$

or

$$MB = \frac{I_o}{\forall} \qquad (2.44)$$

The metacentric height MG is now given by

$$MG = MB \mp GB$$

or

$$MG = \frac{I_o}{\forall} \mp GB \qquad (2.45)$$

EXAMPLE 2.16

For the barge of Example 2.13, compute (a) the metacentric height and (b) the righting moment when the angle of heel is 6°.

Solution (a) The metacentric height MG is calculated using Eq. (2.45).

$$MG = \frac{I_o}{\forall} - GB$$

where

$$I_o = \frac{15 \times (4)^3}{12} = 80 \text{ m}^4$$

$$\forall = (15 \times 4 \times 2.71) = 162.6 \text{ m}^3$$

$$GB = \left(\frac{3}{2} - \frac{2.71}{2}\right) = 0.145 \text{ m}$$

Therefore:

$$MG = \frac{80}{162.6} - 0.145 = 0.347 \text{ m}$$

The metacentric height is 0.347 m.

(b) The righting moment is calculated using Eq. (2.38):

$$M = W \times MG \sin\theta$$

where

$$W = 1739.2 \text{ kN} \qquad \text{(from Ex. 2.13)}$$

$$MG = 0.347 \text{ m}$$

$$\theta = 6°$$

Therefore:

$$M = 1739.2 \times 0.347 \times \sin 6^\circ$$
$$= 63.1 \text{ kN} \cdot \text{m}$$

The righting moment is 63.1 kN · m.

2.8 RELATIVE EQUILIBRIUM

In this section, we will study two instances when the methods of fluidstatics can be used to study the behavior of fluids in motion. One is the case of uniform linear acceleration when the entire fluid is moving as a solid with no relative movement of one layer with respect to another. The other is the case when a fluid is rotated uniformly about a vertical axis. Fluids in such motion are said to be in *relative equilibrium.* In these cases, the planes of equal pressure are not horizontal but parallel to the fluid surface whose slope and shape are determined by the magnitude and the nature of the uniform acceleration. The two cases of uniform acceleration are now described in detail.

Uniform Linear Acceleration

Consider a mass of fluid moving with a uniform acceleration in space as shown in Fig. 2.36a. Once the fluid has reached equilibrium, it moves as a solid having no shear forces within itself. Let us select a two-dimensional rectangular coordinate system in such a way that the acceleration vector a lies in the $x - y$ plane (Fig. 2.36b). Then, the acceleration vector will have no component in the z-direction. Consider an elemental volume of fluid at a point P as shown in Fig. 2.36a. Since the laws of fluidstatics apply, the forces

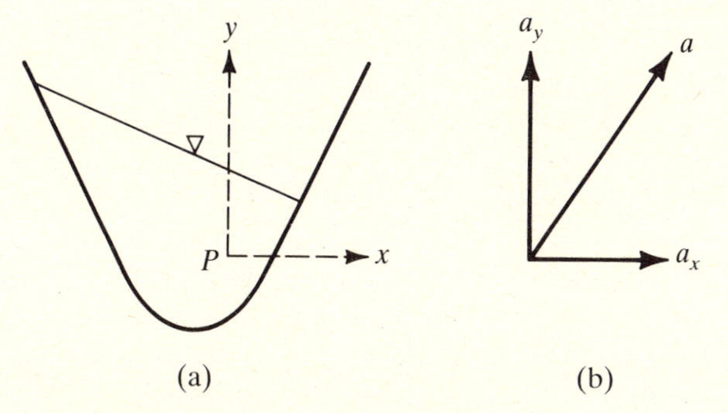

FIGURE 2.36 Linear acceleration with free surface.

acting on the elemental volume are identical to the forces shown in Fig. 2.6b in Section 2.3. From Eq. (2.11)

$$\Sigma F_x = -\frac{\partial p}{\partial x}\,\delta x \delta y \delta z$$

According to Newton's second law of motion

$$\Sigma F_x = m a_x$$

Therefore

$$-\frac{\partial p}{\partial x}(\delta x \delta y \delta z) = \frac{\gamma}{g}(\delta x \delta y \delta z) a_x$$

$$\frac{\partial p}{\partial x} = -\frac{\gamma a_x}{g} \tag{2.46}$$

Also, from Eq. (2.15)

$$\Sigma F_y = -\frac{\partial p}{\partial y}(\delta x \delta y \delta z) - \gamma(\delta x \delta y \delta z)$$

and

$$\Sigma F_y = m a_y$$

Hence:

$$-\frac{\partial p}{\partial y}(\delta x \delta y \delta z) - \gamma(\delta x \delta y \delta z) = \frac{\gamma}{g}(\delta x \delta y \delta z) a_y$$

$$-\frac{\partial p}{\partial y} - \gamma = \frac{\gamma}{g} a_y$$

$$\frac{\partial p}{\partial y} = -\gamma\left(1 + \frac{a_y}{g}\right) \tag{2.47}$$

Since pressure p is a function of position x, y, its total differential is given by:

$$dp = \frac{\partial p}{\partial x} dx + \frac{\partial p}{\partial y} dy$$

Substituting from Eqs. (2.46) and (2.47), we obtain

$$dp = -\frac{\gamma}{g} a_x dx - \gamma\left(1 + \frac{a_y}{g}\right) dy$$

Integrating, we obtain

$$p = -\gamma \frac{a_x}{g} x - \gamma\left(1 + \frac{a_y}{g}\right) y + C \tag{2.48}$$

where C is a constant of integration. To evaluate C, the origin may be placed at a point where pressure is known. Let $p = p_0$ when $x = 0$, $y = 0$. Then

$$C = p_0$$

Substituting in Eq. (2.48), we obtain

$$p = p_0 - \gamma \frac{a_x}{g} x - \gamma\left(1 + \frac{a_y}{g}\right)y \tag{2.49}$$

Solving Eq. (2.49) for y gives

$$y = -\frac{a_x}{a_y + g} x + \frac{p_0 - p}{\gamma\left(1 + \dfrac{a_y}{g}\right)}$$

This is the equation of a straight line whose slope is given by

$$m = -\frac{a_x}{(a_y + g)} \tag{2.50}$$

Hence in the case of linear acceleration, the lines of equal pressure are inclined lines having slope given by Eq. (2.50).

EXAMPLE 2.17

A tank 6 m long, 2 m wide, and 3 m deep containing water to a depth of 1.5 m is accelerated at a rate of 3 m/s^2 in a direction making an angle of 30° with the horizontal. Find the maximum pressure in the tank.

Solution The shape of the water surface at equilibrium is shown in Fig. 2.37. Since the liquid behaves as a static liquid, the pressure will be maximum at the point where the liquid has its maximum height. Obviously, for this problem the point of maximum pressure is at A. To find the pressure at A, first we must find y_2. From Eq. (2.50)

$$m = -\frac{a_x}{a_y + g}$$

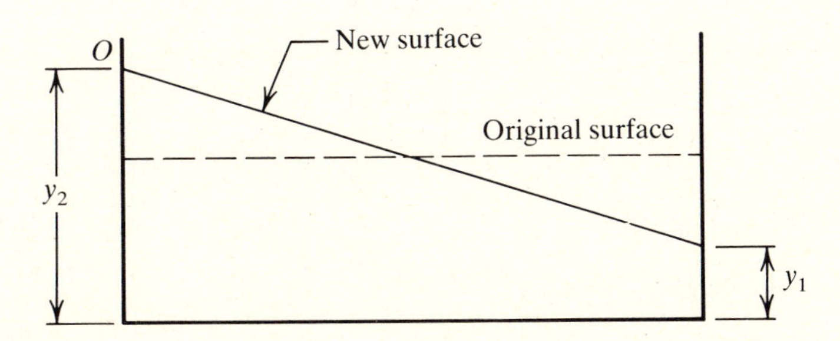

FIGURE 2.37 Example 2.17.

where

$$a_x = a \cos\theta = 3 \times \cos 30^\circ = 2.6 \text{ m/s}^2$$

$$a_y = a \sin\theta = 3 \times \sin 30^\circ = 1.5 \text{ m/s}^2$$

Therefore

$$m = -\frac{2.6}{1.5 + 9.81} = -0.23$$

The minus sign indicates that the slope is negative.

$$y_2 = y_1 + m(6)$$

$$= y_1 + 0.23 \times 6 = y_1 + 1.38$$

or

$$y_2 = y_1 + 1.38 \qquad (1)$$

Also, the volume in the tank in the new position is the same as the volume in the old position, so that

$$\left(\frac{y_1 + y_2}{2}\right)6 \times 2 = 6 \times 2 \times 1.5$$

$$y_1 + y_2 = 3.0 \qquad (2)$$

Solving Eqs. (1) and (2) simultaneously yields

$$y_1 = 0.81 \text{ m}$$

$$y_2 = 2.19 \text{ m}$$

The maximum pressure would occur somewhere on the bottom of the tank. The pressure at any point in the fluid is given by Eq. (2.49), according to which

$$p = p_0 - 9810\frac{2.6}{9.81}x - 9810\left(1 + \frac{1.5}{9.81}\right)y$$

$$p = p_0 - 2600x - 11310y$$

If we place the origin at O, then p_0 is zero; hence

$$p = -2600x - 11310y$$

At the bottom of the tank, $y = -2.19$ m; hence the equation of pressure for the bottom is given by

$$p = -2600x - 11310(-2.19)$$

or

$$p = 24{,}769 - 2600x$$

It is evident from the above equation that pressure would be maximum when $x = 0$. Therefore, maximum pressure would occur at the left bottom corner of the tank and is equal to 24,800 N.

Uniform Rotation About a Vertical Axis

In the case of uniform rotation about a vertical axis (also known as *forced vortex motion*), every particle of the fluid has the same angular velocity. Therefore, there is no relative movement between the fluid layers and no shear forces. Since there are no shear forces, the laws of fluidstatics apply.

Let a jar containing a fluid be rotated about its vertical axis with a constant angular velocity of ω rad/s (Fig. 2.38). Consider an elemental volume of fluid of length δr and depth δy at a distance r from the axis as shown in Fig. 2.38. The forces acting on the various faces of the elemental volume are also shown in Fig. 2.38. Summing up the forces in the vertical direction yields:

$$\Sigma F_y = -\frac{\partial p}{\partial y}(\delta A \delta y) - \gamma(\delta A \delta y)$$

Since there is no acceleration in the y-direction, ΣF_y must be zero. Hence:

$$-\frac{\partial p}{\partial y}(\delta A \delta y) - \gamma(\delta A \delta y) = 0$$

which yields

$$\frac{\partial p}{\partial y} = -\gamma \tag{2.51}$$

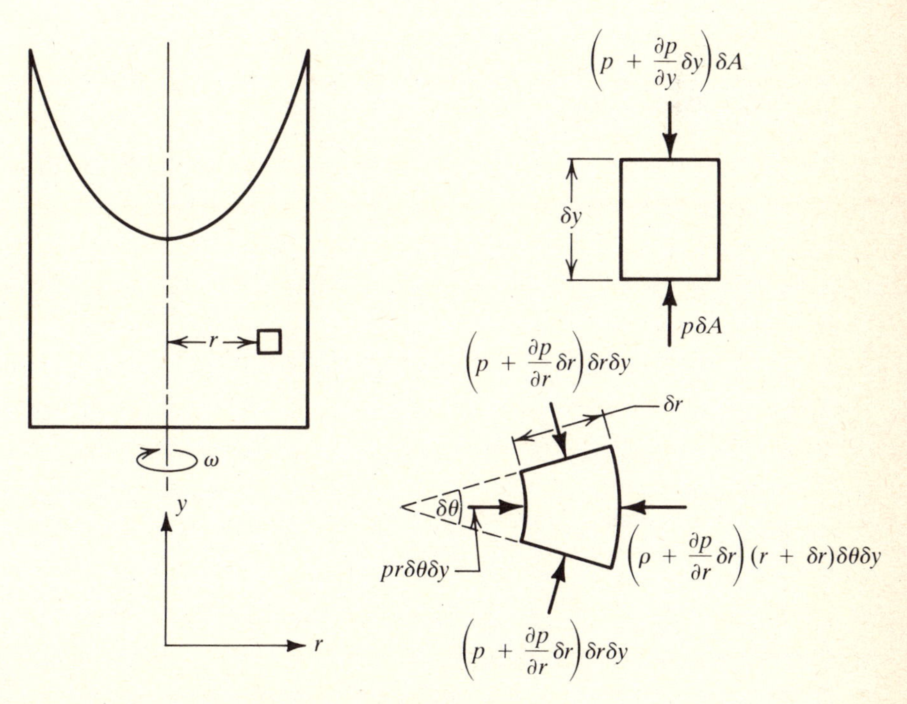

FIGURE 2.38 Uniform rotation about a vertical axis.

Summing up the forces in the radial direction yields:

$$\Sigma F_r = pr\,\delta\theta\delta y - \left(p + \frac{\partial p}{\partial r}\,\delta r\right)(r + \delta r)\delta\theta\delta y$$
$$+ 2\left(p + \frac{\partial p}{\partial r}\,\delta r\right)\delta r\delta y \sin\left(\frac{\delta\theta}{2}\right)$$

Simplifying, neglecting second order of small quantities and assuming $\sin \delta\theta/2 \equiv \delta\theta/2$, we obtain

$$\Sigma F_r = -\frac{\partial p}{\partial r}\, r\delta r\delta\theta\delta y$$

From Newton's second law of motion, $\Sigma F_r = ma_r$, where a_r is the radial acceleration. Also, $a_r = -\omega^2 r$. Hence

$$-\frac{\partial p}{\partial r}\, r\delta r\delta\theta\delta y = \frac{\gamma}{g}(r\delta r\delta\theta\delta y)(-\omega^2 r)$$

or

$$\frac{\partial p}{\partial r} = \frac{\gamma}{g}\,\omega^2 r \tag{2.52}$$

Since the pressure p is a function of position r, y, its total differential is given by:

$$dp = \frac{\partial p}{\partial r}\,dr + \frac{\partial p}{\partial y}\,dy$$

Substituting from Eqs. (2.51) and (2.52), we obtain

$$dp = \gamma\frac{\omega^2}{g}\,rdr - \gamma dy$$

Integration gives

$$p = \gamma\frac{\omega^2}{g}\frac{r^2}{2} - \gamma y + C$$

To evaluate C, the origin may be placed at a point where pressure is known. Let $p = p_0$ when $r = 0$, $y = 0$. Then, $C = p_0$. Hence,

$$p = p_0 + \gamma\frac{\omega^2}{2g}\,r^2 - \gamma y \tag{2.53}$$

Let the coordinates of two points 1 and 2 be (r_1, y_1) and (r_2, y_2) respectively from a common origin. Then from Eq. (2.53)

$$p_1 = p_0 + \frac{\gamma\omega^2}{2g}\,r_1^2 - \gamma y_1$$

$$p_2 = p_0 + \frac{\gamma\omega^2}{2g}\,r_2^2 - \gamma y_2$$

Subtracting expression for p_2 from that for p_1, we obtain

$$p_1 - p_2 = \frac{\gamma\omega^2}{2g}(r_1^2 - r_2^2) - \gamma(y_1 - y_2)$$

or

$$p_1 + \gamma y_1 - \frac{\gamma\omega^2 r_1^2}{2g} = p_2 + \gamma y_2 - \frac{\gamma\omega^2 r_2^2}{2g}$$

or

$$\frac{p_1}{\gamma} + y_1 - \frac{\omega^2 r_1^2}{2g} = \frac{p_2}{\gamma} + y_2 - \frac{\omega^2 r_2^2}{2g} \tag{2.54}$$

Also $r_1\omega = V_1$ and $r_2\omega = V_2$. Then

$$\frac{p_1}{\gamma} + y_1 - \frac{V_1^2}{2g} = \frac{p_2}{\gamma} + y_2 - \frac{V_2^2}{2g} \tag{2.55}$$

This indicates that in a fluid rotating about its vertical axis with a uniform acceleration the expression $(p/\gamma + y - V^2/2g)$ is constant. It is also seen from Eq. (2.53), that the surfaces of equal pressure are paraboloids of revolution.

Consider a cylinder containing a liquid to a height y_0 as shown in Fig. 2.39. If we rotate this cylinder with an angular velocity ω, the water surface will attain a new position. At the equilibrium, let the height of the water at the

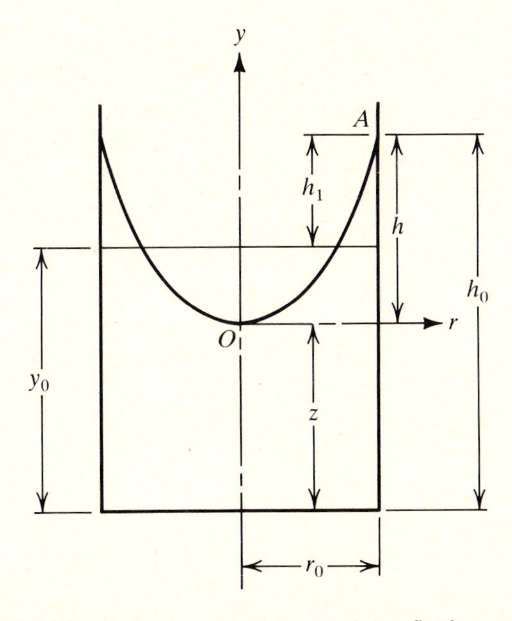

FIGURE 2.39 A circular cylinder containing fluid rotating about its vertical axis.

wall measured from the vertex be h. Then the pressure at point A (the origin being at the vertex) is given by:

$$p_A = p_0 + \frac{\gamma\omega^2}{2g} r_A^2 - \gamma y_A$$

But $p_A = p_0 = 0$, hence

$$0 = 0 + \frac{\gamma\omega^2 r_0^2}{2g} - \gamma h$$

$$h = \frac{\omega^2 r_0^2}{2g} \tag{2.56}$$

Since a paraboloid of revolution has a volume equal to one-half its circumscribing cylinder, $h_1 = h/2$.

$$h_1 = \frac{\omega^2 r_0^2}{4g}$$

Also

$$h_0 = y_0 + h_1$$

$$h_0 = y_0 + \frac{\omega^2 r_0^2}{4g} \tag{2.57}$$

This equation relates the height of the paraboloid of revolution to the original height of the fluid in the cylinder. The height of the vertex from the bottom of the cylinder is given by:

$$z = h_0 - h$$

$$= y_0 + \frac{\omega^2 r_0^2}{4g} - \frac{\omega^2 r_0^2}{2g}$$

$$z = y_0 - \frac{\omega^2 r_0^2}{4g} \tag{2.58}$$

EXAMPLE 2.18

A cylindrical jar 1 ft in diameter and 2 ft high contains water to a height of 1 ft. The jar is rotated about its vertical axis at 150 rpm. Determine (a) the height of the paraboloid of revolution from the base of the cylinder, (b) the maximum pressure and its location, and (c) the pressure at a point 0.2 ft from the center and 0.25 ft from the base.

Solution (a)

$$\omega = \frac{2\pi N}{60} = 2\pi \times \frac{150}{60} = 15.71 \text{ rad/s.}$$

From Eq. (2.57)

$$h_0 = y_0 + \frac{\omega^2 r_0^2}{4g}$$

$$y_0 = 1 \text{ ft}$$

$$r_0 = 0.5 \text{ ft}$$

$$h_0 = 1 + \frac{(15.71)^2(0.5)^2}{4 \times 32.2}$$

$$= 1.48 \text{ ft}$$

(b) The maximum pressure will correspond to the maximum height. Hence:

$$p_{\max} = \gamma h_0 = 62.4 \times 1.48$$

$$= 92.4 \text{ lbs/ft}^2 \text{ at the bottom corners.}$$

(c) From Eq. (2.58)

$$z = y_0 - \frac{\omega^2 \gamma_0^2}{4g}$$

$$= 1 - \frac{(15.71)^2(0.5)^2}{4 \times 32.2}$$

$$= 0.52 \text{ ft}$$

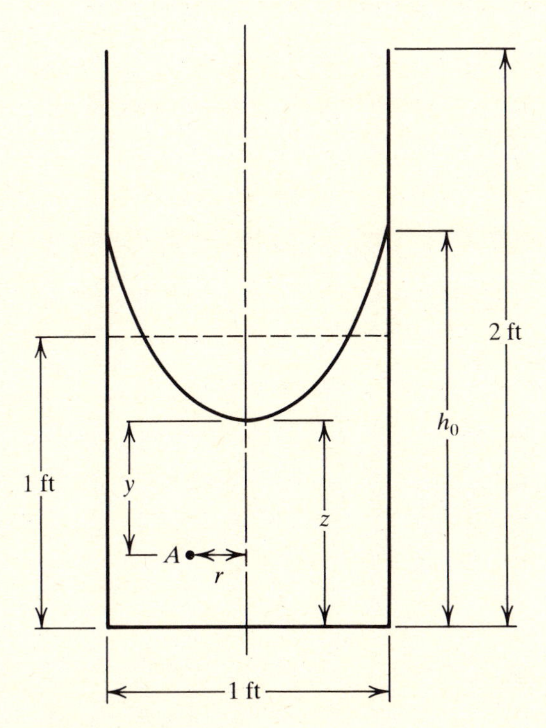

FIGURE 2.40 Example 2.18.

If we place the origin at the vertex, the coordinates of point A, where pressure is required, are: $r = 0.2$, $y = -(0.52 - 0.25) = -0.27$

$$p = p_0 + \frac{\gamma\omega^2}{2g} r^2 - \gamma y$$

$$= 0 + \frac{62.4(15.71)^2(0.2)^2}{2 \times 32.2} - 62.4(-0.27)$$

$$p = 26.4 \text{ lbs/ft}^2$$

PROBLEMS

p.2.1 In a static fluid, pressure p and depth h are related by

a. $p = \rho h$.
b. $p = \gamma h$.
c. $h = \gamma p$.
d. $h = \rho p$.
e. $p = (\rho/g)h$.

p.2.2 The relationship $p = \gamma h$ is valid only for those fluids which are

a. compressible and at rest.
b. incompressible and at rest.
c. ideal.
d. incompressible and in motion.
e. none of these answers.

p.2.3 Pressure at the surface of water in a tank is 5 psi. The pressure in psi at a point 5.0 ft below the surface is

a. 7.16.
b. 71.6.
c. 317.
d. 54.
e. none of these answers.

p.2.4 A 3 m column of oil of specific gravity 0.85 rests on a column of water in a tall cylinder. The pressure at a point 2 m below the water-oil interface in meters of water is

a. 5.
b. 4.55.
c. 5.53.
d. 4.7.
e. 5.35.

p.2.5 A gauge pressure of 300 kN/m^2 is equivalent to a water column height of (in meters)

a. 30.58.
b. 0.03.
c. 4.81.
d. 15.
e. none of these answers.

p.2.6 A gauge pressure of 250 kN/m^2 is equivalent to an oil (sp. gr. 0.8) column height of (in meters)

a. 15.85.
b. 20.39.
c. 31.85.
d. 0.032.
e. none of these answers.

p.2.7 The pressure in psi which is equivalent to 5.0 in of mercury (sp. gr. 13.55) plus 15 in of oil (sp. gr. 0.9) is

a. 5054.4.
b. 35.1.
c. 7.38.
d. 2.93.
e. none of these answers.

p.2.8 A pressure gauge attached to an air duct reads 130 kPa. The atmospheric pressure at this place is 760 mm of mercury. The absolute pressure of the air duct in kPa is

a. 137.45.
b. 140.3.
c. 29.0.
d. 231.02.
e. none of these answers.

p.2.9 The magnitude of force acting on one side of a plane area submerged in a liquid is

a. independent of the orientation of the area.
b. depends on the orientation of the area.
c. given by $\rho A\bar{y}$.
d. given by $\gamma A\bar{y}\sin\theta$.
e. none of these answers.

p.2.10 The center of pressure is located.

a. at the centroid of the plane area.
b. always above the centroid of the area.
c. always below the centroid of the area.
d. sometimes above the centroid and sometimes below the centroid of the area.
e. none of these answers.

p.2.11 A plane area 50 × 50 cm is immersed vertically in water with its upper side 10 cm below the water. The force acting on one side of the area in N is

a. 613.12.
b. 5.46.
c. 858.4.
d. 0.86.
e. none of these answers.

p.2.12 For the plane area of problem 2.11 the vertical distance of the center of pressure from the water surface, in meters, is

a. 0.35.
b. 0.06.
c. 1.06.
d. 0.407.
e. none of these answers.

p.2.13 A circular gate 3 m in diameter holds water with the free surface 6 m above its center. The force acting on the gate in kN is

a. 2646.5.
b. 416.0.
c. 1664.2.
d. 850.0.
e. none of these answers.

p.2.14 For the gate of problem 2.13 the depth of the center of pressure from the water surface, in meters, is

a. 6.0.
b. 3.5.
c. 6.094.
d. 6.9.
e. none of these answers.

p.2.15 The center of buoyancy is located at.

a. the center of gravity of the submerged body.
b. below the center of gravity of the submerged body.

c. the center of gravity of the volume of fluid displaced by the body.
d. midway between the bottom end and the center of gravity of the body.
e. none of these answers.

p.2.16 In the hydraulic jack shown in Fig. 2.41, a force P of 150 N is applied. What weight W can be supported?

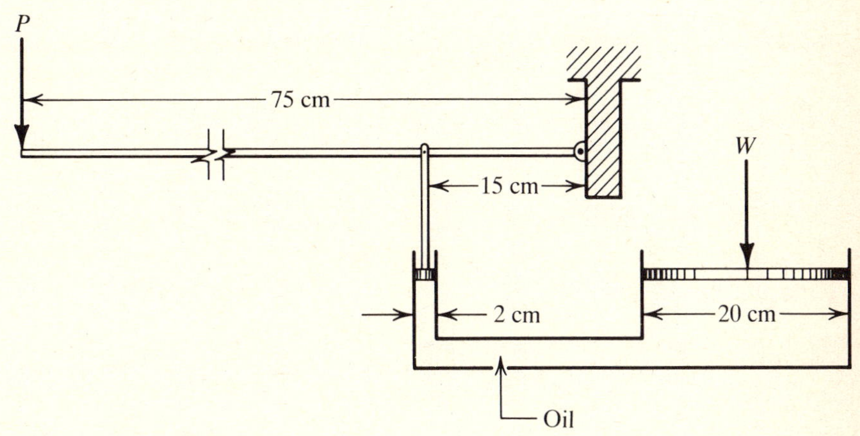

FIGURE 2.41 Problem 2.16.

p.2.17 A hydraulic press has a ram-to-piston area ratio of 20. How much weight can be lifted if a force of 10 lbs is applied on the piston?

p.2.18 Express a pressure of 150 kN/m^2 in terms of the

a. height of a water column at 20°C.
b. height of a glycerine column.
c. height of a castor oil column.

p.2.19 Calculate the pressures at points A, B, C and D in the system shown in Fig. 2.42.

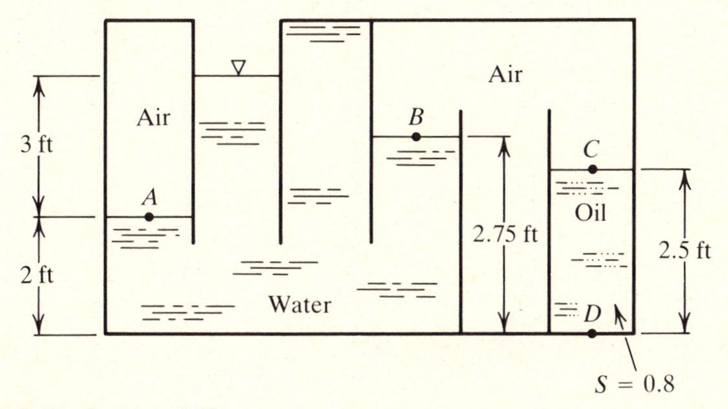

FIGURE 2.42 Problem 2.19.

p.2.20 The pressure in the air space above an oil surface is 10 psi. Calculate the pressure at a point 10 ft below the oil surface. The specific gravity of oil is 0.85.

p.2.21 Determine the pressure acting on a diver if he reaches 1000 m below the surface of the ocean.

p.2.22 Express a pressure of 14.7 psi in

a. inches of mercury.
b. feet of water.
c. feet of oil of specific gravity 0.85.

p.2.23 Determine the atmospheric pressure at an altitude of 9000 ft if the temperature and pressure at sea level are 68°F and 14.7 psi, respectively. Assume polytropic atmosphere.

p.2.24 A person is interested in developing a windmill at an elevation of 6000 ft above sea level. How much kinetic energy will be present in the wind at that level for a wind speed of 10 mi/h? Assume U.S. Standard atmosphere.

p.2.25 Find the difference of pressure between the points *A* and *B* in the manometer shown in Fig. 2.43.

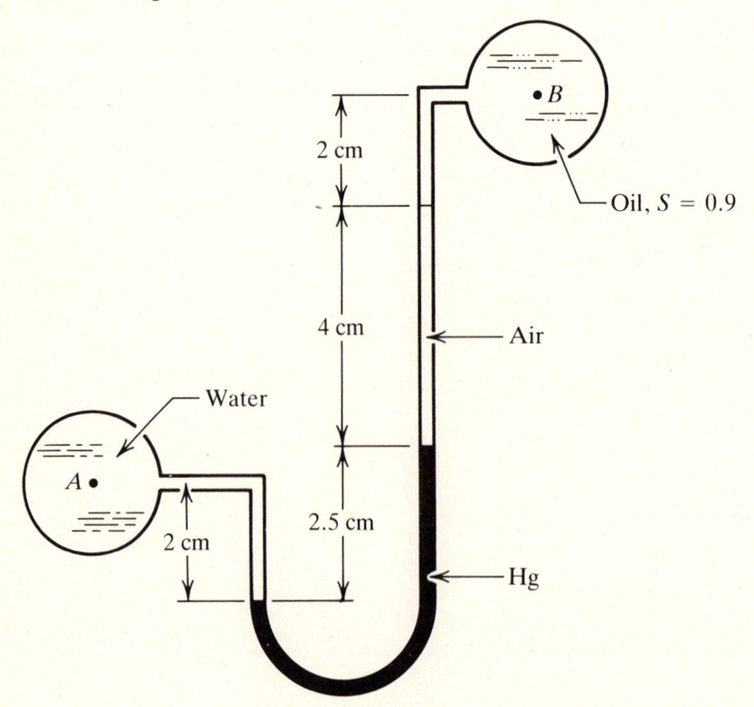

FIGURE 2.43 Problem 2.25.

p.2.26 Determine the specific gravity of the liquid in the left leg of the manometer tube shown in Fig. 2.44.

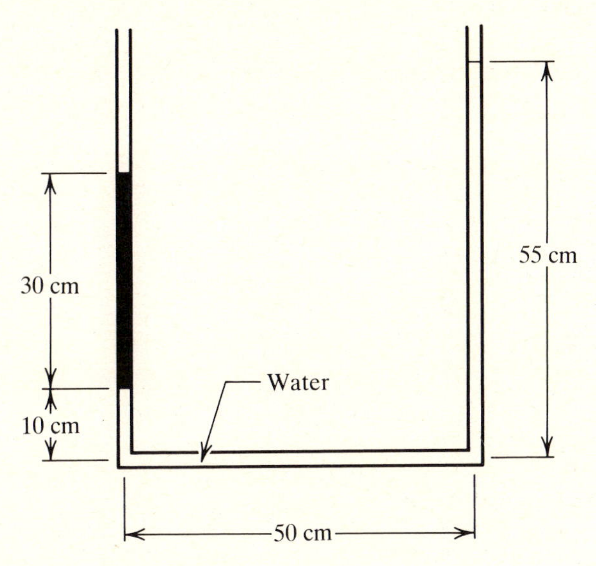

FIGURE 2.44 Problem 2.26.

p.2.27 Determine the pressure at the center of pipe *A*, shown in Fig. 2.45.

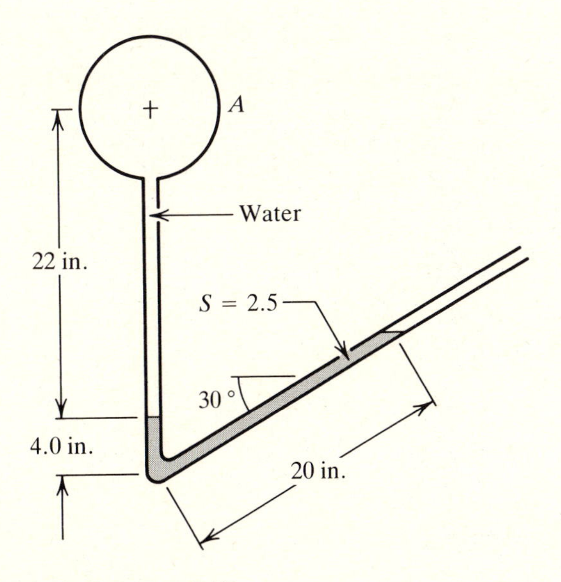

FIGURE 2.45 Problem 2.27.

p.2.28 The inclined manometer shown in Fig. 2.46 is filled with a liquid of specific gravity 0.9. How much pressure should be applied on the surface of the liquid in the cistern to cause the fluid in the tube to move 30 cm?

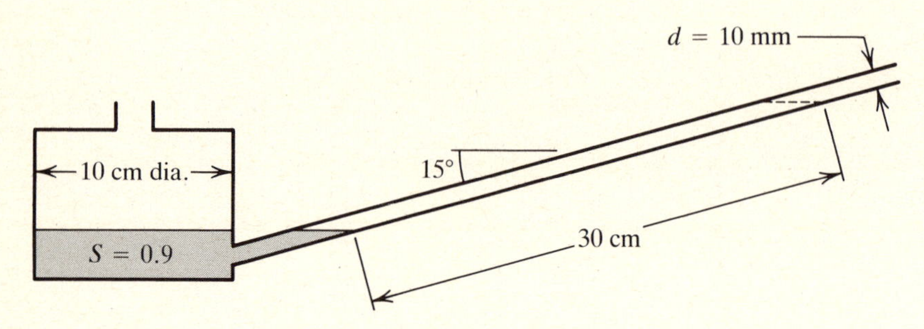

FIGURE 2.46 Problem 2.28.

p.2.29 Find the difference of pressure between the points A and B for the cases shown in Figs. 2.47, 2.48, and 2.49.

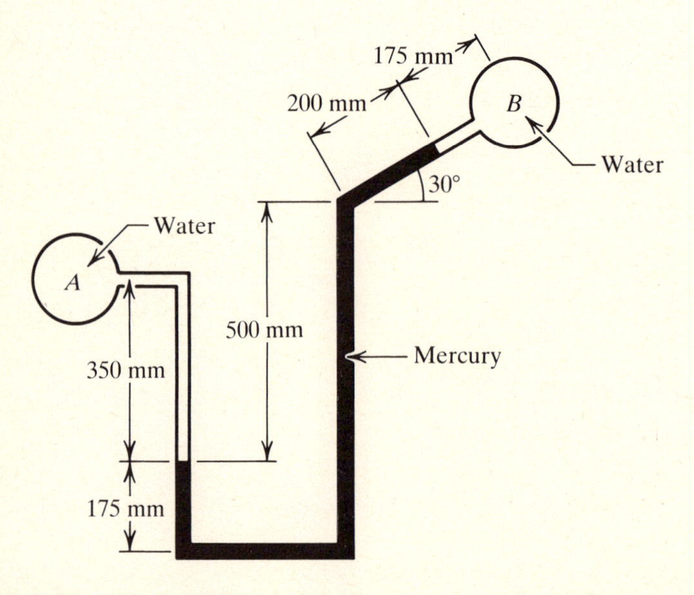

FIGURE 2.47 Problem 2.29.

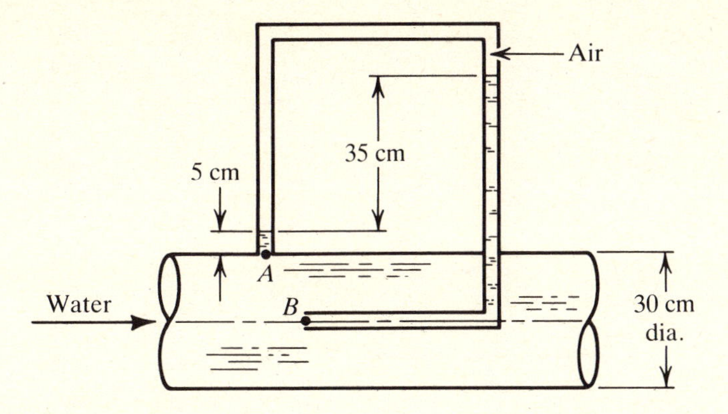

FIGURE 2.48 Problem 2.29.

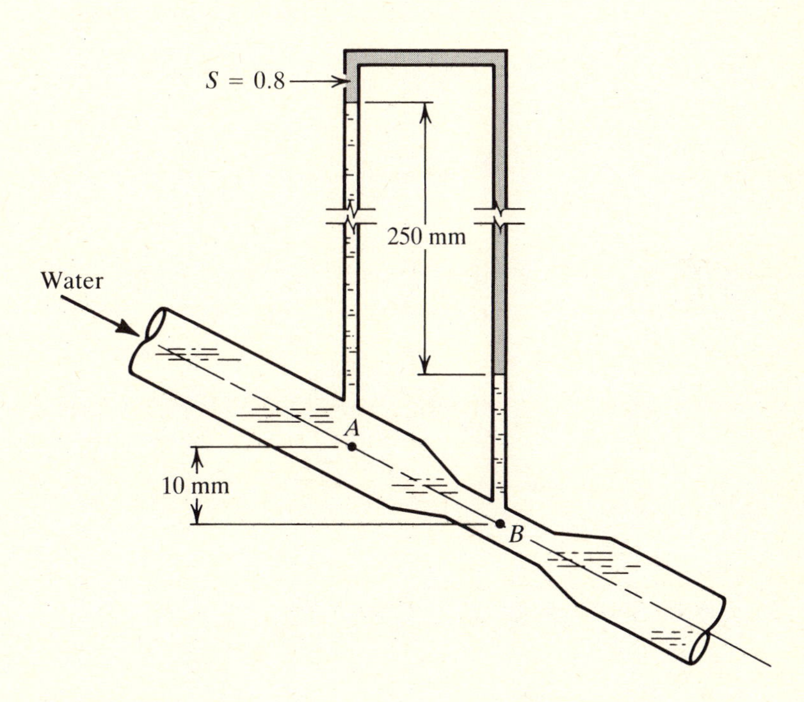

FIGURE 2.49 Problem 2.29.

p.2.30 A plane triangular area is immersed in water as shown, in Fig. 2.50. Determine (a) The force acting on one side of the area and (b) The location of the center of pressure.

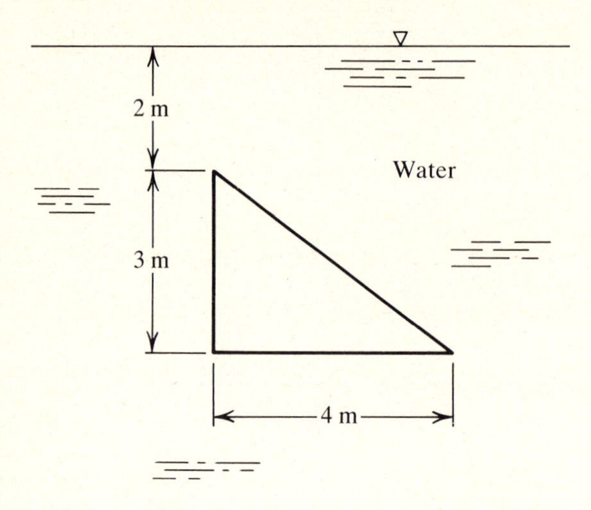

FIGURE 2.50 Problem 2.30.

p.2.31 Determine the magnitude of the force F required to open the gate shown in Fig. 2.51. The gate is hinged at point A. The fluid is water at 20°C.

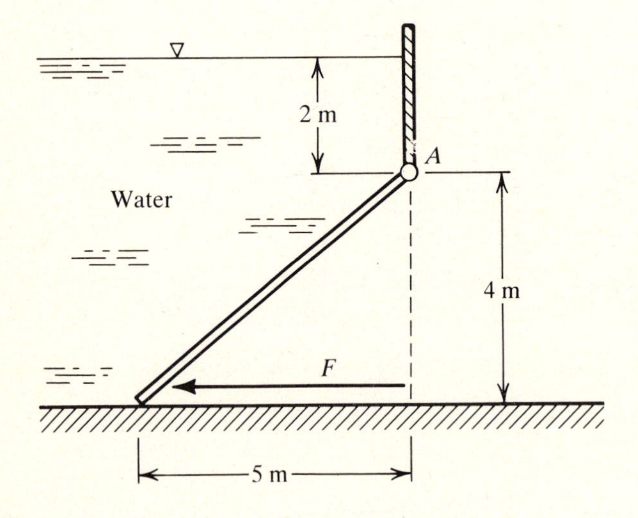

FIGURE 2.51 Problem 2.31.

p.2.32 Determine the magnitude of the force F required to open the gate shown in Fig. 2.52. The gate is hinged at point A. The fluid on both sides is water of specific gravity 1.09.

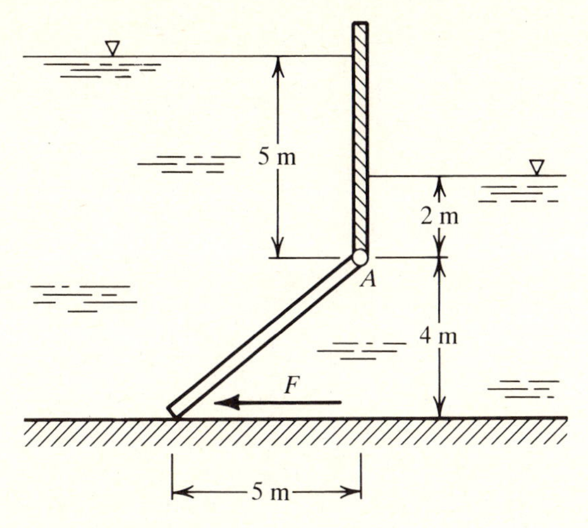

FIGURE 2.52 Problem 2.32.

p.2.33 A concrete dam of rectangular section 70 m high and 30 m wide has the water level at its top. Determine

a. the total hydrostatic force per meter length of the dam.
b. the depth of the center of pressure.
c. the point at which the resultant of the hydrostatic force and the weight of the dam cuts the base.

The specific gravity of concrete is 2.4.

p.2.34 Design a lock gate for a waterway of width 20 m. The gate is to be 10 m high with water elevation of 9 m and 3 m on its two sides. Use a safety factor of 2.0. The available bearings have a maximum allowable thrust of 500,000 N, 600,000 N, and 900,000 N. Use the computer program listed in Appendix B.

p.2.35 Determine the magnitude of the force F required to open the curved gate AB, shown in Fig. 2.53.

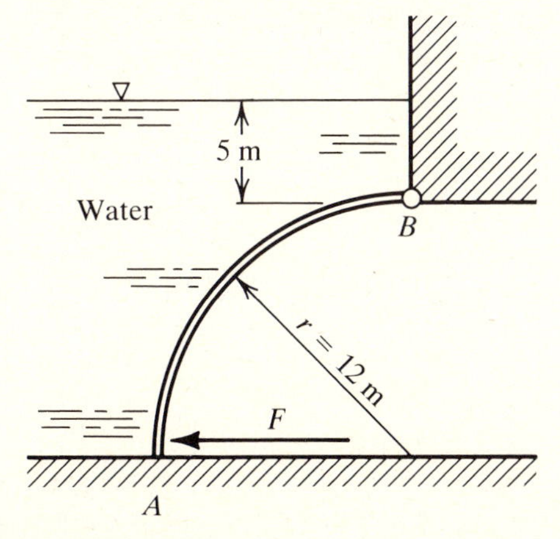

FIGURE 2.53 Problem 2.35.

p.2.36 A log in water and oil is shown in Fig. 2.54. Determine

a. the force pushing the log against the wall.
b. the weight of the log per unit width.
c. the specific gravity of the log.

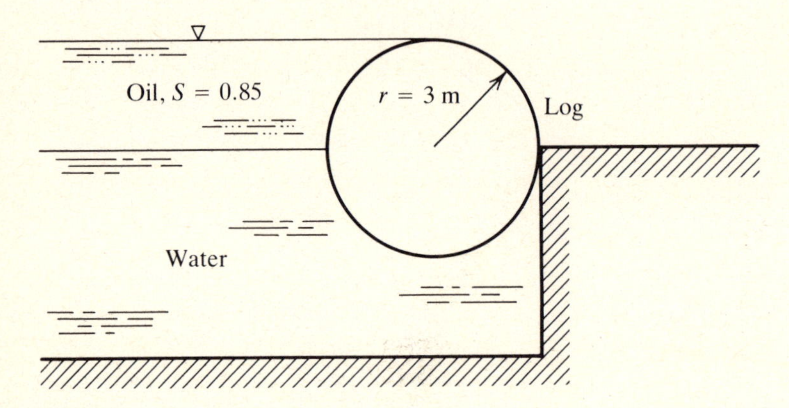

FIGURE 2.54 Problem 2.36.

p.2.37 Calculate the force F required to hold the gate closed in Fig. 2.55. The oil has specific gravity of 0.9.

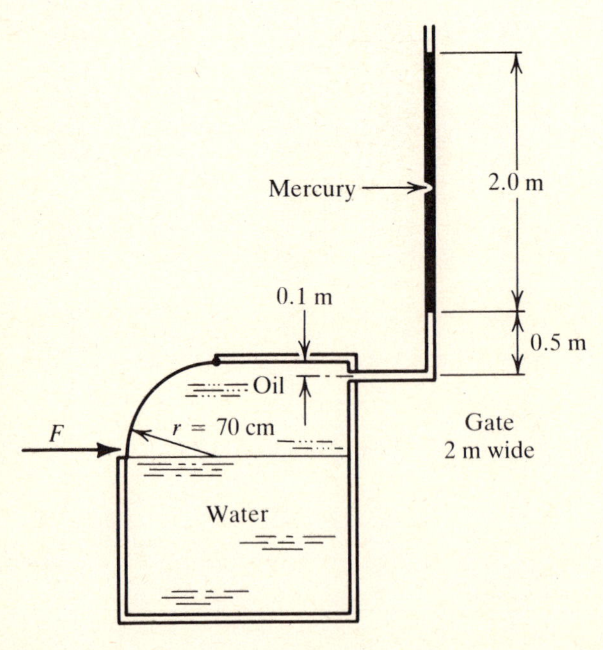

FIGURE 2.55 Problem 2.37.

p.2.38 Design a 3 m wide radial sluice gate to retain a maximum head of 4 m of water of specific weight 9840 N/m^3. Three sizes of thrust bearings are available which can withstand a maximum of 125,000 N, 150,000 N and 200,000 N of thrust. Use a safety factor of 2. Use the computer program listed in Appendix B.

p.2.39 A body floats in stable equilibrium when

a. M is above G.
b. M is below G.
c. M coincides with G.
d. center of buoyancy coincides with G.
e. none of these answers.

p.2.40 A cylinder 2 ft in diameter and 3 ft high weighs 25 lbs. It contains oil of specific gravity 0.8. Will the cylinder sink or float if dropped on its end in water of specific gravity 1.0? If it floats, determine the height to which the water would rise. Is the cylinder floating in stable equilibrium?

p.2.41 A rectangular barge is 20 m long, 5 m wide and 2 m deep. The specific gravity of the barge is 0.8. Determine (a) the metacentric height, and (b) the righting moment when the angle of heel is 8°. The water is of specific gravity 1.09.

p.2.42 The tank shown in Fig. 2.56 is filled with water and accelerated as shown. Determine the pressure at the points B and C. The tank is open at point A.

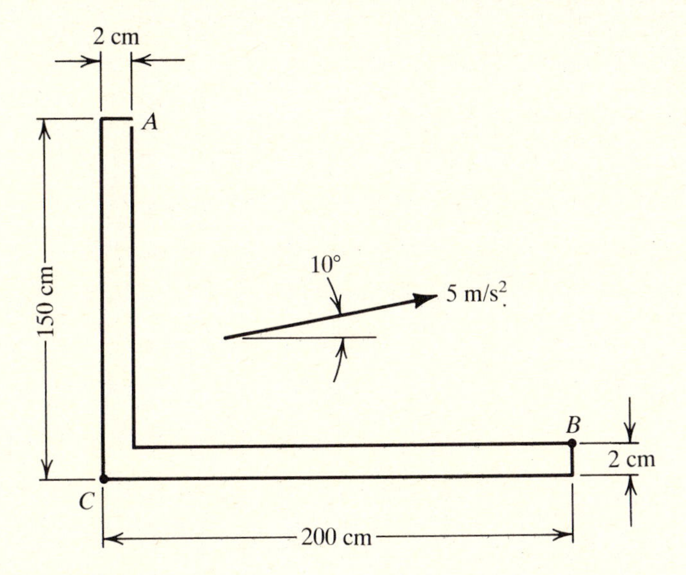

FIGURE 2.56 Problem 2.42.

p.2.43 If the tank shown in Fig. 2.56 is accelerated in the x-direction only, determine the acceleration a_x such that the pressure at B is zero.

p.2.44 The box shown in Fig. 2.57 is accelerated such that $a_x = 2.15\ \text{m/s}^2$, and $a_y = 0$. Find the pressures at A, B and C.

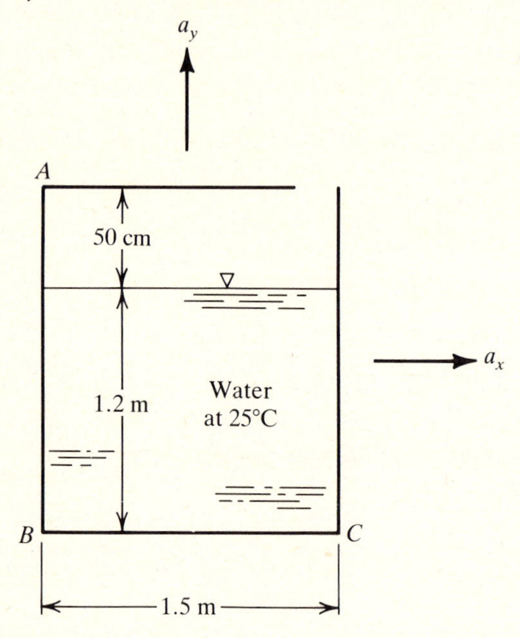

FIGURE 2.57 Problem 2.44, 2.45.

p.2.45 The box shown in Fig. 2.57 is accelerated such that $a_y = 3.00\ \text{m/s}^2$, $a_x = 9.87\ \text{m/s}^2$. Find the pressures at A, B and C.

p.2.46 The tube shown in Fig. 2.58 contains oil of specific gravity 0.8. It is rotated about AA at the rate of 10 rad/s. Find the pressure at the point B.

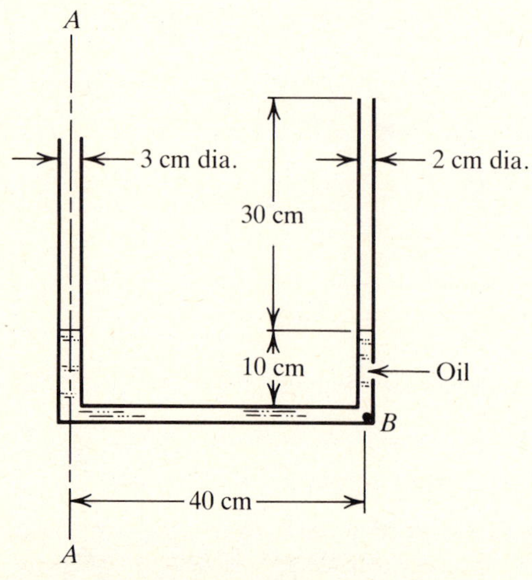

FIGURE 2.58 Problem 2.46.

p.2.47 The tube shown in Fig. 2.59 contains water. It is rotated about a vertical axis passing through A at the rate of 6 rad/s. Determine the pressure at A and B.

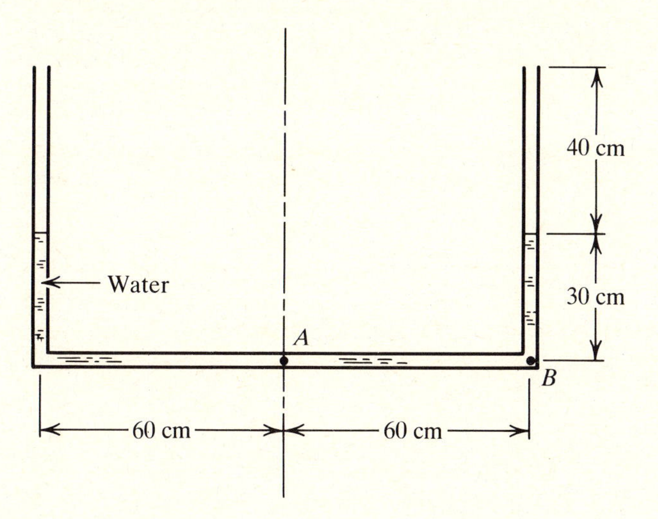

FIGURE 2.59 Problem 2.47.

Chapter Three

FLUID KINEMATICS

In Chapter Two, we considered the equilibrium of a mass of fluid which was at rest or in uniform motion as a solid with no relative movement between various fluid layers. In this chapter, we will consider a mass of fluid in motion by examining the geometry of the motion of its particles. No attention, however, will be paid to the forces which cause the motion of the fluid mass. These forces will be considered in Chapter Four.

3.1 FUNDAMENTAL CONCEPTS OF FLUID FLOW

Fluid Velocity and Fluid Motion

The fluid velocity at a point (using the principle of continuum), is the instantaneous velocity of the center of gravity of the *representative elementary volume* (REV) surrounding the point.

The motion of a fluid may be described using two methods. The first is the *Lagrangian Method* in which attention is focused on motion of a single particle as it occupies various positions in the flow field. The second is the *Eulerian Method* in which attention is focused on a fixed point in the flow field and the velocities of different particles are examined as they pass through that point.

For an explanation of the Lagrangian Method, consider a particle of fluid (Fig. 3.1) which occupies a position (x_1, y_1, z_1) at an instant of time t_1 and has a velocity V_1. At another instant of time t_2, the particle moves to another position (x_2, y_2, z_2) and acquires a velocity V_2. It is clear that V_2 depends on the elapsed time $(t_2 - t_1)$ and on the position $(x_2\ y_2, z_2)$ which, in turn, depends on the original position (x_1, y_1, z_1). Mathematically, Lagrangian velocity may be expressed as

$$V = V[x(t), y(t), z(t), t] \tag{3.1}$$

Clearly, the Lagrangian Method is limited to the study of one particle only.

For the explanation of the Eulerian Method, consider a point (x_1, y_1, z_1) in a flow field (Fig. 3.2). Different particles occupy point (x_1, y_1, z_1) at different times. The velocity of a fluid particle passing the point at time t_1 may be V_1 and the velocity of another fluid particle passing the same point at time t_2 may be V_2.

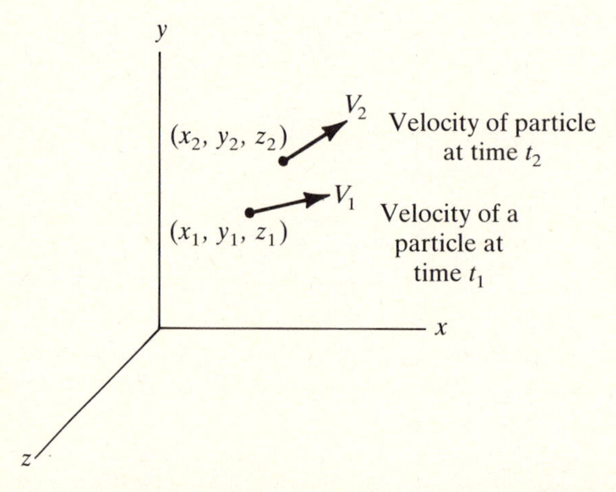

FIGURE 3.1 Lagrangian approach to the movement of a fluid particle.

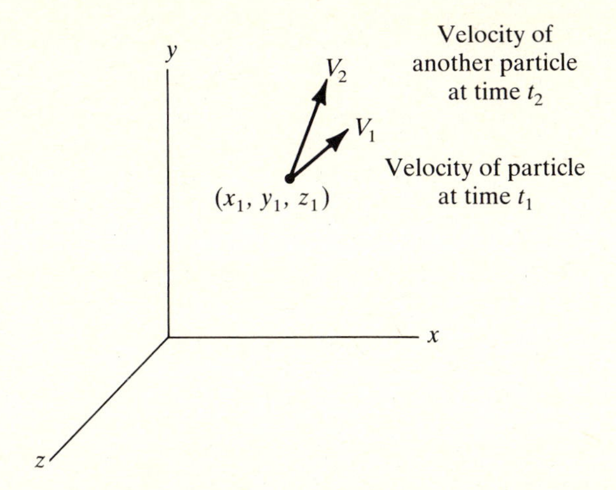

FIGURE 3.2 Eulerian approach to the movement of a fluid particle.

Mathematically, Eulerian velocity may be expressed as

$$V = V(x, y, z, t) \tag{3.2}$$

The Eulerian Method is not limited to the study of one fluid particle; rather, it completely describes a flow field giving information on velocity variation from point to point.

In terms of rectangular components, the velocity vector may be written as

$$\mathbf{V} = u\mathbf{i} + v\mathbf{j} + w\mathbf{k} \tag{3.3}$$

where u, v and w are, respectively, the components of the velocity in the x, y and z directions.

The flow of a fluid is termed as *one-*, *two-* or *three-dimensional* depending on the number of space coordinates required to specify the velocity field.

EXAMPLE 3.1

The velocity distribution of a fluid is given by $\mathbf{V} = 4x(\mathrm{m})\mathbf{i} - 4y(\mathrm{m})\mathbf{j}$, where $\mathbf{V}$ is the velocity in m/s. Determine (a) the dimension of the flow field, (b) velocity components at the point (1, 2), and (c) the magnitude and the direction of the velocity at the point (1, 2).

Solution

(a) From the given velocity, it is evident that

$$u = 4x$$

$$v = -4y$$

Since, two space coordinates x and y are required to specify the velocity field, the flow is two-dimensional.

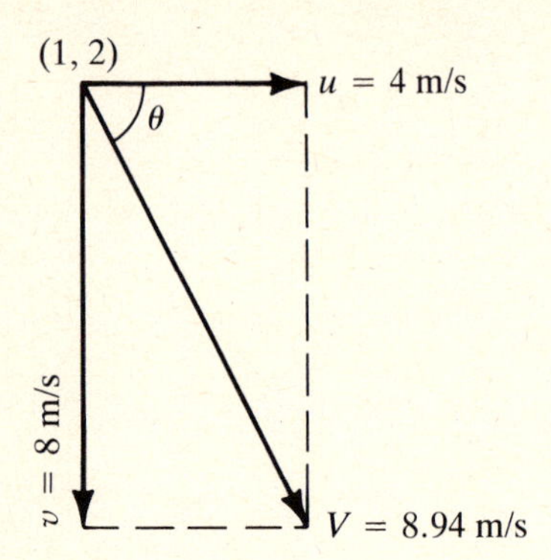

FIGURE 3.3 Example 3.1.

(b) The velocity components at the point (1, 2) are

$$u = 4x = 4 \text{ m/s}$$

$$v = -4y = -8 \text{ m/s}$$

(c) The magnitude of velocity at point (1, 2)

$$= \sqrt{4^2 + 8^2} = 8.94 \text{ m/s}$$

The angle with the horizontal

$$= \theta = \tan^{-1}\frac{v}{u} = \tan^{-1}\left(\frac{-8}{4}\right)$$

$$= -63.4°$$

Rate of Flow

The volume rate of flow, Q, often simply called *rate of flow* or *discharge* is the volume of fluid passing through a cross-section in a unit time. Mathematically

$$Q = AV \tag{3.4}$$

where A is the cross-sectional area through which flow is occurring and V is the mean or average velocity across the cross-section. Here the word "mean" or "average velocity" is used because the velocity, in general, varies across the cross-section. More precisely, the rate of flow may be calculated using integration. Then

$$Q = \int_A u\,dA \tag{3.5}$$

where u is the velocity of fluid passing through an elemental area dA. In the above equation, it has been assumed that the velocity vector is normal to the

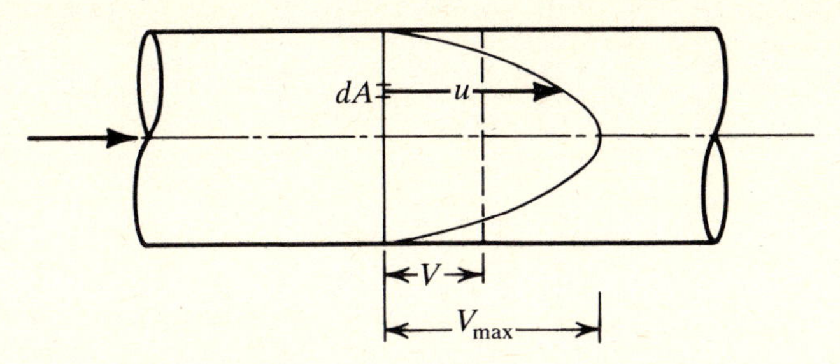

FIGURE 3.4 Velocity distribution in a pipe.

cross-sectional area. If the velocity vector is not normal to the area but inclined at an angle θ to the normal, then

$$Q = \int_A u \cos \theta dA$$

$$Q = \int_A u \cdot dA \tag{3.6}$$

The *mass rate of flow*, $\dot{m}$, is defined as the quantity of mass flowing past a cross-section in a unit time. For a fluid with constant density across the section

$$\dot{m} = \rho Q \tag{3.7}$$

Average or *mean velocity of flow*, V, at a cross-section is given by

$$V = \frac{Q}{A} \tag{3.8}$$

For flow of a fluid in a pipe, the local velocity u, the average velocity V, and the maximum velocity V_{max} are shown in Fig. 3.4.

EXAMPLE 3.1

Water at 25°C is flowing through a pipe, 30 cm in diameter, with an average velocity of 15 m/s. Determine (a) the discharge, and (b) mass rate of flow.

Solution From Table A-1, for water at 25°C

$$\rho = 997.1 \text{ kg/m}^3$$

(a)

$$Q = AV$$

where

$$A = \frac{\pi(0.3)^2}{4} = 0.071 \text{ m}^2$$

$$V = 15 \text{ m/s (given)}$$

Hence

$$Q = 0.071 \times 15 = 1.06 \text{ m}^3/\text{s}$$

(b)

$$\dot{m} = \rho Q$$

$$= 997.1 \times 1.06 = 1056.9 \text{ kg/s}$$

EXAMPLE 3.3

The velocity of flow in a circular pipe is given by

$$u = V_0\left(\frac{r_0^2 - r^2}{r_0^2}\right)^n$$

where r_0 is the radius of the pipe, V_0 is the velocity at the centerline, u is the velocity at a distance r from the centerline and n is a constant. Determine the mean velocity when $V_0 = 10$ m/s, $r_0 = 10$ cm, n = 1.

Solution For the given conditions

$$u = 10\left(\frac{0.1^2 - r^2}{0.1^2}\right)$$

$$= 10(1 - 100r^2)$$

or

$$u = 10 - 1000r^2$$

Now:

$$Q = \int_A u\,dA$$

Consider an elemental circular ring of thickness dr at a distance r from the center of the pipe (Fig. 3.5). The area of the elemental ring is given by $2\pi rdr$. Then:

$$Q = \int_0^{r_0} u2\pi rdr = 2\pi\int_0^{0.1} (10 - 1000r^2)rdr$$

$$= 2\pi\left[\frac{10r^2}{2} - \frac{1000r^4}{4}\right]_0^{0.1} = 2\pi[5(0.1)^2 - 250(0.1)^4]$$

$$= 0.157 \text{ m}^3/\text{s}$$

$$V = \frac{Q}{A} = \frac{0.157}{\pi(0.1)^2} = 5.0 \text{ m/s}$$

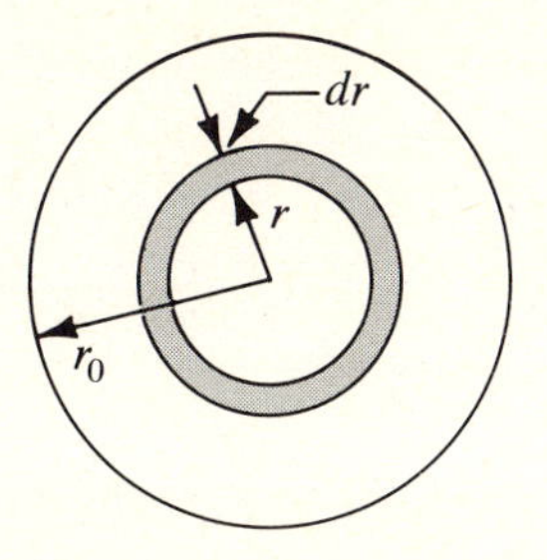

FIGURE 3.5 Example 3.3.

Steady, Unsteady, Uniform, Nonuniform Flows

The fluid property at a point in a flow field is a function of its position as well as time. This can be expressed as:

$$E = E(x, y, z, t) \tag{3.9}$$

where E is any fluid property such as density, velocity, temperature, etc.

The flow of a fluid is said to be *steady* if its properties at a point in the flow field do not change with time. However, the properties may vary from point to point. Hence, for a steady flow

$$\frac{\partial E}{\partial t} = 0 \qquad \text{or} \qquad E = E(x, y, z)$$

In certain types of flow (such as in *turbulent* flow, which will be described later) there are always small fluctuations of velocity at a point. Under such conditions, the flow is said to be *steady* if the temporal mean velocity given by

$$V_t = \frac{1}{t}\int_0^t V\,dt$$

does not change with time.

The flow, on the other hand, is *unsteady* if the properties do vary at a point with respect to time (Fig. 3.6).

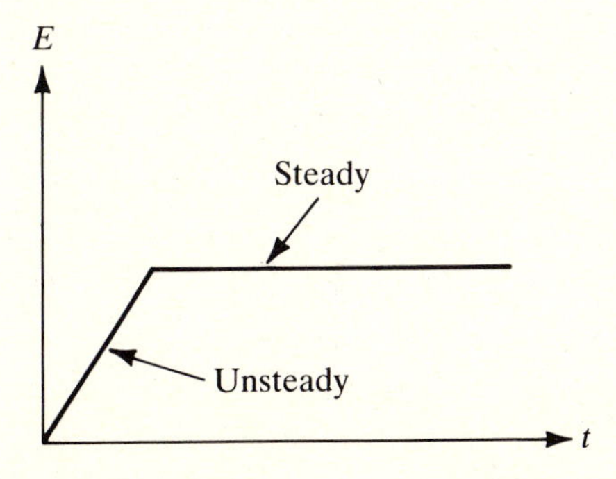

FIGURE 3.6 Variation of a property E of a fluid in steady and unsteady flows.

A practical case of steady and unsteady flow is encountered when a domestic water tap is used. Initially, when the tap is closed, the velocity at a given point in the pipe is zero. When the tap is opened, the velocity gradually increases from zero. Since the velocity is changing with respect to time, the flow is unsteady. When the tap has been opened to a desired position and then left alone, the flow stabilizes and becomes steady.

Contrary to the steady flow which is defined with respect to time, *uniform* flow is defined with respect to space at any instant of time. The flow of a fluid is said to be *uniform* if the fluid properties do not change with respect to space at any instant of time. However, the properties may vary from time to time. Thus, in a uniform flow

$$\frac{\partial E}{\partial s} = 0 \qquad \text{or} \qquad E = E(t)$$

where s refers to the position in space. The flow of a fluid through a pipe of constant diameter is an example of a uniform flow. On the other hand, a nonuniform flow is one in which properties vary from point to point at the same instant of time. The flow of a fluid through a pipe of variable diameter is an example of a *nonuniform* flow.

Streamline, Stream-tube, Stream Function

A *streamline* is a line drawn through the flow field at an instant of time in such a way that a tangent drawn at any point on the line gives the direction of the velocity at that point. Three streamlines drawn in a converging pipe are shown in Fig. 3.7. The portion of the flow field where streamlines are closer together has higher velocity than the portion where the streamlines are farther apart.

Consider a streamline in a two-dimensional flow field as shown in Fig. 3.8. A tangent drawn at point P on the streamline gives the direction of the velocity V at that point. The velocity V, in terms of its rectangular components, can be written as

$$\mathbf{V} = u\mathbf{i} + v\mathbf{j}$$

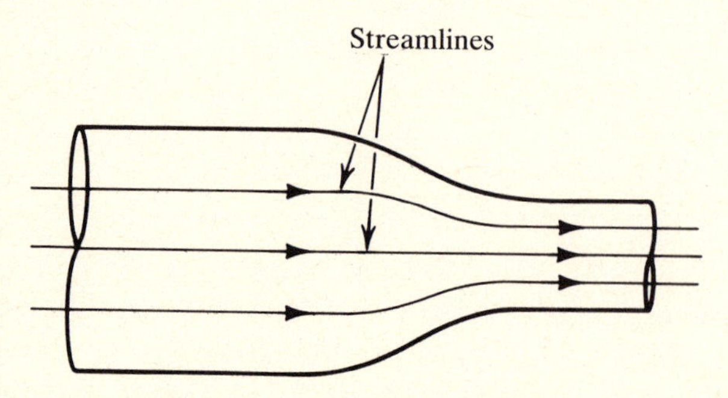

FIGURE 3.7 Streamlines in a converging flow.

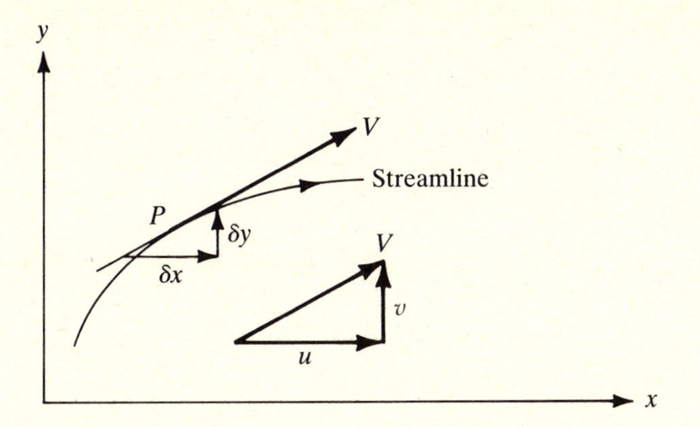

FIGURE 3.8 A streamline in a two-dimensional flow field.

Also, the displacement δs of the fluid particle at point P may be written as

$$\boldsymbol{\delta}\mathbf{s} = \delta x\mathbf{i} + \delta y\mathbf{j}$$

where δx and δy are respectively, the rectangular components of the displacement in the x- and y-directions.

Then for $\boldsymbol{\delta}\mathbf{s}$ and $\mathbf{V}$ to have the same direction, we must have

$$\frac{\delta x}{u} = \frac{\delta y}{v}$$

Or, in the differential form:

$$\frac{dx}{u} = \frac{dy}{v}$$

and

$$vdx - udy = 0 \tag{3.10}$$

This gives the equation of a streamline in a two-dimensional flow field. In three dimensions, the equations would become

$$\frac{dx}{u} = \frac{dy}{v} = \frac{dz}{w} \tag{3.11}$$

In a steady flow, since the direction of velocity is fixed at any point, the streamline is also fixed in space and coincides with the path of movement of a particle. In an unsteady flow, the direction of velocity is not fixed with time; therefore, a streamline may shift in space from time to time. A particle, then, may follow one streamline at one instant of time and another one at another instant of time. Since the velocity vector is tangent to a streamline, it must have no component normal to it. *Therefore, there is no flow across a streamline.*

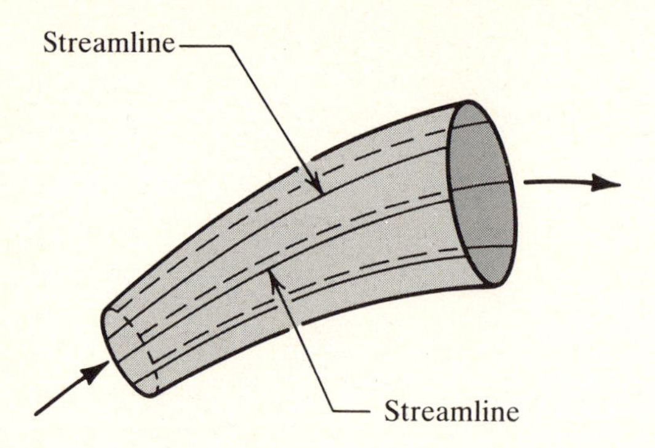

FIGURE 3.9 A stream tube.

A *stream-tube* is a volume of fluid enclosed by a cylindrical surface formed by streamlines as shown in Fig. 3.9. Since the velocity component normal to a streamline is zero, there can be no flow across a stream-tube. A stream-tube of very small cross-section with no velocity variation over it is sometimes called a *stream filament.*

An individual particle of fluid does not always follow a streamline. The path of a single particle of fluid as it moves through a flow field is called a *path line.* The path line of a fluid particle may be determined by injecting a tracer dye in a flow field and then taking a series of photographs over time t.

A *streakline* is an instantaneous picture of the line joining all particles which have emerged from the same source and which have passed through the same point in a flow field. If a dye is injected at a point in a flow field, the dye would follow the path of the fluid particles passing through that point. A photograph of flow taken at any instant of time would show a streak of dye which is a streakline. If the flow is steady, photographs taken at various times would be identical. But if the flow is unsteady, different photographs taken at different times would show different streaklines.

In a steady flow, a path line, a streakline and a streamline coincide with each other if they pass through the same point in space.

In a two-dimensional flow, the equations of streamlines may be described by *stream function* ψ, with different values of stream function representing different streamlines. To understand the concept of stream function, consider two points A and D (Fig. 3.10) in a two-dimensional flow field. Point A is fixed, but point D can assume any position. The point A can be connected to point D in various ways. Two such ways are shown as ACD and ABD. It can be seen that, for an incompressible fluid, the rate of inflow across ABD into the region $ABDCA$ is equal to the rate of outflow across ACD irrespective of the shape of the line ACD. In other words, the point A may be connected to D via ACD, $AC'D$, $AC''D$, etc; the rate of flow across it will be the same and equal to the one across ABD. Hence, the rate of flow across ACD depends only on the position of the end points A and D. Since A is fixed, the rate of flow across ACD is a function of the position of D only. This function is

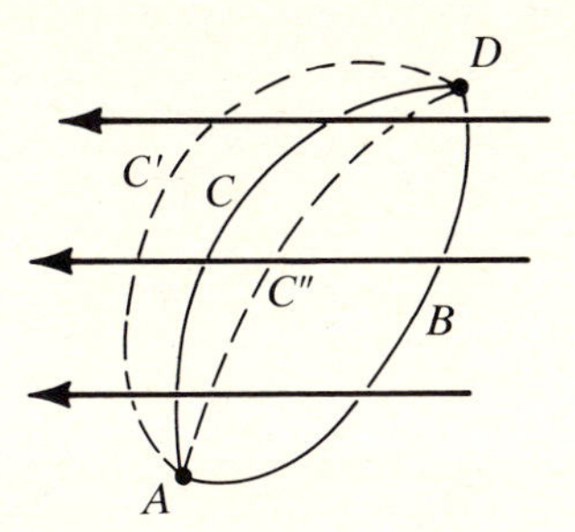

FIGURE 3.10 A fluid flowing between two points A and D.

known as *the stream function* ψ. The value of ψ at the fixed point A is zero, and the value of ψ at any other point, such as D, represents the rate of flow across any line joining A to D.

Now, consider a streamline passing through D (Fig. 3.11). For a fixed point A, the rate of flow across AD' must be the same as across AD. Since no flow takes place across a streamline, the value of ψ must be the same at D and D' or at any other point taken on the streamline. Thus, in a two-dimensional flow, the equations of streamlines may be described by ψ with a different value of ψ representing a different streamline. Since ψ is a function of the location of point D, in general, we can say that $\psi = \psi(x, y)$. Then

$$d\psi = \frac{\partial \psi}{\partial x} dx + \frac{\partial \psi}{\partial y} dy$$

Since ψ is constant along a streamline, the equation of a streamline in terms of ψ will, then be

$$\frac{\partial \psi}{\partial x} dx + \frac{\partial \psi}{\partial y} dy = 0 \tag{3.12}$$

We also developed the equation of a streamline in terms of its rectangular components u and v in the form of Eq. (3.10). Comparing the two equations, we obtain

$$v = \frac{\partial \psi}{\partial x}, \qquad \text{and} \qquad u = -\frac{\partial \psi}{\partial y} \tag{3.13}$$

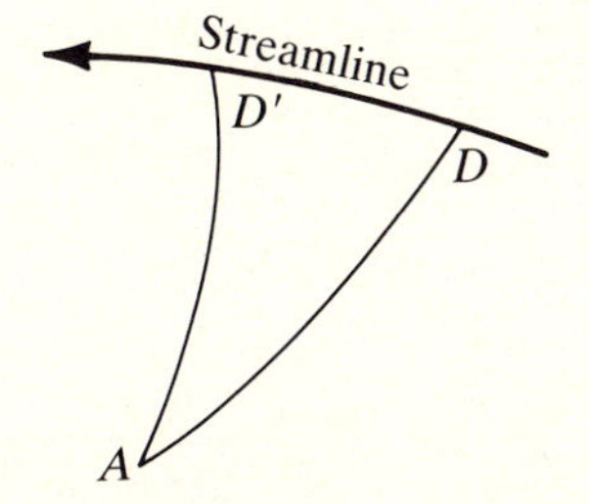

FIGURE 3.11 Definition of stream function.

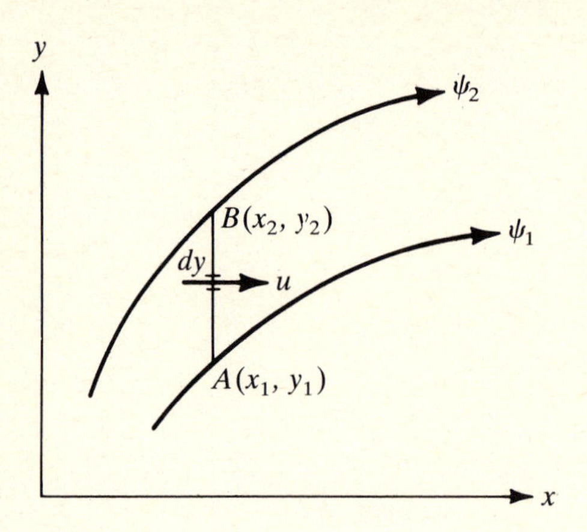

FIGURE 3.12 Discharge between two streamlines ψ_1 and ψ_2.

For a unit depth, discharge between two streamlines ψ_1 and ψ_2 (Fig. 3.12) may be calculated using a vertical plane *AB*. Consider an elemental area *dy* on the line *AB*. Then

$$Q = \int_{y_1}^{y_2} u dy = \int_{y_1}^{y_2} \frac{\partial \psi}{\partial y} dy$$

But, for flow in the *x*-direction only, from Eq. (3.13):

$$\frac{\partial \psi}{\partial y} dy = d\psi$$

Hence:

$$Q = \int_{\psi_1}^{\psi_2} d\psi$$

or

$$Q = \psi_2 - \psi_1 \tag{3.14}$$

Thus, the discharge per unit width between two streamlines is given by the difference between the constant values of ψ defining the two streamlines.

EXAMPLE 3.4

Velocity in a flow field is given by

$$\mathbf{V} = 4y(\text{m})\mathbf{i} + 2(\text{m})\mathbf{J}.$$

Determine the stream function and sketch the streamlines in the first quadrant. Assume the constant of integration to be zero.

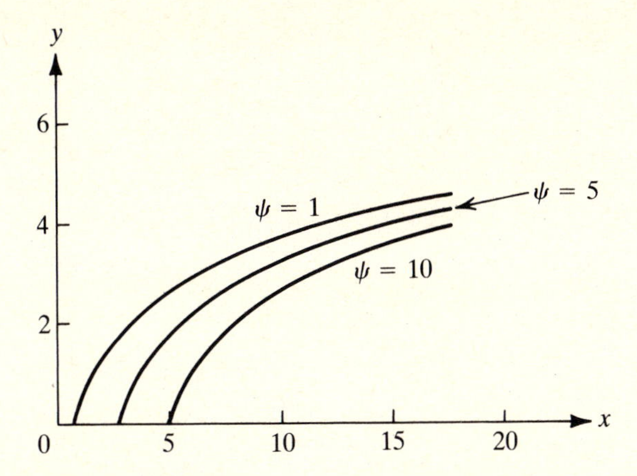

FIGURE 3.13 Example 3.4.

Solution From the given velocity field:

$$\frac{\partial \psi}{\partial x} = v = 2$$

$$\frac{\partial \psi}{\partial y} = -u = -4y$$

$$d\psi = \frac{\partial \psi}{\partial x} dx + \frac{\partial \psi}{\partial y} dy$$

$$= 2dx - 4ydy$$

Integrating:

$$\psi = 2x - 2y^2 + C$$

where C is a constant of integration. It is given that $C = 0$. Hence, the stream function is

$$\psi = 2x - 2y^2$$

The streamlines can be plotted for various values of ψ, as shown in Fig. 3.13.

EXAMPLE 3.5

Given the stream function $\psi = -4xy$, determine the magnitude and direction of velocity at point (2, 4) in the flow field. The units are SI.

Solution

$$\psi = -4xy$$

$$u = -\frac{\partial \psi}{\partial y} = 4x$$

$$v = \frac{\partial \psi}{\partial x} = -4y$$

At point (2, 4)

$$u = 4 \times 2 = 8 \text{ m/s}$$

$$v = -4 \times 4 = -16 \text{ m/s}$$

$$V = \sqrt{8^2 + 16^2} = 17.89 \text{ m/s}$$

$$\theta = \tan^{-1}\left(-\frac{16}{8}\right)$$

$$= -63.4^\circ$$

Hence, the magnitude of the velocity at point (2, 4) is 17.89 m/s and it makes an angle of 63.4° with the x-axis in the fourth quadrant.

Fluid Acceleration

Acceleration of a fluid particle is the rate of change of its velocity as it moves along a streamline. Let a particle of fluid move from a position s at time t to a position $s + ds$ in time dt, as shown in Fig 3.14.

If the velocity of the particle at time t is V, then $V = V(s, t)$ and the acceleration along the streamline, a_s, is given by

$$a_s = \frac{dV(s, t)}{dt}$$

$$a_s = \frac{\partial V}{\partial s}\frac{ds}{dt} + \frac{\partial V}{\partial t}$$

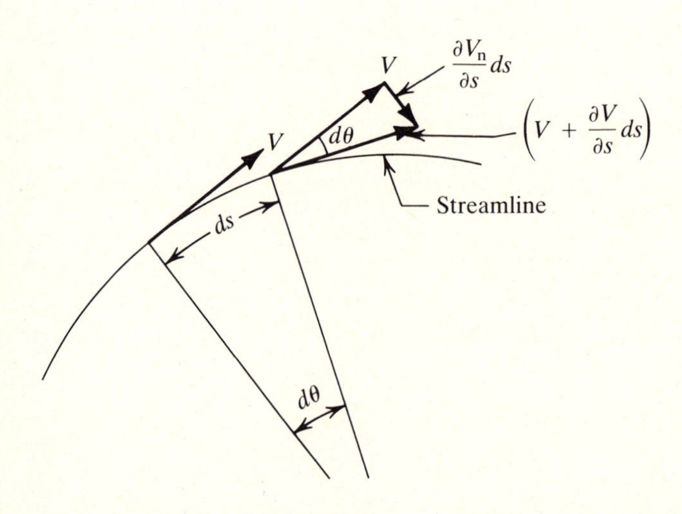

FIGURE 3.14 Acceleration of a fluid particle.

Now ds/dt is the local velocity V. Hence,

$$a_s = V\frac{\partial V}{\partial s} + \frac{\partial V}{\partial t} \tag{3.15}$$

The above expression for acceleration contains two types of terms. The first term, $V(\partial V/\partial s)$, describes change of velocity with respect to position while the second term, $\partial V/\partial t$, describes the change of velocity with respect to time. The first term is known as the *convective* or *spatial acceleration* and the second term is known as the *local* or *temporal acceleration.*

If the streamlines are curved, they are subjected to an additional acceleration in the direction normal to the streamline. Let n be a coordinate, normal to the streamline and directed towards the center of curvature (Fig. 3.14), and let V_n be the velocity along the n-direction. Then the acceleration in the n-direction a_n, is given by

$$a_n = \frac{\partial V_n}{\partial s}\frac{ds}{dt} + \frac{\partial V_n}{\partial t} \tag{3.16}$$

From Fig. (3.14)

$$\frac{\partial V_n}{\partial s}\,ds = \left(V + \frac{\partial V}{\partial s}\,ds\right)\sin d\theta$$

Since $d\theta$ is small, $\sin d\theta = d\theta = ds/r$, where r is the radius of curvature of the streamline.
Then:

$$\frac{\partial V_n}{\partial s}\,ds = \left(V + \frac{\partial V}{\partial s}\,ds\right)\frac{ds}{r}$$

$$\frac{\partial V_n}{\partial s}\,ds = \frac{V}{r}\,ds + \frac{1}{r}\frac{\partial V}{\partial s}(ds)^2$$

Now, $(ds)^2$ is a second order of small quantity, and can be neglected. Therefore

$$\frac{\partial V_n}{\partial s} = \frac{V}{r}$$

Substituting in Eq. (3.16), and noting that $ds/dt = V$, we obtain

$$a_n = \frac{V^2}{r} + \frac{\partial V_n}{\partial t} \tag{3.17}$$

In terms of rectangular components, the acceleration vector **a** may be written

$$\mathbf{a} = a_x\mathbf{i} + a_y\mathbf{j} + a_z\mathbf{k}$$

where a_x, a_y, and a_z are, respectively, the x-, y- and z-components of the acceleration.

Then

$$a_x = \frac{du}{dt};\ a_y = \frac{dv}{dt};\ a_z = \frac{dw}{dt}$$

Since $u = u(x, y, z, t)$, then

$$a_x = \frac{\partial u}{\partial x}\frac{dx}{dt} + \frac{\partial u}{\partial y}\frac{dy}{dt} + \frac{\partial u}{\partial z}\frac{dz}{dt} + \frac{\partial u}{\partial t}$$

But

$$\frac{dx}{dt} = u; \quad \frac{dy}{dt} = v; \quad \text{and} \quad \frac{dz}{dt} = w;$$

thus acceleration in the x-direction is given by

$$a_x = u\frac{\partial u}{\partial x} + v\frac{\partial u}{\partial y} + w\frac{\partial u}{\partial z} + \frac{\partial u}{\partial t} \tag{3.18a}$$

Similarly:

$$a_y = u\frac{\partial v}{\partial x} + v\frac{\partial v}{\partial y} + w\frac{\partial v}{\partial z} + \frac{\partial v}{\partial t} \tag{3.18b}$$

$$a_z = u\frac{\partial w}{\partial x} + v\frac{\partial w}{\partial y} + w\frac{\partial w}{\partial z} + \frac{\partial w}{\partial t} \tag{3.18c}$$

In cylindrical coordinates (r, θ, z) with velocity components V_r, V_θ, and V_z, the acceleration components are:

$$a_r = V_r\frac{\partial V_r}{\partial r} + \frac{V_\theta}{r}\frac{\partial V_r}{\partial \theta} + V_z\frac{\partial V_r}{\partial z} - \frac{V_\theta^2}{r} + \frac{\partial V_r}{\partial t} \tag{3.19a}$$

$$a_\theta = V_r\frac{\partial V_\theta}{\partial r} + \frac{V_\theta}{r}\frac{\partial V_\theta}{\partial \theta} + V_z\frac{\partial V_\theta}{\partial z} + \frac{V_r V_\theta}{r} + \frac{\partial V_\theta}{\partial t} \tag{3.19b}$$

$$a_z = V_r\frac{\partial V_z}{\partial r} + \frac{V_\theta}{r}\frac{\partial V_z}{\partial \theta} + V_z\frac{\partial V_z}{\partial z} + \frac{\partial V_z}{\partial t} \tag{3.19c}$$

If the convective acceleration is zero in a flow field, the velocity does not change from point to point and the flow is uniform. If the local acceleration is zero, the velocity does not change with respect to time and the flow is steady. In other words, in a uniform flow field the convective acceleration is zero and in a steady flow field the local acceleration is zero.

EXAMPLE 3.6

Water of density 1000 kg/m^3 flows through the nozzle shown in Fig. 3.15 at a constant rate of 25.0 kg/s. Determine the (a) velocity, and (b) convective acceleration at a point 10 cm from the left end of the pipe.

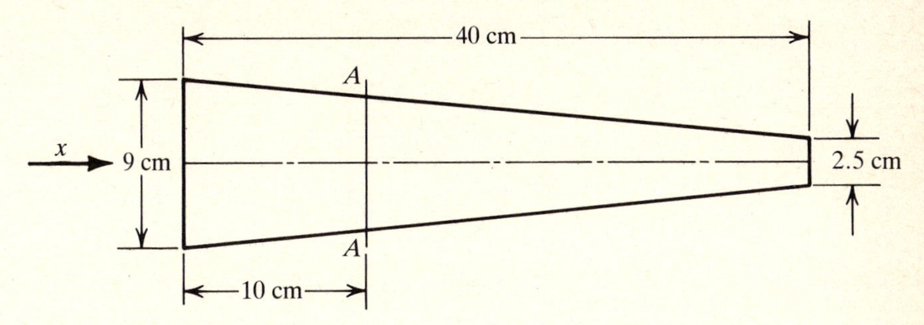

FIGURE 3.15 Example 3.6.

Solution Define the section at which the velocity and the convective acceleration are to be determined by AA. First find the cross-sectional area at AA.

$$\text{Area of left section} = \frac{\pi(9)^2}{4} = 63.62 \text{ cm}^2$$

$$\text{Area of right section} = \frac{\pi(2.5)^2}{4} = 4.91 \text{ cm}^2$$

Assuming a linear variation in area:

$$\text{Area of Section } AA = 63.62 - \frac{63.62 - 4.91}{40} \times 10$$

$$= 63.62 - 14.68$$

$$= 48.94 \text{ cm}^2 = 48.94 \times 10^{-4} \text{ m}^2$$

Now

$$\dot{m} = \rho Q = \rho A V$$

Hence:

$$V = \frac{\dot{m}}{\rho A} = \frac{25 \times 10^4}{1000 \times 48.94} = 5.1 \text{ m/s}$$

(b) Area of cross-section at distance X from the left section

$$= A_x = 63.62 - \frac{63.62 - 4.91}{40} X$$

$$= 63.62 - 1.468X \text{ cm}^2 = (63.62 - 1.468X)10^{-4} \text{ m}^2$$

Thus, the velocity at distance X

$$= V = \frac{Q}{\rho A_X}$$

$$= \frac{25 \times 10^4}{1000 \times (63.62 - 1.468X)}$$

$$= \frac{250}{63.62 - 1.468X}$$

Differentiating with respect to X, we obtain

$$\frac{dV}{dX} = \frac{367}{(63.62 - 1.468X)^2}$$

$$(dV/dX) \text{ at } 0.1\text{m} = 0.09 \text{ s}^{-1}$$

$$A_{cx} = V\frac{dV}{dX} = 5.1 \times 0.09 = 0.47 \text{ m/s}^2$$

System and Control Volume

For the purpose of fluid analysis, sometimes terms such as *system* and *control volume* are used. A *system* is a specific mass of fluid which is isolated for experimentation. A system may change its shape or position but its mass always remains constant. This concept is useful in the consideration of changes which occur in a given mass of fluid such as the case of an air-gasoline mixture in the cylinder of an automobile. Figure 3.16a shows a piston-cylinder assembly containing an air-gasoline mixture at some time $t = t_0$. If a spark is brought in contact with the mixture, the air would expand and the piston would move to the right. Figure 3.16b shows the position of the piston at some other time $t = t_1$. In this case, the boundary of the system has moved, but its mass remains the same. However, heat and work may cross the boundaries of the system.

A *control volume* is a specific region of the flow system in space through which fluid flows and which has been isolated to study the effect of forces and other agents on the fluid. The physical boundary of the control volume is called the *control surface*. Figure 3.17 shows a section of a pipe through which a fluid is flowing. A part of the pipe has been isolated as shown by the broken lines. This is a control volume and the broken lines are the control surface. A section of a stream-tube may be used as a control volume. In this case, flow can move into and out of the control volume through the end cross-sections only, since no flow can take place across a stream-tube.

The concept of a system is useful when studying fluids from the Lagrangian point of view, while the concept of control volume is useful when studying fluids from the Eulerian point of view.

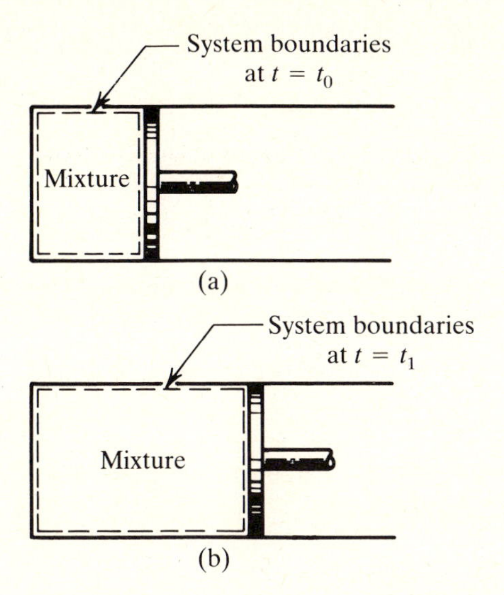

FIGURE 3.16 A piston moving in a circular cylinder.

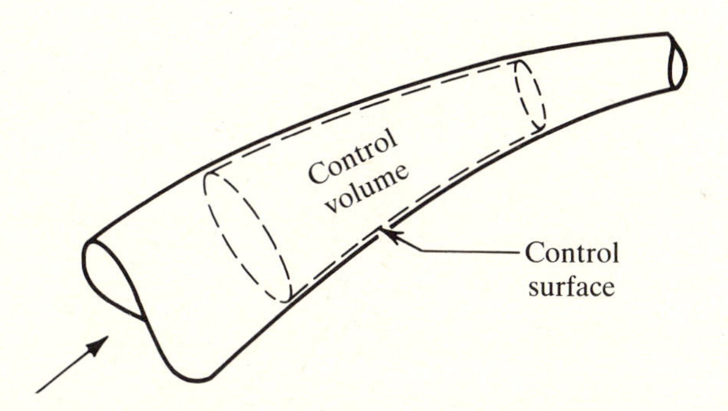

FIGURE 3.17 Control volume.

3.2 BASIC EQUATION FOR A CONTROL VOLUME

In general, the control volume approach to studying fluid behavior is preferred over the system approach. There are two basic reasons for this. First, it is always more convenient to study the behavior of a moving fluid as it flows past a fixed region than to study the behavior of a fluid as it goes from place to place. Second, in general, we are not interested in the motion of a given mass of fluid, but rather in the effect of such a motion on the surrounding objects.

A general equation for the motion of a fluid mass across a control volume will now be developed. As a fluid mass passes through a control volume, it also carries with it its properties such as energy and momentum across the control volume. We would like to develop an equation for a general property rather than a specific one. Let us use the symbol, N, to designate any property (mass, energy, momentum) of the fluid. Also let η be the amount of this property per unit mass throughout the fluid. Then

$$N = \int_{\text{mass}} \eta dm = \int_{\text{volume}} \eta \rho d\forall \tag{3.20}$$

Let us assume that a fixed mass of a fluid totally occupies the control volume at time t (Fig. 3.18a). At another time $t + \Delta t$, the fluid moves and occupies a new position as shown in Fig. 3.18b. The control volume is fixed relative to the reference coordinates x, y, and z.

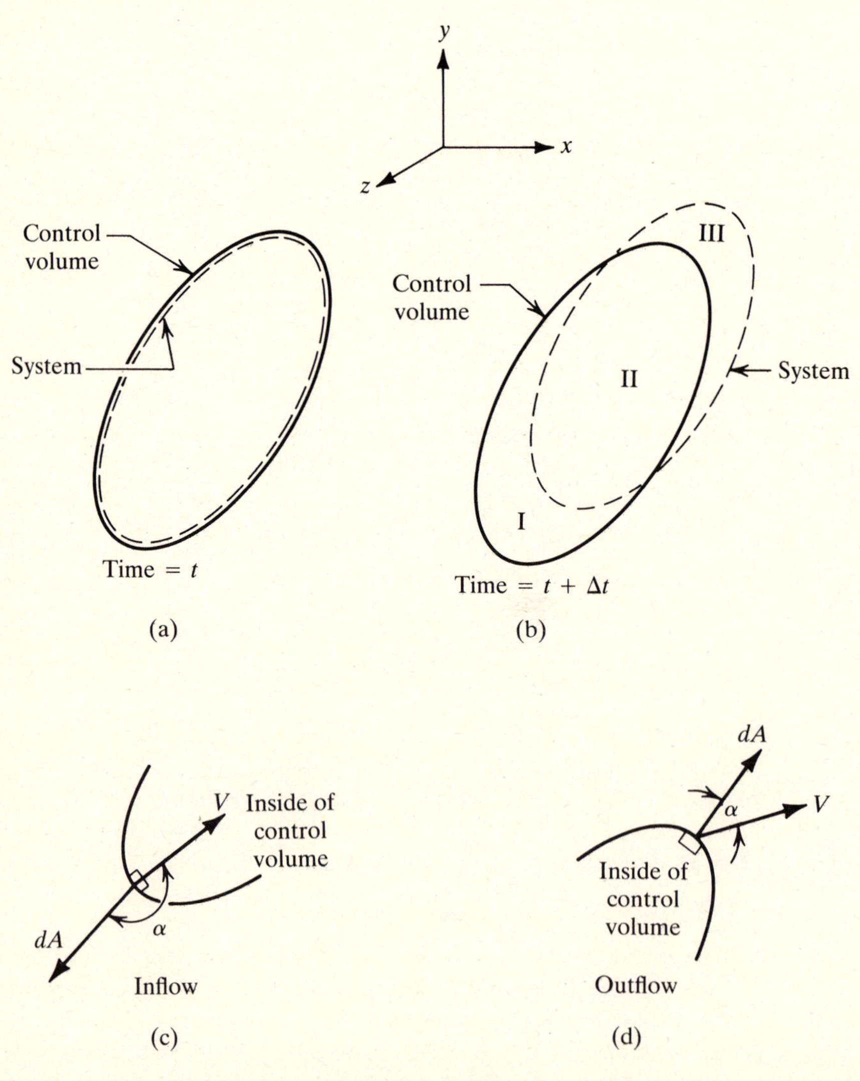

FIGURE 3.18 A system moving through a control volume.

Then:

$$N_t = (N_{cv})_t$$

and

$$N_{t+\Delta t} = (N_{II} + N_{III})_{t+\Delta t} = (N_{cv} - N_I + N_{III})_{t+\Delta t}$$

where the subscripts of N on the right-hand side of the above equations refer to the subregions shown in Fig. 3.18b.

From the above equations, under limiting conditions when Δt approaches zero, we obtain

$$\lim_{\Delta t \to 0} \frac{N_{t+\Delta t} - N_t}{\Delta t} = \lim_{\Delta t \to 0} \frac{(N_{cv})_{t+\Delta t} - (N_{cv})_t}{\Delta t} - \lim_{\Delta t \to 0} \frac{(N_I)_{t+\Delta t}}{\Delta t} + \lim_{\Delta t \to 0} \frac{(N_{III})_{t+\Delta t}}{\Delta t} \quad (3.21)$$

Let us now examine the above four terms more closely.

1. $\lim_{\Delta t \to 0} (N_{t+\Delta t} - N_t)/\Delta t$ is simply dN/dt, or the rate of change of the property N of the fluid as it passes the control volume.
2. $\lim_{\Delta t \to 0} [(N_{cv})_{t+\Delta t} - (N_{cv})_t]/\Delta t$ is the rate of change of the property N within the control volume $= (\partial/\partial t)(N_{cv})$, which, after substitution from Eq. (3.20), is equal to $\partial/\partial t \int_{cv} \eta \rho d\forall$.
3. $\lim_{\Delta t \to 0} (N_I)_{t+\Delta t}/\Delta t$ is the time rate of flow of N into the control volume, which is equal to $-\int_{\text{inflow surface}} \eta \rho \mathbf{V} \cdot \mathbf{dA}$. (Fig. 3.18c).

   Notice that the area vector is drawn normal to the surface in the outward direction. Hence, the angle α for this term will be greater than $\pi/2$.
4. $\lim_{\Delta t \to 0} (N_{III})_{t+\Delta t}/\Delta t$ is the time rate of flow of N out of the control volume, which is equal to $\int_{\text{outflow surface}} \eta \rho \mathbf{V} \cdot \mathbf{dA}$. (Fig. 3.18d).

   The angle α for this case will be less than $\pi/2$. From the above discussion, it follows that

$$\frac{dN}{dt} = \frac{\partial}{\partial t} \int_{cv} \eta \rho d\forall + \int_{\text{inflow surface}} \eta \rho \mathbf{V} \cdot \mathbf{dA} + \int_{\text{outflow surface}} \eta \rho \mathbf{V} \cdot \mathbf{dA}$$

or

$$\frac{dN}{dt} = \frac{\partial}{\partial t} \int_{cv} \eta \rho d\forall + \int_{cs} \eta \rho \mathbf{V} \cdot \mathbf{dA} \quad (3.22)$$

The above equation states that the time rate of change of N of a fixed mass of fluid passing through a control volume is equal to the rate of change of N within the control volume plus the net rate of efflux of N across the control surface.

3.3 MOTION OF A FLUID ELEMENT

Let us now investigate the motion of a small fluid element in a flow field without considering the forces which caused the motion. Consider a small fluid element of sides δx, δy and δz in an x-, y-, z-coordinate system (Fig. 3.19a).

As the fluid moves, the motion of the small fluid element can take place in four different ways. They are: *translation, linear deformation, angular deformation,* and *rotation.*

Translation A fluid element is said to have been translated if it only shifts in position without undergoing any deformation (Fig. 3.19b). In the case of translation, the motion of the fluid element takes place parallel to the original axes—i.e., the planes of the element which were originally perpendicular remain perpendicular. There are no shear stresses. There is no change in the volume of the element. In time dt, the motion in the x-direction is udt, in the

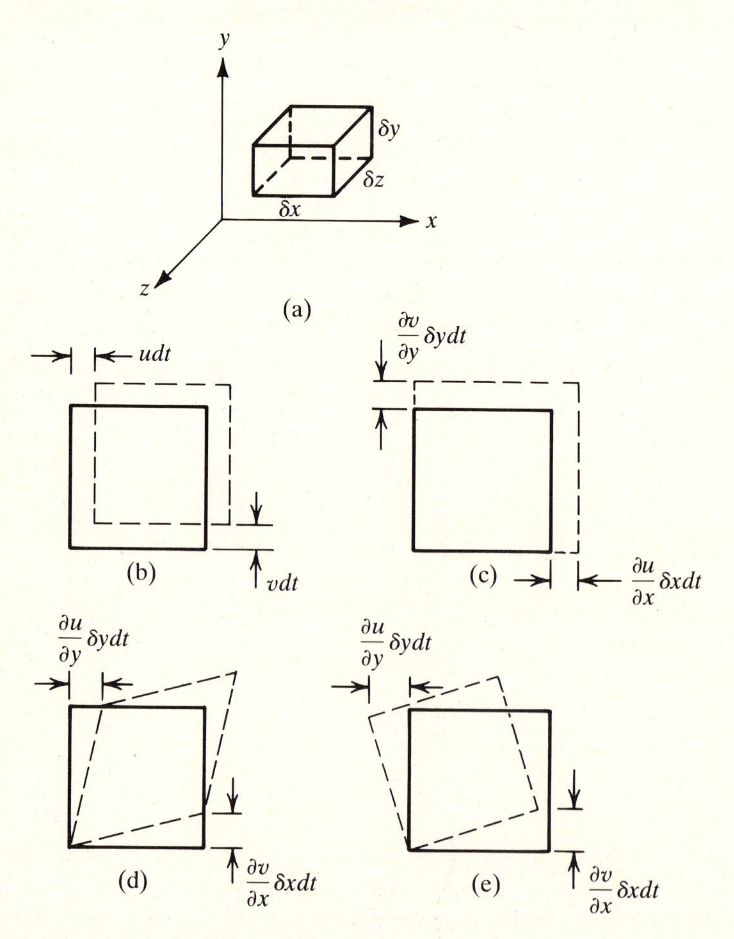

FIGURE 3.19 Deformation of a fluid element.

y-direction it is vdt and in the z-direction it is wdt. Since there is no motion of one fluid layer relative to another, no heat is generated and no energy is dissipated through translation of a fluid element.

Linear Deformation A fluid element is said to have been linearly deformed if it has expanded or contracted parallel to the original axes (Fig. 3.19c). In this case, there is a change in the volume of the element. In time dt, the linear deformations in the x-, y-, and z-directions are $\partial u/\partial x\ \delta x dt$, $\partial v/\partial y\ \delta y dt$ and $\partial w/\partial z\ \delta z dt$, respectively. The relative change in volume with respect to the original volume is

$$\frac{\left(\delta x + \frac{\partial u}{\partial x}\delta x dt\right)\left(\delta y + \frac{\partial v}{\partial y}\delta y dt\right)\left(\delta z + \frac{\partial w}{\partial z}\delta z dt\right) - \delta x \delta y \delta z}{\delta x \delta y \delta z dt}$$

$$= \frac{\partial u}{\partial x} + \frac{\partial v}{\partial y} + \frac{\partial w}{\partial z} \qquad \text{[ignoring the terms containing } (dt)^2 \text{ and } (dt)^3\text{]}$$

$$= \nabla \cdot \mathbf{V}$$

Shear stresses are present, due to the existence of relative motion between fluid layers. Heat is generated and energy is dissipated in linear deformation.

Angular Deformation A fluid element is said to have been angularly deformed if the sides of the element have deformed in the angular direction (Fig. 3.19d). In this case, there is a change in the shape of the fluid element. Referring to Fig. 3.19d, and considering counterclockwise motion as positive, the angular deformation θ, of the sides δx and δy in time dt is given by:

$$\theta_{xy} = \left(\frac{\frac{\partial v}{\partial x}\delta x dt}{\delta x}\right) - \left(\frac{-\frac{\partial u}{\partial y}\delta y dt}{\delta y}\right)$$

$$\theta_{xy} = \left(\frac{\partial v}{\partial x} + \frac{\partial u}{\partial y}\right)dt$$

The rate of angular deformation $\dot{\theta}_{xy}$, in the x-y plane is then given by

$$\dot{\theta}_{xy} = \left(\frac{\partial v}{\partial x} + \frac{\partial u}{\partial y}\right) \tag{3.23a}$$

Similarly, the rate of angular deformation in the y-z plane is

$$\dot{\theta}_{yz} = \left(\frac{\partial v}{\partial z} + \frac{\partial w}{\partial y}\right) \tag{3.23b}$$

and the rate of angular deformation in the x-z plane is

$$\dot{\theta}_{xz} = \left(\frac{\partial w}{\partial x} + \frac{\partial u}{\partial z}\right) \tag{3.23c}$$

The *rate of distortion* is defined as half the rate of angular deformation. Heat is generated and mechanical energy is dissipated through angular deformation. Shear stresses are present.

Rotation A fluid element is said to have been rotated if the sides of the element have rotated through an angle without any change in the shape of the fluid element (Fig. 3.19e). Hence, there is no change in volume of the fluid element and no shear stresses are present. Referring to Fig. 3.19e and considering counterclockwise motion as positive, the rotation in the x-y plane is the average of the rotation of sides δx and δy. Also, the rotation in the x-y plane takes place about an axis parallel to the z-axis.

Hence

$$(\text{rotation})_z = \frac{1}{2}\left(\frac{\partial v}{\partial x} - \frac{\partial u}{\partial y}\right)dt$$

The rate of rotation is the angular velocity.

$$\omega_z = \frac{1}{2}\left(\frac{\partial v}{\partial x} - \frac{\partial u}{\partial y}\right) \tag{3.24a}$$

where ω_z is the angular velocity about the z-axis. Similarly:

$$\omega_x = \frac{1}{2}\left(\frac{\partial w}{\partial y} - \frac{\partial v}{\partial z}\right) \tag{3.24b}$$

and

$$\omega_y = \frac{1}{2}\left(\frac{\partial u}{\partial z} - \frac{\partial w}{\partial x}\right) \tag{3.24c}$$

The resultant angular velocity $\boldsymbol{\omega}$ of the fluid element is then given by

$$\boldsymbol{\omega} = \omega_x \mathbf{i} + \omega_y \mathbf{j} + \omega_z \mathbf{k}$$

$$\boldsymbol{\omega} = \frac{1}{2}\begin{vmatrix} \mathbf{i} & \mathbf{j} & \mathbf{k} \\ \dfrac{\partial}{\partial x} & \dfrac{\partial}{\partial y} & \dfrac{\partial}{\partial z} \\ u & v & w \end{vmatrix} = \tfrac{1}{2}\,\text{curl}\,\mathbf{V} = \tfrac{1}{2}\nabla \times \mathbf{V}$$

No heat is generated and no mechanical energy is dissipated through rotation.

Irrotational Flow

A flow is said to be *irrotational* if the resultant angular velocity is zero, or ω_x, ω_y, and ω_z are equal to zero. Setting ω_x, ω_y, and ω_z in Eq. (3.24) equal to zero yields the conditions for an irrotational flow. They are:

in the x-y plane,

$$\frac{\partial v}{\partial x} = \frac{\partial u}{\partial y} \tag{3.25a}$$

in the y-z plane, $$\frac{\partial w}{\partial y} = \frac{\partial v}{\partial z} \tag{3.25b}$$

in the x-z plane, $$\frac{\partial u}{\partial z} = \frac{\partial w}{\partial x} \tag{3.25c}$$

Vorticity (ζ) is defined as twice the angular velocity

$$\zeta = 2\omega \tag{3.26}$$

Vorticity is analogous to velocity. As is the case with velocity, vorticity is also a function of space and time. Similar to a streamline, a *vortex line* is defined as a line drawn through the flow field in such a way that a tangent drawn to it at any point is the direction of the vorticity vector at that point. A vortex tube is a tube bounded by vortex lines.

Circulation

Circulation (Γ) is defined as the line integral of the tangential component of velocity around a closed curve. Mathematically, in a system of rectangular axes

$$\Gamma = \oint (udx + vdy + wdz) \tag{3.27}$$

The vorticity at a point within a closed curve is the limit of the circulation per unit area enclosed by the curve.

$$\zeta = \underset{A \to 0}{\text{limit}} \frac{\Gamma}{A} \tag{3.28}$$

According to Eq. (3.28), circulation is analogous to discharge. Discharge is the product of velocity and area while circulation is the product of vorticity and area.

In cylindrical coordinates, the circulation may be obtained by considering the rotation of a fluid element *abcd* (Fig. 3.20).

Circulation $= \Gamma = Vr\delta\theta - (V + \partial V/\partial n\, \delta n)(r - \delta n)\delta\theta$

$$= \left(\frac{V}{r} - \frac{\partial V}{\partial n}\right)\delta n r \delta\theta$$

$$= \left(\frac{V}{r} - \frac{\partial V}{\partial n}\right)\delta A$$

Since vorticity is circulation per unit area

$$\zeta = \frac{V}{r} - \frac{\partial V}{\partial n}$$

$$\omega = \tfrac{1}{2}\zeta = \frac{1}{2}\left(\frac{V}{r} - \frac{\partial V}{\partial n}\right)$$

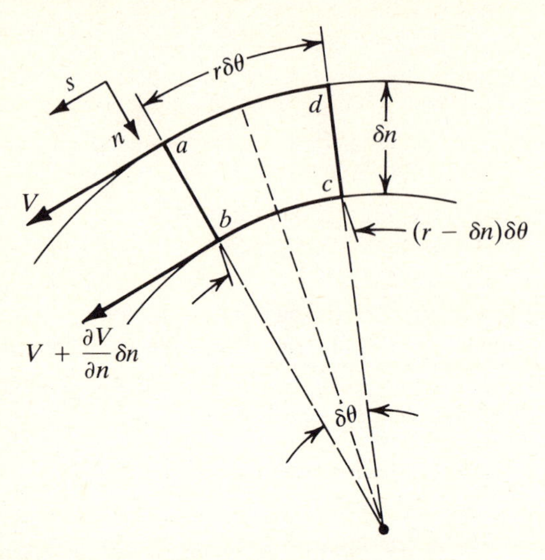

FIGURE 3.20 A fluid element *abcd* in a flow field.

For irrotational flow $\omega = 0$, hence:

$$\frac{V}{r} - \frac{\partial V}{\partial n} = 0 \tag{3.29}$$

Velocity Potential

Let us define a function ϕ such that a negative derivative of ϕ in any direction gives the velocity vector in that direction. In terms of x-y components

$$u = -\frac{\partial \phi}{\partial x} \tag{3.30a}$$

$$v = -\frac{\partial \phi}{\partial y} \tag{3.30b}$$

The function ϕ is known as the velocity potential or the potential function. The minus sign in the above equations appears as a matter of convention to indicate that the velocity potential decreases in the direction of flow. The flow described by the potential function is irrotational. This can be proved by using the conditions of irrotational flow given by Eq. (3.25), according to which for a flow to be irrotational in the x-y plane, $\partial v/\partial x = \partial u/\partial y$. From Eq. (3.30)

$$\frac{\partial v}{\partial x} = -\frac{\partial^2 \phi}{\partial y \partial x} = -\frac{\partial^2 \phi}{\partial x \partial y}$$

$$\frac{\partial u}{\partial y} = -\frac{\partial^2 \phi}{\partial x \partial y}$$

It can be seen from the above expressions that $\partial v/\partial x = \partial u/\partial y$, hence, the flow described by the velocity potential is always irrotational. For this reason, irrotational flow is also known as potential flow. Velocity potential is not a physical quantity which can be measured. It is, rather, a mathematical concept.

Let us consider a line of $\phi(x, y) = \text{constant}$. Such a line of constant velocity potential is known as an *equipotential* line. The total differential of $\phi(x, y)$ is given by

$$d\phi = \frac{\partial \phi}{\partial x} dx + \frac{\partial \phi}{\partial y} dy = 0$$

Substituting from Eq. (3.31), we obtain

$$d\phi = -udx - vdy = o$$

or

$$\frac{dy}{dx} = -\frac{u}{v} \tag{3.31}$$

It can be seen from Eq. (3.10), that for a streamline

$$\frac{dy}{dx} = \frac{v}{u} \tag{3.32}$$

It is evident that the product of the two slopes, given by Eqs. (3.31) and (3.32), is equal to negative one. This means that the streamlines (lines along which ψ is constant) and the equipotential lines (lines along which ϕ is constant) cross each other orthogonally. In a flow field, the network formed by the streamlines and the equipotential lines is known as a flow net.

It may be pointed out here that a stream function ψ can be defined for a rotational or an irrotational flow, but a velocity potential ϕ exists only for an irrotational flow.

The Flow Net

A graphical representation of streamlines of $\psi = \text{constant}$ and equipotential lines of $\phi = \text{constant}$ within a two-dimensional flow field is called a flow net. A portion of a flow net with two streamlines and two equipotential lines is shown in Fig. (3.21). In this figure Δn denotes the distance between two adjacent streamlines and Δs denotes the distance between the two adjacent equipotential lines. In practice Δn is made equal to Δs, then in the limits when $\Delta s = \Delta n \to 0$, the region A (Fig. 3.21) becomes a perfect square. A flow net may be drawn through trial and error by first sketching a number of streamlines and then drawing equipotential lines in such a way that approximate squares are drawn. A flow net for a reservoir-pipe system is shown in Fig. 3.22.

Flow nets can provide valuable information about the flow patterns in a flow field.

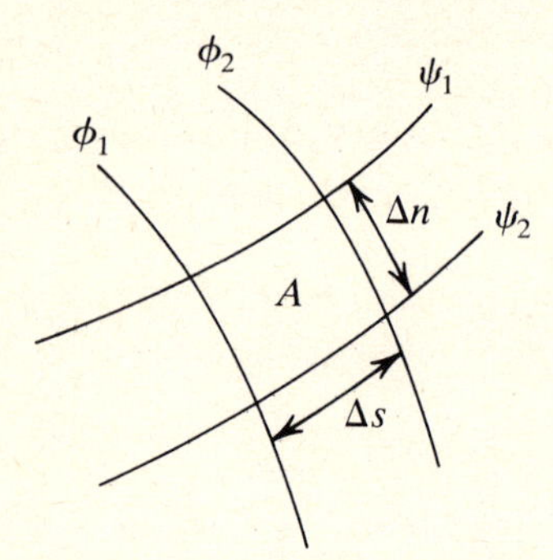

FIGURE 3.21 Streamlines and equipotential lines.

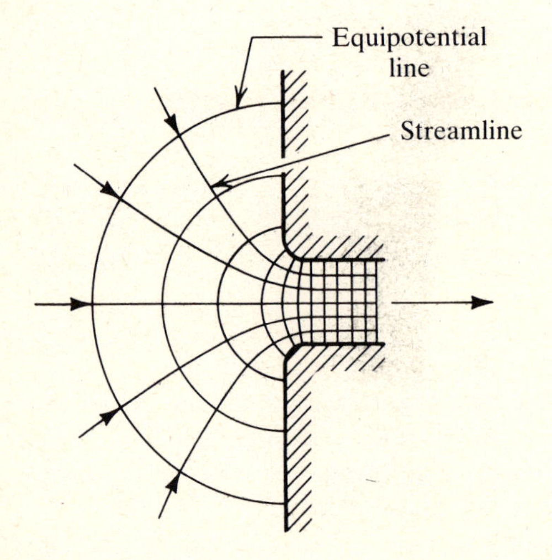

FIGURE 3.22 A flow net.

EXAMPLE 3.7

A flow can be described by the following equations:

$$u = x^2 \cos y \qquad \text{and} \qquad v = -2x \sin y$$

Determine whether the flow is rotational or irrotational.

Solution The flow is two-dimensional and is taking place in the x-y plane. To determine whether it is rotational or irrotational, its vorticity about the z-axis must be examined. To calculate vorticity about the z-axis $\partial u/\partial y$ and $\partial v/\partial x$ are required.

$$u = x^2 \cos y$$

Hence:

$$\frac{\partial u}{\partial y} = -x^2 \sin y$$

Also:

$$v = -2x \sin y$$

$$\frac{\partial v}{\partial x} = -2 \sin y$$

Now

$$\frac{\partial v}{\partial x} - \frac{\partial u}{\partial y} = -2 \sin y + x^2 \sin y \neq 0$$

Therefore, the flow is rotational.

EXAMPLE 3.8

Is the flow given by the velocity distribution

$$\mathbf{V} = (16y - 12x)\mathbf{i} + (12y - 9x)\mathbf{j}$$

irrotational?

Solution The flow is two-dimensional in the x-y plane, and if it is irrotational, Eq. (3.25a) must be satisfied.

$$u = 16y - 12x$$

$$v = 12y - 9x$$

$$\frac{\partial v}{\partial x} = -9$$

$$\frac{\partial u}{\partial y} = 16$$

$$\frac{\partial v}{\partial x} - \frac{\partial u}{\partial y} = -9 - 16 = -25 \neq 0$$

Therefore, the flow is rotational.

3.4 CONSERVATION OF MASS AND THE CONTINUITY EQUATION

The Law of Conservation of Mass states that in the absence of a nuclear reaction, matter can neither be created nor destroyed. When applied to a control volume, it would mean that:

> Rate of mass flowing (i.e., mass flux) into the control volume − Rate of mass flowing out of the control volume = Rate of mass accumulating in the control volume.

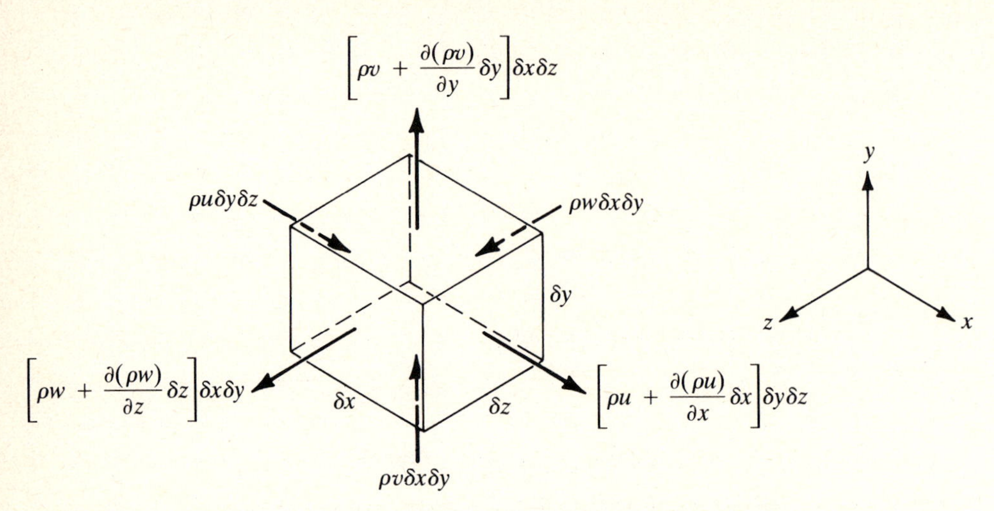

FIGURE 3.23 Elemental fluid volume used to derive continuity equation.

The Equation of Continuity is a mathematical statement of the Law of Conservation of Mass.

The Continuity equation is now derived for an elementary rectangular control volume of a fluid shown in Fig. 3.23. Conservation of the fluid mass requires that the mass flux entering the control volume minus the mass flux leaving the control volume must equal the rate of change of mass inside the control volume. Figure (3.23) shows the mass flux across the six faces of the control volume. The net mass flux in the x-direction, $\dot{m}_x$, is given by

$$\dot{m}_x = \rho u\delta y\delta z - \left[\rho u + \frac{\partial(\rho u)}{\partial x}\delta x\right]\delta y\delta z$$

$$\dot{m}_x = -\frac{\partial(\rho u)}{\partial x}\delta x\delta y\delta z$$

Similarly:

$$\dot{m}_y = -\frac{\partial(\rho v)}{\partial y}\delta x\delta y\delta z$$

$$\dot{m}_z = -\frac{\partial(\rho w)}{\partial z}\delta x\delta y\delta z$$

The rate of change of mass inside the control volume is given by

$$\frac{\partial m}{\partial t} = \frac{\partial}{\partial t}(\rho\delta x\delta y\delta z)$$

Since $\delta x\delta y\delta z$ is constant

$$\frac{\partial m}{\partial t} = \delta x\delta y\delta z\frac{\partial \rho}{\partial t}$$

According to the Law of Conservation of Mass:

$$\dot{m}_x + \dot{m}_y + \dot{m}_z = \frac{\partial m}{\partial t}$$

$$-\frac{\partial(\rho u)}{\partial x}\,\delta x\delta y\delta z - \frac{\partial(\rho v)}{\partial y}\,\delta x\delta y\delta z - \frac{\partial(\rho w)}{\partial z}\,\delta x\delta y\delta z = \frac{\partial \rho}{\partial t}\,\delta x\delta y\delta z$$

Dividing throughout by $-\delta x\delta y\delta z$, we obtain

$$\frac{\partial(\rho u)}{\partial x} + \frac{\partial(\rho v)}{\partial y} + \frac{\partial(\rho w)}{\partial t} = -\frac{\partial \rho}{\partial t} \tag{3.33}$$

This is the general form of the continuity equation applicable to compressible and incompressible fluids under steady and unsteady flow conditions. For an incompressible fluid, the density may be assumed constant. Also, for steady flow $\partial\rho/\partial t = 0$. Therefore, the continuity equation for the steady flow of an incompressible fluid may be written as

$$\frac{\partial u}{\partial x} + \frac{\partial v}{\partial y} + \frac{\partial w}{\partial z} = 0 \tag{3.34}$$

or

$$\nabla \cdot \mathbf{V} = 0 \tag{3.35}$$

The continuity equation for a section of a pipe or a duct may be derived by setting up a control volume as shown in Fig. 3.24. Assume that the fluid enters the control volume through section 1 and leaves it through section 2. There is no flow across the control volume as there is no flow across the pipe.

According to the Law of Conservation of Mass

$$\dot{m}_1 - \dot{m}_2 = \frac{dm}{dt} \tag{3.36}$$

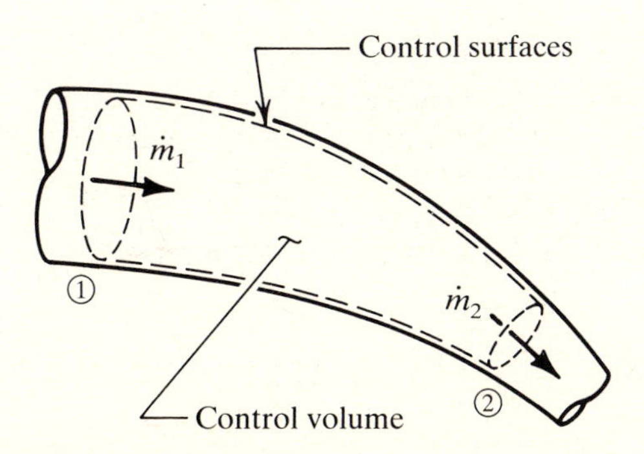

FIGURE 3.24 Control volume in a pipe to develop continuity equation.

where $\dot{m}_1$ = mass rate of flow entering the control volume, $\dot{m}_2$ = mass rate of flow leaving the control volume, and m = mass of fluid contained within the control volume. For a steady flow, $dm/dt = 0$, hence:

$$\dot{m}_1 - \dot{m}_2 = 0$$

$$\dot{m}_1 = \dot{m}_2$$

$$\rho_1 Q_1 = \rho_2 Q_2 \tag{3.37}$$

where ρ_1 and Q_1 are, respectively, the density and the volume rate of flow at section 1, and ρ_2 and Q_2 are, respectively, the density and the volume rate of flow at section 2.

For an incompressible fluid, the density may be assumed constant. Therefore:

$$Q_1 = Q_2$$

$$A_1 V_1 = A_2 V_2 \tag{3.38}$$

where A_1 and V_1 are, respectively, the area of cross-section and the average velocity at section 1, and A_2 and V_2 are, respectively, the area of cross-section and the average velocity at section 2.

Alternatively, the continuity equation for pipe flow may be obtained using the control volume equation derived earlier and given by Eq. (3.22). Then N is the mass m, and η is mass per unit mass—i.e., $\eta = 1$. Also, from the general principle of conservation of mass, the mass within a system remains constant with time—i.e., $dm/dt = 0$. Substituting in Eq. (3.22), we obtain

$$\frac{dm}{dt} = \frac{\partial}{\partial t}\int_{cv} \rho d\forall + \int_{cs} \rho \mathbf{V} \cdot \mathbf{dA} = 0$$

For steady flow

$$\frac{\partial}{\partial t}\int_{cv} \rho d\forall = 0$$

Hence:

$$\int_{cs} \rho \mathbf{V} \cdot \mathbf{dA} = 0$$

which states that the net mass outflow from the control volume must be zero.

Consider a section of a pipe with a single stream filament shown inside it (Fig. 3.25a). The same stream filament is shown highly exaggerated in Fig. 3.25b, with the velocity and the area vectors at its ends.

At section 1, the net mass outflow is $\rho_1 \mathbf{V}_1 \cdot \mathbf{dA}_1 = -\rho_1 V_1 dA_1$ and at section 2, it is $\rho_2 \mathbf{V}_2 \cdot \mathbf{dA}_2 = \rho_2 V_2 dA_2$. Since there is no flow across the stream filament

$$\int_{cs} \rho \mathbf{V} \cdot \mathbf{dA} = -\rho_1 V_1 dA_1 + \rho_2 V_2 dA_2 = 0$$

or

$$\rho_1 V_1 dA_1 = \rho_2 V_2 dA_2$$

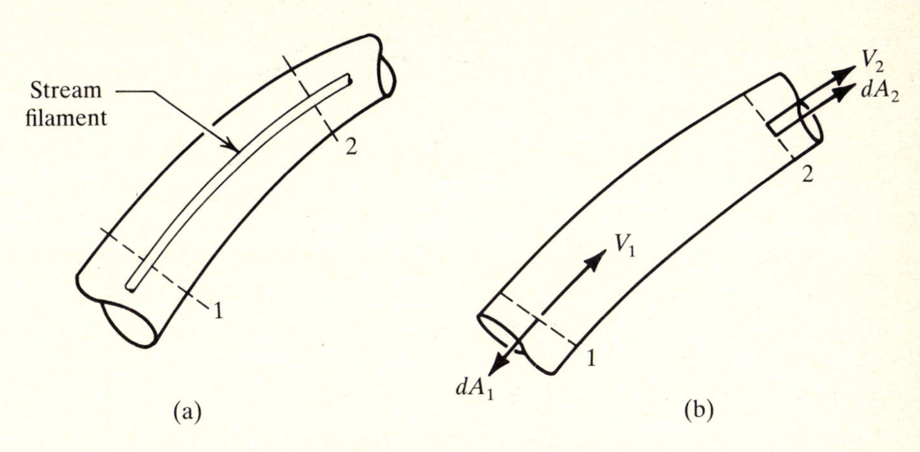

FIGURE 3.25 Control volume consisting of stream filaments.

The entire section of the pipe shown in Fig. 3.25a, can be assumed to consist of an infinite number of such stream filaments. Then, for the given section of the pipe, the continuity equation becomes

$$\int \rho_1 V_1 dA_1 = \int \rho_2 V_2 dA_2$$

Assuming that the density and the velocity at a section are constant, the above equation becomes:

$$\rho_1 V_1 A_1 = \rho_2 V_2 A_2$$

where V is the average velocity at cross-sectional area A. For an incompressible fluid, the density is constant and

$$V_1 A_1 = V_2 A_2$$

This equation is identical to the one obtained before, as Eq. (3.38).

EXAMPLE 3.9

A velocity field for steady flow of water is given by $\mathbf{V} = (x^2 + y^2)\mathbf{i} + 2xy\mathbf{j} - 4xz\mathbf{k}$. Does the flow satisfy continuity?

Solution From the given equation:

$$u = x^2 + y^2; \qquad \frac{\partial u}{\partial x} = 2x$$

$$v = 2xy; \qquad \frac{\partial v}{\partial y} = 2x$$

$$w = -4xz; \qquad \frac{\partial w}{\partial z} = -4x$$

For flow to be continuous:

$$\frac{\partial u}{\partial x} + \frac{\partial v}{\partial y} + \frac{\partial w}{\partial z} = 0$$

In this problem:

$$2x + 2x - 4x = 0$$

Therefore, the flow satisfies continuity.

EXAMPLE 3.10

Water is flowing at constant rate through the pipe shown in Fig. 3.26. The diameter of the smaller section is 10 cm and the diameter of the larger section is 30 cm. The average velocity in the smaller section is measured to be 30 m/s. Determine the velocity in the larger section.

Solution From Eq. (3.38):

$$A_1 V_1 = A_2 V_2$$

$$V_2 = \left(\frac{A_1}{A_2}\right) V_1$$

Now

$$\frac{A_1}{A_2} = \left(\frac{\pi D_1^2}{4}\right) \bigg/ \left(\frac{\pi D_2^2}{4}\right) = \left(\frac{D_1}{D_2}\right)^2$$

Hence:

$$V_2 = \left(\frac{D_1}{D_2}\right)^2 V_1$$

$$= \left(\frac{10}{30}\right)^2 30$$

$$= 3.33 \text{ m/s}$$

Therefore, the velocity at the larger section is 3.33 m/s.

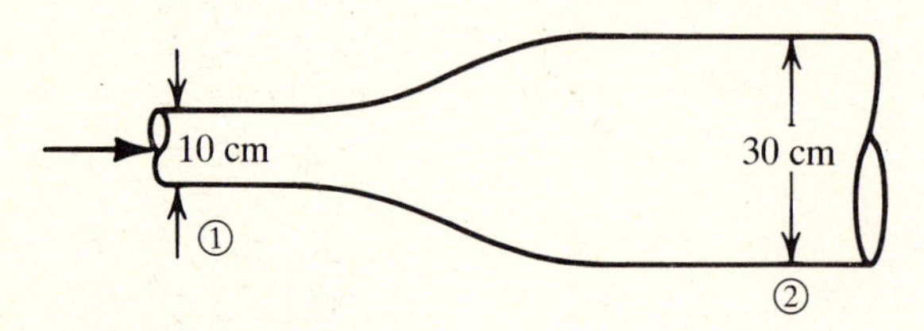

FIGURE 3.26 Example 3.10.

PROBLEMS

p.3.1 Steady flow occurs when
a. $\partial V/\partial t$ is constant.
b. $\partial V/\partial s = 0$.
c. conditions change steadily with time.
d. conditions are constant with respect to time at a point.
e. none of these answers.

p.3.2 Uniform flow occurs when.
a. $\partial V/\partial s$ is constant.
b. $\partial V/\partial s = 0$.
c. $\partial V/\partial t$ is constant.
d. conditions change uniformly with time.
e. the flow is steady.

p.3.3 Water is flowing through a uniformly tapering pipe with a constant discharge. The flow is
a. steady-uniform,
b. steady-nonuniform,
c. unsteady-uniform,
d. unsteady-nonuniform,
e. none of these answers.

p.3.4 A streamline is
a. always the path of a fluid particle.
b. the path of a fluid particle only if the flow is steady.
c. the path of a fluid particle only if the flow is uniform.
d. always fixed in space.
e. none of these answers.

p.3.5 Select the correct statement.
a. Flow can take place across a stream-tube under steady conditions.
b. Flow can take place across a stream-tube under uniform conditions.
c. Flow can take place across a stream-tube under unsteady conditions.
d. Flow can take place across a stream-tube under non-uniform conditions.
e. Flow cannot take place across a stream-tube under any condition.

p.3.6 The velocity in a flow field is given by $\mathbf{V} = 7\mathbf{i} + (x + y^2)\mathbf{j} + 5xy\mathbf{k}$. Determine acceleration at point (1, 4, 5) and at (−1, −4, −5).

p.3.7 Water at 20°C is flowing through the nozzle shown in Fig. 3.27. What is the convective acceleration at a distance of 30 cm from the left section if the rate of flow is 10 l/s?

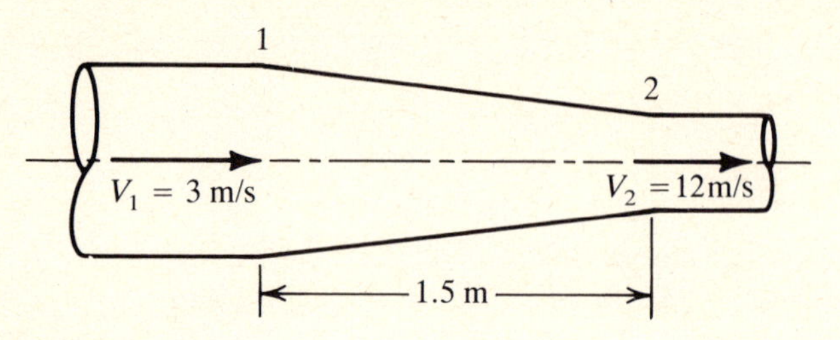

FIGURE 3.27 Problem 3.7.

p.3.8 The velocity of water as it approaches the stagnation point at the leading edge of a cylinder of radius R (Fig. 3.28) held normal to the stream is given by

$$u = u_s\left(1 - \frac{R^2}{x^2}\right)$$

where u_s is the free stream velocity. What is the acceleration at

a. $x = -4R$?
b. $x = -2R$?
c. $x = -R$?

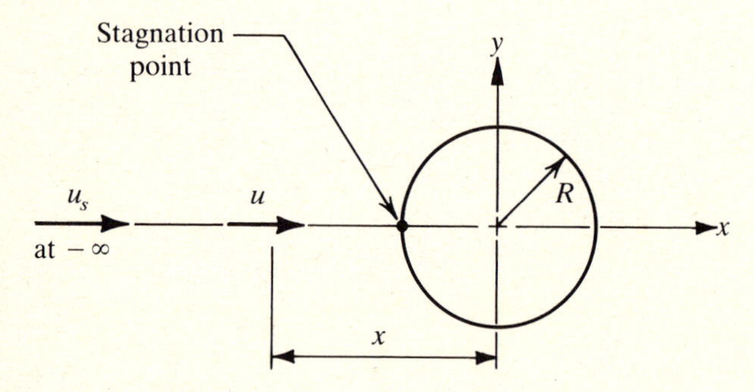

FIGURE 3.28 Problem 3.8.

p.3.9 Given $u = x^2$ and $v = 4$.

a. What is the stream function for this flow?
b. Sketch the streamlines for the first two quadrants.

p.3.10 A stream function is given by $\psi = y^2 + 2y + x^2$. Determine the flow rate passing through a straight line connecting the points (3, 1) and (6, 2) per unit width in the z-direction.

p.3.11 The velocity profile for laminar flow between two parallel plates is given by

$$u = u_m\left(1 - \frac{4r^2}{r_0^2}\right)$$

Where u_m is the centerline velocity, r_0 is the distance between the plates and u is the velocity at a distance r from the centerline. Determine the average velocity if $u_m = 8.5$ m/s, and $r_0 = 30$ cm.

p.3.12 The velocity in a flow field is given by

$$\mathbf{V} = (2x^2 + y^2)\mathbf{i} + 2xy\mathbf{j}$$

Is the flow irrotational?

p.3.13 The velocity in a flow field is given by

$$u = \ln x + y$$

$$v = xy - \frac{y}{x}$$

Is the flow irrotational?

p.3.14 Which of the following flows satisfies continuity?

a. $\mathbf{V} = (x^2 \cos y)\mathbf{i} - (2x \sin y)\mathbf{j}$
b. $\mathbf{V} = (-x^2 + y^2)\mathbf{i} + 2xy\mathbf{j}$
c. $\mathbf{V} = (10x)\mathbf{i} + (10y)\mathbf{j}$
d. $\mathbf{V} = [y(x^2 + y^2)^{-3/2}]\mathbf{i} - [x(x^2 + y^2)^{-3/2}]\mathbf{j}$

p.3.15 Water flows from a 30 cm diameter pipe at a rate of 0.2 m^3/s, into a 50 cm diameter pipe. What is the velocity of flow in the 50 cm diameter pipe?

p.3.16 Water flowing with a velocity of 5 m/s in a 20 cm diameter pipe and alcohol flowing at a rate of 0.15 m^3/s meet in a Y-connection and flow as a mixture in a 30 cm diameter pipe. What is the velocity of flow of the mixture in the 30 cm diameter pipe?

p.3.17 What is the velocity of flow of water in the 15 cm diameter pipe shown in Fig. 3.29?

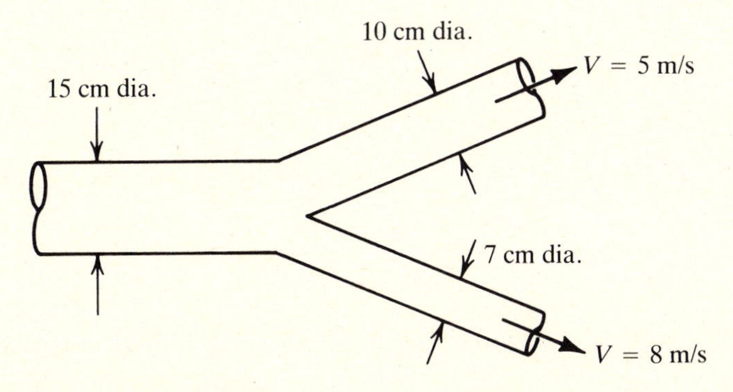

FIGURE 3.29 Problem 3.17.

p.3.18 At what rate is the water level rising or falling in the tank shown in Fig. 3.30?

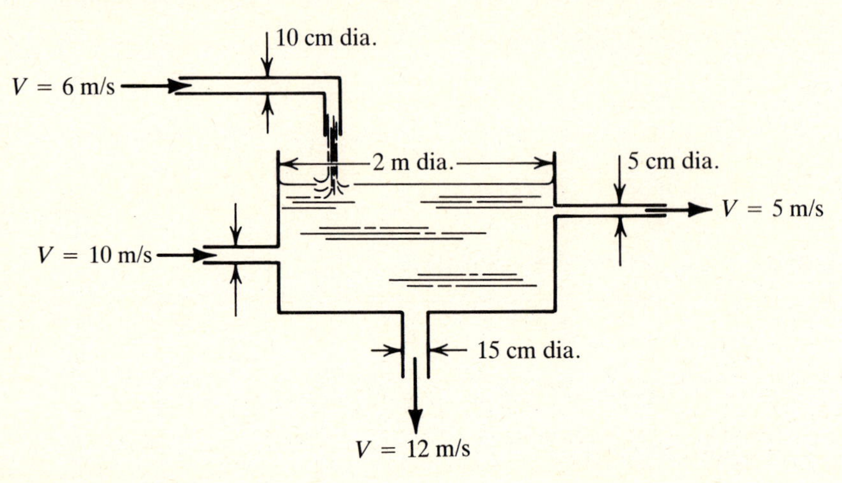

FIGURE 3.30 Problem 3.18.

Chapter Four

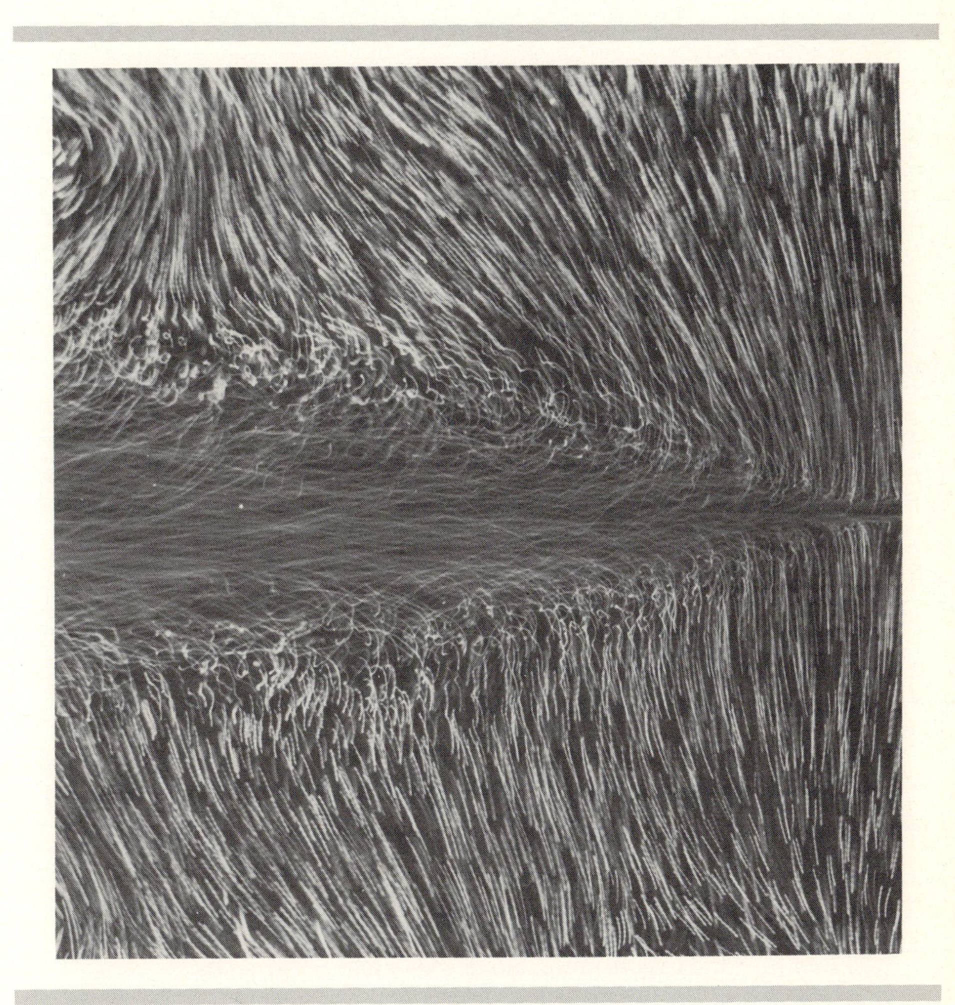

FLUID DYNAMICS

In Chapter Three, we considered the velocities and accelerations of fluid particles without considering the forces which caused the fluid to move. In this chapter, we will consider the forces also. First we will study the motion of a particle of an inviscid (non-viscous) fluid leading to the development of Euler and Bernoulli equations. Then the energy equation for a real fluid will be developed.

4.1 THE EULER AND BERNOULLI EQUATIONS

In general, the velocity of a fluid particle varies from point to point in a flow field. Since force is required to change velocity, the pressure must vary from point to point. The relationship between the changes in pressure and velocity in a flow field is described by the Euler equation, named after the Swiss mathematician Leonhard Euler (1707–83).

Consider a fluid element of unit thickness having a length ds along the streamline and a length dn normal to the streamline, the coordinates s and n being directed along and normal to the streamline respectively (Fig. 4.1a). We will develop *two equations* of motion, one *along the streamline* and the other *normal to it.* Only the pressure and the gravity forces will be considered. Other forces, such as those due to viscosity, surface tension, chemical and nuclear reactions, etc., are assumed to be negligible.

Along the Streamline

Let the pressure at the upstream face of the element be p. Then, using only the first two terms in the Taylor series expansion, the pressure on the downstream face of the element is $[p + (\partial p/\partial s)ds]$. The weight of the fluid element is $(\rho g ds dn)$ and acts vertically downward. The resultant force acting on the element in the s-direction is then given by

$$\Sigma F_s = p dn - \left(p + \frac{\partial p}{\partial s} ds\right) dn - \rho g ds dn \cos\theta$$

$$\Sigma F_s = -\frac{\partial p}{\partial s} ds dn - \rho g ds dn \cos\theta$$

Taking z as the vertical axis, the relationship between the directions s, n and z is given in Fig. 4.1b. From this, $\cos\theta = z/s$, or, in the differential form $\cos\theta = \partial z/\partial s$. Then, substituting for $\cos\theta$, we obtain:

$$\Sigma F_s = -\frac{\partial p}{\partial s} ds dn - \rho g ds dn \frac{\partial z}{\partial s}$$

From Eq. (3.15), acceleration along a streamline is given by

$$a_s = V\frac{\partial V}{\partial s} + \frac{\partial V}{\partial t}$$

Also, along the streamline, $\Sigma F_s = m a_s$. Hence:

$$-\frac{\partial p}{\partial s} ds dn - \rho g ds ds \frac{\partial z}{\partial s} = \rho ds dn\left(V\frac{\partial V}{\partial s} + \frac{\partial V}{\partial t}\right)$$

Dividing by $(\rho ds dn)$ and rearranging, we obtain:

$$\frac{1}{\rho}\frac{\partial p}{\partial s} + g\frac{\partial z}{\partial s} + V\frac{\partial V}{\partial s} + \frac{\partial V}{\partial t} = 0 \qquad (4.1)$$

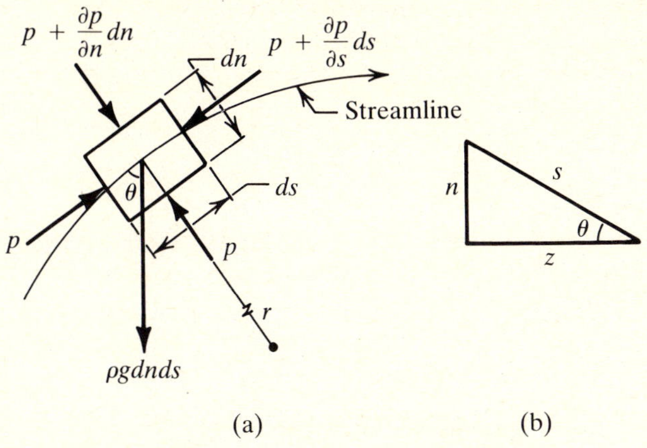

FIGURE 4.1 Motion of a fluid element along a streamline.

This is the *Euler equation of motion* of a fluid particle *along* a streamline. The flow may be steady or unsteady and the fluid compressible or incompressible. However, it has been assumed that the fluid has zero viscosity.

Normal to the Streamline

Let the pressure at the inner face be p (Fig. 4.1a). Then, the pressure at the outer face is $[p + (\partial p/\partial n)dn]$. Therefore, the resultant force acting on the element in the n-direction is given by

$$\Sigma F_n = pds - \left(p + \frac{\partial p}{\partial n}\, dn\right)ds - \rho gdnds \sin\theta$$

$$\Sigma F_n = -\frac{\partial p}{\partial n}\, dnds - \rho gdnds \sin\theta$$

From Fig. (4.1b), $\sin\theta = n/s$, or, in the differential form $\sin\theta = \partial n/\partial s$. Substituting for $\sin\theta$, we obtain

$$\Sigma F_n = -\frac{\partial p}{\partial n}\, dnds - \rho gdnds \frac{\partial n}{\partial s}$$

From Eq. (3.17), acceleration normal to a streamline is given by

$$a_n = \frac{V^2}{r} + \frac{\partial V_n}{\partial t}$$

Also, normal to the streamline $\Sigma F_n = ma_n$. Hence:

$$-\frac{\partial p}{\partial n}\, dnds - \rho gdnds \frac{\partial n}{\partial s} = \rho dnds\left(\frac{V^2}{r} + \frac{\partial V_n}{\partial t}\right)$$

Dividing by ($\rho dnds$) and rearranging, we obtain

$$\frac{1}{\rho}\frac{\partial p}{\partial n} + g\frac{\partial n}{\partial s} + \frac{V^2}{r} + \frac{\partial V_n}{\partial t} = 0 \tag{4.2}$$

This is the *Euler equation of motion* for a *fluid* particle *normal* to a streamline. It is applicable to compressible and incompressible fluids under steady or unsteady flow conditions. However, the fluid is assumed to have zero viscosity.

The Bernoulli Equation

The Bernoulli equation describes the motion of a fluid particle *along a streamline* when the flow is steady and the fluid is incompressible and non-viscous. It can be obtained from Eq. (4.1) by substituting the conditions for a steady flow and an incompressible fluid. For a steady flow $\partial V/\partial t = 0$. Substituting in Eq. (4.1), we obtain

$$\frac{1}{\rho}\frac{\partial p}{\partial s} + g\frac{\partial z}{\partial s} + V\frac{\partial V}{\partial s} = 0$$

In this equation, p, z and V are functions of s only; therefore, partial derivatives may be replaced by the total derivatives. Hence:

$$\frac{1}{\rho}\frac{dp}{ds} + g\frac{dz}{ds} + V\frac{dV}{ds} = 0 \tag{4.3}$$

Integrating along the streamline, we obtain

$$\int \frac{dp}{\rho} + gz + \frac{V^2}{2} = \text{constant} \tag{4.4}$$

Or, integrating between two points 1 and 2 along the streamline, we obtain

$$\int_1^2 \frac{dp}{\rho} + g(z_2 - z_1) + \frac{(V_2^2 - V_1^2)}{2} = 0 \tag{4.5}$$

If the fluid is incompressible, the density remains constant and $\int dp/\rho$ becomes p/ρ. Hence, for the steady flow of a non-viscous incompressible fluid:

$$\frac{p}{\rho} + gz + \frac{V^2}{2} = \text{constant} = C \tag{4.6}$$

This is commonly known as the *Bernoulli equation.* Another widely used form of the Bernoulli equation is obtained by dividing Eq. (4.6) by g. Then

$$\frac{p}{\gamma} + z + \frac{V^2}{2g} = \text{constant} = H \tag{4.7}$$

In this equation, each term has units of length. They may also be interpreted as energy per unit weight of the fluid. Then, the first term p/γ is the energy per unit weight of the fluid due to its pressure and is called the *pressure head.* The second term z is the energy per unit weight of the fluid due to its elevation and is called the *elevation head.* The third term $V^2/2g$ is the energy per unit weight of the fluid due to its velocity, and is called the *velocity head.*

According to Eq. (4.7), the sum of pressure head, elevation head and velocity head remains constant (equal to H) along a single streamline, but it may vary from streamline to streamline.

However, in cases where all the streamlines start from, or pass through, the same conditions of pressure, velocity, and elevation (for example, the surface of a reservoir), the sum of the three terms is equal for all the streamlines.

Let us now examine the condition which must be fulfilled so that the constant H may be applied throughout a flow field. For a steady flow, Eq. (4.2) becomes (since $\partial V_n/\partial t = 0$),

$$\frac{1}{\rho}\frac{\partial p}{\partial n} + g\frac{\partial z}{\partial n} + \frac{V^2}{r} = 0 \tag{4.8}$$

From Eq. (4.7)

$$\frac{p}{\rho} + gz + \frac{V^2}{2} = gH$$

Differentiating with respect to n, we obtain

$$\frac{1}{\rho}\frac{\partial p}{\partial n} + g\frac{\partial z}{\partial n} + V\frac{\partial V}{\partial n} = g\frac{\partial H}{\partial n} \tag{4.9}$$

Subtracting Eq. (4.9) from Eq. (4.8), we obtain

$$\frac{V^2}{r} - V\frac{\partial V}{\partial n} = -g\frac{\partial H}{\partial n}$$

$$\frac{V}{r} - \frac{\partial V}{\partial n} = -\left(\frac{g}{V}\right)\frac{\partial H}{\partial n}$$

But H is constant along s; hence $\partial H/\partial n = dH/dn$.
Then:

$$\frac{V}{r} - \frac{\partial V}{\partial n} = -\left(\frac{g}{V}\right)\frac{dH}{dn}$$

For H to be constant in the direction normal to the streamline, dH/dn must be zero. Therefore, the condition which must be satisfied so that the Bernoulli equation may be applied throughout a flow field is given by:

$$\frac{V}{r} - \frac{\partial V}{\partial n} = 0$$

This expression is identical to the one given by Eq. (3.29), which indicates an irrotational flow. Therefore, it is concluded that the *Bernoulli equation can be applied throughout a flow field provided that the flow is irrotational.* In other words, the expression $(p/\gamma + z + V^2/2g)$ remains constant from streamline to streamline in an irrotational flow field.

Since it is assumed that the fluid is non-viscous, it is implied that the velocity distribution over a cross-section is uniform. Hence, if a point is taken

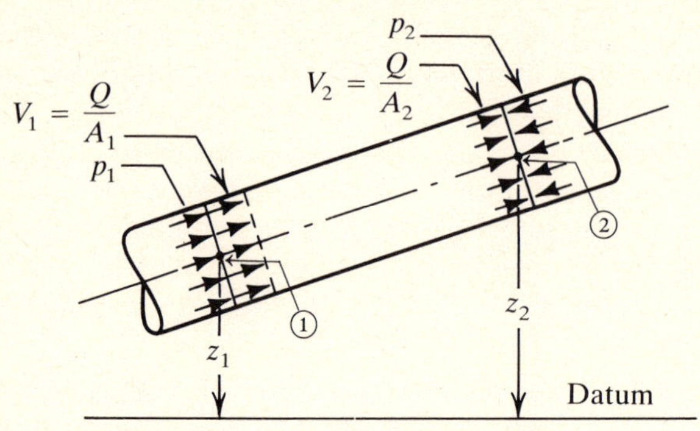

FIGURE 4.2 Fluid flowing through an inclined pipe.

on the centerline of a pipe, the average velocity over that cross-section is used to compute the velocity head. For example, for points 1 and 2 in Fig. 4.2, the Bernoulli equation will have the form

$$\frac{p_1}{\gamma} + z_1 + \frac{V_1^2}{2g} = \frac{p_2}{\gamma} + z_2 + \frac{V_2^2}{2g}$$

where for points (1) and (2):

p_1, p_2 = average pressure

V_1, V_2 = average velocity

z_1, z_2 = elevation above a datum.

Compressible Fluid

If a fluid is compressible, the density may vary from point to point along a streamline and the term $\int dp/\rho$ in the Euler equation cannot be evaluated unless the relationship between density and pressure is known. For gases, density ρ and pressure p are related by the polytropic formula

$$\frac{p}{\rho^n} = \text{constant}$$

where n depends on the mode of compression or expansion. For isothermal compression, $n = 1$ and for isentropic compression, $n = k = (c_p/c_v)$. Now, for isentropic compression

$$\frac{p}{\rho^k} = \frac{p_1}{\rho_1^k} \qquad \text{or} \qquad \rho = \left(\frac{\rho_1^k}{p_1}p\right)^{1/k}$$

Then:

$$\int_1^2 \frac{dp}{\rho} = \frac{1}{\rho_1}\int_1^2 \left(\frac{p_1}{p}\right)^{1/k} dp$$

or

$$\int_1^2 \frac{dp}{\rho} = \frac{k}{(k-1)}\left(\frac{p_1}{\rho_1}\right)\left[\left(\frac{p_2}{p_1}\right)^{(k-1)/k} - 1\right] \tag{4.10}$$

Equation (4.10) when substituted in Eq. (4.5) would yield the Euler equation between two points 1 and 2 for a nonviscous compressible fluid. Or

$$\left(\frac{k}{k-1}\right)\left(\frac{p_1}{\rho_1}\right)\left[\left(\frac{p_2}{p_1}\right)^{(k-1)/k}-1\right]+g(z_2-z_1)+\frac{V_2^2-V_1^2}{2}=0 \quad (4.11)$$

EXAMPLE 4.1

Water is flowing down a vertical tapering pipe 6 m long. The diameters of the pipe at the top and the bottom sections are 10 cm and 5 cm, respectively. If the discharge through the pipe is 100 l/s, find the difference of pressure between the top and the bottom ends of the pipe. Neglect the head losses.

Solution Let us select two points 1 and 2 at center of the top and bottom cross-sections as shown in Fig. 4.3

$$Q = 100 \text{ l/s} = 0.1 \text{ m}^3/\text{s}$$

$$A_1 = \frac{\pi(10)^2}{4} = 78.54 \text{ cm}^2$$

$$= 78.54 \times 10^{-4} \text{ m}^2$$

$$A_2 = \frac{\pi(5)^2}{4} = 19.63 \text{ cm}^2$$

$$= 19.63 \times 10^{-4} \text{ m}^2$$

$$V_1 = \frac{Q}{A_1} = \frac{0.1}{78.54 \times 10^{-4}} = 12.73 \text{ m/s}$$

$$V_2 = \frac{Q}{A_2} = \frac{0.1}{19.63 \times 10^{-4}} = 50.94 \text{ m/s}$$

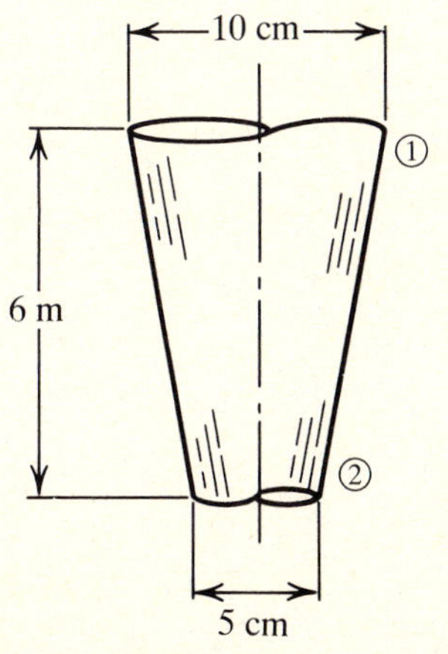

FIGURE 4.3 Example 4.1.

Applying the Bernoulli equation to points 1 and 2 and taking the datum through point 2, we have

$$\frac{p_1}{\gamma} + z_1 + \frac{V_1^2}{2g} = \frac{p_2}{\gamma} + z_2 + \frac{V_2^2}{2g}$$

$$\frac{p_1}{\gamma} + 6 + \frac{(12.73)^2}{2 \times 9.81} = \frac{p_2}{\gamma} + 0 + \frac{(50.94)^2}{2 \times 9.81}$$

$$\frac{p_2}{\gamma} + 6 + 8.26 = \frac{p_2}{\gamma} + 0 + 132.26$$

$$\left(\frac{p_1}{\gamma} - \frac{p_2}{\gamma}\right) = 132.26 - 6 - 8.26$$

$$= 118 \text{ m of water}$$

or

$$(p_1 - p_2) = 118 \times 9810 = 1158 \text{ kN/m}^2$$

EXAMPLE 4.2

Water is flowing from a tank through an orifice 2.5 cm in diameter under a constant head of 2.0 m. Determine the theoretical discharge from the orifice. What is the actual discharge, if the coefficient of discharge is 0.65?

Solution The tank and the orifice set is shown in Fig. 4.4. Consider points 1 and 2 as shown. Point 1 is located at the surface of water in the tank and point 2 is located slightly outside the tank on the centerline of the orifice.

Applying the Bernoulli equation to points 1 and 2 and taking the centerline of the orifice as the datum, we obtain

$$\frac{p_1}{\gamma} + z_1 + \frac{V_1^2}{2g} = \frac{p_2}{\gamma} + z_2 + \frac{V_2^2}{2g} \tag{1}$$

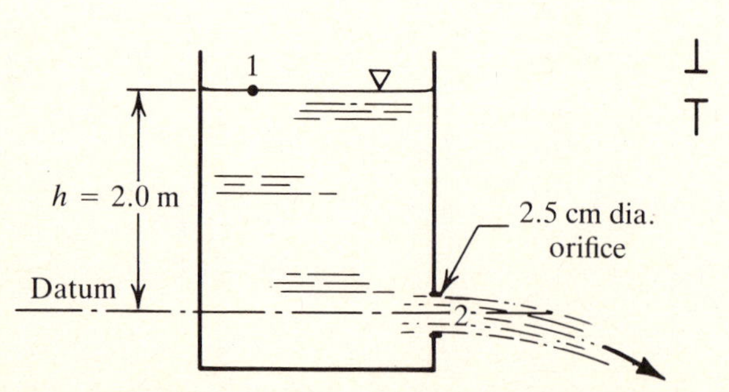

FIGURE 4.4 Example 4.2.

Now

$\frac{p_1}{\gamma} = 0$, or the pressure at the surface of the water is atmospheric and equal to zero gauge pressure

$z_1 = h$ or the height of point 1 above the datum line

$V_1 = 0$, or velocity at the surface of the tank is zero

$\frac{p_2}{\gamma} = 0$, or the pressure around and inside the jet is atmospheric and equal to zero gauge pressure

$z_2 = 0$, or point 2 lies on the datum line

$V_2 =$ unknown, or this is what we want to determine

Substituting in Eq. (1), we obtain:

$$0 + h + 0 = 0 + 0 + \frac{V_2^2}{2g}$$

$$V_2 = \sqrt{2gh}$$

Also,

$$Q = A_0 V_2$$

where

$$A_0 = \text{area of the orifice}$$

Hence

$$Q_{th} = A_0\sqrt{2gh}$$

This last equation is the theoretical discharge. Due to frictional losses, the actual discharge will be less than the theoretical discharge. Or

$$Q = C_d A_0 \sqrt{2gh} \tag{2}$$

where $C_d =$ coefficient of discharge. The coefficient of discharge is determined experimentally. It varies with the head and the type of the orifice. Usually its value is between 0.6 and 0.65.

In this problem

$$h = 2.0 \text{ m} = 200 \text{ cm}$$

$$d = 2.5 \text{ cm}$$

$$A_0 = \frac{\pi(2.5)^2}{4} = 4.91 \text{ cm}^2$$

Therefore:

$$Q = 0.65 \times 4.91\sqrt{2 \times 981 \times 200}$$
$$= 2.0 \text{ l/s}$$

EXAMPLE 4.3

For the large orifice shown in Fig. 4.5

$$H_1 = 3 \text{ m}, H_2 = 4 \text{ m}, b = 50 \text{ cm}, C_d = 0.65.$$

Determine the discharge through the orifice.

Solution If a vertical orifice is as large as shown in Fig. 4.5, the velocity of liquid flowing through the orifice will not be the same along the vertical axis.

Let
H_1 = height of liquid above the top of the orifice
H_2 = height of liquid above the bottom of the orifice
b = width of orifice

Consider an elemental strip of thickness dh, located at depth h from the liquid surface as shown in Fig. 4.5. Then the velocity V, through the strip is given by

$$V = \sqrt{2gh}$$

Let dQ_t = theoretical discharge through the elemental strip. Then

$$dQ_t = V \times \text{area of the strip}$$
$$= \sqrt{2gh}bdh$$
$$= b\sqrt{2g}h^{1/2}dh$$

$$\text{Total theoretical discharge} = Q_t = \int_{H_1}^{H_2} b\sqrt{2g}h^{1/2}dh$$
$$= b\sqrt{2g} \times \tfrac{2}{3}[h^{3/2}]_{H_1}^{H_2}$$
$$Q_t = \tfrac{2}{3}b\sqrt{2g}[H_2^{3/2} - H_1^{3/2}]$$

Let C_d = coefficient of discharge of the orifice. Then

$$\text{Actual discharge} = Q = C_d\tfrac{2}{3}b\sqrt{2g}[H_2^{3/2} - H_1^{3/2}] \quad (1)$$

In this problem:

$$H_1 = 3 \text{ m}$$
$$H_2 = 4 \text{ m}$$
$$b = 50 \text{ cm} = 0.5 \text{ m}$$
$$C_d = 0.65$$

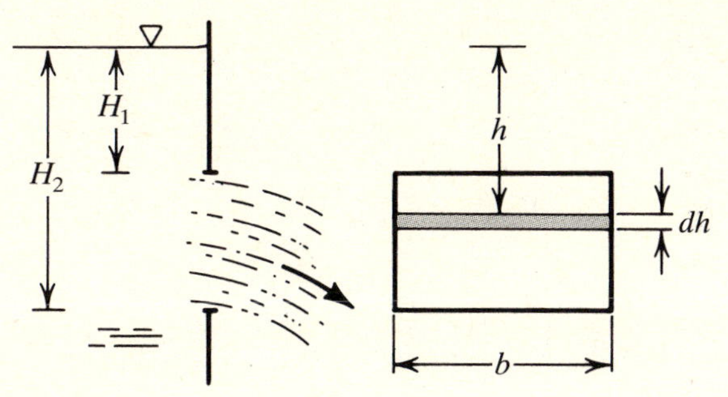

FIGURE 4.5 Example 4.3.

Hence:

$$Q = 0.65 \times \tfrac{2}{3} \times 0.5\sqrt{2 \times 9.81}[4^{3/2} - 3^{3/2}]$$
$$= 2.69 \text{ m}^3/\text{s}$$

EXAMPLE 4.4

An orifice in the side of a tank is rectangular in shape, 1 m broad and 0.5 m deep. The water level on one side of the orifice is 2 m above the top edge; the water level on the other side of the orifice is 20 cm below the top edge. Find the discharge through the orifice, if $C_d = 0.6$.

Solution If an orifice discharges into a liquid, rather than the atmosphere, and if the liquid head on the discharge side is above the top of the orifice, the orifice is known as *drowned* or *submerged.* If the discharge side is partially submerged, the orifice is known as *partially drowned* or *partially submerged.*

The discharge through a drowned orifice may be calculated using the same equation as for a free orifice except that the head causing the flow will be the difference between the heads on either side of the orifice.

The discharge through a partially drowned orifice may be found by treating the lower portion as a drowned orifice and the upper portion as a free orifice and by adding together the two discharges thus found.

The orifice for this problem is large and partially drowned. The lower portion may be treated as a drowned orifice and the upper half as a free orifice.

For the upper portion, Eq. (1) of Example 4.3 applies.
Hence:

$$Q_1 = C_d \tfrac{2}{3} b\sqrt{2g}[H_2^{3/2} - H_1^{3/2}]$$

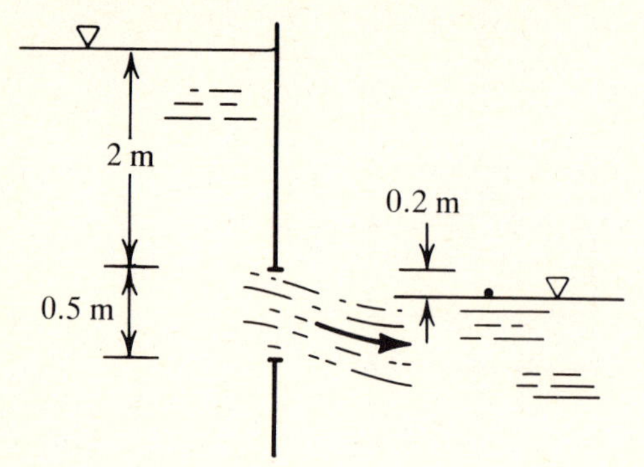

FIGURE 4.6 Example 4.4.

Here

$$C_d = 0.6$$
$$b = 1 \text{ m}$$
$$g = 9.81 \text{ m/s}^2$$
$$H_1 = 2 \text{ m}$$
$$H_2 = 2 + 0.2 = 2.2 \text{ m}$$

Therefore:

$$Q_1 = 0.6 \times \tfrac{2}{3} \times 1\sqrt{2 \times 9.81}[2.2^{3/2} - 2^{3/2}]$$
$$= 0.77 \text{ m}^3/\text{s}$$

For the lower portion, the head causing the flow is the difference between the heads on either side of the orifice.

Hence:

$$H = 2.0 + 0.2 = 2.2 \text{ m}$$

and is constant over the vertical cross-section of the submerged portion.

Therefore, from Eq. (2) of Example 4.2:

$$Q_2 = C_d\sqrt{2gH}A$$

Here

$$C_d = 0.6$$
$$H = 2.2$$
$$A = \text{area of the submerged portion only}$$
$$= 1 \times 0.3 = 0.3 \text{ m}^2$$
$$Q_2 = 0.6\sqrt{2 \times 9.81 \times 2.2} \times 0.3$$
$$= 1.18 \text{ m}^3/\text{s}$$
$$\text{Total discharge} = Q_1 + Q_2$$
$$= 0.77 + 1.18 = 1.95 \text{ m}^3/\text{s}$$

EXAMPLE 4.5

Determine the rate of flow of water through the venturimeter shown in Fig. 4.7. Assume $C_d = 0.95$.

Solution Consider two points 1 and 2 located on the centerline and directly below the manometer tubes. Neglecting head losses, the Bernoulli equation between points 1 and 2 may be written as:

$$\frac{p_1}{\gamma} + z_1 + \frac{V_1^2}{2g} = \frac{p_2}{\gamma} + z_2 + \frac{V_2^2}{2g}$$

Substituting for individual terms and taking the centerline as datum, we get

$$(30 + \underbrace{2}_{\text{radius of the pipe}}) + 0 + V_1^2/2g = (20 + \underbrace{1}_{\text{radius of the throat}}) + 0 + V_2^2/2g$$

or

$$V_2^2/2g - V_1^2/2g = 32 - 21 = 11$$

$$V_2^2 - V_1^2 = 2 \times 981 \times 11 = 21582 \qquad (1)$$

From continuity

$$A_1 V_1 = A_2 V_2$$

or

$$V_1 = \left(\frac{A_2}{A_1}\right) V_2 = \left(\frac{D_2}{D_1}\right)^2 V_2 = \left(\frac{20}{40}\right)^2 V_2$$

or

$$V_1 = 0.25 V_2$$

Substituting in Eq. (1), we get

$$V_2^2 - (0.25 V_2)^2 = 21582$$

$$0.9375 V_2^2 = 21582$$

$$V_2 = 151.73 \text{ cm/s} = 1.52 \text{ m/s}$$

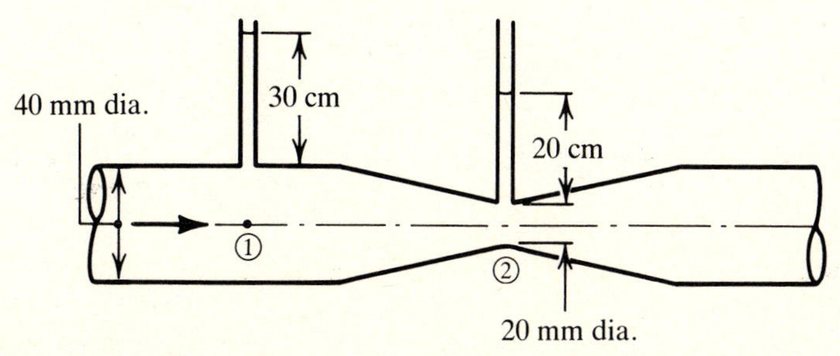

FIGURE 4.7 Example 4.5.

Therefore, the velocity of flow through the throat of the venturimeter is 1.52 m/s.

$$Q = VA = \frac{1.52\pi(0.02)^2}{4} = 4.775 \times 10^{-4} \text{ m}^3/\text{s}$$

$$\begin{aligned} \text{Actual Discharge} &= C_d Q \\ &= 0.95 \times 4.775 \times 10^{-4} \\ &= 4.53 \times 10^{-4} \text{ m}^3/\text{s} \end{aligned}$$

Therefore, the rate of flow $= 4.53 \times 10^{-4}$ m^3/s.

EXAMPLE 4.6

A pitot tube was lowered in a flowing stream to a depth of 1 ft. The water rose in the tube 6 in above the surface of the stream. Determine the velocity of the stream at the point of insertion.

Solution The Pitot tube is an instrument by which velocity may be measured at any desired point in a flowing liquid. In its simplest form it consists of a glass tube with the lower end bent through 90° as shown in Fig. 4.8. It is placed in the moving liquid with the lower opening facing the direction of flow. The liquid flows up the tube until all of its kinetic energy is converted to potential energy. The velocity of the liquid may then be estimated by the application of the Bernoulli equation. Let 1 and 2 be two points taken such that the point 1 lies on the centerline of the tube and slightly outside it, while point 2 lies on the free surface of the liquid inside the tube as shown. Let

Δx = height of liquid in the tube above the water surface

h = depth of liquid above point 1

V = velocity of the liquid at point 1

Applying the Bernoulli equation to points 1 and 2 and taking a line passing through point 1 as datum:

$$\frac{p_1}{\gamma} + z_1 + \frac{V_1^2}{2g} = \frac{p_2}{\gamma} + z_2 + \frac{V_2^2}{2g}$$

$$h + 0 + \frac{V_1^2}{2g} = 0 + (h + \Delta x) + 0$$

$$\begin{aligned} \frac{V_1^2}{2g} &= \Delta x + h - h \\ &= \Delta x \end{aligned}$$

$$V_1 = \sqrt{2g\Delta x}$$

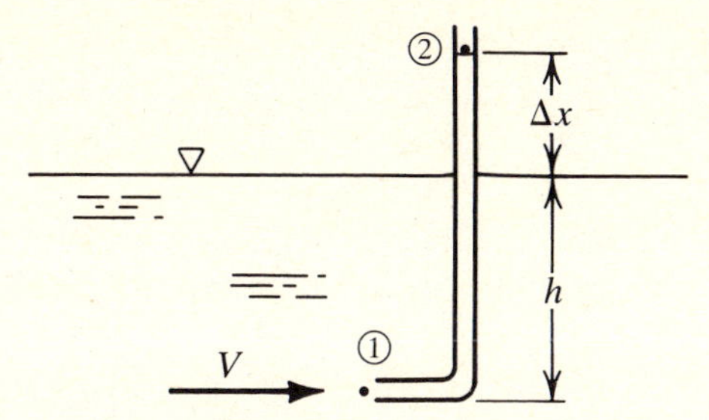

FIGURE 4.8 Example 4.6.

Let K = coefficient of the Pitot tube. Then, the actual velocity = $K\sqrt{2g\Delta x}$. In well-constructed instruments, K is equal to unity.

For the given problem:

$$\Delta x = 6 \text{ in} = 0.5 \text{ ft}$$

Assume $K = 1$. Then the velocity at a depth of 1 ft =

$$V = \sqrt{2 \times 32.2 \times 0.5}$$
$$= 5.67 \text{ ft/s}$$

4.2 GENERAL STEADY STATE ENERGY EQUATION

The general energy equation for the steady flow of a fluid may be developed using the first law of thermodynamics, which states that for any mass in a closed system, the net heat added to the system equals the increase in the energy of the system plus the energy used in doing any work, or

$$\delta Q = \delta E + \delta W \tag{4.12}$$

where δQ is the heat added to the system; δE is the change in the energy of the system and δW is the work done by the system.

The total energy of a mass of fluid consists of:

1. *Kinetic energy* or energy due to the motion of the fluid.
2. *Potential energy* or energy due to the position of the fluid, usually gravitational energy but it may also be electrical or magnetic.
3. *Internal energy* or energy due to the kinetic and potential energies of the molecules of the fluid. It is a function of the internal state of the fluid system only and does not depend on the position and the velocity of the system as a whole.

Let us now apply the first law of thermodynamics to a closed system of a fluid shown in Fig. 4.9

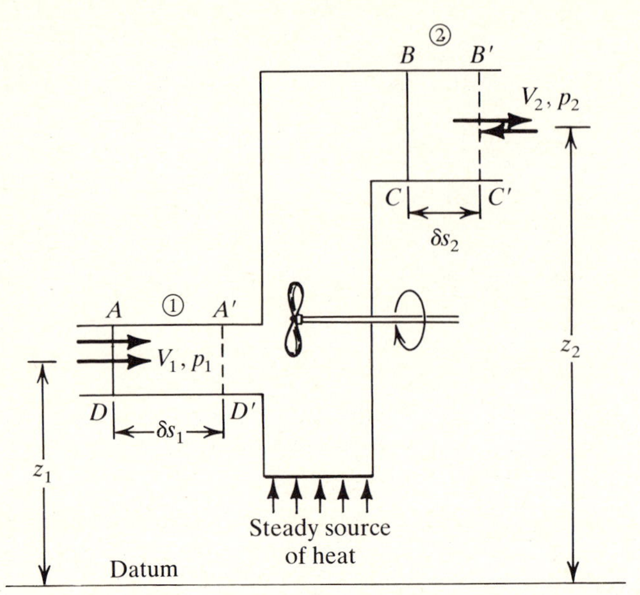

FIGURE 4.9 A closed fluid system used to develop energy equation.

The system consists of a device through which a fluid is flowing steadily. While flowing through the device, the fluid rotates a shaft and performs mechanical work. Fluid at pressure p_1 and with a velocity V_1 enters the device through section 1 and leaves the device at pressure p_2 and with a velocity V_2 through section 2. The elevation of section 1 is z_1 and that of section 2 is z_2 measured from a common datum. The device is being supplied with heat at a steady rate.

Consider a mass of fluid bounded between the planes *AD*, *BC* and the device boundaries at any time t. After a short time δt, the fluid moves forward and occupies a new position bounded by the planes *A′D′*, *B′C′* and the device boundaries. It is evident that during time δt, a small mass of fluid, δm, contained within the planes *AD* and *A′D′* has moved in the device and an equal mass of fluid, δm, contained within the planes *BC* and *B′C′* has moved out of the device. The mass δm at entry has kinetic energy $\frac{1}{2}\delta m V_1^2$, potential energy $\delta m g z_1$ and internal energy $\delta m u_1$ (where u is the internal energy per unit mass). If the energy of the system within the planes *A′D′*, *BC* and the device boundaries is E, then the energy of the fluid within the planes *AD*, *BC* and the device boundaries is $E + \delta m u_1 + \frac{1}{2}\delta m V_1^2 + \delta m g z_1$. Hence, energy of the system at time t is:

$$(\text{Energy})_t = E + \delta m(u_1 + V_1^2/2 + gz_1) \tag{4.13}$$

At time $t + \delta t$, the fluid occupies the position bounded between the planes *A′D′*, *B′C′* and the device boundaries. The energy in this position is equal to the energy between the planes *A′D′*, *BC* and the device boundaries (which is E) plus the energy of mass δm contained between the planes *BC* and *B′C′*

(which is the sum of $\delta m u_2$, $\delta m V_2^2/2$, and $\delta m g z_2$). Hence, energy of the system at time $t + \delta t$ is:

$$(\text{Energy})_{t+\delta t} = E + \delta m(u_2 + V_2^2/2 + gz_2) \tag{4.14}$$

The increase in energy during time δt is then given by:

$$\delta E = (\text{Energy})_{t+\delta t} - (\text{Energy})_t$$

or

$$\delta E = \delta m[(u_2 - u_1) + \tfrac{1}{2}(V_2^2 - V_1^2) + g(z_2 - z_1)] \tag{4.15}$$

Let us assume that during time δt, an amount of heat δQ was supplied to the system at a steady rate and the fluid did work $\delta W'$ on the shaft also at a steady rate. In addition to the work done on the shaft, the fluid also did work, $\delta W''$, against the pressure while moving from its first position to the second. The work done at the inlet is $p_1 A_1 \delta s_1$ (where A is the cross-sectional area and δs is the distance moved) and at the outlet it is $p_2 A_2 \delta s_2$. Hence:

$$\delta W'' = p_2 A_2 \delta s_2 - p_1 A_1 \delta s_1$$

The minus sign arises because the force $p_1 A_1$ is in the opposite direction to the displacement δs_1. Thus, total work done δW, is the sum of $\delta W'$ and $\delta W''$, or

$$\delta W = \delta W' + p_2 A_2 \delta s_2 - p_1 A_1 \delta s_1 \tag{4.16}$$

Substituting Eqs. (4.15) and (4.16) in Eq. (4.12), we obtain

$$\delta Q = \delta m[(u_2 - u_1) + \tfrac{1}{2}(V_2^2 - V_1^2) + g(z_2 - z_1)] + \delta W' + p_2 A_2 \delta s_2 - p_1 A_1 \delta s_1$$

Dividing by δm and noting that $\delta m = \rho_1 A_1 \delta s_1 = \rho_2 A_2 \delta s_2$, we get

$$\frac{\delta Q}{\delta m} = (u_2 - u_1) + \tfrac{1}{2}(V_2^2 - V_1^2) + g(z_2 - z_1) + \frac{\delta W'}{\delta m} + \frac{p_2}{\rho_2} - \frac{p_1}{\rho_1}$$

or

$$q = \left(\frac{p_2}{\rho_2} + \frac{V_2^2}{2} + gz_2\right) - \left(\frac{p_1}{\rho_1} + \frac{V_1^2}{2} + gz_1\right) + (u_2 - u_1) + W' \tag{4.17}$$

where q is the heat added to the system per unit mass and W' is the shaft work done by the fluid per unit mass. Equation (4.17) is known as the *steady state energy equation.* The equation is applicable to liquids, gases and vapors subject to meeting of the basic assumptions employed in its derivation, which are: (a) the flow is steady and continuous; (b) conditions within the device are steady; (c) the application of heat is at a steady rate; (d) the work done on the shaft is steady; (e) the velocity distribution is uniform at the cross-sections, and (f) that energy due to electricity, magnetism, surface tension or nuclear reaction is absent.

For many practical applications, Eq. (4.17) may be simplified by dropping various terms which are simply not present, or if present, are negligible.

If there is no heat transfer, q may be taken as zero, and the energy equation then becomes:

$$\frac{p_1}{\rho_1} + \frac{V_1^2}{2} + gz_1 = \frac{p_2}{\rho_2} + \frac{V_2^2}{2} + gz_2 + u_2 - u_1 + W' \tag{4.18}$$

For an ideal incompressible fluid with no shaft work, u_2, u_1 and W' are zeros, and Eq. (4.18) transforms to the Bernoulli equation.

Real fluids have viscosity and they do work when in motion to overcome viscous forces. The energy required to overcome friction is transformed into heat energy, which, in turn, increases the temperature and the internal energy of the fluid, and, in general, there is a heat transfer from the fluid to its surroundings. Let it be $-q$ per unit mass. This energy, coupled with the energy used in increasing the internal energy from u_1 to u_2, is actually lost as useless energy. Hence, the total energy lost, E_1 from the system in overcoming viscous resistance is:

$$E_1 = u_2 - u_1 - q \tag{4.19}$$

For an incompressible fluid, this loss of energy is usually expressed in terms of energy loss per unit weight and designated by HL. Hence:

$$HL = \frac{u_2 - u_1 - q}{g} \tag{4.20}$$

Dividing Eq. (4.17) by g and then substituting from Eq. (4.20) and putting $\rho_1 = \rho_2$ for an incompressible fluid, we get

$$\frac{p_1}{\rho g} + \frac{V_1^2}{2g} + z_1 = \frac{p_2}{\rho g} + \frac{V_2^2}{2g} + z_2 + HL + h_s \tag{4.21}$$

where $h_s = W'/g$ and is the shaft work done per unit weight of the fluid. If there is a pump in the system, then the work is done on the fluid and the energy is added to the system. In this case, h_s will be negative.

The final form of the energy equation for the steady flow of an incompressible real fluid with a source or sink of energy within the system may be written as:

$$\frac{p_1}{\gamma} + z_1 + \frac{V_1^2}{2g} = \frac{p_2}{\gamma} + z_2 + \frac{V_2^2}{2g} + HL \pm h_s \tag{4.22}$$

For a compressible fluid, the terms p_1/γ and p_2/γ cannot be obtained unless the density is known as a function of pressure. Hence, the energy equation for a compressible flow may be written as

$$z_1 + \frac{V_1^2}{2g} = z_2 + \frac{V_2^2}{2g} + \int_1^2 \frac{dp}{\rho g} + HL \pm h_s \tag{4.23}$$

The Kinetic Energy Correction Factor

While deriving the energy equation, it was assumed that the velocity did not vary over the cross-section. In practice, the velocity does vary over the cross-section due to the viscous forces which cause the velocity to be zero at the solid boundaries. A factor, called the *kinetic energy correction factor* (α), is used to correct this deficiency so that $\alpha V^2/2g$ is the average kinetic energy per unit weight passing the cross-section. Figure 4.10 shows a cross-section with actual velocity variation and average velocity V. The actual kinetic energy passing the cross-section per unit time is given by:

$$KE = \gamma \int \frac{u^2}{2g} u dA$$

This kinetic energy must be equal to the kinetic energy used in the energy equation multiplied by the correction factor α.
Hence:

$$\alpha \frac{V^2}{2g} \gamma V A = \gamma \int_A \frac{u^2}{2g} u dA$$

Solving for α:

$$\alpha = \frac{1}{A} \int_A \left(\frac{u}{V}\right)^3 dA \tag{4.24}$$

Let us now compute the kinetic energy correction factor for turbulent flow through a pipe.

The velocity distribution in turbulent flow in a smooth pipe is given by Prandtl's one-seventh power law, according to which

$$\frac{u}{u_{max}} = \left(\frac{R - r}{R}\right)^{1/7}$$

where u is the velocity at a distance r from the centerline of the pipe; u_{max} is the velocity at the centerline and R is the radius of the pipe. The average velocity in this case is given by (see Example 3.3)

$$V = \frac{98}{120} u_{max}$$

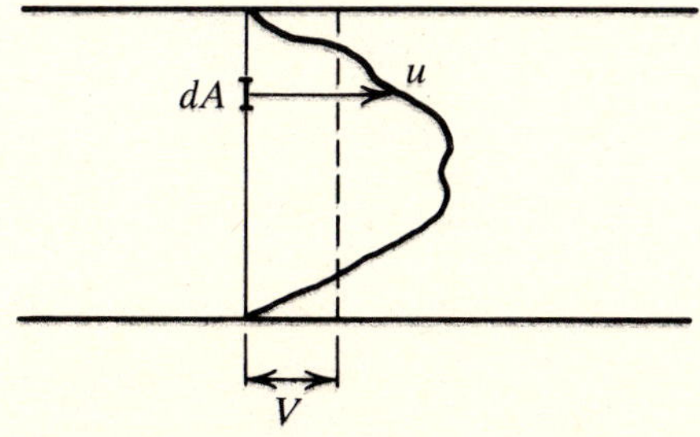

FIGURE 4.10 An arbitrary velocity variation over a cross section.

Then:

$$\frac{u}{V} = \frac{120}{98}\left(\frac{R-r}{R}\right)^{1/7}$$

Substituting in Eq. (4.24), we obtain:

$$\alpha = \frac{1}{\pi R^2}\int_0^R 2\pi r\left(\frac{120}{98}\right)^3\left(\frac{R-r}{R}\right)^{3/7} dr = 1.06$$

Therefore, using this particular velocity distribution, the kinetic energy will be about 6% less if the average velocity is used and the correction factor is not employed. This is not significant and can be ignored in practical problems except where great accuracy is required.

EXAMPLE 4.7

Determine the discharge through the system shown in Fig. 4.11. Assume that the head loss in the pipe is given by $4V_1^2/2g$ and the head loss in the nozzle is given by $3V_2^2/2g$, where V_1 and V_2 are the velocities in the pipe and the nozzle respectively.

Solution Consider two points 1 and 2, point 1 being at the top of the reservoir and point 2 being slightly outside the exit of the nozzle. Applying the energy equation to points 1 and 2 and noting that $h_s = 0$, we obtain:

$$\frac{p_1}{\gamma} + z_1 + \frac{V_1^2}{2g} = \frac{p_2}{\gamma} + z_2 + \frac{V_2^2}{2g} + HL_{(1-2)}$$

or

$$0 + 25 + 0 = 0 + 0 + \frac{V_2^2}{2g} + HL_{(1-2)}$$

or

$$\frac{V_2^2}{2g} + HL_{(1-2)} = 25 \tag{1}$$

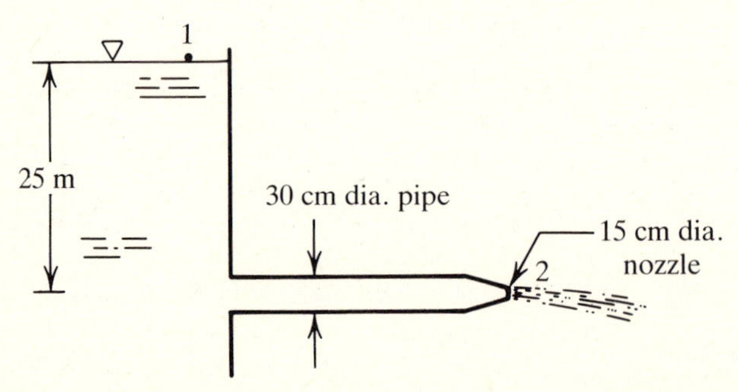

FIGURE 4.11 Example 4.7.

The head losses between the points 1 and 2 consist of head losses through the pipe given by $4V_1^2/2g$ and head losses through the nozzle given by $3V_2^2/2g$. Then

$$HL_{(1-2)} = 4\frac{V_1^2}{2g} + 3\frac{V_2^2}{2g}$$

Substituting in Eq. (1), we obtain:

$$\frac{V_2^2}{2g} + 4\frac{V_1^2}{2g} + 3\frac{V_2^2}{2g} = 25 \tag{2}$$

From continuity

$$A_1V_1 = A_2V_2$$

$$V_1 = \left(\frac{A_2}{A_1}\right)V_2 = \left(\frac{D_2}{D_1}\right)^2 V_2$$

$$V_1 = \left(\frac{15}{30}\right)^2 V_2 = 0.25V_2$$

Substituting in Eq. (2) we obtain:

$$\frac{V_2^2}{2g} + \frac{4(0.25V_2)^2}{2g} + 3\frac{V_2^2}{2g} = 25$$

$$\frac{1}{2g}(V_2^2 + 0.25V_2^2 + 3V_2^2) = 25$$

or

$$V_2 = \sqrt{\frac{2 \times 9.81 \times 25}{4.25}}$$

$$= 10.74 \text{ m/s}$$

$$A_2 = \text{area of nozzle}$$

$$= \frac{\pi(15)^2}{4} = 176.7 \text{ cm}^2 = 0.0177 \text{ m}^2$$

Hence:

$$Q = A_2V_2$$

$$= 0.0177 \times 10.74 = 0.19 \text{ m}^3\text{/s}$$

$$= 190 \text{ l/s}$$

EXAMPLE 4.8

Find the necessary power input to the pump (P) for the flow conditions shown in Fig. 4.12. The pump is 80% efficient and the head losses from points 1 to 2 are equal to 30 m. The fluid is water with $\gamma = 9810\ \text{N/m}^3$.

Solution From continuity:

$$A_1 V_1 = A_2 V_2$$

or

$$V_1 = \left(\frac{A_2}{A_1}\right) V_2 = \left(\frac{D_2}{D_1}\right)^2 V_2$$

$$= \left(\frac{2}{6}\right)^2 40 = 4.44\ \text{m/s}$$

Also:

$$Q = A_1 V_1 = 0.013\ \text{m}^3/\text{s}$$

Applying the energy equation to points 1 and 2 and noting that there is a pump between them, we have:

$$\frac{p_1}{\gamma} + z_1 + \frac{V_1^2}{2g} = \frac{p_2}{\gamma} + z_2 + \frac{V_2^2}{2g} + HL - h_s$$

Taking datum through point 1 and substituting for various terms, we obtain:

$$\frac{-50}{9810} + 0 + \frac{(4.44)^2}{2.0 \times 9.81} = 0 + 100 + \frac{(40)^2}{2.0 \times 9.81} + 0 + 30 - h_s$$

$$-0.005 + 0 + 1.005 + h_s = 0 + 100 + 81.55 + 0 + 30$$

or

$$h_s = 210.6\ \text{m}$$

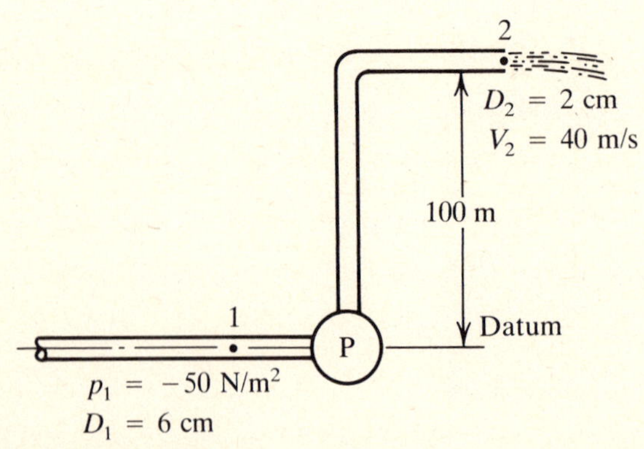

FIGURE 4.12 Example 4.8.

Therefore the pump must be able to provide a head of 210.6 meters. Let P = power output of the pump in watts.
Then:

$$P = \gamma Q h_s$$
$$= 9.810 \times 0.013 \times 210.6$$
$$= 26.9 \text{ kW}$$

$$\text{Power input} = \frac{\text{Power output}}{\text{efficiency}}$$
$$= \frac{26.9}{0.8} = 33.6 \text{ kW}$$

4.3 HYDRAULIC AND ENERGY GRADE LINES

When a fluid flows through a pipe, it is subjected to frictional resistance depending on the velocity of flow, the nature of the fluid and the nature of the pipe surface. The energy lost in overcoming the frictional resistance is known as *the head lost in friction* and is expressed in terms of the height of a fluid column.

Consider water flowing through the system of Fig. 4.13. If a piezometer is inserted into the pipe, the water would rise to a height p/γ above the pipe, giving the pressure head at that point. If the pressure heads at all points along the length of the pipe are plotted as vertical ordinates using the centerline of the pipe as the baseline, a straight sloping line *ab* would be obtained. This line is called *the hydraulic grade line* (*HGL*). Therefore, a hydraulic grade line is a line which joins all the points p/γ distance from the centerline of the pipe, or $(p/\gamma + z)$ distance from any arbitrary datum.

Another line, *cd*, drawn in such a way that every point on it is $(p/\gamma + V^2/2g)$ from the centerline of the pipe is called the *energy grade line* (*EGL*). The energy grade line would always slope down uniformly in the

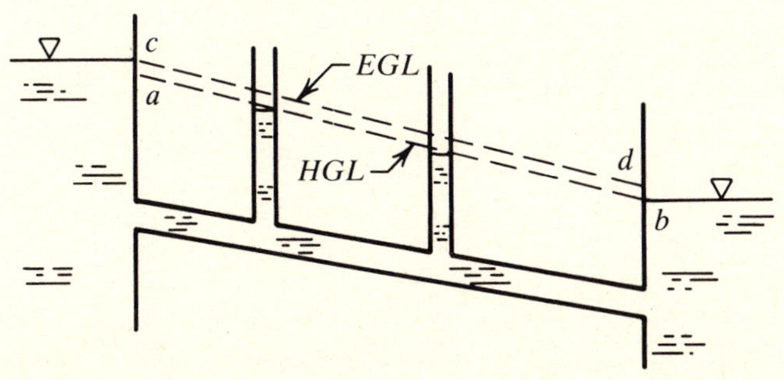

FIGURE 4.13 Hydraulic grade line (*HGL*) and energy grade line (*EGL*) in a flow field.

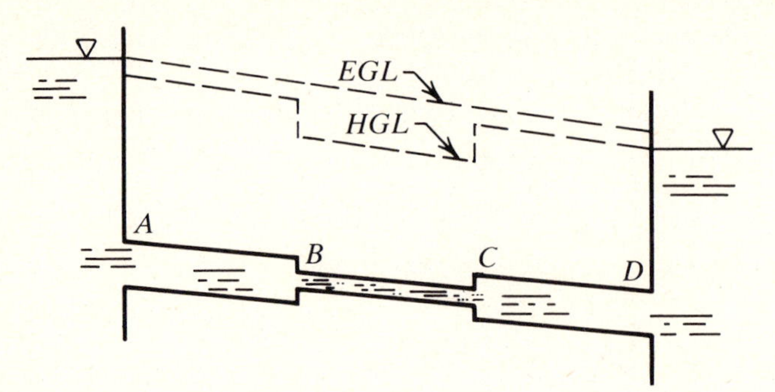

FIGURE 4.14 *HGL* and *EGL* for pipes of different sizes connected in series.

direction of flow, although the hydraulic grade line may go up or down depending on the variation of pressure in the direction of the flow. To illustrate this, consider the flow system shown in Fig. 4.14. At cross-section *B*, the pipe diameter suddenly decreases, resulting in a proportionate increase in velocity. Since total energy at *B* is constant, when the velocity increases, the pressure must decrease. This causes the hydraulic grade line to suddenly drop at *B*. At cross-section *C*, the diameter of the pipe suddenly increases causing a slower velocity and a larger pressure. This causes the *HGL* to suddenly rise as shown in Fig. 4.14. The *EGL*, however, slopes uniformly in the direction of flow irrespective of the changes in pipe diameter.

The energy grade line may go up only if a pump is installed in the line as shown in Fig. 4.15. The abrupt rise is equal to the head generated by the pump.

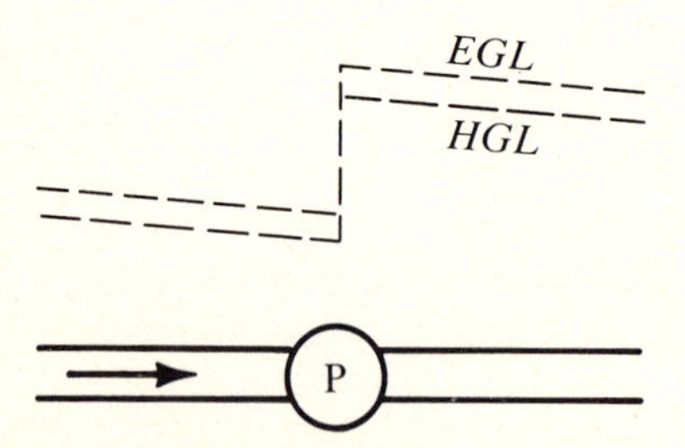

FIGURE 4.15 *HGL* and *EGL* for a pump.

Similarly, an abrupt drop in the energy grade line can be caused if a turbine is present in the system.

EXAMPLE 4.9

For the flow system shown in Fig. 4.16, the flow rate is 200 l/s. Assume that the head losses in the pipe are given by $0.02(L/D)(V^2/2g)$ where L is the length of the pipe and D is the diameter of the pipe. Determine the pressure head at points A and B and draw the hydraulic and the energy grade lines.

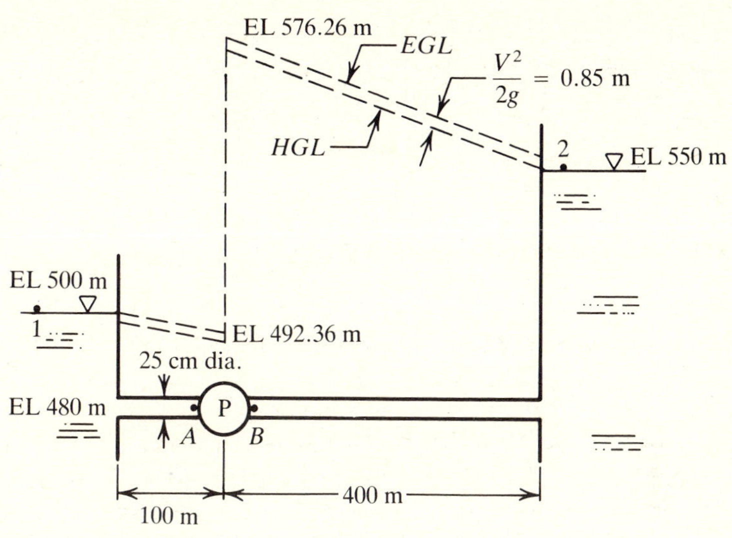

FIGURE 4.16 Example 4.9.

Solution

$$\text{Discharge through the system} = Q = 0.2 \text{ m}^3/\text{s}$$

$$\text{Area of pipe} = A = \frac{\pi(25)^2}{4} = 491 \text{ cm}^2 = 0.049 \text{ m}^2$$

$$\text{Velocity in the pipe} = \frac{Q}{A} = \frac{0.2}{0.049} = 4.08 \text{ m/s}$$

Pressure head at A

To find the pressure head at A, take point 1 on the surface of the lower reservoir and apply the energy equation to points 1 and A.

$$\frac{p_1}{\gamma} + z_1 + \frac{V_1^2}{2g} = \frac{p_A}{\gamma} + z_A + \frac{V_A^2}{2g} + HL_{(1-A)}$$

$$0 + 500 + 0 = \frac{p_A}{\gamma} + 480 + \frac{(4.08)^2}{2 \times 9.81} + 0.02\left(\frac{100}{0.25}\right)\frac{(4.08)^2}{2 \times 9.81}$$

$$500 = \frac{p_A}{\gamma} + 480 + 0.85 + 6.79$$

$$\frac{p_A}{\gamma} = 500 - 480 - 0.85 - 6.79$$

$$= 12.36 \text{ m}$$

Elevation of the HGL at $A = 480 + 12.36 = 492.36$ m.

Pressure head at B

The pressure head at B is equal to the head generated by the pump (h_s). To find h_s, take point 2 on the surface of the higher reservoir, and apply the energy equation to points 1 and 2.

Then:

$$\frac{p_1}{\gamma} + z_1 + \frac{V_1^2}{2g} + h_s = \frac{p_2}{\gamma} + z_2 + \frac{V_2^2}{2g} + HL_{(1-2)}$$

$$0 + 500 + 0 + h_s = 0 + 550 + 0 + 0.02\left(\frac{500}{0.25}\right)\frac{(4.08)^2}{2 \times 9.81}$$

$$500 + h_s = 550 + 33.9$$

$$h_s = 83.9 \text{ m}$$

Elevation of the HGL at $B = 492.36 + 83.9 = 576.26$ m.

The hydraulic grade line (HGL) and the energy grade line (EGL) are drawn on Fig. 4.16. The head losses in the pump itself are neglected. While drawing the HGL and EGL, points A and B have been assumed to be close to each other. The velocity head in the pipe is $V^2/2g = 0.85$ m. Therefore, the EGL everywhere is 0.85 m above the HGL.

PROBLEMS

p.4.1 The correct form of Bernoulli equation is

a. $\frac{p}{\rho} + z + \frac{V^2}{g} = C.$

b. $\frac{p}{\gamma} + z + \frac{V^2}{g} = C.$

c. $\frac{p}{\gamma} + z + \frac{V^2}{2g} = C.$

d. $\frac{p}{g} + z + \frac{V^2}{2\gamma} = C.$

e. none of these.

p.4.2 Bernoulli equation in the English system has the units of

a. ft · lb/s.
b. lb.
c. ft · lb/slug.
d. ft.
e. none of these.

p.4.3 The pressure head term in Bernoulli's equation is

a. z.

b. $\frac{V^2}{2g}$.

c. $\frac{p}{\gamma}$.

d. $\frac{p}{2g}$.

e. none of these.

p.4.4 Water flows through an orifice under a constant head of 5 ft of water. The theoretical velocity in ft/s is

a. 332.
b. 17.9.
c. 9.9.
d. 12.65.
e. none of these.

p.4.5 The discharge through a sharp-edged circular orifice, 1 in in diameter under a constant head of 4.0 ft is 0.054 ft^3/s. The coefficient of discharge is

a. 0.92.
b. 0.62.
c. 1.2.
d. 0.35.
e. none of these.

p.4.6 The discharge (in ft^3/s) through a large rectangular vertical orifice, 6.0 ft wide and 4.0 ft deep, when the water level is 10 ft above the top edge of the orifice and $C_d = 0.61$ is

a. 300.
b. 403.
c. 150.
d. 806.
e. none of these.

p.4.7 The hydraulic grade line is always above the centerline of the pipe by a distance equal to

a. $\frac{p}{\gamma} + z + \frac{V^2}{2g}$.
b. $\frac{p}{\gamma} + z$.
c. $\frac{p}{\gamma}$.
d. $z + \frac{V^2}{2g}$.
e. none of these answers.

p.4.8 The energy grade line is always above the centerline of the pipe by a distance equal to

a. $\frac{p}{\gamma} + z + \frac{V^2}{2g}$.
b. $\frac{p}{\gamma} + z$.
c. $\frac{p}{\gamma}$.
d. $z + \frac{V^2}{2g}$.
e. none of these answers.

p.4.9 At a point in a flow system, the diameter of a pipe is suddenly reduced from 30 cm to 15 cm. The *HGL* at this point would

a. have a sudden vertical rise.
b. have a sudden vertical drop.
c. have no sudden rise or fall, but would fall gradually.
d. coincide with the energy grade line.
e. none of these answers.

p.4.10 At a point in a flow system, the diameter of a pipe is suddenly increased from 15 cm to 30 cm. The *EGL* at this point would

a. have a sudden vertical rise.
b. have a sudden vertical drop.
c. have no sudden rise or fall but would fall gradually.
d. coincide with the hydraulic grade line.
e. none of these answers.

p.4.11 The velocity at point 2 in Fig. 4.17 is known to be 10 m/s. Neglecting all the head losses, determine the pressure at point 2.

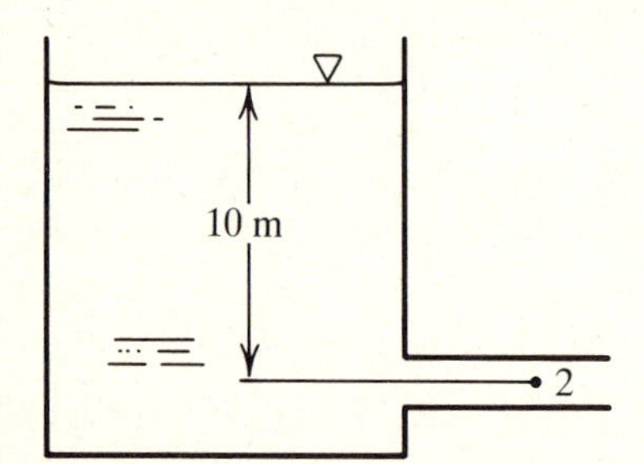

FIGURE 4.17 Problem 4.11.

p.4.12 Neglecting all the head losses, determine the discharge through the orifice, in Fig. 4.18.

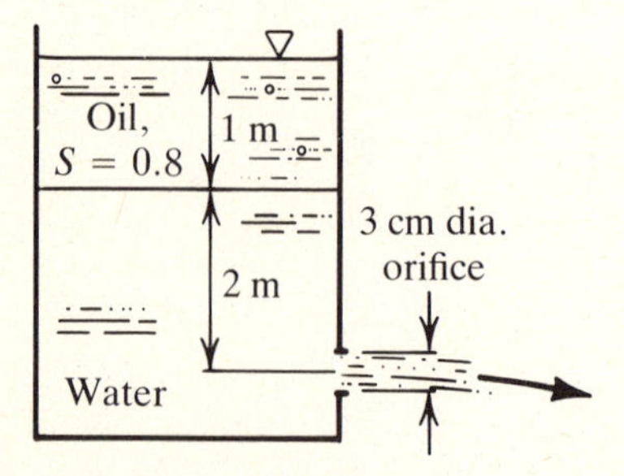

FIGURE 4.18 Problem 4.12.

p.4.13 How long will it take for the water surface in the tank shown in Fig. 4.19 to drop from $h = 4$ m to $h = 1$ m?

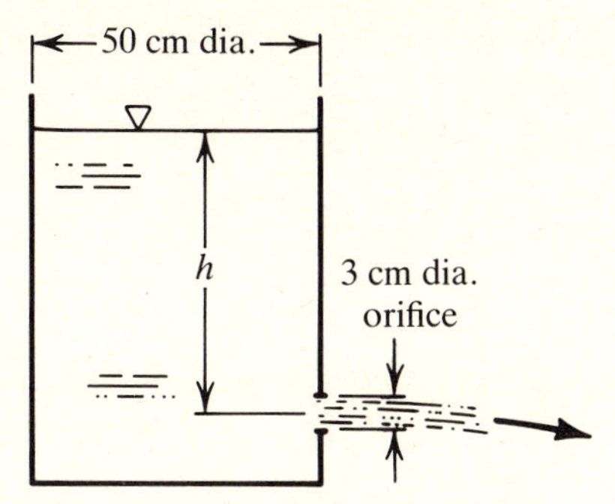

FIGURE 4.19 Problem 4.13.

p.4.14 Water flows through a venturimeter. The diameter of the pipe is 6 in and the diameter of the throat is 3 in. The head loss between the entrance and the throat is 1.5 ft. Determine the discharge if $C_d = 0.95$.

p.4.15 Neglecting all the head losses, determine the discharge through the venturimeter shown in Fig. 4.20. The fluid in the venturimeter is water.

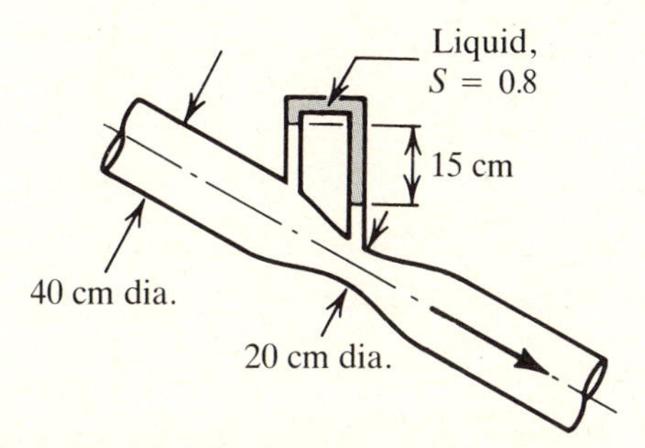

FIGURE 4.20 Problem 4.15.

p.4.16 Water is flowing through the pipeline shown in Fig. 4.21. The rate of flow is 30 l/s. Determine the water level in the tube A. Neglect all head losses.

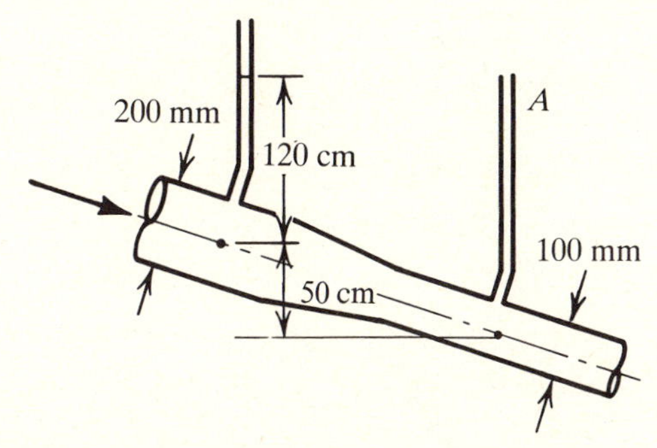

FIGURE 4.21 Problem 4.16.

p.4.17 Water is flowing through the hydraulic system shown in Fig. 4.22. Calculate the pressure at point A and the distance H. Neglect all head losses.

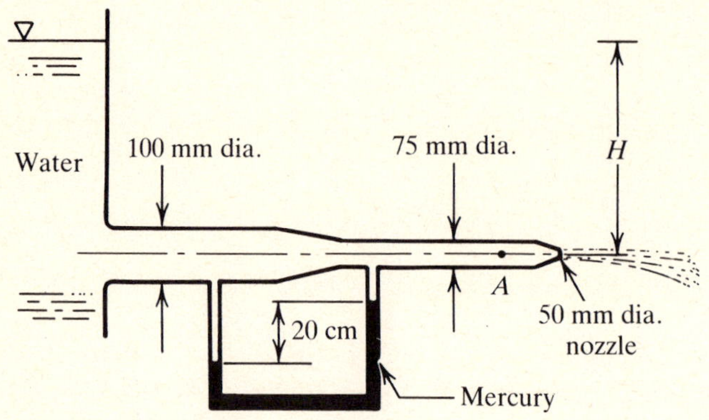

FIGURE 4.22 Problem 4.17.

p.4.18 The diameter of a pipe changes gradually from 6 in at a point A, 20 ft above datum, to 3 in at B, 10 ft above datum. The pressure at A is 15 psi, and the velocity of flow 12 ft/s. Neglecting losses between A and B, determine the pressure at B. The fluid is water.

p.4.19 A pipe 400 m long is laid on a slope of 2% and tapers from 120 cm in diameter at the high end to a 60-cm diameter at the low. The rate of flow is 1.5 m^3/s. If the pressure at the high end is 250 kPa, find the pressure at the low end. Neglect friction. The fluid is water.

p.4.20 Determine the time required to lower the water level from the top to 2 m below the top in Fig. 4.23. Assume $C_d = 0.65$.

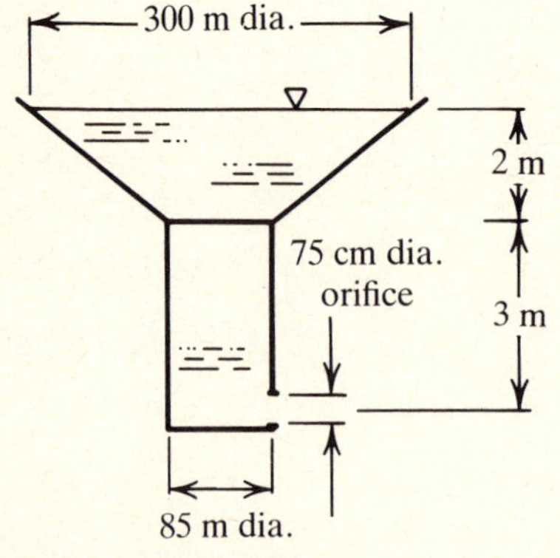

FIGURE 4.23 Problem 4.20.

p.4.21 Determine the discharge through the system shown in Fig. 4.24 when $H = 10$ m and the losses are given by $3V^2/2g$. The fluid is water.

p.4.22 For a flow of 1000 gpm and losses of $10V^2/2g$, determine H for the system shown in Fig. 4.24. The fluid is water.

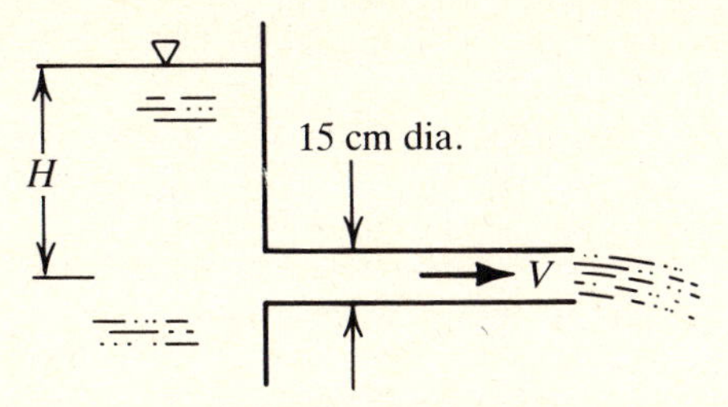

FIGURE 4.24 Problems 4.21, 4.22, 4.23.

p.4.23 For a flow of 100 l/s and H of 10 m, determine the coefficient K in the expression $KV^2/2g$ for the system shown in Fig. 4.24. The fluid is water.

p.4.24 Determine the discharge through the system of Fig. 4.25 and the pressure at A, if the losses up to section A are $6V_1^2/2g$ and the losses beyond section A up to the end are $3V_2^2/2g$. Use $H = 30$ m. The fluid is water.

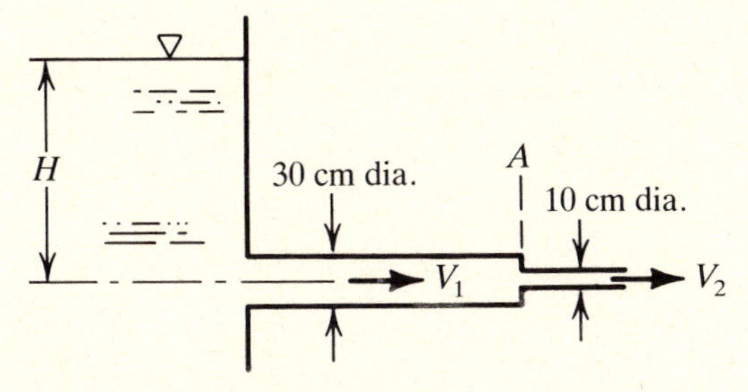

FIGURE 4.25 Problems 4.24, 4.25.

p.4.25 For the system in Fig. 4.25 determine H when the pressure at A is 300 kPa. Use expressions for head losses as given in Problem 4.24.

p.4.26 Calculate the discharge through the system of Fig. 4.26 and pressures at points A, B, C and D. The head losses up to point D are given by $0.1V^2/2g$ per

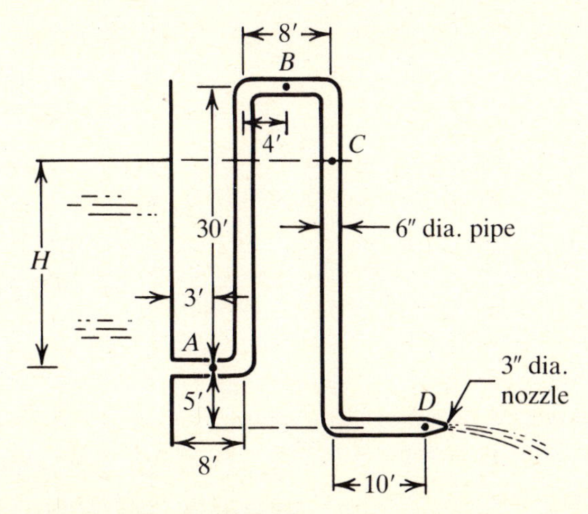

FIGURE 4.26 Problems 4.26, 4.27.

foot length of the pipe and the losses beyond the point D up to the end are $3V^2/2g$. Use $H = 25$ ft. The fluid is water.

p.4.27 Calculate the discharge through the system of Fig. 4.26 and pressures at points A, B, C and D neglecting all the head losses. Use $H = 25$ ft.

p.4.28 Determine the power input to the pump when the discharge through the system of Fig. 4.27 is 0.1 m³/s. The total head loss through the system is $12V^2/2g$ and $H = 20$ m. Pump efficiency is 80%. The fluid is water.

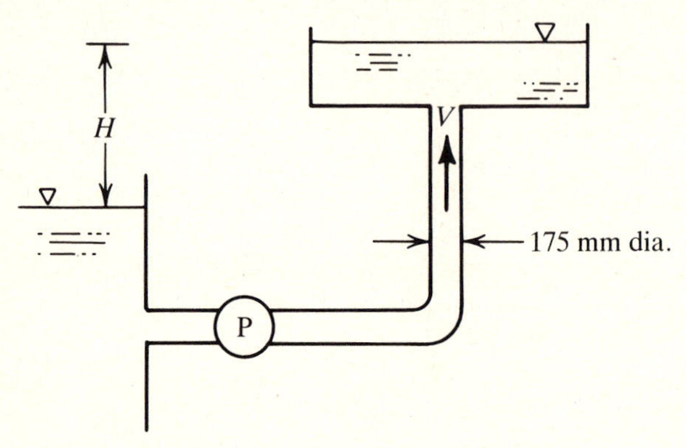

FIGURE 4.27 Problem 4.28.

p.4.29 Draw the hydraulic and energy grade lines for the flow system shown in Fig. 4.28. The head loss through the pipe is given by $0.015(L/D)(V^2/2g)$. Neglect the head loss through the pump. The flow rate is 1.5 ft³/s. The fluid is water.

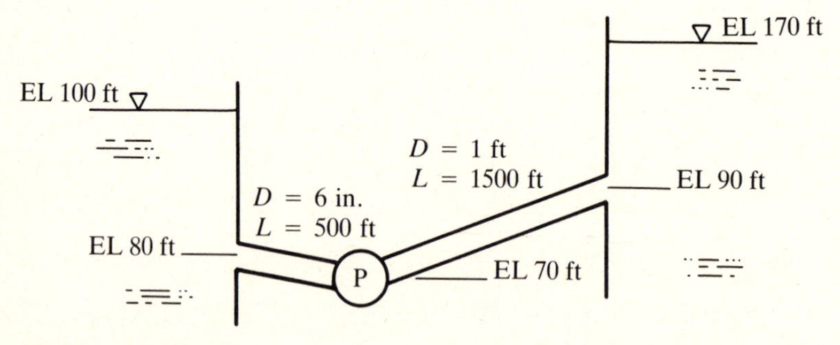

FIGURE 4.28 Problem 4.29.

p.4.30 For the system shown in Fig. 4.29, the head loss in the pipe is given by 0.012 $(L/D)(V^2/2g)$. The pipe diameter is 20 cm. The pump is 80% efficient and the power input is 20 KW. The fluid is water.

a. Determine the rate of flow through the system.
b. Draw the hydraulic and the energy grade lines.
c. Locate and calculate the maximum and the minimum pressures in the system.

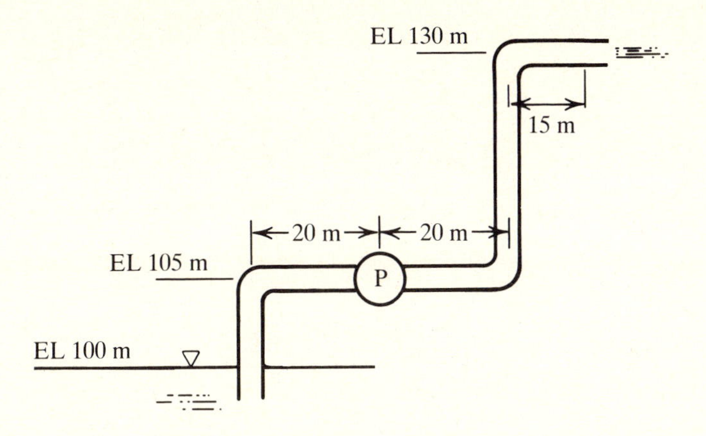

FIGURE 4.29 Problem 4.30.

p.4.31 In the flow system shown in Fig. 4.30, B is a hydraulic machine. Between points C and D, there is a pipe.

a. Is the hydraulic machine at B a pump or a turbine?
b. Is the pipe CD of the same diameter as BC? If not, is pipe CD of smaller or larger diameter?
c. Which is the point of minimum pressure in the system?
d. Which is the point of maximum pressure in the system?
e. Is the pressure negative at any point in the system? If yes, where?

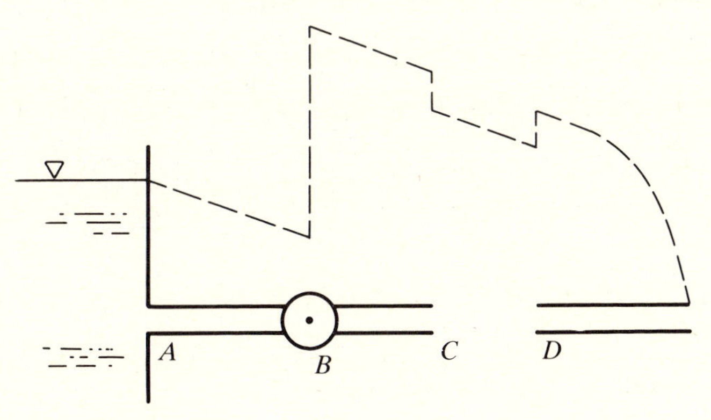

FIGURE 4.30 Problem 4.31.

Chapter Five

FLUID MOMENTUM

In the previous chapters we applied Newton's second law of motion to a particle of fluid moving along a streamline and derived the Euler and Bernoulli equations.

In this chapter, we will apply Newton's second law of motion to a mass of particles of fluid to obtain the momentum equation. The momentum equation has wide application in fluid mechanics. For example, the momentum equation may be used to analyze forces exerted on a solid surface when a jet of fluid impinges on it; aerodynamic forces on the wings of an aeroplane; the thrust on a propeller; forces in pipe-bends, etc.

5.1 THE MOMENTUM EQUATION

Newton's second law of motion, when applied to a mass of particles of a fluid, states that the resultant force, acting on the mass of particles, is equal to the rate of change of momentum. Thus:

$$\Sigma F = \frac{\Delta M}{\Delta t} = \Delta \dot{M} \tag{5.1}$$

where ΣF is the vector sum of all the external forces acting on the mass of particles and ΔM is the change in momentum in time Δt. $\Delta \dot{M}$ is the change in momentum flux.

Now, let us derive the momentum equation by applying it to a control volume shown in Fig. 5.1. Suppose that at time t, the mass of particles of fluid is contained within the boundaries of the control volume $ABCD$, and that after a time Δt, it occupies a new position $A'B'C'D'$. Let us assume that the velocity V and the pressure force F act uniformly over the control surfaces AB and CD. The change in momentum of the system in time Δt is then given by:

$$\begin{aligned}\Delta M &= M(A'B'C'D')_{t+\Delta t} - M(ABCD)_t \\ &= M(A'B'CD)_{t+\Delta t} + M(CDC'D')_{t+\Delta t)} \\ &\quad - M(ABA'B')_t - M(A'B'CD)_t\end{aligned}$$

For steady flow:

$$M(A'B'CD)_{t+\Delta t} = M(A'B'CD)_t$$

Hence:

$$\Delta M = M(CDC'D')_{t+\Delta t} - M(ABA'B')_t$$

$$\Delta M = M_O - M_I \tag{5.2}$$

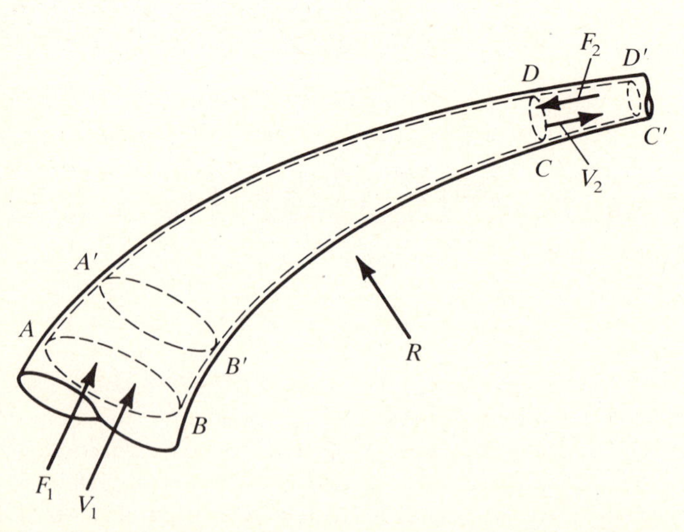

FIGURE 5.1 Control volume ABCD for deriving momentum equation.

where

$$M_O = \text{momentum leaving the control volume}$$

$$M_I = \text{momentum entering the control volume.}$$

The rate of change of momentum is given by

$$\frac{\Delta M}{\Delta t} = \frac{M_O}{\Delta t} - \frac{M_I}{\Delta t}$$

$$\Delta \dot{M} = \dot{M}_O - \dot{M}_I$$

From Eq. (5.1), $\Delta \dot{M} = \Sigma F$, and therefore

$$\Sigma F = \dot{M}_O - \dot{M}_I \tag{5.3}$$

Since use of Eq. (5.3) involves vectors, it is easier to arrange calculations in the direction of rectangular axes. In terms of three-dimensional cartesian axes, Eq. (5.3) may be written as

$$\Sigma F_x = \dot{M}_{Ox} - \dot{M}_{Ix} \tag{5.4a}$$

$$\Sigma F_y = \dot{M}_{Oy} - \dot{M}_{Iy} \tag{5.4b}$$

$$\Sigma F_z = \dot{M}_{Oz} - \dot{M}_{Iz} \tag{5.4c}$$

where

ΣF_x = sum of the x-components of all the external forces

ΣF_y = sum of the y-components of all the external forces

ΣF_z = sum of the z-components of all the external forces

$\dot{M}_{Ox}, \dot{M}_{Oy}, \dot{M}_{Oz}$ = respectively, x-, y-, and z-components of the momentum rate leaving the control volume

$\dot{M}_{Ix}, \dot{M}_{Iy}, \dot{M}_{Iz}$ = respectively, x-, y-, and z-components of the momentum rate entering the control volume.

Also:

$$\dot{M} = \dot{m}V$$

$$\dot{M} = \rho Q V$$

Substituting in Eqs. (5.4), we obtain

$$\Sigma F_x = (\rho Q V)_{Ox} - (\rho Q V)_{Ix} \tag{5.5a}$$

$$\Sigma F_y = (\rho Q V)_{Oy} - (\rho Q V)_{Iy} \tag{5.5b}$$

$$\Sigma F_z = (\rho Q V)_{Oz} - (\rho Q V)_{Iz} \tag{5.5c}$$

For an incompressible fluid, $\rho_O = \rho_I = \rho$.
Hence:

$$\Sigma F_x = \rho(Q_O V_{Ox} - Q_I V_{Ix}) \tag{5.6a}$$

$$\Sigma F_y = \rho(Q_O V_{Oy} - Q_I V_{Iy}) \tag{5.6b}$$

$$\Sigma F_z = \rho(Q_O V_{Oz} - Q_I V_{Iz}) \tag{5.6c}$$

These equations may be used to calculate the force exerted on a moving fluid. According to these equations, in a steady flow, the net force on the fluid contained within the control volume is equal to the net momentum flux leaving the control volume. Of course, the force and the momentum must have the same direction. It may be noticed that only the conditions at the inlet and the outlet to the control volume enter into the calculations. Whatever occurs inside the control volume does not concern us.

It should be kept in mind that the derivation of the momentum equation is based on the assumption that the density, velocity and pressure vary uniformly over the cross-section.

Alternatively, the momentum equation may be derived using the control volume equation developed in Chapter Three and given by Eq. (3.23). Then N is the linear momentum, or $N = mV$, and η is the linear momentum per unit mass, or $\eta = mV/m = V$. Substituting in Eq. (3.23), we obtain

$$\frac{d(mV)}{dt} = \frac{\partial}{\partial t}\int_{cv} \rho V \cdot d\forall + \int_{cs} \rho V \mathbf{V} \cdot \mathbf{dA}$$

According to Newton's second law of motion, the rate of change of momentum is equal to the net external force, or $d(mV/dt) = \Sigma F$. Also, for a steady flow:

$$\frac{\partial}{\partial t}\int_{cv} \rho V d\forall = 0.$$

Hence:

$$\Sigma F = \int_{cs} \rho V \mathbf{V} \cdot \mathbf{dA} \tag{5.7}$$

Consider the control volume shown in Fig. 5.2, where the area and the velocity vectors on the end sections have been drawn.

At section 1:

$$\int \rho V \mathbf{V} \cdot \mathbf{dA} = -\rho_1 V_1 V_1 A_1 = -\rho_1 Q_1 V_1 = -\dot{M}_1$$

At section 2:

$$\int \rho V \mathbf{V} \cdot \mathbf{dA} = \rho_2 V_2 V_2 A_2 = \rho_2 Q_2 V_2 = \dot{M}_2$$

Substituting in Eq. (5.7), we obtain

$$\Sigma F = \dot{M}_2 - \dot{M}_1$$

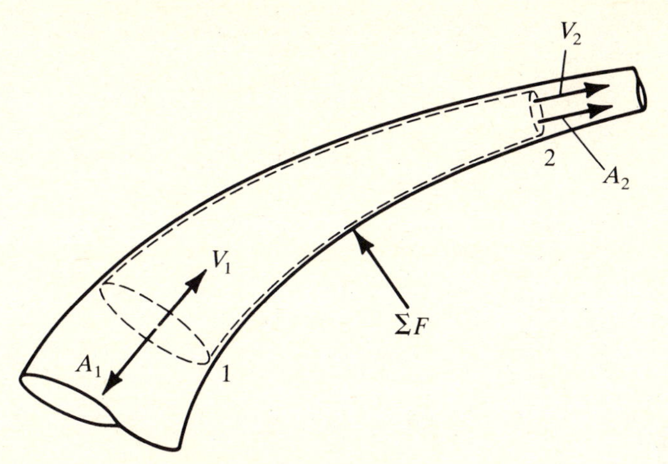

FIGURE 5.2 Control volume within a pipe.

Or, in general:

$$\Sigma F = \dot{M}_O - \dot{M}_I$$

This equation is identical to Eq. (5.3) derived earlier.

The Momentum Correction Factor

While deriving the momentum equation, it was assumed that the velocity and the momentum did not vary over the cross-section. In the case of a real fluid, the velocity does vary over the cross-section due to viscous forces. A factor, called the *momentum correction factor* (β), is used to correct this deficiency so that $\beta\dot{M}$ is the average momentum passing through the section.

A cross-section BB with actual velocity variation and average velocity V is shown in Fig. 5.3. Let u be the velocity of the flow through the elemental area dA. Then, the momentum rate of fluid passing through dA is given by

$$d\dot{M} = \rho u^2 dA$$

Total momentum rate passing through the entire section is given by:

$$\dot{M} = \int_A \rho u^2 dA$$

Using the average velocity V and momentum correction factor β, the momentum passing through the entire section will be given by:

$$\dot{M} = \beta\rho A V^2$$

Therefore:

$$\beta\rho A V^2 = \int_A \rho u^2 dA$$

$$\beta = \frac{1}{A}\int_A \left(\frac{u}{V}\right)^2 dA \tag{5.8}$$

For turbulent flow given by Prandtl's one-seventh law, β is calculated to be equal to 4/3.

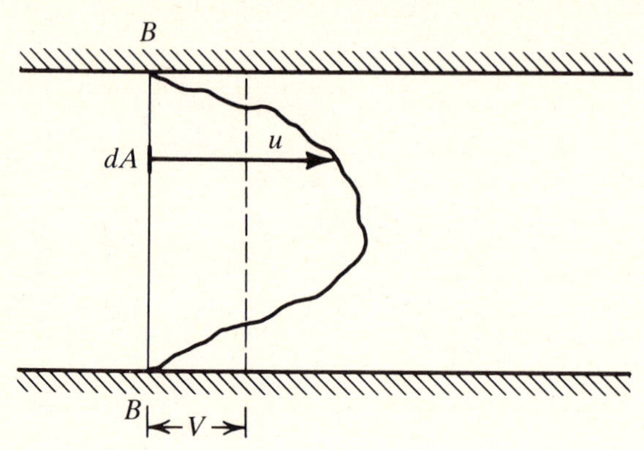

FIGURE 5.3 An arbitrary velocity variation over a cross section.

Comments

1. When solving momentum problems in SI units, care should be exercised so that the forces are expressed in Newtons (N) and density in kg/m^3. A common error is made by students in using forces in kilonewtons (kN) and density in kg/m^3.
2. When using FPS units, the forces should be in lbs and the density in slugs/ft^3.
3. Since forces and velocities are vectors, use of proper signs is important.
4. Remember that the forces at the end sections of the control volume act *towards* the sections.

EXAMPLE 5.1

Water flows through a 45° reducing elbow at a rate of 300 l/sec. The diameter at inlet and outlet are 30 cm and 15 cm, respectively. The pressure is 175 kPa at the inlet and 160 kPa at the outlet. Determine the resultant force on the elbow. Assume the bend to be in the horizontal plane.

Solution The elbow with the control volume is shown in Fig. 5.4

$$Q = 300 \text{ l/sec} = 0.3 \text{ m}^3\text{/sec}$$

$$D_1 = 30 \text{ cm}, \; A_1 = \frac{\pi(30)^2}{4} = 706.8 \text{ cm}^2 = 0.0707 \text{ m}^2$$

$$D_2 = 15 \text{ cm}, \; A_2 = \frac{\pi(15)^2}{4} = 176.7 \text{ cm}^2 = 0.01767 \text{ m}^2$$

$$F_1 = p_1 A_1 = 175 \times 0.0707 = 12.372 \text{ kN} = 12372 \text{ N}$$

$$F_2 = p_2 A_2 = 160 \times 0.01767 = 2.827 \text{ kN} = 2827 \text{ N}$$

$$V_1 = 0.3/0.0707 = 4.24 \text{ m/s}$$

$$V_2 = 0.3/0.01767 = 16.98 \text{ m/s}$$

$$\rho = 1000 \text{ kg/m}^3$$

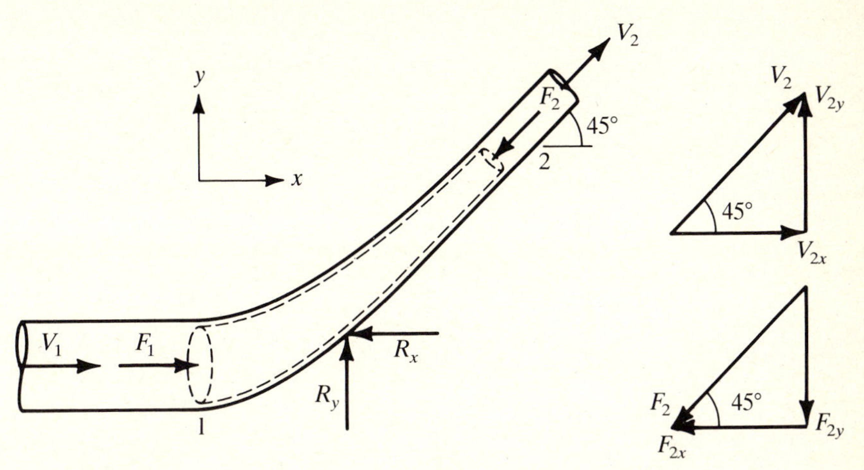

FIGURE 5.4 Example 5.1.

Let R_x and R_y be the x- and y-components of the force experienced by the elbow. Applying the momentum equation in the x-direction:

$$\Sigma F_x = \dot{M}_{Ox} - \dot{M}_{Ix}$$

$$F_1 - F_{2x} - R_x = \rho_2 Q_2 V_{2x} - \rho_1 Q_1 V_{1x}$$

For water:

$$\rho_1 = \rho_2$$

Also

$$Q_1 = Q_2$$

Hence:

$$F_1 - F_{2x} - R_x = \rho Q(V_{2x} - V_{1x})$$

$$12{,}372 - 2{,}827 \cos 45^\circ - R_x = 1{,}000 \times 0.3(16.98 \cos 45^\circ - 4.24)$$

$$R_x = 8043 \text{ N}$$

Applying the momentum equation in the y-direction:

$$\Sigma F_y = \dot{M}_{Oy} - \dot{M}_{Iy}$$

$$R_y - F_{2y} = \rho Q(V_{2y} - V_{1y})$$

$$R_y - 2827 \sin 45^\circ = 1{,}000 \times 0.3(16.98 \sin 45^\circ - 0)$$

$$R_y = 5600 \text{ N}$$

The resultant

$$R = \sqrt{R_x^2 + R_y^2}$$

$$= \sqrt{(8043)^2 + (5600)^2}$$

$$= 9800 \text{ N}$$

$$\theta = \tan^{-1} R_y/R_x$$

$$= \tan^{-1} (5600/8043)$$

$$= 34.8°$$

EXAMPLE 5.2

A liquid jet strikes a curved vertical blade (Fig. 5.5) and is deflected through an angle of 60°. The jet velocity at entrance is 80 ft/s, the jet area is 0.1 ft^2, and $\rho = 1.94$ slugs/ft^3. What force is exerted by the jet on the vane?

Solution The blade with the control volume is shown in Fig. 5.5. Since the blade is in the vertical plane, the velocity at section 2 would be less than the velocity at section 1. The velocity at section 2 may be found by the application of the Bernoulli equation.

$$V_1 = 80 \text{ ft/s}$$

$$A_1 = 0.1 \text{ ft}^2$$

$$Q = A_1 V_1 = 0.1 \times 80 = 8 \text{ ft}^3\text{/s}$$

Applying the Bernoulli equation to points 1 and 2, taking datum at point 1 and ignoring head losses, we get:

$$\frac{p_1}{\gamma} + z_1 + \frac{V_1^2}{2g} = \frac{p_2}{\gamma} + z_2 + \frac{V_2^2}{2g}$$

$$0 + 0 + \frac{(80)^2}{2 \times 32.2} = 0 + 3 + \frac{V_2}{2g}$$

$$V_2 = 78.8 \text{ ft/s}$$

Notice that the pressures at the end sections of the control volume are atmospheric and equal to zero.

Hence:

$$F_1 = F_2 = 0$$

Applying the momentum equation in the x-direction

$$-R_x = \rho Q(V_{2x} - V_{1x})$$

$$= (1.94)(8)(78.8 \cos 60° - 80)$$

$$R_x = 630 \text{ lbs}$$

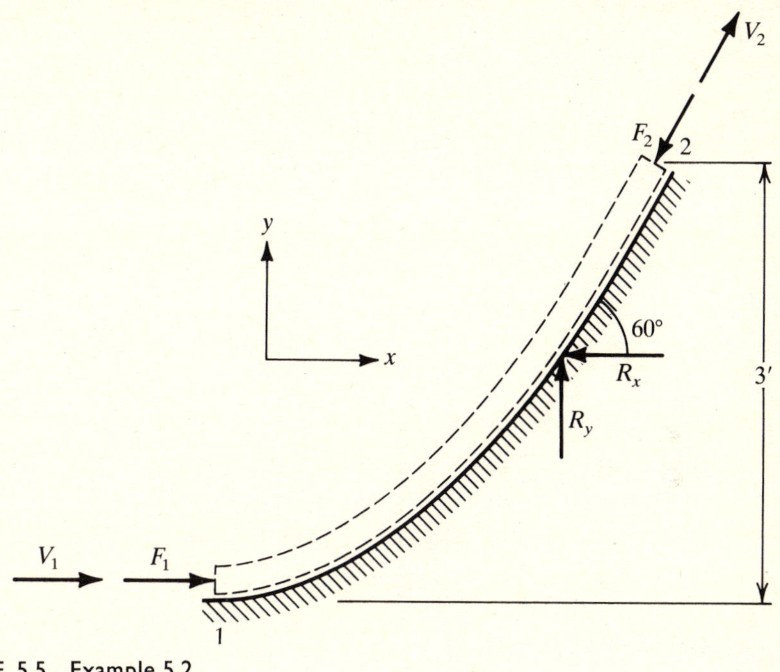

FIGURE 5.5 Example 5.2.

Applying the momentum equation in the y-direction:

$$R_y = \rho Q(V_{2y} - V_{1y})$$

$$= (1.94)(8)(78.8 \sin 60° - 0)$$

$$R_y = 1059 \text{ lbs}$$

$$\text{Resultant force} = R = \sqrt{(630)^2 + (1059)^2} = 1232 \text{ lbs}$$

$$\theta = \tan^{-1}\left(\frac{1059}{630}\right) = 59.3°$$

EXAMPLE 5.3

Neglecting losses, determine the force needed to hold the Y shown in Fig. 5.6 in place. Assume the Y to be in the horizontal plane.

Solution Consider the control volume shown. The water enters the control volume through section 1 and leaves through sections 2 and 3. The following

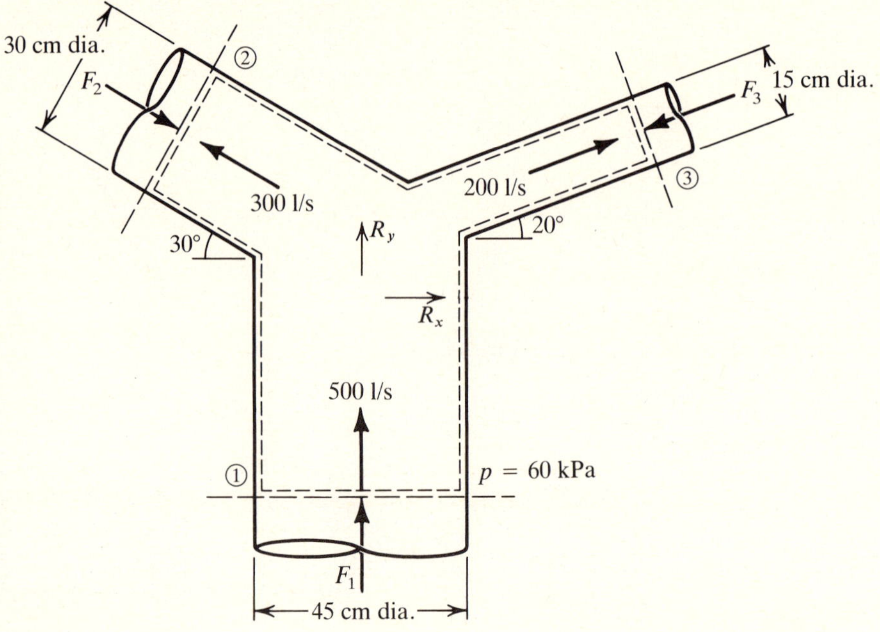

FIGURE 5.6 Example 5.3.

information is given (subscripts 1, 2 and 3 refer to sections 1, 2, and 3 respectively):

$$Q_1 = 500\ \text{l/s} = 0.5\ \text{m}^3/\text{s}$$

$$Q_2 = 300\ \text{l/s} = 0.3\ \text{m}^3/\text{s}$$

$$Q_3 = 200\ \text{l/s} = 0.2\ \text{m}^3/\text{s}$$

$$D_1 = 45\ \text{cm},\ A_1 = \frac{\pi(45)^2}{4} = 1590\ \text{cm}^2 = 0.159\ \text{m}^2$$

$$D_2 = 30\ \text{cm},\ A_2 = \frac{\pi(30)^2}{4} = 706.8\ \text{cm}^2 = 0.07068\ \text{m}^2$$

$$D_3 = 15\ \text{cm},\ A_3 = \frac{\pi(15)^2}{4} = 176.7\ \text{cm}^2 = 0.01767\ \text{m}^2$$

$$p_1 = 60\ \text{kPa} = 60{,}000\ \text{N/m}^2$$

Using the given information

$$V_1 = Q_1/A_1 = 0.5/0.159 = 3.14\ \text{m/s}$$

$$V_2 = Q_2/A_2 = 0.3/0.07068 = 4.24\ \text{m/s}$$

$$V_3 = Q_3/A_3 = 0.2/0.01767 = 11.32\ \text{m/s}$$

the pressures at sections 2 and 3 may be found using the Bernoulli equation.

To find p_2, apply the Bernoulli equation to points 1 and 2.

$$\frac{p_1}{\gamma} + z_1 + \frac{V_1^2}{2g} = \frac{p_2}{\gamma} + z_2 + \frac{V_2^2}{2g}$$

$$\frac{60{,}000}{9810} + 0 + \frac{(3.14)^2}{2 \times 9.81} = \frac{p_2}{\gamma} + 0 + \frac{(4.24)^2}{2 \times 9.81}$$

$$\frac{p_2}{\gamma} = 5.7$$

$$p_2 = 9810 \times 5.7 = 55{,}941 \text{ N/m}^2$$

To find p_3, apply the Bernoulli equation to points 1 and 3.

$$\frac{p_1}{\gamma} + z_1 + \frac{V_2^2}{2g} = \frac{p_3}{\gamma} + z_3 + \frac{V_3^2}{2g}$$

$$\frac{60{,}000}{9810} + 0 + \frac{(3.14)^2}{2 \times 9.81} = \frac{p_3}{\gamma} + 0 + \frac{(11.32)^2}{2 \times 9.81}$$

$$\frac{p_3}{\gamma} = 0.0875$$

$$p_3 = 9810 \times 0.0875 = 858 \text{ N/m}^2$$

Also:

$$F_1 = p_1 A_1 = 60{,}000 \times 0.159 = 9540 \text{ N}$$

$$F_2 = p_2 A_2 = 55{,}941 \times 0.07068 = 3954 \text{ N}$$

$$F_3 = p_3 A_3 = 858 \times 0.01767 = 15 \text{ N}$$

To find R_x, apply the momentum equation in the x-direction, assuming $\rho_1 = \rho_2 = \rho_3 = \rho$

$$\Sigma F_x = \dot{M}_{Ox} - \dot{M}_{Ix}$$

$$F_{2x} - F_{3x} + R_x = \rho[(Q_2 V_{2x} + Q_3 V_{3x}) - Q_1 V_{1x}]$$

$$(3954 \cos 30°) - (15 \cos 20°) + R_x$$
$$= 1000[(0.3 \times 4.24 \cos 30° - 0.2 \times 11.32 \cos 20°) - 0]$$

or

$$R_x = -4436 \text{ N}$$

Hence, R_x should be directed toward the left. Or:

$$R_x = -4436 \text{ N} \leftarrow$$

To find R_y, apply the momentum equation in the y-direction.

$$\Sigma F_y = \dot{M}_{Oy} - \dot{M}_{Iy}$$

$$-F_{2y} - F_{3y} + F_1 + R_y = \rho[(Q_2 V_{2y} + Q_3 V_{3y}) - Q_1 V_{1y}]$$

$$(3954 \sin 30°) - (15 \sin 20°) + 9540 + R_y$$
$$= 1000[(0.3 \times 4.24 \sin 30° + 0.2 \times 11.32 \sin 20°) - (0.5 \times 3.14)]$$

or

$$R_y = 11{,}671 \text{ N}$$

Hence, R_y should be directed downwards. Or:

$$R_y = 11{,}671 \text{ N}\downarrow$$

$$\text{Resultant force} = R = \sqrt{R_x^2 + R_y^2}$$

$$= \sqrt{(4436)^2 + (11671)^2} = 12{,}485 \text{ N}$$

$\theta = \tan^{-1}(11671/4436) = 69.2°$ with the horizontal.

Force on a Stationary Vertical Flat Plate

If a jet of fluid strikes normally at a flat plate, the plate would experience a force which can be calculated using the momentum equation.

Consider a jet of fluid striking a plate as shown in Fig. 5.7. Establish a control volume as shown and let

V_1 = velocity of the jet at section 1

Q_1 = discharge at section 1

V_2 = velocity of the jet after deflection at section 2

Q_2 = discharge at section 2

V_3 = velocity of the jet after deflection at section 3

Q_3 = discharge at section 3

R_x = horizontal force required to keep the plate in equilibrium

R_y = vertical force required to keep the plate in equilibrium

Applying the momentum equation in the x-direction:

$$\Sigma F_x = \dot{M}_{Ox} - \dot{M}_{Ix}$$

$$-R_x = (\rho_2 Q_2 V_{2x} + \rho_3 Q_3 V_{3x}) - (\rho_1 Q_1 V_{1x})$$

Since the jet is horizontal and the plate is vertical

$$V_{2x} = V_{3x} = 0$$

Hence:

$$-R_x = -\rho_1 Q_1 V_{1x}$$

$$R_x = \rho Q V = \frac{\gamma}{g} Q V \tag{5.9}$$

where

Q = rate of flow of the jet

V = velocity of the jet.

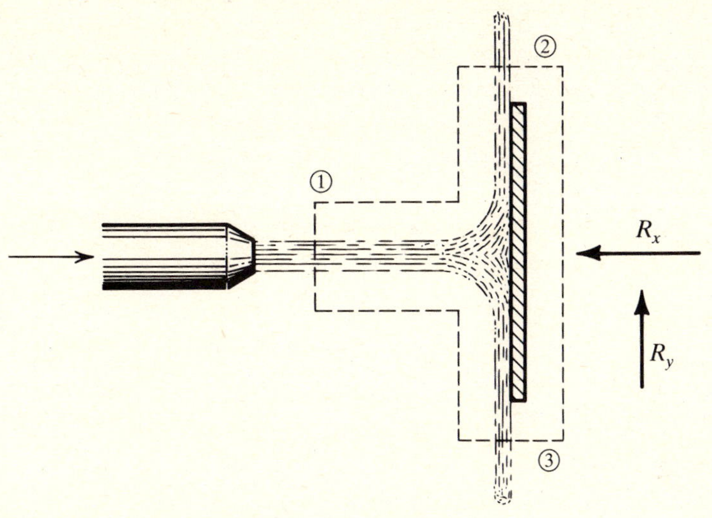

FIGURE 5.7 A jet striking a stationary vertical flat plate.

Since a force equal to R_x must be applied to keep the plate in equilibrium, it follows that a force equal to R_x must be experienced by the plate.

Hence:

$$\text{Force on plate} = \rho QV$$

There will be no force in the y-direction, since $V_{2y} = -V_{3y}$ and $V_{1y} = 0$.

EXAMPLE 5.4

A jet of water 2 in in diameter impinges normally on a fixed plate and has a velocity of 100 ft/s. Find the normal force on the plate.

Solution

$$\text{Diameter of jet} = 2 \text{ in}$$

$$\text{Area of jet} = A = \frac{\pi(2)^2}{4 \times 144} = 0.0218 \text{ ft}^2$$

$$\text{Velocity of jet} = V = 100 \text{ ft/s}$$

$$\text{Discharge} = AV$$

$$= 0.0218 \times 100 = 2.18 \text{ ft}^3/\text{s}$$

Using Eq. (5.9)

$$R_x = \frac{\gamma}{g} QV$$

$$= \frac{62.4}{32.2} \times 2.18 \times 100 = 422 \text{ lbs}$$

Force on a Stationary Inclined Flat Plate

If the plate is inclined to the jet, the plate will experience a force in the horizontal, as well as in the vertical direction. Let a jet of fluid strike a plate as shown in Fig. 5.8. The problem can be easily solved by orienting the coordinate axes, and establishing a control volume as shown. Now applying the momentum equation in the x-direction:

$$\Sigma F_x = \dot{M}_{Ox} - \dot{M}_{Ix}$$

$$-R_x = (\rho_2 Q_2 V_{2x} + \rho_3 Q_3 V_{3x}) - (\rho_1 Q_1 V_{1x})$$

But:

$$V_{2x} = V_{3x} = 0$$

Hence:

$$R_x = \rho Q V_{1x}$$

$$R_x = \rho Q V \sin\theta \tag{5.10}$$

Similarly, it can be shown that

$$R_y = \rho Q V \cos\theta \tag{5.11}$$

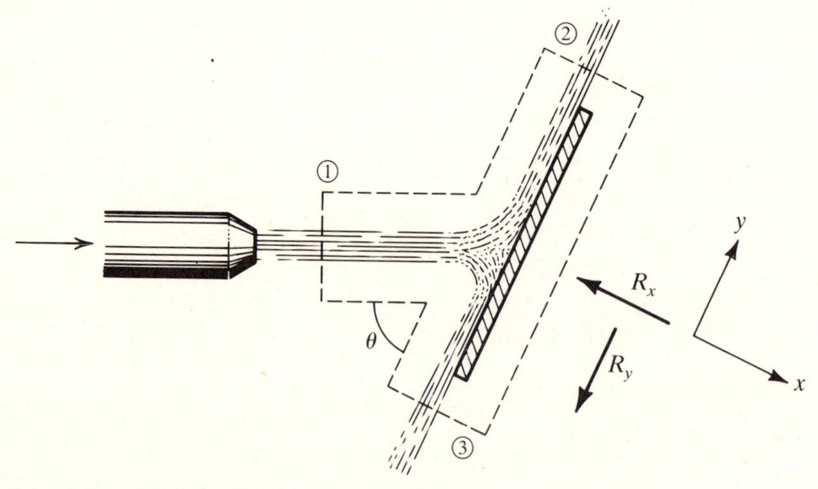

FIGURE 5.8 A jet striking a stationary inclined flat plate.

Force on a Moving Flat Plate

If a jet of fluid strikes a plate which is moving, the strike velocity will be the relative velocity between the jet and the plate. Hence, if the plate is moving in the same direction as the jet, the strike velocity will be the difference between the jet and the plate velocities, and if the plate is moving in the opposite direction of the jet, the strike velocity will be the sum of the jet and the plate velocities. A practical example may be a wheel, with flat plates fixed radially

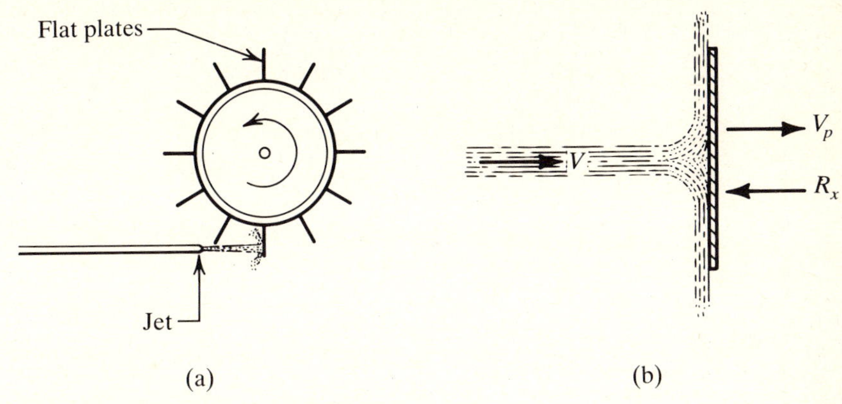

FIGURE 5.9 A jet striking a moving flat plate.

around the circumference at regular intervals, moving under the force of a jet (Fig. 5.9a). Then, for a single flat plate, for the case when the plate is moving in the same direction as the jet (Fig. 5.9b):

$$R_x = \frac{\gamma}{g} Q(V - V_p) \tag{5.12}$$

where V = velocity of the jet, and V_p = tangential velocity of the plate.

$$\text{Work done per second on the plate} = R_x V_p = \frac{\gamma}{g} Q(V - V_p)V_p$$

Also:

$$\text{Energy supplied by the jet} = \text{Kinetic Energy (KE) of the jet}$$

$$= \frac{\gamma Q V^2}{2g}$$

$$\text{Efficiency of the plate} = \eta = \frac{\text{work done}}{\text{KE of jet}}$$

$$= \frac{\frac{\gamma}{g} Q(V - V_p)V_p}{\frac{\gamma}{2g} QV^2}$$

$$\eta = \frac{2(V - V_p)V_p}{V^2} \tag{5.13}$$

The maximum efficiency can be found by differentiating η with respect to V_p and equating to zero, or

$$\frac{d\eta}{dV_p} = V - 2V_p = 0$$

or

$$V_p = \frac{V}{2} \tag{5.14}$$

Then

$$\eta_{max} = \frac{2\left(V - \frac{V}{2}\right)\frac{V}{2}}{V^2} = \frac{1}{2}$$

Hence, maximum possible efficiency of such a device will be only 50%.

EXAMPLE 5.5

A jet of water 7.5 cm in diameter and having velocity of 15 m/s strikes a series of flat plates normally. If the plates are moving in the same direction as the jet with a velocity of 10 ft/s, find the force on the plates, the work done per second, and the efficiency.

Solution

$$D = 7.5 \text{ cm}$$

$$A = \frac{\pi(7.5)^2}{4} = 44.18 \text{ cm}^2 = 44.18 \times 10^{-4} \text{ m}^2/\text{s}$$

$$V = 15 \text{ m/s}$$

Hence:

$$Q = AV$$

$$= 44.18 \times 10^{-4} \times 15 = 66.27 \times 10^{-3} \text{ m}^3/\text{s}$$

$$V_p = 10 \text{ m/s}$$

Using Eq. (5.12)

$$\text{Force on the plate} = R_x = \frac{\gamma}{g} Q(V - V_p)$$

$$= \frac{9810}{9.81} \times 66.27 \times 10^{-3}(15 - 10)$$

$$= 331.3 \text{ N}$$

$$\text{Work done/sec} = \text{force} \times \text{velocity of plates} = 331.3 \times 10 = 3313 \text{ m} \cdot \text{N/s}$$

Using Eq. (5.13)

$$\text{Efficiency} = \frac{2(V - V_p)V_p}{V^2}$$

$$= \frac{2(15 - 10)10}{15^2}$$

$$= 0.444 = 44.4\%$$

Force on a Fixed Curved Vane

Consider a curved vane as shown in Fig. 5.10.

Let α = an angle with the x-axis, at which the jet strikes the vane
β = an angle with the x-axis, at which the jet leaves the vane

Draw a control volume and identify control surfaces 1 and 2 as shown. Applying the momentum equation in the x-direction, we have

$$\Sigma F_x = \dot{M}_{Ox} - \dot{M}_{Ix}$$
$$= \rho_2 Q_2 V_{2x} - \rho_1 Q_1 V_{1x}$$

If the fluid is incompressible

$$\rho_1 = \rho_2 = \rho$$

Also, for this particular case:

$$Q_1 = Q_2 = Q$$

Hence:

$$\Sigma F_x = \rho Q (V_{2x} - V_{1x})$$
$$-R_x = \rho Q (-V_2 \cos\beta - V_1 \cos\alpha)$$

Hence:

$$R_x = \rho Q (V_2 \cos\beta + V_1 \cos\alpha) \tag{5.15}$$

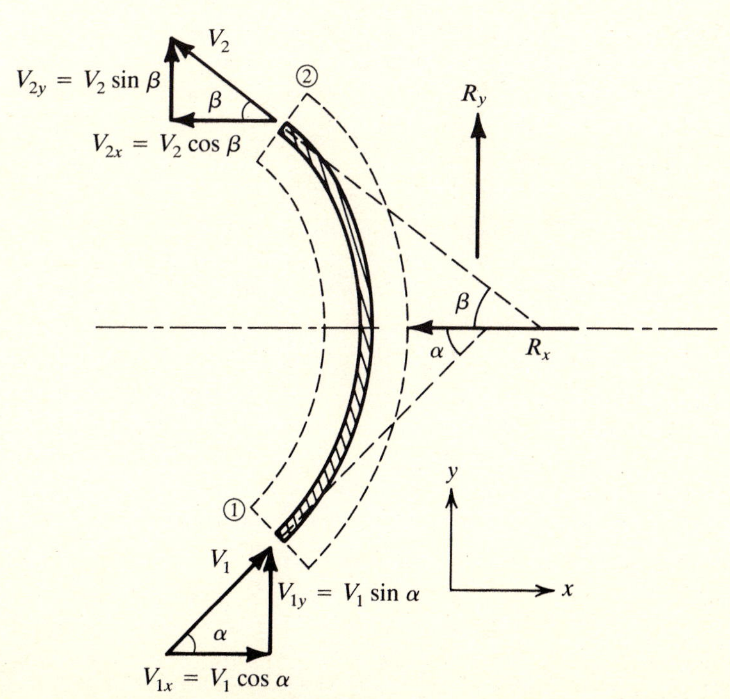

FIGURE 5.10 A jet striking a fixed curved vane.

Applying momentum equation in the y-direction, we have

$$\Sigma F_y = \dot{M}_{Oy} - \dot{M}_{Iy}$$

or, as described above

$$\Sigma F_y = \rho Q(V_{2y} - V_{1y})$$

$$R_y = \rho Q(V_2 \sin \beta - V_1 \sin \alpha) \tag{5.16}$$

EXAMPLE 5.6

A jet of water 3 in in diameter and discharging 2.0 ft^3/s strikes a fixed curved vane. The angles of inlet and outlet of the vane are 30° each. Find the force exerted on the plate. Assume that the vane is in the horizontal plane and that the jet has constant speed.

Solution For the given problem

$$\text{Diameter of jet} = D = 3 \text{ in}$$

$$\text{Area of jet} = A = \frac{\pi(3)^2}{4 \times 144} = 4.91 \times 10^{-2} \text{ ft}^2$$

$$Q = 2.0 \text{ ft}^3/\text{s}$$

Hence:

$$V = \frac{2}{4.91 \times 10^{-2}} = 40.7 \text{ ft/s}$$

$$\alpha = 30^\circ$$

$$\beta = 30^\circ$$

Since jet has constant speed, $V_1 = V_2 = V = 40.7$ ft/s
Using Eq. (5.15)

$$\begin{aligned} R_x &= \rho Q(V_2 \cos \beta + V_1 \cos \alpha) \\ &= 1.94 \times 2(40.7 \cos 30^\circ + 40.7 \cos 30^\circ) \\ &= 273.5 \text{ lbs} \\ R_x &= 273.5 \text{ lbs} \end{aligned}$$

Using Eq. (5.16)

$$\begin{aligned} R_y &= \rho Q(V_2 \sin \beta - V_1 \sin \alpha) \\ &= 1.94 \times 2(40.7 \sin 30^\circ - 40.7 \sin 30^\circ) \\ &= 0 \\ R_y &= 0.0 \text{ lbs} \end{aligned}$$

Force on Moving Curved Vane

If the curved vane of Fig. 5.11 is moving in the x-direction along the jet with a velocity V_p, the forces on the vane can be calculated as described below.

Let

V_1 = velocity of jet at section 1

V_{1x} = component of V_1 in the x-direction

V_{1y} = component of V_1 in the y-direction

V_{1r} = relative velocity of fluid striking the vane. This is obtained by subtracting vector V_p from vector V_1.

V_{2r} = relative velocity of fluid leaving the vane.

V_2 = actual velocity of jet at exit. This is obtained by adding vector V_p to vector V_{2r}.

V_{2x} = x-component of V_2

V_{2y} = y-component of V_2

V_p = velocity of vane

α, β = as defined previously

θ = angle between relative velocity and direction of motion at inlet

ϕ = angle between relative velocity and direction of motion at outlet

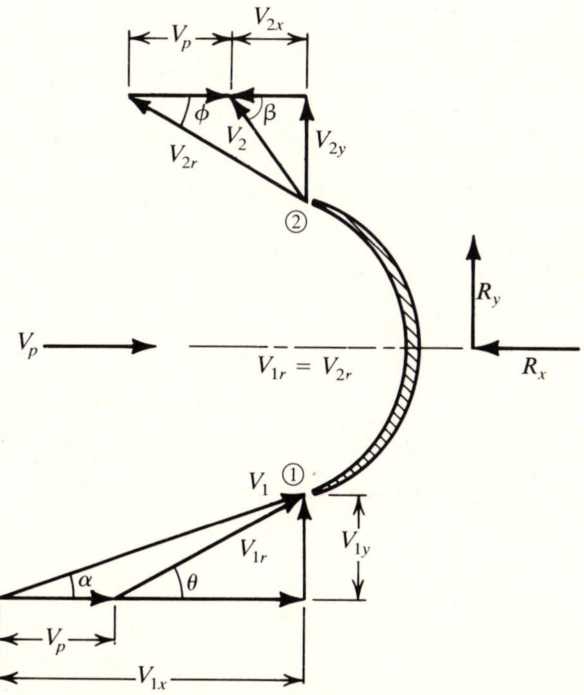

FIGURE 5.11 A jet striking a moving curved vane.

Applying the momentum equation in the x-direction, we have

$$\Sigma F_x = \dot{M}_{Ox} - \dot{M}_{Ix}$$

$$-R_x = \rho Q[(-V_{2x} + V_p) - (V_{1x} + V_p)]$$

$$= \rho Q(-V_{2x} - V_{1x})$$

$$R_x = \rho Q(V_{2x} + V_{1x})$$

$$R_x = \rho Q(V_2 \cos \beta + V_1 \cos \alpha) \tag{5.17}$$

$V_2 \cos \beta$ is called *the velocity of whirl at the exit* and $V_1 \cos \alpha$ is called *the velocity of whirl at the entrance.* Let

$$V_{1w} = V_1 \cos \alpha$$

$$V_{2w} = V_2 \cos \beta$$

Then

$$R_x = \rho Q(V_{2w} + V_{1w}) \tag{5.18}$$

Applying the momentum equation in the y-direction it is found that

$$R_y = \rho Q(V_2 \sin \beta - V_1 \sin \alpha) \tag{5.19}$$

Since the fluid enters and leaves the vane at angles of θ and ϕ respectively, it follows that if the fluid is to enter and leave the vane without shock, the angles of the blade at inlet and outlet must be made equal to θ and ϕ respectively.

Also:

$$\text{Work done on vane/unit time} = F_x V_p$$

$$= \rho Q(V_{2w} + V_{1w})V_p$$

$$\text{Work done/unit weight of fluid} = (V_{2w} + V_{1w})\frac{V_p}{g} \tag{5.20}$$

The work done is also equal to the change in kinetic energy of the jet per second, or

$$\text{work done} = \tfrac{1}{2}\rho Q(V_1^2 - V_2^2)$$

$$\text{KE of jet} = \tfrac{1}{2}\rho Q V_1^2$$

$$\text{The efficiency of the vane} = \frac{\text{work done}}{\text{KE of jet}}$$

$$= \frac{\tfrac{1}{2}\rho Q(V_1^2 - V_2^2)}{\tfrac{1}{2}\rho Q V_1^2}$$

$$\eta = 1 - \left(\frac{V_2}{V_1}\right)^2 \tag{5.21}$$

Also, if the friction between the fluid and the vane is neglected, the relative velocity at exit equals the relative velocity at entrance, or

$$V_{2r} = V_{1r} \tag{5.22}$$

EXAMPLE 5.7

A vane has a velocity 20 m/s. Water impinges on the vane at an angle of 30° and leaves at an angle of 160° to the direction of motion. If the entering water has an absolute velocity of 40 m/s, find (a) the angles of the blade tips at inlet and outlet for no shock conditions; (b) the work done on the vane, and (c) the efficiency of the vane.

Solution Referring to Fig. 5.11

$$V_1 = 40 \text{ m/s (given)}$$

$$V_p = 20 \text{ m/s (given)}$$

$$\alpha = 30°$$

$$\beta = 180° - 160° = 20°$$

(a) From the diagram of velocities at inlet

$$V_{1x} = V_{1w} = V_1 \cos \alpha = 40 \cos 30° = 34.64 \text{ m/s}$$

$$V_{1y} = 40 \sin 30° = 20.0 \text{ m/s}$$

$$\tan \theta = \frac{V_{1y}}{V_{1w} - V_p} = \frac{20}{34.64 - 20.0} = 1.366$$

$$\theta = \tan^{-1} 1.366 = 53.8°$$

$$V_{1r} = \frac{V_{1y}}{\sin \theta} = \frac{20}{\sin 53.8°} = 24.78 \text{ m/s}$$

From the diagram of velocities at the outlet, using Eq. 5.22

$$V_{2r} = V_{1r} = 24.78 \text{ m/s}$$

$$\tan \beta = \frac{V_{2y}}{V_{2x}}$$

$$= \frac{V_{2r} \sin \phi}{V_{2r} \cos \phi - V_p}$$

$$\tan 20° = \frac{24.78 \sin \phi}{24.78 \cos \phi - 20}$$

This equation can be solved by trial and error using various trial values for ϕ. Then,

$$\phi = 4°$$

Also

$$V_2 = \frac{V_{2y}}{\sin \beta} = \frac{V_{2r} \sin \phi}{\sin \beta}$$

$$V_2 = \frac{24.78 \sin 4°}{\sin 20°}$$

$$= 5.05 \text{ m/s}$$

$$V_{2w} = V_2 \cos \beta = 5.05 \cos 20° = 4.74 \text{ m/s}$$

(b) Using Eq. (5.20)

$$\text{Work done per N of fluid per second} = (V_{2w} + V_{1w})\frac{V_p}{g}$$

$$= (4.74 + 34.64)\frac{20}{9.81} = 80.3\ \text{N}\cdot\text{m}$$

(c) Using Eq. (5.21)

$$\text{Efficiency} = 1 - (V_2/V_1)^2 = 1 - \left(\frac{5.05}{40}\right)^2$$

$$= 0.984 = 98.4\%$$

5.2 MOMENTUM THEORY OF PROPELLERS

A propeller uses the torque of a shaft to increase the momentum of the fluid in which it is submerged to create a force which is then used for propulsion.

A propeller in a stream of fluid is shown in Fig. 5.12. Consider the control volume bounded by the slipstream boundaries and the end sections. The

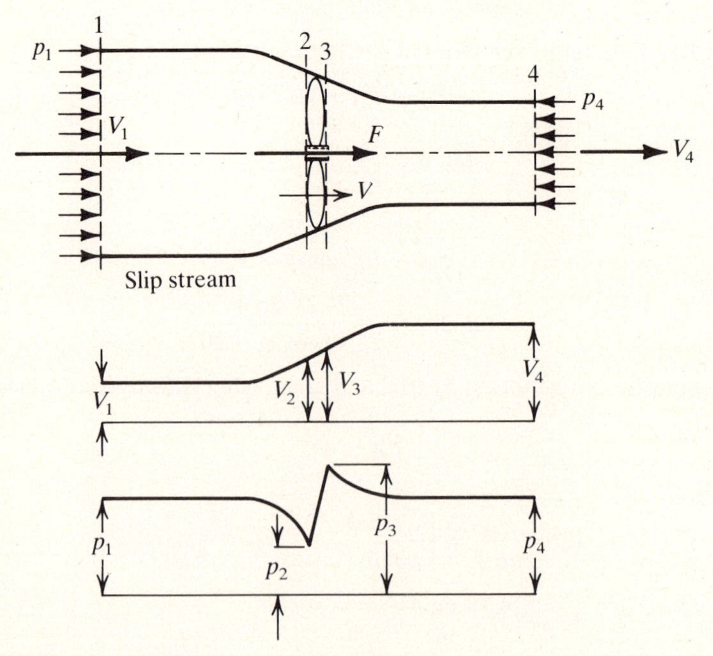

FIGURE 5.12 A propeller in a stream of fluid with pressure and velocity variation diagrams.

slipstream is the fluid which is directly affected by the action of the propeller. At entrance to the control volume at section 1, the flow is undisturbed with velocity V_1 and pressure p_1. At exit to the control volume at section 4, the pressure p_4 is equal to pressure p_1 but the velocity V_4 is larger than V_1 due to the reduced section. At section 2, the pressure is p_2 and the velocity V_2, while at section 3, the pressure increases to p_3 and the velocity is V_3.

It is immaterial whether the propeller is stationary and the fluid is moving through it (for example, an electric fan) or the propeller is moving and the fluid is stationary (for example, an airplane). In both cases, the pattern of flow relative to the propeller is the same.

For simplicity of mathematical analysis, the following assumptions are made:

1. The propeller does not impart any rotary motion to the fluid. In actuality, the fluid leaving the propeller has some rotary motion in addition to the axial motion. The rotary motion, however, does not contribute to the process of propulsion.
2. Conditions (i.e., velocity, pressure) on each side of the propeller are uniform. In actuality, the conditions may not be uniform unless the propeller consists of an infinite number of blades.
3. The density of the fluid is constant.
4. The velocity of the fluid across the propeller remains constant (i.e., $V_2 = V_3 = V$).
5. The fluid is frictionless.

The pressure all around the control volume is the same and there are no shear forces according to assumption 5; therefore, the only force in the axial direction is the thrust F produced by the propeller. Applying the momentum equation to the control volume in the axial direction, we obtain

$$F = \rho Q(V_4 - V_1)$$

$$F = \dot{m}(V_4 - V_1) = \frac{\gamma}{g} AV(V_4 - V_1) \tag{5.23}$$

where A is the area of the propeller and V is the fluid velocity through the propeller.

Applying the Bernoulli equation to sections 1 and 4 and ignoring the head losses through the propeller, we obtain

$$\frac{p_1}{\gamma} + z_1 + \frac{V_1^2}{2g} + h_p = \frac{p_4}{\gamma} + z_4 + \frac{V_4^2}{2g}$$

where h_p is the head generated by the propeller. Notice that $p_1 = p_4$, $z_1 = z_4$ and $h_p = F/\gamma A$.

Hence

$$\frac{V_1^2}{2g} + \frac{F}{\gamma A} = \frac{V_4^2}{2g}$$

$$F = \frac{\gamma A}{2g}(V_4^2 - V_1^2)$$

$$F = \frac{\gamma A}{2g}(V_4 + V_1)(V_4 - V_1) \tag{5.24}$$

Comparing Eqs. (5.23) and (5.24),

$$\frac{\gamma}{g} AV(V_4 - V_1) = \frac{\gamma A}{2g}(V_4 + V_1)(V_4 - V_1)$$

$$V = \left(\frac{V_1 + V_4}{2}\right) \tag{5.25}$$

Thus, the velocity through the propeller is the average of the velocities upstream and downstream from it.

If the fluid is stationary and the propeller moves through it with a velocity V_1, then, the rate of work done by the propeller is given by FV_1. From Eq. (5.23), $F = \dot{m}(V_4 - V_1)$. Hence:

$$\text{work done} = \dot{m}(V_4 - V_1)V_1$$

The rate of power input to the propeller is equal to the rate of change in kinetic energy of the fluid passing through the propeller.
Hence:

$$\text{power input} = \frac{\dot{m}(V_4^2 - V_1^2)}{2}$$

The theoretical efficiency of the propeller is the ratio of the work done to the power input.
Hence:

$$\eta_t = \frac{\text{work done}}{\text{power input}}$$

$$= \frac{\dot{m}(V_4 - V_1)V_1}{\dot{m}(V_4^2 - V_1^2)/2} = \frac{V_1}{(V_4 + V_1)/2} = \frac{V_1}{V}$$

$$\eta_t = \frac{V_1}{V} \tag{5.26}$$

The pressure difference across the propeller, $p_3 - p_2$, may be found by applying the Bernoulli equation between sections 1 and 2 and between sections 3 and 4 and then solving for $p_3 - p_2$. This yields

$$p_3 - p_2 = \tfrac{1}{2}\rho(V_4^2 - V_1^2) \tag{5.27}$$

From Eq. (5.23) it follows that for a propulsive force to exist, $(V_4 - V_1)$ must be nonzero, or, in other words, V_4 must be greater than V_1. Then, from Eq. (5.25), V will be larger than V_1. If V is larger than V_1, the theoretical efficiency, according to Eq. (5.26), cannot approach 100% even for an ideal fluid. In general, best efficiency is obtained when the quantity of air involved is large and the increase in velocity, $(V_4 - V_1)$, is small. Under optimum operating conditions, the actual efficiency of an aircraft propeller is about 0.85 to 0.9 times the theoretical efficiency given by Eq. (5.26).

The foregoing analysis is based on, among others, two important assumptions: (1) there is no rotary motion of the fluid, and (2) conditions are uniform over the cross-section of the propeller. These assumptions lead to inaccuracies when applied to actual propellers. A more detailed analysis involving aerofoil theory is needed to evaluate the performance of an actual propeller.

Comment

For a still fluid, V_1 is the velocity of the propelled object. For example, if an airplane is moving through still air, V_1 is the velocity of the airplane.

EXAMPLE 5.8

An airplane is traveling through still air at 220 mph. The density of air is 0.0022 slugs/ft^3. The airplane has a 6 ft in diameter propeller and discharges 12000 ft^3/s of air. Calculate (a) the thrust on the plane, (b) the propeller efficiency, (c) the theoretical horsepower required to drive the propeller, and (d) the pressure difference across the blades.

Solution The air is still, hence V_1 is equal to the velocity of the plane

$$V_1 = \frac{220 \times 5280}{3600} = 322.6 \text{ ft/s}$$

$$V = \frac{12{,}000}{\pi 6^2/4} = 424.4 \text{ ft/s}$$

(a) From Eq. (5.25)

$$V_4 = 2V - V_1$$
$$= 2 \times 424.4 - 322.6$$
$$V_4 = 526.2 \text{ ft/s}$$

From Eq. (5.23), the thrust on the plane is

$$F = \dot{m}(V_4 - V_1)$$
$$= 0.0022 \times 12{,}000(526.2 - 322.6)$$
$$= 5375 \text{ lbs}$$

(b) Theoretical efficiency

$$= \eta_t = \frac{V_1}{V} = \frac{322.6}{424.4} = 0.76 = 76\%$$

(c) Theoretical horsepower required

$$= \frac{FV_1}{500\eta_t} = \frac{5375 \times 322.6}{500 \times 0.76}$$

$$= 4563 \text{ hp}$$

(d) From Eq. (5.27)

$$p_3 - p_2 = \frac{\rho}{2}(V_4^2 - V_1^2)$$

$$= \frac{0.0022}{2}(526.2^2 - 322.6^2)$$

$$= 190.1 \text{ lbs/ft}^2$$

Jet Propulsion

In jet engines, the thrust is created by taking air into the engine, burning it with a small amount of fuel and then ejecting the gases with a much higher velocity than in a propeller slipstream. If the mass of the fuel is neglected, Eqs. (5.23) to (5.26) which were developed for the propeller system also apply to the jet propulsion. In a jet stream there is no propeller, hence, sections 2 and 3 in Fig. 5.12 are not present. Also, section 4 may now be called section 2. Then, Eqs. (5.23) to (5.26) may be used for jet propulsion problems by changing the subscript from "4" to "2".

The equations may also be easily modified to account for the fuel consumption. Let r be the ratio of the mass of fuel burnt to the mass of air. Then:

$$F = \dot{m}(V_2 - V_1) + r\dot{m}V_2$$
$$F = \dot{m}[(1 + r)V_2 - V_1] \tag{5.28}$$

Work done by the jet is given by FV_1.
Hence:

$$\text{work done} = \dot{m}V_1[(1 + r)V_2 - V_1] \tag{5.29}$$

The change in kinetic energy of the fuel may be expressed:

$$\text{Change in KE} = \frac{\dot{m}}{2}(V_2^2 - V_1^2) + \frac{\dot{m}}{2}rV_2^2$$
$$= \frac{\dot{m}}{2}[(1 + r)V_2^2 - V_1^2] \tag{5.30}$$

The change in kinetic energy is equal to the power input. Therefore, efficiency of the jet can be calculated by taking the ratio of the work done by the jet to the change in kinetic energy.

PROBLEMS

p.5.1 The momentum rate is given by

a. γQV.
b. ρQV.
c. $\rho QV^2/2$.
d. $1/2mV^2$.
e. none of these answers.

p.5.2 A jet of water 5 cm in diameter impinges on a fixed vertical plate with a velocity of 30 m/s. The normal force on the plate in N is

a. 1767.
b. 1.77×10^7.
c. 17335.
d. 465.
e. none of these answers.

p.5.3 A wheel is fixed with flat plates radially around the circumference. The wheel moves under the force of a jet. The maximum efficiency in percent which can be obtained for such a system is

a. 90.
b. 50.
c. 80.
d. 95.
e. none of these answers.

p.5.4 A jet of water 2 in in diameter, having a velocity of 75 ft/s, strikes a flat fixed plate normally. The force on the plate in lbs. is

a. 2380.
b. 238.
c. 7655.
d. 980.
e. none of these answers.

p.5.5 A jet of water 2 in in diameter, having a velocity of 75 ft/s, strikes a flat moving plate normally. The plate is moving in the same direction as the jet with a velocity of 25 ft/s. The force in lbs on the plate is

a. 158.
b. 106.
c. 238.
d. 765.
e. none of these answers.

p.5.6 The work done per second for p.5.5 in ft · lbs is

a. 7950.
b. 5300.
c. 2650.
d. 3950.
e. none of these answers.

p.5.7 A 7.5 cm diameter horizontal jet having a velocity of 30 m/s strikes a flat plate inclined at 60° with the horizontal. The normal pressure in N on the plate, is

a. 199.
b. 1988.
c. 3443.
d. 980.
e. none of these answers.

p.5.8 The normal pressure, in N, on the plate of p.5.7 when the plate is moving away from the jet at a velocity of 15 m/s is

a. 861.
b. 497.
c. 1721.
d. 780.
e. none of these answers.

p.5.9 A 7.5 cm diameter jet having a velocity of 30 m/s strikes a hemispherical cup and is deflected through 180°. The force in N on the cup in the direction of the jet is

a. 3976.
b. 1980.
c. 7952.
d. 1560.
e. none of these answers.

p.5.10 The force on the cup, in N, of p.5.9 when the cup is moving away with the velocity of 10 m/s is

a. 3534.
b. 1767.
c. 5301.
d. 2650.
e. none of these answers.

p.5.11 A jet of water having a velocity of 75 ft/s strikes a series of vanes moving with a velocity of 30 ft/s. The jet makes an angle of 30° with the direction of motion of the vanes at the entrance and leaves at an angle of 120°. Determine

a. the angles of the vane tips for no shock.
b. the efficiency.

p.5.12 The diameter of a jet is 10 cm and the velocity is 50 m/s. Determine the magnitude and the direction of the force required to keep the vane shown in Fig. 5.13 in place. The vane is in the vertical plane.

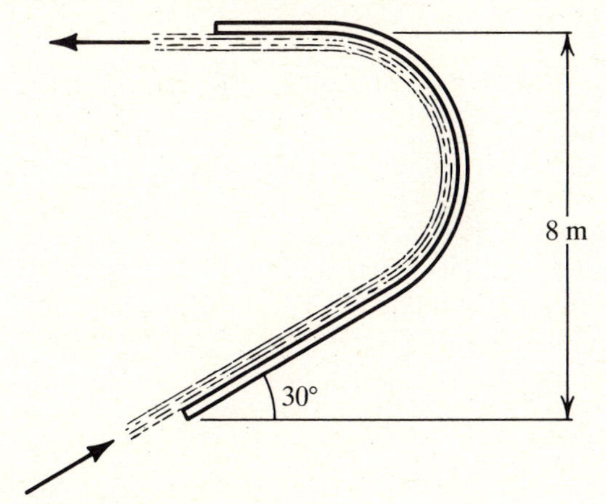

FIGURE 5.13 Problem 5.12.

p.5.13 A jet is 7.5 cm in diameter and has a velocity of 30 m/s. Determine the magnitude and direction of the force required to keep the plate shown in Fig. 5.14 in place.

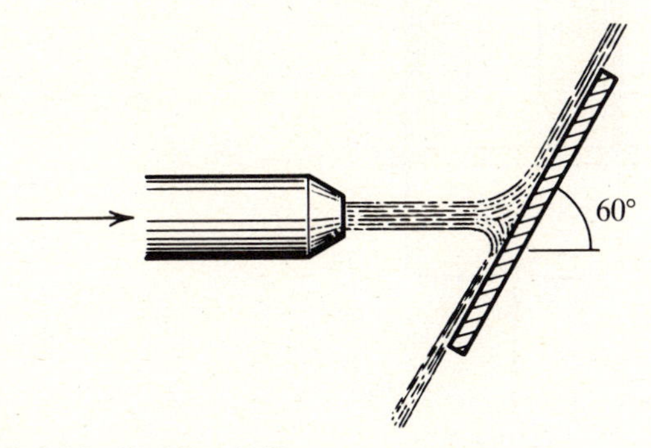

FIGURE 5.14 Problem 5.13.

p.5.14 Water is flowing through the 90° elbow shown in Fig. 5.15 at the rate of 300 l/s. The head loss in the elbow is given by $0.9V^2/2g$. Determine the magnitude and direction of the force required to hold the elbow in place. Assume the system to be in the horizontal plane.

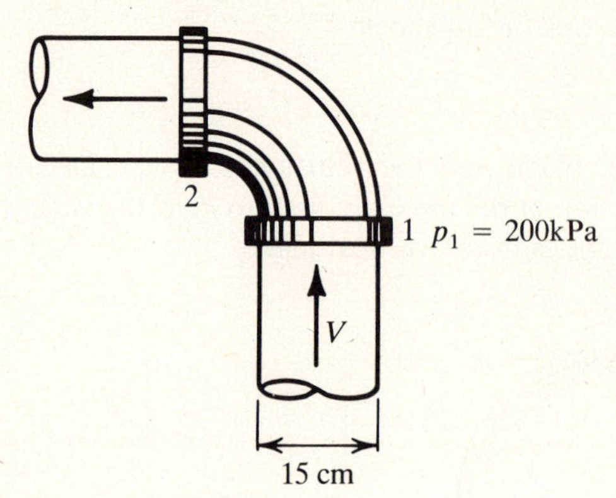

FIGURE 5.15 Problem 5.14.

p.5.15 Determine the magnitude and the direction of the force required to hold the Y shown in Fig. 5.16 in place. Assume the Y to be in the horizontal plane.

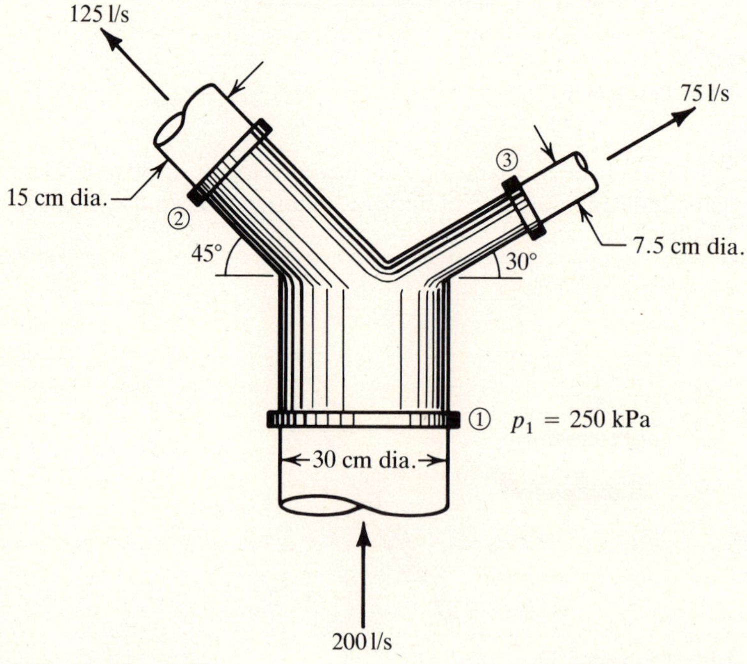

FIGURE 5.16 Problem 5.15.

p.5.16 Water is flowing through the vertical tee shown in Fig. 5.17. Given that $Q_3 = 0.5$ ft^3/s, $Q_1 = 0.2$ ft^3/s, $D_1 = D_3 = 2.0$ in, $D_2 = 1$ in, $p_3 = 40$ psi, determine the force required to keep the tee in place.

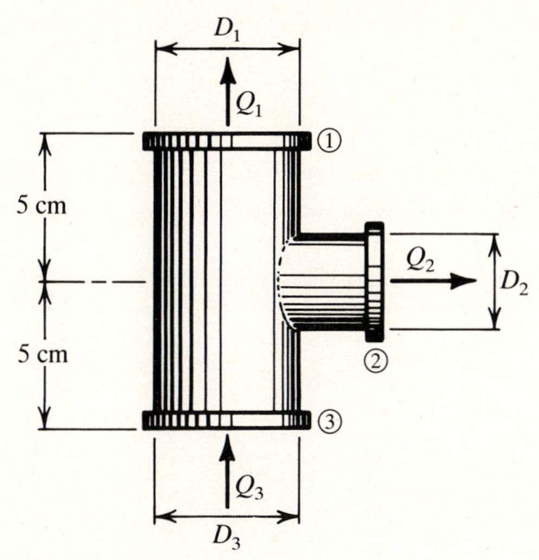

FIGURE 5.17 Problem 5.16.

p.5.17 A motor boat travels upstream in a river at a velocity of 35 km/hr. The stream itself is flowing at 2 m/s. The boat is powered with a jet propulsion system which takes water at the bow and discharges it at the stern with a velocity of 20 m/s relative to the boat. The amount of water discharged is 0.2 m^3/s. The engine produces a power of 25 kW. Determine the thrust and the efficiency.

p.5.18 An aircraft is moving with an absolute velocity of 350 km/hr. The wind velocity at the head is 50 km/hr. The thrust required to maintain this velocity is 12 kN. Assuming a theoretical efficiency of 90%, determine the diameter of the ideal propeller and the power needed to drive it. Assume the density of air to be 1.2 kg/m^3.

p.5.19 An ideal windmill, 36 ft in diameter, operates at a theoretical efficiency of 55% when the wind velocity is 40 ft/s. Determine (a) thrust on the windmill, (b) air velocity through the blade, (c) pressure difference across the blade, and (d) shaft power developed. Assume the density of air to be 1.2 kg/m^3.

p.5.20 An airplane with a 6 ft diameter propeller travels through still air of density 0.0022 $slugs/ft^3$ at 190 mph. The speed of air through the propeller is 260 mph relative to the airplane. Determine (a) the thrust on the plane, (b) the theoretical horsepower required to drive the propeller, (c) the propeller efficiency, and (d) the pressure difference across the blade.

p.5.21 A jet propelled airplane travelling at 1100 km/hr takes in 45 kg/s of air and discharges it at 575 m/s relative to the airplane. Neglecting the weight of the fuel, determine the thrust produced.

Chapter Six

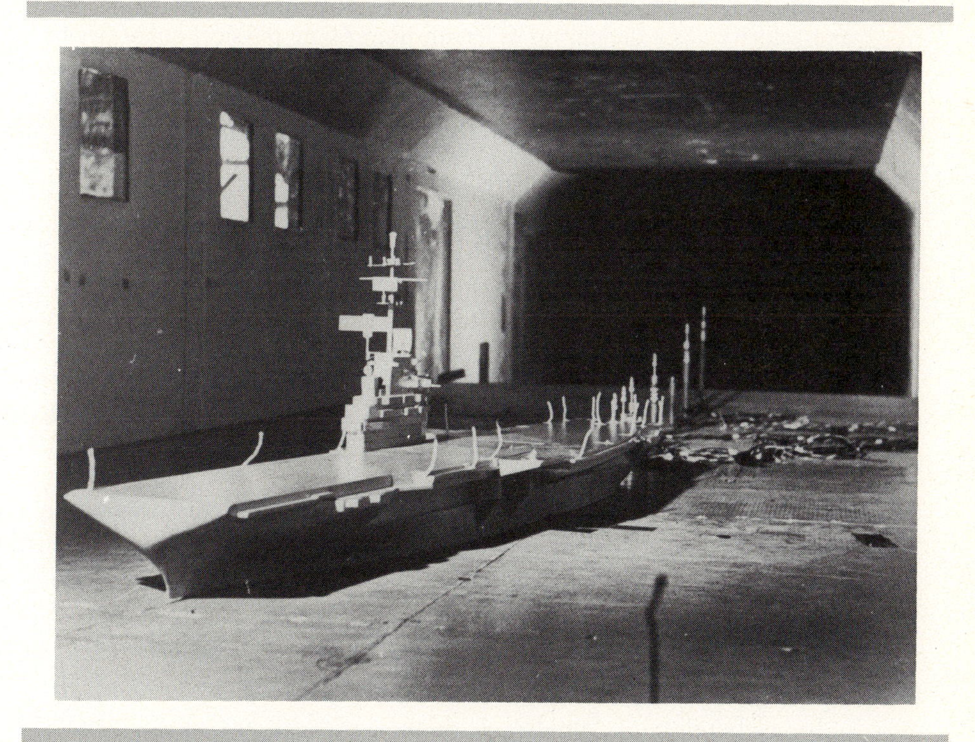

DIMENSIONAL ANALYSIS AND DYNAMIC SIMILITUDE

In this chapter, we shall consider the dimensions of the physical quantities encountered in fluid mechanics. We shall also learn the technique for grouping various physical quantities into dimensionless numbers. Dimensionless numbers play an important role in experimental fluid mechanics.

In this chapter, we shall also examine the technique of physical modeling of engineering structures, machines and vehicles.

6.1 INTRODUCTION

Engineers are most often interested in knowing the relationships between various physical quantities which affect a fluid phenomenon. Most physical quantities have dimensions; for example, distance has a dimension of length and velocity has a dimension of length/time. Some physical quantities have no dimensions; for example, angular displacement θ is dimensionless.

Dimensional analysis is a technique whereby physical quantities may be arranged in dimensionless parameters. Dimensionless parameters play an important role in understanding the flow of fluid in the laboratory—i.e., under experimental conditions. For example, in a hydraulic press (Example 2.1), the ratio of piston-to-ram diameters, a dimensionless number, determines the mechanical advantage regardless of the overall size of the press. Using the dimensionless parameters, limited experimental results can be applied to actual physical conditions involving much larger dimensions, and often different fluid properties. Also, the concept of dimensional analysis, coupled with the understanding of the mechanics of flow, may be used to generalize the experimental data. This would mean that limited and selective experimental results could be applied to a broad category of larger physical problems.

To illustrate the usefulness of dimensionless numbers, consider the problem of determining the discharge through a small diameter round tube under laminar flow conditions. Let it be known that the discharge depends on the difference of pressure at the two ends of the tube, Δp, length of the tube, l, diameter of the tube, D, and the viscosity of the fluid, μ. That is:

$$Q = f(\Delta p, l, D, \mu)$$

In a laboratory, a considerable amount of work and money would be required to establish relationships between Q and the remaining variables taken one at a time. For example, to determine a curve of Q vs Δp, Δp will be varied while l, D and μ remain constant, and to obtain Q vs μ, μ will be varied while Δp, l, and D remain constant. The latter case would require use of different fluids. Not only the development, but also the use and interpretation of such curves becomes a challenging task.

It will be shown later in Section 6.3 that the variables Q, Δp, l, D and μ can be arranged in two dimensionless numbers, one being l/D and the other $QD^3\mu/\Delta p$. In the laboratory, a relationship between l/D and $QD^3\mu/\Delta p$ can be obtained at comparatively low expense of time and money. This curve can then be used to compute Q when all other variables are known.

6.2 FUNDAMENTAL DIMENSIONS

Two types of dimensions, length (L) and time (T) were mentioned in Section 6.1. The kinetic relationships, such as acceleration, may be expressed in terms of these two types of dimensions.

In dealing with dynamic problems, a further dimension, that of mass (M), is required because no combination of length and time will lead to the

Table 6.1 DIMENSIONS FOR CERTAIN PHYSICAL QUANTITIES

Physical quantity	Dimension	Physical quantity	Dimension
Mass	M	Force	MLT^{-2}
Length	L	Work	ML^2T^{-2}
Time	T	Energy	ML^2T^{-2}
Area	L^2	Power	ML^2T^{-3}
Volume	L^3	Angular momentum	ML^2T^{-1}
Linear velocity	LT^{-1}	Torque	ML^2T^{-2}
Angular velocity	T^{-1}	Pressure and stress	$ML^{-1}T^{-2}$
Linear acceleration	LT^{-2}	Bulk modulus	$ML^{-1}T^{-2}$
Angular acceleration	T^{-2}	Surface tension	MT^{-2}
Angle	$M^\circ L^\circ T^\circ$	Density	ML^{-3}
Volumetric flow rate	L^3T^{-1}	Dynamic viscosity	$ML^{-1}T^{-1}$
Mass flow rate	MT^{-1}	Kinematic viscosity	L^2T^{-1}
Linear momentum	MLT^{-1}		

dimension of mass. Thus it follows that in mechanics, mass (M), length (L) and time (T) are three independent dimensions. All physical quantities may be expressed in terms of these three dimensions. In thermodynamics, however, an additional dimension, that of temperature is also required.

The dimensions of various physical quantities normally encountered in fluid mechanics are given in Table 6.1.

It should be noted here that force (F) may be selected as a fundamental dimension in place of mass (M). In that case, all the physical quantities stated in Table 6.1. can be expressed in terms of force (F), length (L), and time (T), noting that $M = FT^2L^{-1}$.

6.3 BUCKINGHAM Π THEOREM

E. Buckingham, in 1915, presented a theorem based on model experiments. According to the theorem, if a physical problem includes n quantities (such as pressure, viscosity, density, etc.) in which there are m dimensions, then these quantities can be arranged into $(n - m)$ independent dimensionless parameters. Let $A_1, A_2, A_3, \ldots, A_n$ be the n quantities involved, such that

$$\phi(A_1, A_2, A_3, \ldots, A_n) = 0 \tag{6.1}$$

Then these quantities can be arranged into $(n - m)$ independent parameters $\Pi_1, \Pi_2, \Pi_3, \ldots, \Pi_{(n-m)}$, in such a way that

$$f(\Pi_1\ \Pi_2, \Pi_3, \ldots, \Pi_{(n-m)}) = 0 \tag{6.2}$$

The procedure for determining the Π parameters is now described.

Let us assume that the rate of flow, Q, through a tube of small diameter, D, depends upon the pressure drop, Δp, between its two ends, the length of the tube, l, and the viscosity of the fluid, μ. Then

$$\phi(Q, \Delta p, l, D, \mu) = 0 \tag{6.3}$$

The dimensionless parameters can now be determined through the following step-by-step procedure.

1. After determining the quantities involved in the physical phenomenon under study, count the number of quantities. In this case, five quantities are involved; therefore, $n = 5$.

   It is evident that some understanding of the physical phenomenon under investigation is necessary to list all the pertinent quantities. If there is any difficulty about selecting the pertinent quantities, it is wise to use a generous number of them. Missing a quantity which should have been included may give a distorted view of the phenomenon; however, including a quantity which should not have been included would probably result in an extra dimensionless variable which eventually would be eliminated in the experimental process.
2. List the dimensions of all the variables. For our problem, the dimensions are:

Symbol	Dimensions
Q	L^3T^{-1}
Δp	$ML^{-1}T^{-2}$
l	L
D	L
μ	$ML^{-1}T^{-1}$

3. Count the total number of fundamental dimensions (e.g., M, L, T) involved. In this case, all three dimensions appear; therefore, $m = 3$.
4. Determine the number of independent dimensionless parameters which can be obtained by subtracting m from n. For this case, number of Πs $= 5 - 3 = 2$. According to Buckingham's theorem, Eq. 6.3 can be replaced by the equation

$$f(\Pi_1, \Pi_2) = 0 \tag{6.4}$$

5. Select m number of quantities with different dimensions, that contain among them the m dimensions, as repeating variables. It is essential that no one of the m quantities selected as repeating variables be derivable from the other repeating variables. For example, do not use both length (L) and moment of inertia of an area (L^4) as repeating variables. For this problem, Q, Δp and D may be used as repeating variables.
6. Set up dimensional equations combining the repeating variables with each of the remaining variables to form $(n - m)$ dimensionless groups. For this problem

$$\Pi_1 = Q^a \Delta p^b D^c l$$

$$\Pi_2 = Q^a \Delta p^b D^c \mu$$

   Here the repeating variables have been raised to the powers a, b and c. Our next task will be to determine the numerical values of a, b and c.

7. Form linear equations by equating the powers of M, L and T equal to zero and determine the values of a, b and c by solving the linear equations. Now

$$\Pi_1 = Q^a \Delta p^b D^c l$$

Substituting in the dimensions gives

$$\Pi_1 = (L^3T^{-1})^a(ML^{-1}T^{-2})^b(L)^c(L)^1 = M^0L^0T^0$$

Comparing the exponents of L on both sides of the equation yields

$$3a - b + c + 1 = 0 \qquad (6.5)$$

Similarly for T:

$$-a - 2b = 0 \qquad (6.6)$$

and for M:

$$b = 0 \qquad (6.7)$$

Simultaneous solution of Eqs. (6.5), (6.6) and (6.7) yields $a = 0$, $b = 0$, $c = -1$.

Hence:

$$\Pi_1 = \frac{l}{D}$$

Repeat the procedure for the remaining Πs. Therefore,

$$\Pi_2 = Q^a \Delta p^b D^c \mu$$

$$(L^3T^{-1})^a(ML^{-1}T^{-2})^b(L)^c(ML^{-1}T^{-1}) = M^0L^0T^0$$

$$3a - b + c - 1 = 0$$

$$-a - 2b - 1 = 0$$

$$b + 1 = 0$$

From which $a = 1$, $b = -1$, $c = -3$.
Hence:

$$\Pi_2 = \frac{Q\mu}{D^3 \Delta p}$$

8. After all the dimensionless parameters have been determined, then

$$f\left[\left(\frac{l}{D}\right), \left(\frac{Q\mu}{D^3 \Delta p}\right)\right] = 0 \qquad (6.8)$$

or

$$\Delta p = f\left(\frac{Q\mu l}{D^4}\right)$$

Equation (6.8) is only one of a number of possible forms of solution to Eq. (6.3), since the resulting form is determined by the choice of the repeating variables.

EXAMPLE 6.1

The phenomenon of drag is of considerable importance to engineers involved in the design of submarines, aeroplanes, cars, etc. The drag F on a fully submerged smooth sphere of diameter D depends on its velocity V, the density of the fluid, ρ, and the viscosity of the fluid μ. Arrange these variables into independent dimensionless numbers.

Solution On the basis of the information provided in the problem, we may write that

$$\phi(F, \rho, V, D, \mu) = 0\text{: } n = 5$$

An examination of the dimensions of these variables reveals that all three basic dimensions—i.e., M, L and T, are present, or $m = 3$. Therefore, number of Πs $= 5 - 3 = 2$, or

$$f(\Pi_1, \Pi_2) = 0$$

Let us choose ρ, V and D as repeating variables.

$$\Pi_1 = \rho^a V^b D^c F$$

$$\Pi_2 = \rho^a V^b D^c \mu$$

To determine Π_1:

$$(ML^{-3})^a (LT^{-1})^b (L)^c (MLT^{-2}) = M^0 L^0 T^0$$

Equating the exponents of M, L and T, we obtain

$$a + 1 = 0$$

$$-3a + b + c + 1 = 0$$

$$-b - 2 = 0$$

from which $a = -1$; $b = -2$; $c = -2$.

Hence:

$$\Pi_1 = \left(\frac{F}{\rho V^2 D^2}\right)$$

To determine Π_2:

$$(ML^{-3})^a (LT^{-1})^b (L)^c (ML^{-1}T^{-1}) = M^0 L^0 T^0$$

Equating the exponents of M, L and T, we obtain

$$a + 1 = 0$$

$$-3a + b + c - 1 = 0$$

$$-b - 1 = 0$$

From which $a = -1$; $b = -1$; $c = -1$.

Hence:

$$\Pi_2 = \left(\frac{\mu}{VD\rho}\right)$$

Now, according to the Buckingham Π theorem,

$$f\left(\frac{F}{\rho V^2 D^2}, \frac{\mu}{VD\rho}\right) = 0$$

$$\frac{F}{\rho V^2 D^2} = f\left(\frac{\mu}{VD\rho}\right)$$

$$\frac{F}{\rho V^2 D^2} = f\left(\frac{VD\rho}{\mu}\right) = f(\mathbf{R})$$

It may be noticed here that the reciprocal of a dimensionless quantity is also a dimensionless quantity.

6.4 SOME COMMON DIMENSIONLESS NUMBERS

Several dimensionless numbers have been identified in fluid mechanics. The understanding of these numbers is important for a better insight into the subject of fluid mechanics. These numbers are now described.

The Reynolds Number (R)

Named after the British scientist Osborne Reynolds, the Reynolds number provides the criterion for laminar and turbulent flows. The general form of the Reynolds number is given by:

$$\mathbf{R} = \frac{VL\rho}{\mu}$$

where V = average velocity of flow, L = a characteristic length, ρ = density of the fluid, μ = viscosity of the fluid.

By rearranging, the Reynolds number may be written as

$$\mathbf{R} = \frac{\rho VL}{\mu} = \frac{\rho V^2 L^2}{(\mu V/L)L^2}$$

where

$$\rho V^2 L^2 \simeq \text{dynamic pressure} \times \text{area}$$
$$\simeq \text{inertia force}$$

$$\frac{\mu V}{L} L^2 \simeq \text{viscous stress} \times \text{area}$$
$$\simeq \text{viscous force}$$

Therefore:

$$\mathbf{R} \simeq \frac{\text{inertia forces}}{\text{viscous forces}}$$

Thus the Reynolds number may also be intrepreted as a ratio of inertia forces to viscous forces.

The Mach Number (M)

Named after the Austrian physicist Ernst Mach (1836–1916), the Mach number is a key parameter in the study of compressible fluids. The Mach number is given by:

$$\mathbf{M} = \frac{V}{c}$$

where V = average velocity of flow, and c = local velocity of sound.

The speed of sound in a liquid may also be written as:

$$c = \sqrt{\beta/\rho}$$

where β = bulk modulus of elasticity.

The Mach number may be written as

$$\mathbf{M} = \frac{V}{\sqrt{\beta/\rho}}$$

By rearranging, the Mach number may be written as:

$$\mathbf{M} = \frac{V}{\sqrt{\beta/\rho}} = \frac{\sqrt{\rho V^2 L^2}}{\sqrt{\beta L^2}}$$

where

$$\rho V^2 L^2 \simeq \text{inertia force}$$

$$\beta L^2 \simeq \text{elastic force}$$

Therefore:

$$\mathbf{M} \simeq \frac{\text{inertia forces}}{\text{elastic forces}}$$

Thus, the Mach number may be interpreted as a ratio of inertia forces to elastic forces. It may also be shown to be a ratio of the kinetic energy of the flow to the internal energy of the fluid.

Effects of compressibility become an important consideration at Mach numbers exceeding 0.4. The Mach number has significant application in connection with high speed aircraft, missiles, propellers and rotary compressors.

The Froude Number (F)

Named after naval architect William Froude (1810–79), the Froude number provides the criterion for critical and noncritical flows. The Froude number is given by:

$$\mathbf{F} = \frac{V}{\sqrt{gL}}$$

where V = average velocity of flow, g = acceleration due to gravity, and L = a characteristic length.

By rearranging, the Froude number may be written as:

$$\mathbf{F} = \frac{\sqrt{\rho V^2 L^2}}{\sqrt{\rho g L^3}}$$

where

$$\rho V^2 L^2 \simeq \text{inertia forces}$$

$$\rho g L^3 \simeq \text{gravity forces}$$

Therefore:

$$\mathbf{F} \simeq \frac{\text{inertia forces}}{\text{gravity forces}}$$

Thus, the Froude number may be interpreted as the ratio of inertia forces to gravity forces.

The Weber Number (W)

Named after the German naval architect Moritz Weber (1871–1951), the Weber number is given by

$$\mathbf{W} = \frac{\rho V^2 L}{\sigma} = \frac{\rho V^2 L^2}{\sigma L}$$

where

$$\rho V^2 L^2 \simeq \text{inertia force}$$

$$\sigma L \simeq \text{surface tension force}$$

Therefore:

$$\mathbf{W} \simeq \frac{\text{inertia forces}}{\text{surface tension forces}}$$

Thus, the Weber number may be interpreted as the ratio of the inertia forces to the surface tension forces. The Weber number is important in those engineering applications where surface tension forces are important, such as in the behavior of small jets formed under low heads, and in the flow of a thin sheet of liquid over a solid surface.

The Pressure Coefficient (c_p)

The pressure coefficient is defined as

$$c_p = \frac{\Delta p}{\rho V^2/2}$$

where Δp = local pressure minus the freestream pressure, ρ = density of the fluid, and V = average velocity of flow.

By rearranging, the pressure coefficient may be written as:

$$c_p = \frac{\Delta p}{\rho V^2/2} = \frac{\Delta p L^2}{\rho V^2 L^2/2}$$

where

$$\Delta p L^2 \simeq \text{pressure force}$$

$$\rho V^2 L^2 \simeq \text{inertia force}$$

Therefore:

$$c_p \simeq \frac{\text{pressure forces}}{\text{inertia forces}}$$

Thus, the pressure coefficient is the ratio of the pressure forces to the inertia forces. The pressure coefficient has application in aerodynamic and other model testing.

6.5 PHYSICAL MODELING

In engineering practice, mathematical procedures and equations used in the design of hydraulic structures, machines, turbines, pumps, aircraft, etc. are based on numerous simplifying assumptions. On many occasions, a lot of subjective judgment on the part of a designer is required when rigorous mathematical analysis of the flow conditions is not available. For these reasons and others, it is customary in engineering practice to conduct model studies to ascertain the correct behavior of the prototype under actual flow conditions by observing the behavior of the model under similar flow conditions in the laboratory. For a model to correctly predict the behavior of the prototype, certain conditions of similarity between the model and the prototype must be met. These similarities include *geometric similarity*, *kinematic similarity* and *dynamic similarity*.

Geometric Similarity

Geometric similarity requires that the ratio of any length in the model to the corresponding length in the prototype be the same everywhere. This ratio is usually known as *the scale factor* (λ_1).

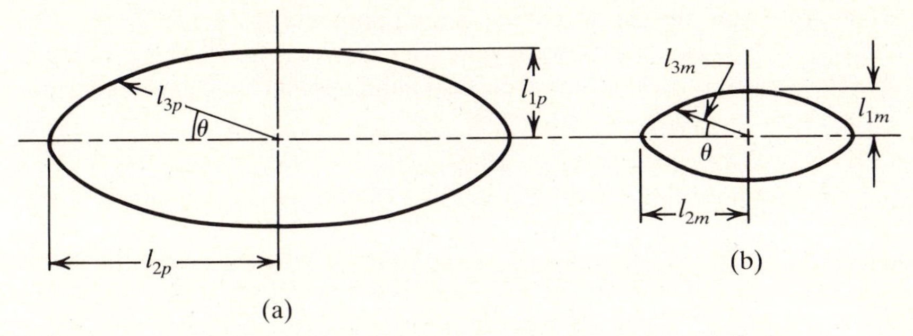

FIGURE 6.1 A bridge pier and its model showing geometric similarity.

A bridge pier and its model are shown in Figs. 6.1a and 6.1b, respectively. To meet the condition of geometric similarity of model and prototype, the following equation must be satisfied:

$$\frac{l_{1p}}{l_{1m}} = \frac{l_{2p}}{l_{2m}} = \frac{l_{3p}}{l_{3m}} = \frac{l_{ip}}{l_{im}} = \lambda_1$$

where l_{ip} is the ith dimension of the prototype, l_{im} is the ith dimension of the model, and λ_1 is the scale factor.

Perfect geometric similarity is not always possible. For example, in a model, not only the physical dimensions of the prototype but also the height of the roughness projections must be reduced by the scale factor. This sometimes may become extremely difficult to achieve. In such a case, a distorted model will result.

Kinematic Similarity

The term *kinematic similarity* means that at corresponding points in the prototype and in the model, the velocities and the accelerations should have a constant ratio.

The pier model-prototype system discussed before is shown in Fig. 6.2 with water flowing around it. Consider two points defined by (r_{1p}, θ_1) and (r_{2p}, θ_2)

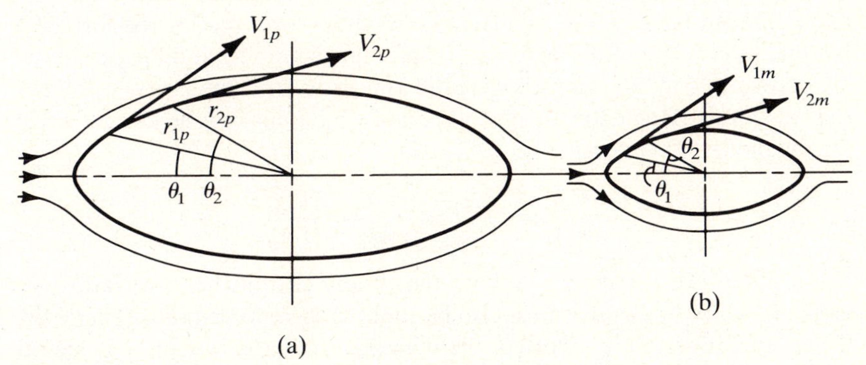

FIGURE 6.2 A bridge pier and its model showing kinematic similarity.

in the prototype, and let the corresponding points in the model be (r_{1m}, θ_1) and (r_{2m}, θ_2). Then for kinematic similarity between model and prototype, the velocities and accelerations at corresponding points must have a constant ratio. Thus

$$\frac{V_{1p}}{V_{1m}} = \frac{V_{2p}}{V_{2m}} = \frac{V_{ip}}{V_{im}} = \lambda_V$$

and

$$\frac{a_{1p}}{a_{1m}} = \frac{a_{2p}}{a_{2m}} = \frac{a_{ip}}{a_{im}} = \lambda_a$$

where λ_V and λ_a are the velocity and acceleration scale factors, respectively. To ensure kinematic similarity, it is a must that geometric similarity be present.

Dynamic Similarity

The term *dynamic similarity* means that at corresponding points in the model and the prototype, the forces should have a constant ratio. A fluid may be subjected to various types of forces. Some possible types of forces are:

1. *Pressure* forces, caused by the differences of pressure between various points in the fluid.
2. *Viscous* forces, caused by the viscosity of the fluid.
3. *Elastic* forces, caused by the compressibility of the fluid.
4. *Surface tension* forces.
5. *Gravity* forces.
6. *Inertia* forces. From Newton's second law of motion, inertia forces are equal to $-ma$—i.e., a hypothetical force (equal to $-ma$) is required to bring the acceleration of the fluid particle to zero.

Let us now investigate the conditions of dynamic similarity.

In the model-prototype system shown in Fig. 6.3, consider a particle of fluid of mass m, in the model as well as in the prototype. This particle is acted upon by the pressure and viscous forces. Let these forces be, respectively, F_p and F_v. From Newton's second law, the resultant force (i.e., inertia force, F_i) is equal to ma. The resulting force triangles for the model and the prototype are shown in Fig. 6.3. From kinematic and geometric similarity, it follows that the force triangle for the prototype must be similar to the force triangle for the model. Therefore:

$$\frac{(F_p)_p}{(F_p)_m} = \frac{(F_v)_p}{(F_v)_m} = \frac{(ma)_p}{(ma)_m} = \frac{(F_i)_p}{(F_i)_m}$$

From the above equation it follows that if any two of the force ratios are equal, the third force ratio must also be equal to these force ratios. Therefore, if only three forces are involved, the dynamic similarity would be achieved when any two of the forces are in direct proportion. For example, in a flow

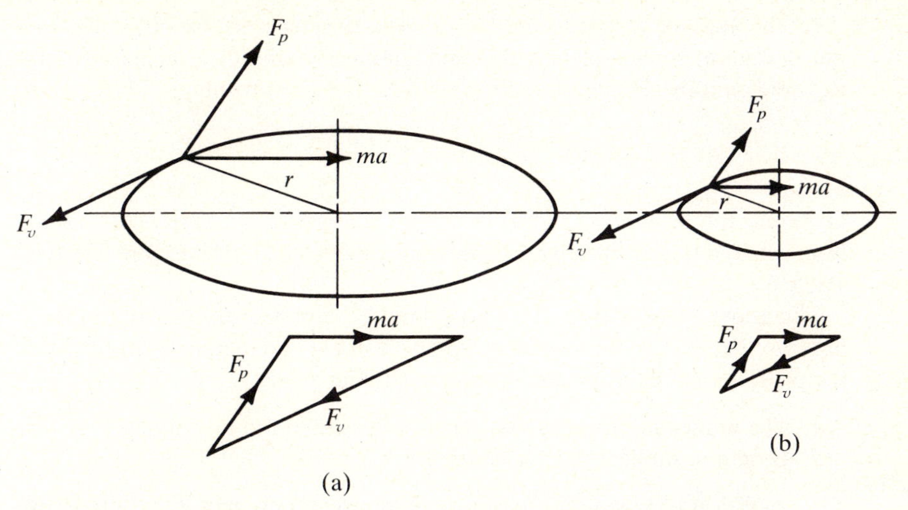

FIGURE 6.3 A bridge pier and its model showing dynamic similarity.

phenomenon, presume inertia and viscous forces are involved; then for dynamic similarity:

$$\frac{(F_v)_p}{(F_v)_m} = \frac{(F_i)_p}{(F_i)_m}$$

or

$$\left(\frac{F_i}{F_v}\right)_p = \left(\frac{F_i}{F_v}\right)_m$$

There is no need to consider the third force, because the similarity condition for this force is automatically satisfied.

The acceleration of particles and, hence, the inertia forces play an important role in all fluid flow phenomena, and for this reason it has become customary to include inertia force as one of the two forces considered for dynamic similarity. The other force may be one of the pressure, viscous, elastic, surface tension or gravity forces. We recall from Section 6.4 that the ratio of inertia force to one of the other forces is represented by a specific number; for example, the ratio of the inertia force to the gravity force is represented by the Froude number. Therefore, it follows that dynamic similarity can be achieved by making the Mach, Reynolds, Froude or Weber number the same (as the case may be) in both the model and the prototype.

For example, in an open channel flow, the inertia forces and gravity forces are dominant forces; therefore, dynamic similarity between model and prototype would be achieved if the Froude number at a point is the same for the model and the prototype, or

$$\left(\frac{V}{\sqrt{gL}}\right)_m = \left(\frac{V}{\sqrt{gL}}\right)_p$$

In the case of a steady flow in a pipe, inertia and viscous forces are the dominant forces. Hence, dynamic similarity would be achieved if the Reynolds number is the same in the model and the prototype. Thus:

$$\left(\frac{VL\rho}{\mu}\right)_m = \left(\frac{VL\rho}{\mu}\right)_p$$

Dynamic similarity automatically ensures kinematic similarity. The conditions of dynamic similarity for various practical conditions are listed in Table 6.2.

The above discussion on similarity may be summed up by stating that in order for the results of model testing to be valid for predicting the behavior of the prototype, the following conditions should be met:

1. The model should be geometrically similar to the prototype.
2. Dynamic similarity must be present.

The condition of kinematic similarity is automatically met if the above two conditions are present.

Table 6.2 CONDITIONS OF DYNAMIC SIMILARITY FOR VARIOUS PRACTICAL PROBLEMS

Significant forces	Practical examples	Condition for dynamic similarity
Viscous Pressure Inertia	1. Pressure flow through pipes 2. Flow of air past a low speed aircraft 3. Flow of water past a submarine deeply submerged so as to produce no surface waves	$(\mathbf{R})_m = (\mathbf{R})_p$
Gravity Pressure Inertia	1. Flow of a liquid in an open channel 2. Wave motion caused by a ship in water 3. Flow over weirs and spillways 4. Flow through an orifice into the atmosphere	$(\mathbf{F})_m = (\mathbf{F})_p$
Surface tension Pressure Inertia	1. Capillary waves 2. Small jets formed under low heads 3. Flow of a thin sheet of liquid over a solid surface	$(\mathbf{W})_m = (\mathbf{W})_p$
Elastic Pressure Inertia	1. High speed aircraft and missiles 2. Propellers and rotary compressers 3. Problems involving isentropic conditions in gases when **M** exceeds 0.4.	$(\mathbf{M})_m = (\mathbf{M})_p$

EXAMPLE 6.2

A pipeline 6 ft in diameter is to be designed to carry crude oil at 20°C at an average velocity of 2.5 m/s. It is decided a one-tenth scale model of the pipeline will be tested in the laboratory to determine the head losses. If the fluid to be used in the model is water at 20°C, at what velocity should the water flow?

Solution For the dynamic similarity in a pipe flow, the Reynolds number must be the same in the model and the prototype, or:

$$\frac{V_m l_m \rho_m}{\mu_m} = \frac{V_p l_p \rho_p}{\mu_p}$$

$$V_m = V_p\left(\frac{l_p}{l_m}\right)\left(\frac{\rho_p}{\rho_m}\right)\left(\frac{\mu_m}{\mu_p}\right)$$

For crude oil at 20°C:

$$\rho_p = 0.86 \times 998.2 = 858.45 \text{ kg/m}^3$$

$$\mu_p = 8 \times 10^{-3} \text{ N}\cdot\text{s/m}^2$$

For water at 20°C:

$$\rho_m = 998.2 \text{ kg/m}^3$$

$$\mu_m = 1.005 \times 10^{-3} \text{ N}\cdot\text{s/m}^2$$

Hence:

$$V_m = 2.5(10)\left(\frac{858.45}{998.2}\right)\left(\frac{1.005 \times 10^{-3}}{8 \times 10^{-3}}\right) = 2.7 \text{ m/s}$$

Hence, the model should be tested at a velocity of 2.7 m/s. This velocity in the model is called the *corresponding velocity*.

EXAMPLE 6.3

A concrete spillway has a length of 300 ft, a maximum discharge of 100,000 cfs and a coefficient of discharge of 3.8. The structure is to be studied in the laboratory on a one-fiftieth scale model. Determine the maximum head on the crest and the corresponding discharge for dynamically similar conditions. Same fluid at the same temperature is to be used for model testing.

Solution In flow over a spillway, gravity forces are the dominant forces; hence, the Froude number should be the same in the model and the prototype.

Therefore:

$$\frac{V_p}{\sqrt{l_p g_p}} = \frac{V_m}{\sqrt{l_m g_m}}$$

$$V_m = V_p \sqrt{\frac{l_m}{l_p}} \qquad \text{(assuming } g_p = g_m\text{)}$$

Now

$$Q = VA = Vl^2$$

$$V = \frac{Q}{l^2}$$

Hence:

$$\frac{Q_m}{l_m^2} = \frac{Q_p}{l_p^2}\sqrt{\frac{l_m}{l_p}}$$

$$Q_m = Q_p\left(\frac{l_m}{l_p}\right)^2\left(\frac{l_m}{l_p}\right)^{1/2} = Q_p\left(\frac{l_m}{l_p}\right)^{5/2}$$

For this problem:

$$\frac{l_m}{l_p} = \frac{1}{50}; \ Q_p = 100{,}000 \text{ cfs}$$

Hence:

$$Q_m = 100{,}000\left(\frac{1}{50}\right)^{5/2} = 5.66 \text{ cfs}$$

Discharge over a spillway is given by

$$Q = CLH^{3/2}$$

For this problem:

$$C = 3.8; \ L = 300 \text{ ft}; \ Q = 100{,}000 \text{ cfs}$$

$$H = \left(\frac{100{,}000}{3.8 \times 300}\right)^{2/3} = 19.771 \text{ ft}$$

Head over model = 19.771/50 = 0.3954 ft = 4.75 in

6.6 SOME PRACTICAL APPLICATIONS

It was mentioned earlier that in fluid mechanics, model testing is often done to predict the behavior of the prototype. A well-known example is the design of an aircraft. To avoid costly alterations on the aircraft and possible risk to human life, a model of the aircraft is usually made and tested in a wind tunnel. For the results of the model testing to be useful, the model must be geometrically and dynamically similar to the prototype. Let us assume that we are interested in finding the drag force F on the aircraft. In this case, inertia and viscous forces are involved. The gravity forces are small and may be neglected. There are no surface tension forces due to the lack of an interface between a liquid and another fluid. If the aircraft is a low-speed one, the effects of compressibility are negligible and, therefore, elastic forces may be neglected. The only dominant forces are inertia and viscous forces. Of course, pressure forces are inevitable and are always present. For dynamic similarity—in this case, the Reynolds number—must be the same for the model and the prototype. Hence:

$$\frac{V_p l_p \rho_p}{\mu_p} = \frac{V_m l_m \rho_m}{\mu_m}$$

and

$$V_m = V_p \left(\frac{l_p}{l_m}\right)\left(\frac{\rho_p}{\rho_m}\right)\left(\frac{\mu_m}{\mu_p}\right) \tag{6.9}$$

The velocity V_m is called *the corresponding velocity*. Also in Example 6.1 it was found that in the case of drag forces

$$\frac{F}{\rho V^2 l^2} = f(\mathbf{R})$$

Since the Reynolds number is the same for the model and the prototype, we have

$$\frac{F_p}{\rho_p V_p^2 l_p^2} = \frac{F_m}{\rho_m V_m^2 l_m^2}$$

In model testing, the size of the model is considerably smaller than the size of the prototype. A scale factor of 100 is not uncommon in the case of large aircraft. Using this scale factor and the same testing fluid (air), the corresponding velocity for a prototype of 300 km/h comes out to be 30,000 km/h. It may be difficult to attain such a velocity. Also at such a high velocity the effects of compressibility become very important and the similarity of Mach numbers cannot be ignored. This difficulty in modeling is accentuated if the aircraft is a high-speed one. This problem can be avoided by using a high density fluid for testing of the model. Air compressed to higher density may also be used.

For high-speed aircraft, elastic forces become important and equality of Mach numbers is required for dynamic similarity. Hence, for complete dynamic similarity, the Reynolds number and the Mach number must be the same for the model and the prototype, or

$$\frac{V_p l_p \rho_p}{\mu_p} = \frac{V_m l_m \rho_m}{\mu_m}$$

and

$$\frac{V_p}{c_p} = \frac{V_m}{c_m}$$

For both conditions to be satisfied simultaneously:

$$\left(\frac{l_p}{l_m}\right)\left(\frac{\rho_p}{\rho_m}\right)\left(\frac{\mu_m}{\mu_p}\right) = \left(\frac{V_m}{V_p}\right) = \left(\frac{c_m}{c_p}\right) = \left(\frac{\beta_m/\rho_m}{\beta_p/\rho_p}\right)^{1/2}$$

For all available fluids, the above equation requires that the scale factor be nearly unity, which, in turn, means that the model be of the same size as the prototype. However, similarity of Mach number only does not impose any restriction on the size of the model because no characteristic length appears in the Mach number. In this case, if the same fluid at the same temperature and pressure is used for both the model and the prototype, then $V_m = V_p$. Therefore, the model should be tested at the same velocity as the prototype. However, in such a case, similarity of Reynolds numbers will not be satisfied.

PROBLEMS

p.6.1 Dimensional analysis is a technique which is used to

a. determine the dimensions of a variable.
b. group various quantities in dimensionless parameters.
c. group various quantities into parameters having a single dimension of M, L or T.
d. determine the dimensions of objects in the laboratory.
e. none of these answers.

p.6.2 A dimensionless combination of $VD\rho\mu$ is

a. $\dfrac{VD\mu}{\rho}$.

b. $\dfrac{VD\rho}{\mu}$.

c. $\dfrac{V\rho\mu}{D}$.

d. $\dfrac{D\mu\rho}{V}$.

e. $\dfrac{V\mu}{D\rho}$.

p.6.3 The power P, required to drive a propeller of diameter D depends on the density of the fluid ρ, viscosity of the fluid μ, free stream velocity V, angular velocity of propeller ω, and the velocity of sound in fluid c. The total number of variables involved in the problem is

a. 5.
b. 6.
c. 7.

d. 8.
e. none of these answers.

p.6.4 The number of Πs for p.6.3 is

a. 2.
b. 3.
c. 4.
d. 5.
e. none of these answers.

p.6.5 $VD\rho/\mu$ is the form for

a. the Reynolds number.
b. the Froude number.
c. the Mach number.
d. the Weber number.
e. the pressure coefficient.

p.6.6 $\rho V^2 L/\sigma$ is the form for

a. the Reynolds number.
b. the Froude number.
c. the Mach number.
d. the Weber number.
e. the pressure coefficient.

p.6.7 The Froude number may also be interpreted as the ratio of

a. inertia forces to viscous forces.
b. inertia forces to elastic forces.
c. inertia forces to gravity forces.
d. inertia forces to surface tension forces.
e. none of these answers.

p.6.8 The Mach number may also be interpreted as the ratio of

a. inertia forces to viscous forces.
b. inertia forces to elastic forces.
c. inertia forces to gravity forces.
d. inertia forces to surface tension forces.
e. none of these answers.

p.6.9 In problems involving pressure flow through pipes, the significant force in addition to the pressure and inertia forces is the

a. viscous force.
b. gravity force.
c. surface tension force.
d. elastic force.
e. none of these answers.

p.6.10 In problems involving open channel flow, the significant force in addition to the pressure and inertia forces is the

a. viscous force.
b. gravity force.
c. surface tension force.
d. elastic force.
e. none of these answers.

p.6.11 In problems involving high speed aircraft and missiles, the significant force in addition to the pressure and inertia forces is the

a. viscous force.
b. gravity force.
c. surface tension force.
d. elastic force.
e. none of these answers.

p.6.12 In a turbulent flow, the losses $\Delta h/l$ per unit length of pipe depend upon the diameter of pipe D, the density of the fluid ρ, the viscosity of the fluid μ, the velocity of flow V and the acceleration due to gravity g. Group these quantities into dimensionless parameters.

p.6.13 The head Δh developed by turbomachines depends upon the diameter of rotor D, rotational speed N, discharge through the machine Q, density of the fluid ρ, viscosity of the fluid μ, and acceleration due to gravity g. Determine the general form of the head equation.

p.6.14 A boat moving over the surface of a lake experiences a drag force F which depends upon the velocity of the boat V, length of the boat l, width of the boat W, density of water ρ, viscosity of water μ, and acceleration due to gravity g. Determine the general form of the drag equation.

p.6.15 The pressure drop Δp in a one-dimensional compressible flow of a fluid in a round duct depends upon the diameter of the duct D, length of the duct l, density of the fluid ρ, viscosity of the fluid μ, velocity of the fluid V, and the velocity of sound c. Develop the general form of the pressure drop equation.

p.6.16 A model of a venturimeter is made on a scale ratio of 1:4. The prototype operates at a temperature of 90°C and has a throat diameter of 50 cm. If the velocity at the throat in the prototype is 8 m/s, what rate of flow is needed through the model for similitude? The model operates at 20°C.

p.6.17 A model of a ship is made on a scale ratio of 1:15. The drag due to waves is found to be 15 N at a speed of 3 m/s in the model. What would be the corresponding drag speed and drag on the prototype? The liquid and the temperature are the same in the model and the prototype.

p.6.18 A 1:4 scale model of a pipe system is tested to determine the overall head losses. Air at 20°C and at a pressure of 1 atm is used for model testing. The prototype uses water at 15°C. For a prototype velocity of 5 m/s in a 4 m diameter pipe, determine the air velocity and the quantity needed to operate the model. Also state how you will estimate the losses in the prototype from the losses determined in the model. The viscosity of air is 1.8×10^{-5} N·s/m^2.

p.6.19 An aircraft is to fly at an elevation of 30,000 ft, where the temperature is −55°F and the pressure is 4.9 psi, at a speed of 1200 ft/s. A model of the aircraft is made at a scale ratio of 1:20 to be tested in a pressurized wind tunnel at 60°F. For complete dynamic similarity what pressure and velocity should be used in the wind tunnel? The viscosity of air is 1.8×10^{-5} N·s/m^2.

p.6.20 A submarine is to travel at a speed of 3 m/s when deeply submerged. The torque required to operate the rudder of this submarine is studied with a 1:20 scale model to be operated in a fresh-water tunnel. The torque on the model is found to be 8.5 N.m. What would be the corresponding torque on the prototype?

p.6.21 The flow rate over a spillway is 200 m^3/s. This spillway is to be studied in a laboratory by a suitable model. What scale factor should be used for the model if the maximum available flow rate in the laboratory is 1 m^3/s? On a part of such a model, a force of 3 N is measured. What will be the corresponding force on the prototype?

Chapter Seven

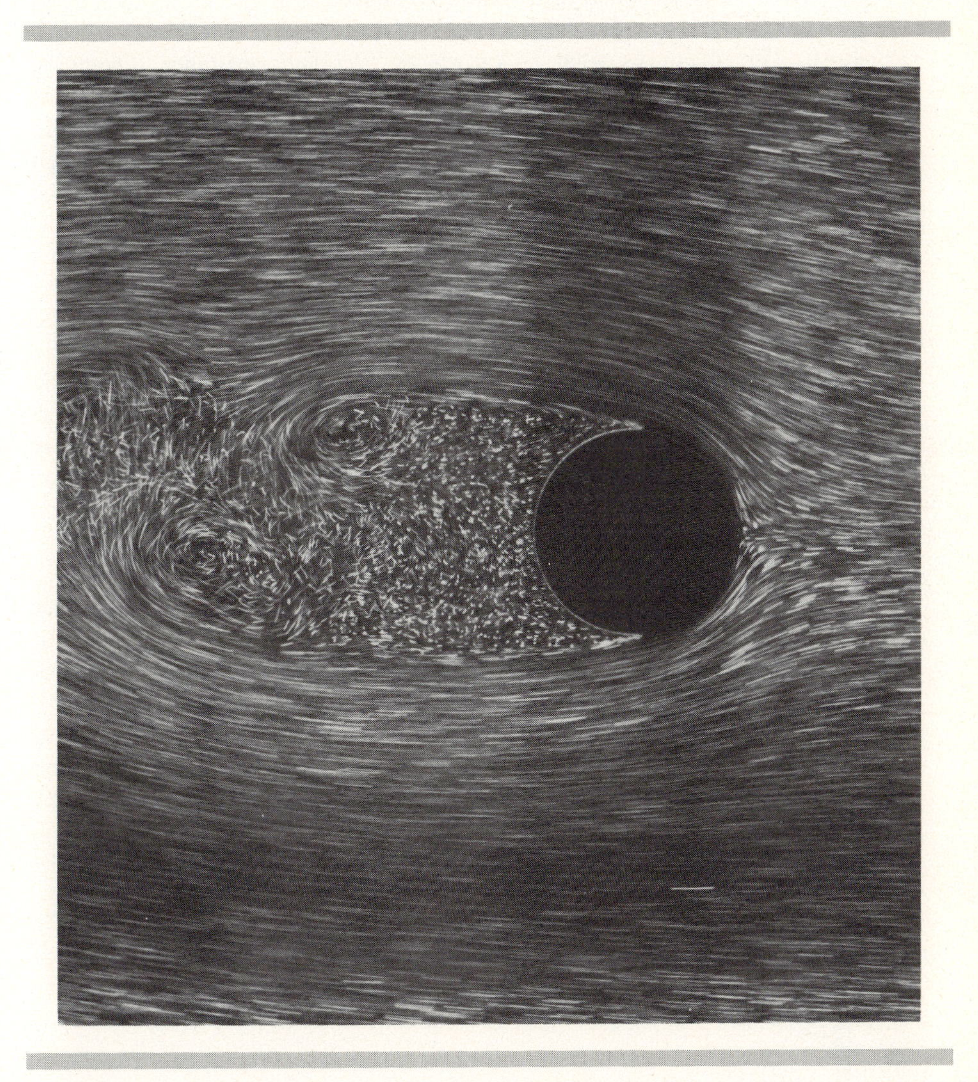

FLUID RESISTANCE

In this chapter, we will study two manners of fluid flow—namely, laminar flow and turbulent flow. The effect of viscosity in producing boundary layer and drag will be investigated. We will also study the phenomenon of separation.

7.1 INTRODUCTION

When a fluid flows past a solid surface, the molecules of fluid in contact with the solid surface remain at rest. Since fluid molecules attract each other, those molecules which pass by the stationary molecules have a tendency to slow down. This produces what is known as *frictional resistance*. Therefore, whenever a fluid flows past a solid surface, it experiences resistance to flow and energy must be expended to overcome this resistance. Similarly, if a solid surface moves through a fluid (for example, motion of an automobile through air), the solid experiences resistance to its motion and energy must be expended to overcome this resistance.

It is found by experiments that the frictional resistance to fluid flow depends on (a) the area of the solid surface in contact with the fluid, called *the wetted surface*, (b) the density of the fluid which in turn depends on the temperature, (c) the surface roughness, (d) the velocity of the fluid, and (e) the viscosity of the fluid. The frictional resistance does not depend on the pressure of the fluid. The factors a, b, d, and e mentioned above are also parameters of the Reynolds number (**R**). Therefore, frictional resistance is related to **R**.

7.2 LAMINAR AND TURBULENT FLOW

Laminar flow is one in which flow occurs in layers with one layer gliding over the other with little or no mixing. The flow generally has a smooth appearance. *Turbulent flow*, on the other hand, is characterized by mixing of the fluid by eddies of varying size within the flow.

The two manners of motion were first reported by Osborne Reynolds in 1883. He conducted his experiments on the flow of water through a glass tube using an apparatus similar to the one shown in Fig. 7.1. A glass tube was mounted horizontally with one end in a tank and the other fitted with a valve. A funnel with a bent tube fitted with a valve was used to inject aniline dye into

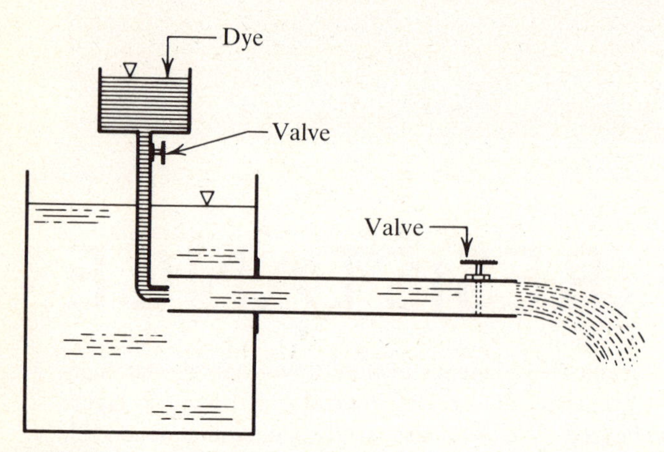

FIGURE 7.1 Reynolds apparatus.

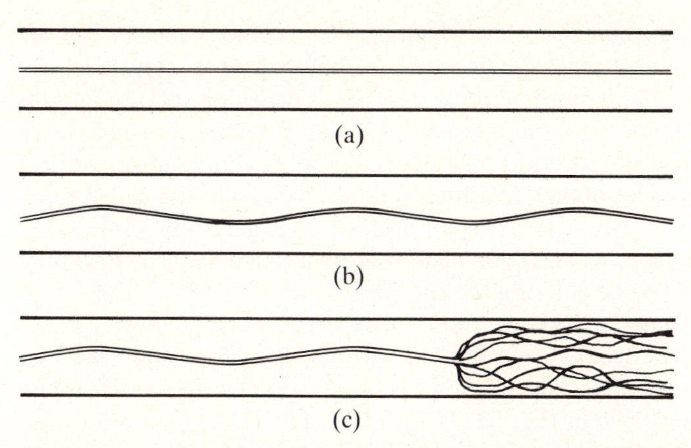

FIGURE 7.2 Various shapes of dye streaks within the pipe of the Reynolds apparatus.

the glass tube. He noticed that for small values of flow, the dye injected along the axis of the glass tube moved in a straight line without diffusion (see Fig. 7.2a). He inferred that for such a condition to exist, the flow must have been taking place in parallel laminae with no mixing. He called this type of flow *laminar flow*. He then slowly increased the rate of flow. As the rate of flow increased, the condition of the dye also changed gradually until a stage was reached when the stream of dye began to waver (Fig. 7.2b). Further increase in the rate of flow caused the wavering to become more intense, and then it suddenly broke up and diffused throughout the lower end of the tube (Fig. 7.2c). At this stage, the flow was no longer taking place in straight lines, but rather crosscurrents, and eddies had developed. He called this type of flow *turbulent flow*. It is of interest to note that in Reynolds's experiments, diffusion always occurred after the dye had travelled a distance about 30 times the diameter of the tube.

Reynolds also found that it was not only the velocity, but also the density and the viscosity of the fluid and the diameter of the tube which determined the flow regime. He proposed a dimensionless number, now known as the *Reynolds number* (**R**), which could be used to determine the flow regime. The Reynolds number is expressed as:

$$\mathbf{R} = \frac{Vl}{\nu}$$

where V = average velocity of flow; l = a characteristic length. In pipe flow, the characteristic length is the diameter of the pipe; ν = kinematic viscosity of the fluid.

After a series of experiments conducted by Reynolds and many others, it is believed that there is no single value of Reynolds number at which the flow changes from laminar to turbulent. There is a range of values of Reynolds numbers over which the flow will change from laminar to turbulent. Therefore, two types of Reynolds numbers are defined. The Reynolds *lower critical number* is the number below which the flow is always *laminar* and the Reynolds *upper critical number* is the number above which the flow is always *turbulent*. The Reynolds upper critical number depends on pipe roughness and changes from pipe to pipe and cannot be assigned a universal single value and, therefore, has no practical significance. The Reynolds lower critical number is independent of pipe roughness and can be assigned a single value and, therefore, is of great practical importance. It is generally agreed that below a Reynolds number of 2,000, the flow is always laminar. The flow may also be considered to be entering the turbulent zone above $\mathbf{R} = 2{,}000$.

Stanton and Pannel carried out a series of experiments on the flow of fluids in round pipes. The results were plotted with log **R** as base and $\tau/\rho V^2$ as ordinate, where τ is the viscous resistance per unit area of wetted surface. The graph obtained is shown in Fig. 7.3. The portion *AB* of the graph represents experimental results on laminar flow, the point *B* being the Reynolds lower critical number. The portion *BC* represents the transition from the laminar to the turbulent flow, the point *C* being the Reynolds upper critical number. The portion *CD* represents the experimental results on the turbulent flow. The Reynolds numbers at points *B* and *C* are found to be 2,000 and 2,500 respectively.

Although Reynolds used water in his original experiments, subsequent data with other fluids revealed that the two regimes of flow occur in all fluids.

The motion of fluid particles in laminar flow is very orderly and a rigorous mathematical analysis is possible without any need for much experimentation. Turbulent flow is very complex because the individual particles neither have a definite frequency nor an orderly pattern of movement. Because of this, an exact mathematical analysis of turbulent flow is impossible and experimental results are widely used in the analysis of turbulent flow problems.

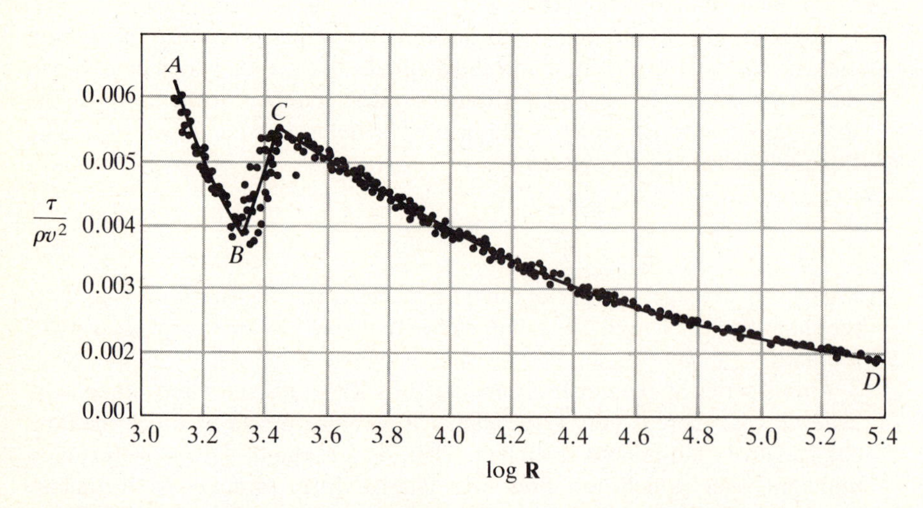

FIGURE 7.3 Results of experiments carried out by Stanton and Pannel.

EXAMPLE 7.1

Water at 30°C is flowing through a pipe 2 cm in diameter with a velocity of 30 cm/s. Is the flow laminar or turbulent? Also classify the flow if the fluid is glycerine.

Solution

$$\nu = 0.804 \times 10^{-6}\ \text{m}^2/\text{s} \qquad \text{(for water, from Table A-1)}$$

$$V = 30\ \text{cm/s} = 0.3\ \text{m/s} \qquad \text{(Given)}$$

$$D = 2\ \text{cm} = 0.02\ \text{m} \qquad \text{(Given)}$$

Hence:

$$\mathbf{R} = \frac{VD}{\nu}$$

$$= \frac{0.3 \times 0.02}{0.804 \times 10^{-6}}$$

$$= 7462$$

Since **R** is greater than 2,000, the flow is turbulent. For glycerine, from Table A-4

$$\mu = 8620 \times 10^{-4}\ \text{N} \cdot \text{s/m}^2$$

$$S = 1.26$$

For water at 30°C, from Table A-1

$$\rho = 995.7\ \text{kg/m}^3$$

Therefore; for glycerine

$$\nu = \frac{\mu}{\rho}$$

$$= \frac{8620 \times 10^{-4}}{1.26 \times 995.7}$$

$$= 6.87 \times 10^{-4}\ \text{m}^2/\text{s}$$

Then:

$$\mathbf{R} = \frac{VD}{\nu}$$

$$= \frac{0.3 \times 0.02}{6.87 \times 10^{-4}}$$

$$= 8.7$$

Since **R** is less than 2,000, the flow is laminar.

7.3 EDDY VISCOSITY: PRANDTL'S MIXING LENGTH THEORY

In a turbulent flow, the particles of fluid move erratically producing eddies and crosscurrents. It is difficult to follow the paths of individual particles; however, the behavior of the fluid can be examined by considering average conditions. We recall from Section 1.5 that the interchange of fluid molecules from one layer to another is, in part, responsible for the viscosity of a fluid. Similarly, transfer of fluid particles much larger than the molecules from one point to another in a turbulent flow produces shear stresses called *apparent shear stresses*. Hence, similar to Newton's law of viscosity, an expression for the apparent shear stresses in a turbulent flow may be written as:

$$\tau = \eta \frac{d\bar{u}}{dy} \tag{7.1}$$

where η is the eddy viscosity and $\bar{u}$ is the velocity at the point under consideration averaged over a considerable length of time. Eq. (7.1) is also known as the *Boussinesq equation* after the French mathematician J. Boussinesq (1842–1929) who first suggested it. There is an important difference between the viscosity of a fluid and the eddy viscosity. While viscosity is the property of a fluid, eddy viscosity is not. It depends on the degree of turbulence of the flow and the location of the point under consideration. The total shear stress in a turbulent flow may be derived from both the viscosity of the fluid and the eddy viscosity. Then:

$$\tau = (\mu + \eta) \frac{d\bar{u}}{dy} \tag{7.2}$$

Since η is not the property of the fluid and depends only on the degree of turbulence, it cannot be measured directly. Therefore, Eq. (7.2) has little practical use.

Reynolds, in 1886, developed an expression for apparent shear stresses by applying the momentum principle to the flow of fluid particles across a small area in a flow field. Consider a small area δA in a turbulent flow field as shown in Fig. 7.4. Let the average velocity in the s-direction be $\bar{u}$. Due to

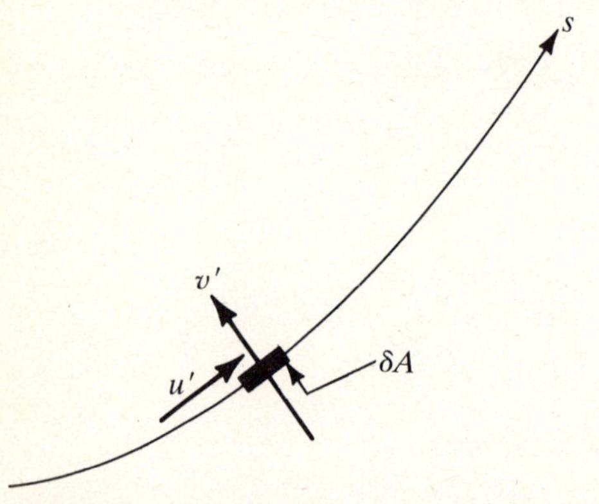

FIGURE 7.4 A streamline in a turbulent flow field.

turbulence, the instantaneous velocity u at a point may be assumed to consist of $\bar{u}$ and a fluctuating component. Let u' be the fluctuating component in the s-direction and v' be the fluctuating component normal to the s-direction. The shear stress δF acting on the fluid of area δA may now be determined by considering the momentum interchange across δA. The mass flux across δA is given by $\rho \delta A v'$. Then the change of momentum across δA is equal to the product of the mass flux and the velocity fluctuation u', across δA. This change of momentum is equal to the shear stress δF acting on the fluid element of area δA. Hence:

$$\delta F = (\rho \delta A v')u'$$

$$\frac{\delta F}{\delta A} = \rho v' u'$$

$$\tau = \rho v' u' \tag{7.3}$$

In the above expression the stress referred to is called *the Reynolds stress.*

In 1925, a German engineer, Ludwig Prandtl (1875–1953) presented a most useful theory of turbulence called *the mixing-length theory.* In the mixing-length theory, expressions for u' and v' are obtained in terms of mixing length distance l and the velocity gradient $d\bar{u}/dy$ in which $\bar{u}$ is as defined before and y is the distance normal to $\bar{u}$. The distance y is usually measured from the boundary. Prandtl assumed that the fluctuations u' and v' are related to l by

$$v' \simeq u' \simeq l \frac{d\bar{u}}{dy}$$

Substituting in Eq. (7.3) and letting l absorb the proportionality we obtain

$$\tau = \rho l^2 \left(\frac{d\bar{u}}{dy}\right)^2 \tag{7.4}$$

Comparison of Eqs. (7.1) and (7.4) yields an expression for the eddy viscosity, which is

$$\eta = \rho l^2 \frac{d\bar{u}}{dy} \tag{7.5}$$

At, or near, a boundary, the exchange of momentum is almost zero; hence, l approaches zero. At any distance y from the wall, l can be calculated as suggested by Von Kármán.

$$l = \kappa \frac{d\bar{u}/dy}{d^2\bar{u}/dy^2} \tag{7.6}$$

where $\kappa =$ the von Kármán constant, also called *the turbulence constant.* In a region close to the wall, the mixing length at a distance y from the wall, may be written as

$$l = \kappa y \tag{7.7}$$

Substituting for l in Eq. (7.4), an expression for shear stress near the wall may be written as:

$$\tau = \rho\kappa^2 y^2 \left(\frac{d\bar{u}}{dy}\right)^2 \tag{7.8}$$

EXAMPLE 7.2

A turbulent flow of water occurs in a pipe 1.75 m in diameter. The velocity profile is given by $u = 8.5 + 0.7 \ln y$ where y is the distance in meters from the wall. The shear stress at 30 cm from the wall is found to be 103 Pa. Determine eddy viscosity, mixing length and turbulence constant at this point.

Solution

$$u = 8.5 + 0.7 \ln y$$

$$\frac{du}{dy} = \frac{0.7}{y}$$

At $y = 0.3$ m, $\dfrac{du}{dy} = \dfrac{0.7}{0.3} = 2.333$

From Eq. (7.8)

$$\kappa^2 = \frac{\tau}{\rho y^2 (du/dy)^2}$$

$$= \frac{103}{1000(0.3)^2(2.333)^2} = 0.21$$

Hence:

$$\kappa = 0.46$$

$$\text{Mixing length} = l = \kappa y$$

$$= 0.46 \times 0.3 = 0.138 \text{ m}$$

or

$$l = 13.8 \text{ cm}$$

$$\text{Eddy viscosity} = \eta = \rho l^2 \left(\frac{du}{dy}\right)$$

$$= 1000 \times (0.138)^2 \times 2.333$$

$$= 44.4 \text{ N} \cdot \text{m/s}^2$$

7.4 THE BOUNDARY LAYER

In a flow regime, whether laminar or turbulent, the effects of viscosity are most pronounced at the point of contact between the fluid and the solid boundary. It is immaterial whether the fluid is moving and the solid is stationary or the solid is moving and the fluid is stationary. Due to the 'no slip' condition at the boundary, the fluid particles in contact with the solid do not move with respect to the boundary. These stationary particles cause the other fluid particles moving nearby to slow down due to molecular attraction. This causes the velocity of flow to increase gradually from zero at the wall to the main stream velocity at some distance away from the wall. The region where the velocity varies from zero to its full stream velocity is called the *boundary layer*. The development of a boundary layer can be visualized by examining Fig. 7.5. The figure shows wind of velocity U passing over a thin plate. When the column of wind reaches the edge of the plate it experiences

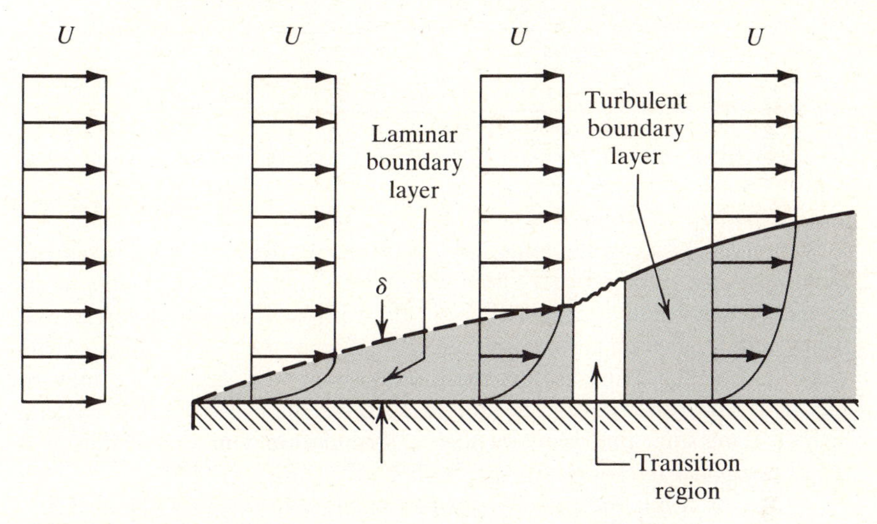

FIGURE 7.5 Development of a boundary layer over a flat plate.

shear resistance which causes the wind velocity to become zero at points where it comes into contact with the plate. At points away from the surface, the velocity increases and ultimately reaches the main stream velocity U. The region in which the velocity varies from zero to U is the boundary layer, its thickness being denoted by δ. The dotted lines in Fig. 7.5 define the outer limits of the boundary layer. As the wind moves along the plate, more and more of it slows down and the thickness of the boundary layer increases. At the boundary, the shear stress is the greatest and the velocity gradient is the steepest. The velocity gradient becomes less steep as one proceeds in the direction of flow.

At first, the boundary layer is laminar. As it grows in thickness, a stage is reached when the flow becomes unstable, develops turbulence and the thickness of the layer increases more rapidly. These changes occur over a relatively short distance called the *transition region*. Downstream of the transition region, the flow is completely turbulent and the boundary layer is called the *turbulent boundary layer*. In the turbulent boundary layer, eddies mix higher velocity fluid into the region close to the wall so that the velocity gradient at the wall now becomes greater than that in the laminar boundary layer just upstream of the transition point.

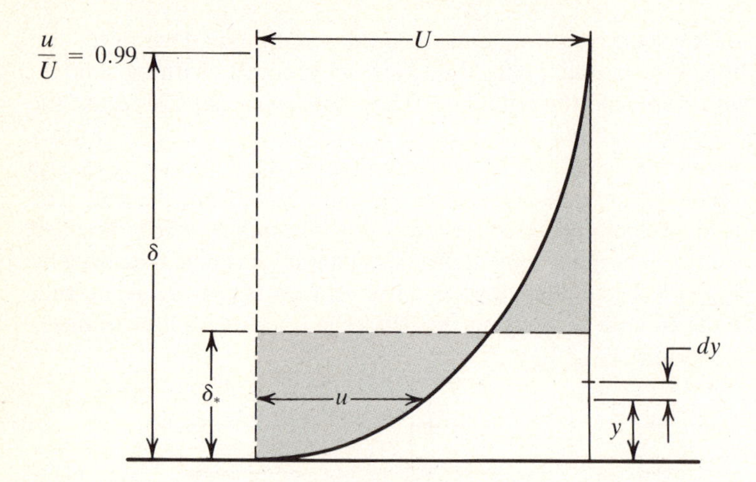

FIGURE 7.6 Definitions of boundary layer thickness and displacement thickness.

Since the velocity in the boundary layer reaches the main stream velocity asymptotically, some arbitrary limit should be assigned to the velocity in the boundary layer to define its thickness. For example, it may be assumed that the boundary layer ends when the velocity becomes 99% of the main stream velocity (Fig. 7.6). For more precise work, a greater percentage may be specified. The thickness of the boundary layer may also be defined as a 'length' using some precise mathematical formulation. One such definition is the *displacement thickness*, δ_*.

In Fig. 7.6, if u is the velocity through an elemental area dy located at a distance y from the plate, the rate of flow through the elemental area is given by udy. However, if there had been no boundary layer, the flow through the elemental area would have been Udy. This amounts to a reduction in flow rate of $(U - u)dy$. The total reduction in flow rate through the entire boundary layer would then be $\int_0^\delta (U - u)dy$. If we draw a line at a distance δ_* such that the shaded areas in Fig. 7.6 are equal, then

$$\delta_* U = \int_0^\delta (U - u)dy$$

$$\delta_* = \frac{1}{U}\int_0^\delta (U - u)dy$$

$$\delta_* = \int_0^\delta \left(1 - \frac{u}{U}\right)dy \tag{7.9}$$

The Momentum Equation Applied to the Boundary Layer

The flow in the boundary layer is highly complex and exact mathematical solutions are difficult to obtain. Von Kármán, however, obtained approximate but useful results by applying the momentum equation to the boundary

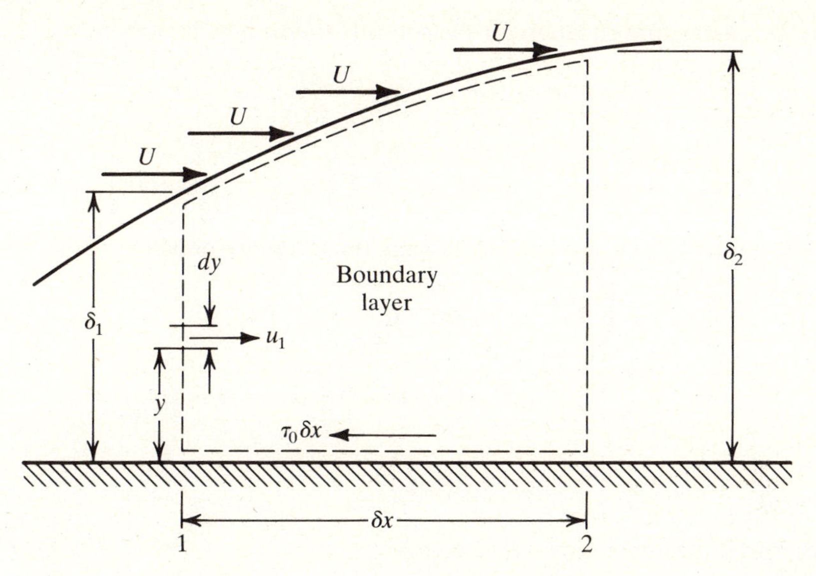

FIGURE 7.7 Control volume for applying momentum equation to the boundary layer.

layer. Consider a boundary layer over a flat plate as shown in Fig. 7.7. Consider two sections 1 and 2, a distance δx apart. Erect a control volume taking the plate, sections 1 and 2, and the top of the boundary layer as control surfaces. Let the shear stress at the plate be τ_0. Assuming that the pressure gradient along the plate is zero, the only force acting on the control volume is the shear force equal to $\tau_0 \delta x$. Consider an elementary strip of thickness dy at a distance y from the plate along section 1. The momentum rate through the strip is given by $\rho u_1^2 dy$. Then, the rate at which the momentum is entering the control volume through section 1 is given by

$$\dot{M}_{IX} = \int_0^{\delta_1} \rho u_1^2 dy$$

Similarly, the momentum leaving the control volume through section 2 is given by

$$\dot{M}_{OX} = \int_0^{\delta_2} \rho u_2^2 dy$$

Any change in momentum within the control volume is caused by the fluid entering from the top. This may be expressed as the product of U and the difference in the mass flow rates past sections 1 and 2. This can be expressed as:

$$\frac{dM}{dt} = U\left[\int_0^{\delta_2} \rho u_2 dy - \int_0^{\delta_1} \rho u_1 dy\right]$$

Now, the momentum equation for the control volume may be written as

$$\sum F_x = \dot{M}_{OX} - \dot{M}_{IX} - \frac{dM}{dt}$$

$$-\tau_0 \delta x = \int_0^{\delta_2} \rho u_2^2 dy - \int_0^{\delta_1} \rho u_1^2 dy - U\left[\int_0^{\delta_2} \rho u_2 dy - \int_0^{\delta_1} \rho u_1 dy\right]$$

Assuming ρ to be constant and rearranging the terms, we obtain

$$-\tau_0 \delta x = \rho \int_0^{\delta_2} (u_2^2 - Uu_2)dy - \rho \int_0^{\delta_1} (u_1^2 - Uu_1)dy$$

$$-\tau_0 = \frac{1}{\delta x}\left[\rho \int_0^{\delta_2} (u_2^2 - Uu_2)dy - \rho \int_0^{\delta_1} (u_1^2 - Uu_1)dy\right]$$

$$-\tau_0 = \frac{\delta}{\delta x}\left[\rho \int_0^{\delta} (u^2 - Uu)dy\right]$$

In the limits, the above equation becomes

$$-\tau_0 = \rho \frac{d}{dx} \int_0^{\delta} (u^2 - Uu)dy \tag{7.10}$$

$$\tau_0 = \rho U^2 \frac{d}{dx} \int_0^{\delta} \frac{u}{U}\left(1 - \frac{u}{U}\right)dy \tag{7.11}$$

Equations (7.10) and (7.11) are two forms of the momentum integral equation of the boundary layer over a flat plate with zero pressure gradient and constant density. It is applicable to laminar, turbulent or transition flow in the boundary layer.

The Laminar Boundary Layer

In laminar flow, the shear stress varies with the gradient of the velocity, or $\tau = \mu du/dy$, where y is measured perpendicular to u. In a laminar boundary layer, if τ_0 is the stress at the plate and y is measured from the plate into the boundary layer, then

$$\tau_0 = \mu \left(\frac{du}{dy}\right)_{y=0}$$

Substituting in Eq. (7.11), we obtain

$$\mu\left(\frac{du}{dy}\right)_{y=0} = \rho U^2 \frac{d}{dx} \int_0^{\delta} \frac{u}{U}\left(1 - \frac{u}{U}\right)dy \tag{7.12}$$

In a boundary layer u varies with y. When $y = 0, u = 0$; when $y = \delta, u = U$. Let us assume that the velocity distribution has the same form at each value of x along the plate—i.e.,

$$\frac{u}{U} = f\left(\frac{y}{\delta}\right) = f(\eta) \text{ where } \eta = \frac{y}{\delta}$$

Substituting $y = \eta\delta$ and $u/U = f(\eta)$ in Eq. (7.12), we obtain

$$\frac{\mu}{\delta} U \left[\frac{df(\eta)}{d\eta}\right]_{\eta=0} = \rho U^2 \frac{d}{dx}\left[\delta \int_0^1 \{1 - f(\eta)\} f(\eta) d\eta\right]$$

Since $f(\eta)$ is independent of x, the expressions $\int_0^1 \{1 - f(\eta)\} f(\eta) d\eta$ and $[df(\eta)/d\eta]_{\eta=0}$ are constants. Let the former be designated by A and the latter by B. Then

$$\frac{\mu}{\delta} UB = \rho U^2 \frac{d}{dx} \delta A$$

$$= \rho U^2 A \frac{d\delta}{dx}$$

$$\mu B = \rho U A \delta \frac{d\delta}{dx}$$

Integrating with respect to x, we obtain

$$\mu B x = \frac{\rho U A \delta^2}{2} + \mathrm{C}$$

where C is a constant of integration. The value of C can be obtained by using the boundary condition that when $x = 0$, $\delta = 0$. This yields $\mathrm{C} = 0$.
Hence:

$$\mu B x = \frac{\rho U A \delta^2}{2}$$

$$\delta^2 = \left(\frac{2B}{A}\right)\left(\frac{\mu x}{\rho U}\right)$$

$$\delta^2 = \left(\frac{2B}{A}\right)\left[\frac{x^2}{(\rho U x/\mu)}\right] \tag{7.13}$$

$$\delta = \sqrt{\frac{2B}{A}}\left[\frac{x}{\sqrt{\mathbf{R}_x}}\right] \tag{7.14}$$

$$\delta = \mathrm{K}_1\left[\frac{x}{\sqrt{\mathbf{R}_x}}\right]$$

where $\mathbf{R}_x = \rho U x/\mu$ is the Reynolds number based on the free stream velocity and the distance x along the plate and $\mathrm{K}_1 = \sqrt{(2B/A)}$ is a constant. If the values of the constants A and B are known, the thickness of the boundary layer at any distance x along the plate can be calculated using Eq. (7.14). The values of the constants A and B depend on the form of the function $f(\eta)$. For the laminar boundary layer, Prandtl assumed that $f(\eta) = \frac{3}{2}\eta - \eta^3/2$ for $0 \leqslant y \leqslant \delta$. Then:

$$A = \int_0^1 \left[1 - \left(\frac{3}{2}\eta - \frac{\eta^3}{2}\right)\right]\left(\frac{3}{2}\eta - \frac{\eta^3}{2}\right) d\eta$$

$$B = \frac{d}{d\eta}\left(\frac{3}{2}\eta - \frac{\eta^3}{2}\right)_{\eta=0}$$

Solving, we obtain $A = 117/840$ and $B = 3/2$ and $K_1 = 4.64$. Substituting in Eq. (7.14), we obtain

$$\delta = 4.64 \frac{x}{\sqrt{\mathbf{R}_x}} \tag{7.15}$$

The above equation gives the thickness of the laminar boundary layer as a function of x. Now, let us develop an expression for the shear stress on one side of the plate. From Eq. (7.11),

$$\tau_0 = \rho U^2 \frac{d}{dx} \int_0^\delta \frac{u}{U}\left(1 - \frac{u}{U}\right) dy$$

Substituting $y = \eta\delta$ and $u/U = f(\eta)$, and simplifying, we obtain

$$\tau_0 = \rho U^2 A \frac{d\delta}{dx} \tag{7.16}$$

From Eq. (7.13)

$$\frac{d\delta}{dx} = \tfrac{1}{2}\sqrt{(2\mu B/\rho U A x)}$$

or

$$\frac{d\delta}{dx} = \sqrt{\frac{B}{2A\mathbf{R}_x}}$$

Substituting in Eq. (7.16), we obtain

$$\tau_0 = \rho U^2 A \sqrt{\frac{B}{2A\mathbf{R}_x}}$$

or

$$\tau_0 = \rho U^2 \sqrt{\frac{AB}{2\mathbf{R}_x}} = \sqrt{\frac{AB}{2}} \frac{\rho U^2}{\sqrt{\mathbf{R}_x}} = K_2 \frac{\rho U^2}{\sqrt{\mathbf{R}_x}}$$

Substituting $A = 117/840$ and $B = 3/2$, we obtain $K_2 = 0.323$.
Hence:

$$\tau_0 = 0.323\, \rho U^2 \frac{1}{\sqrt{\mathbf{R}_x}} \tag{7.17}$$

$$\tau_0 = 0.323 \sqrt{\frac{\mu \rho U^3}{x}} \tag{7.18}$$

Table 7.1 VALUES OF K_1 AND K_2 FOR VARIOUS VALUES OF FUNCTION $F(\eta)$

$\frac{u}{U}$	$K_1 = \sqrt{\frac{2B}{A}}$	$K_2 = \sqrt{\frac{AB}{2}}$
$2\eta - \eta^2$	5.48	0.365
$\frac{3}{2}\eta - \frac{1}{2}\eta^3$	4.64	0.323
$2\eta - 2\eta^3 + \eta^4$	5.84	0.342
$\text{Sin}\left(\frac{\pi}{2}\eta\right)$	4.8	0.327
Blasius's Solution	5.0	0.332

The total shear force per unit width on one side of the plate of length L, called *drag*, will be given by

$$\text{Drag} = \int_0^L \tau_0 dx = 0.646\sqrt{\mu\rho U^3 L} \tag{7.19}$$

Other assumptions about $f(\eta)$ may be made, but it is found that it does not change the values of A and B radically (Table 7.1). An analytical solution obtained by a German engineer, P.R.H. Blasius, (1883–1970) in 1908, for a laminar boundary layer on a flat plate with zero pressure gradient is shown in Fig. 7.8. He obtained $K_1 = 5.0$ and $K_2 = 0.332$.

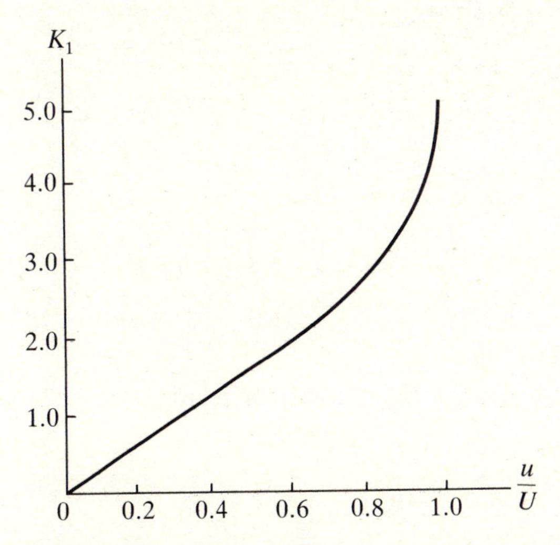

FIGURE 7.8 Laminar boundary layer on a flat plate.

Drag can be expressed in terms of the dynamic pressure of the free stream, $\rho U^2/2$, and a skin friction coefficient C_f. Then:

$$\text{Drag} = C_f \frac{\rho U^2 L}{2}$$

where $C_f = 1.292/\sqrt{\mathbf{R}_L}$. Using $K_2 = 0.332$ from Blasius

$$C_f = 1.328/\sqrt{\mathbf{R}_L}$$

Experiments have shown that the boundary layer remains laminar up to a Reynolds number of 5×10^5. Beyond this it becomes turbulent.

EXAMPLE 7.3

Glycerine at 68°F is flowing past a thin flat plate, 1 m wide and 2 m long, with a velocity of 1 m/s. Determine and plot the boundary layer thickness and the shear stress distribution along the plate. Also find the drag on the plate. Use Blasius's solution.

Solution From Table A-5, for glycerine

$$S = 1.26; \mu = 0.862 \text{ N} \cdot \text{s/m}^2$$

Then $\rho = 1.26 \times 1000 = 1260 \text{ kg/m}^3$

$$\mathbf{R}_x = \frac{Ux\rho}{\mu} = \frac{(1)(x)(1260)}{0.862} = 1462x$$

Using Blasius's solution:

$$\delta(x) = \frac{5x}{\sqrt{\mathbf{R}_x}}$$

$$= \frac{5x}{\sqrt{1462x}} = 0.1307\sqrt{x}$$

Now δ can be calculated for various values of x

x(m)	0	0.1	0.5	1.0	1.5	2.0
δ(cm)	0	4.13	9.24	13.07	16.01	18.48

The shear stress as a function of x is calculated using the relation

$$\tau_0 = K_2 \rho \frac{U^2}{\sqrt{\mathbf{R}_x}}$$

where, from Blasius, $K_2 = 0.332$.

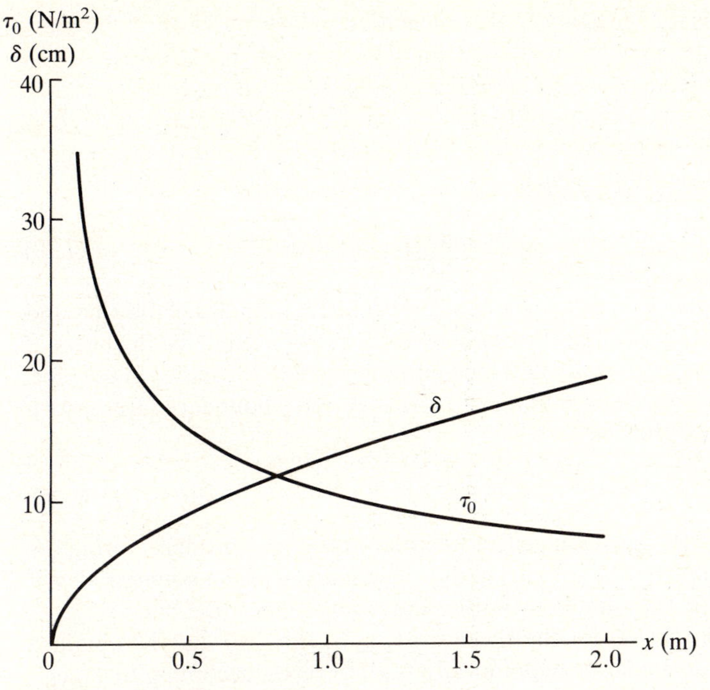

FIGURE 7.9 Example 7.3.

Hence:

$$\tau_0 = \frac{0.332(1260)(1)^2}{\sqrt{1462x}}$$

$$= \frac{10.94}{\sqrt{x}} \text{ N/m}^2$$

Now τ_0 can be calculated for various values of x

x(m)	0	0.1	0.5	1.0	2.0
τ_0(N/m²)	∞	34.60	15.47	10.94	7.74

The drag on the plate is calculated using the relation

$$\text{drag} = 0.664\sqrt{\mu\rho U^3 L} = 0.664\sqrt{0.862 \times 1260 \times 1 \times 2} = 30.95\text{N}$$

The thickness of the boundary layer and the shear stress as a function of distance are given in Fig. 7.9.

Turbulent Boundary Layer

Most of the boundary layers encountered in practice are turbulent. The analysis of the turbulent boundary layer depends heavily on experimental data. The basis for the analysis of turbulent boundary layer over a flat plate comes from the experimental data on flow through pipes. Most of the pioneering work was done by Prandtl who suggested that the boundary layer over a flat plate was similar to that in a pipe. The velocity distribution for a fully developed turbulent flow in a pipe is given by Prandtl's one-seventh power law, according to which

$$\frac{u}{U} = \left(\frac{y}{R}\right)^{1/7} \tag{7.20}$$

where u = velocity at a distance y from the center; R = radius of the pipe, and U = maximum velocity, at the center of the pipe. In a pipe, the thickness of the boundary layer in a fully developed flow is equal to the radius of the pipe. By analogy, the above equation can be written for a boundary layer over a flat plate in the form

$$\frac{u}{U} = \left(\frac{y}{\delta}\right)^{1/7} \tag{7.21}$$

where u = velocity in the turbulent boundary layer at a distance y from the plate; U = main stream velocity, and δ the thickness of the boundary layer. This equation is valid for $Ux/\nu < 10^7$, and is not valid at the boundary.

Again an expression for the shear stress τ_0 at the wall may be obtained from Blasius's formula for hydraulically smooth pipes, according to which

$$\tau_0 = 0.0225\rho U^2\left(\frac{\nu}{UR}\right)^{1/4} \tag{7.22}$$

By analogy, an expression for τ_0 for a flat plate may be obtained by substituting δ for R. This yields

$$\tau_0 = 0.0225\rho U^2\left(\frac{\nu}{U\delta}\right)^{1/4} \tag{7.23}$$

The thickness of the turbulent boundary layer may now be determined using the momentum equation derived earlier and the expressions for u/U and τ_0 developed here. Substituting Eqs. (7.21) and (7.23) in Eq. (7.11), we obtain

$$0.0225\rho U^2\left(\frac{\nu}{U\delta}\right)^{1/4} = \rho U^2 \frac{d}{dx}\int_0^\delta \left(\frac{y}{\delta}\right)^{1/7}\left[1 - \left(\frac{y}{\delta}\right)^{1/7}\right]dy$$

$$= \rho U^2 \frac{d}{dx}\left[\frac{7}{72}\delta\right]$$

$$= \frac{7}{72}\rho U^2 \frac{d\delta}{dx}$$

$$\left(\frac{\nu}{U\delta}\right)^{1/4} = 4.321\frac{d\delta}{dx}$$

$$\delta^{1/4}d\delta = 0.231\left(\frac{\nu}{U}\right)^{1/4} dx$$

Integrating, we obtain

$$\delta^{5/4} = 0.289\left(\frac{\nu}{U}\right)^{1/4} x + C \tag{7.24}$$

where C is a constant of integration. The value of C can be obtained if the thickness of the boundary layer at a distance x is known. This poses a problem because the turbulent boundary layer begins after the transition from the laminar boundary layer and its initial thickness is not known. However, we can assume that the turbulent boundary layer begins when $x = 0$. Prandtl showed that such an assumption does not change the results radically. Using this assumption, the constant of integration becomes zero. Then from Equation (7.24), we obtain

$$\frac{\delta}{x} = 0.37\left(\frac{\nu}{Ux}\right)^{1/5}$$

$$\delta = \frac{0.37x}{(\mathbf{R}_x)^{1/5}} \tag{7.25}$$

This equation gives the thickness of the turbulent boundary layer as a function of x. The shear stress at the plate can be obtained by substituting Eq. (7.25) in Eq. (7.23) which yields

$$\tau_0 = 0.0225\rho U^2\left(\frac{\nu}{U}\right)^{1/4}\left[\frac{1}{0.37}\left(\frac{Ux}{\nu}\right)^{1/20} x^{-1/4}\right]$$

The drag F on one side of the plate of length L is given by:

$$F = \int_0^L \tau_0 \, dx$$

$$= \int_0^L \left\{0.0225\,\rho U^2\left(\frac{\nu}{U}\right)^{1/4}\left[\frac{1}{0.37}\left(\frac{Ux}{\nu}\right)^{1/20} x^{-1/4}\right]\right\} dx$$

$$= 0.036\rho U^2\left(\frac{\nu}{UL}\right)^{1/5} L$$

or

$$F = 0.036\,\frac{\rho U^2 L}{(\mathbf{R}_L)^{1/5}} \tag{7.26}$$

where $\mathbf{R}_L$ is the Reynolds number based on the total length of the plate.

Equation (7.26) may be written in terms of a skin friction coefficient C_f and the stagnation pressure $\frac{1}{2}\rho U^2$. Then:

$$F = \frac{1}{2} C_f \rho U^2 L$$

where

$$C_f = \frac{0.072}{(\mathbf{R}_L)^{1/5}} \tag{7.27}$$

Experimental data show that the above equation gives values of C_f slightly lower than the observed values. The experimental data gives

$$C_f = \frac{0.074}{(\mathbf{R}_L)^{1/5}} \tag{7.28}$$

It may also be recalled that Eqs. (7.25) and (7.27) are based on Blasius's formula for hydraulically smooth pipes which is valid when $5 \times 10^5 < \mathbf{R} < 10^7$. Therefore, these equations are also valid within this range of the Reynolds number. For Reynolds numbers between 10^7 and 10^9 an empirical relation devised by H. Schlichting gives results close to the experimental data. This relationship is given by

$$C_f = \frac{0.455}{[\log_{10}(\mathbf{R}_L)]^{2.58}} \tag{7.29}$$

To summarize:

(i) For laminar drag: $C_f = 1.328(\mathbf{R}_L)^{1/2}$

(ii) For turbulent drag when $5 \times 10^5 \leqslant \mathbf{R}_L \leqslant 10^7$:

$$C_f = \frac{0.074}{(\mathbf{R}_L)^{1/5}}$$

(iii) For turbulent drag when $\mathbf{R}_L > 10^7$:

$$C_f = \frac{0.455}{[\log_{10} \mathbf{R}_L]^{2.58}}$$

EXAMPLE 7.4

Air of density 1.197 kg/m^3 and viscosity 18.22×10^{-6} N·s/m^2 blows over a flat plate 5 m long at a speed of 8 m/s. Sketch the boundary layer and determine the drag.

Solution The boundary layer starts as a laminar boundary layer at the leading edge and then transforms into a turbulent boundary layer at the critical Reynolds number of 5×10^5.

The thickness of the laminar boundary can be determined using the relation

$$\delta = \frac{5.0x}{\sqrt{\mathbf{R}_x}}$$

$$\mathbf{R}_x = \frac{\rho U x}{\mu} = \frac{1.197 \times 8x}{18.22 \times 10^{-6}} = 525576x$$

Hence:

$$\delta = \frac{5.0x}{\sqrt{525576x}} = 0.0069\sqrt{x}$$

Now, let us determine the distance x corresponding to the critical Reynolds number of 5×10^5.

$$525576x = 5 \times 10^5$$

$$x = 0.951 \text{ m}$$

x (m)	0	0.1	0.3	0.6	0.951
δ (cm)	0	0.218	0.378	0.534	0.673

For a turbulent boundary layer (from Eq. 7.25)

$$\delta = \frac{0.37x}{(\mathbf{R}_x)^{1/5}}$$

$$\delta = \frac{0.37x}{(525576x)^{1/5}} = 0.0265x^{4/5} \text{ m for } x \geqslant 0.951 \text{ m}$$

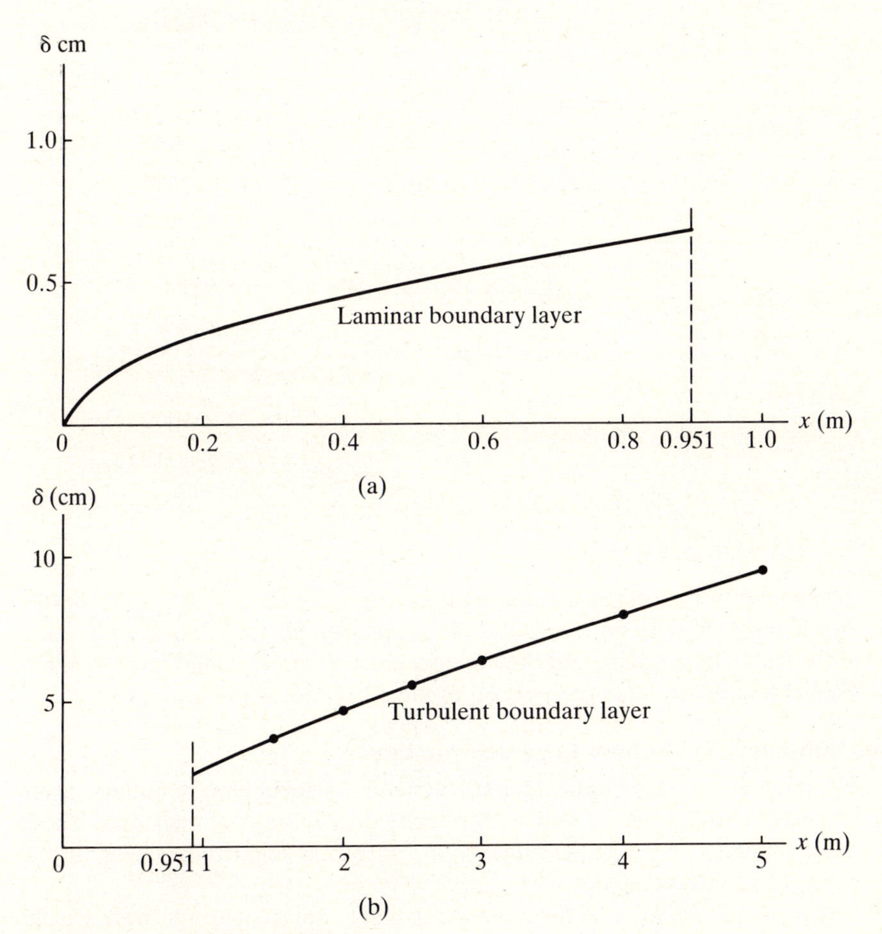

FIGURE 7.10 Example 7.4.

Now, δ for the turbulent boundary layer can be calculated for $0.951 \leqslant x \leqslant 5.0$.

x (m)	0.951	1.5	2.0	2.5	3.0	4.0	5.0
δ (cm)	2.54	3.66	4.61	5.51	6.38	8.03	9.60

The drag on the plate can be determined by summing up the laminar drag and the turbulent drag. However, the turbulent drag equation was obtained assuming that the turbulent boundary layer began at the leading edge of the plate. The effect of this assumption can be discounted by subtracting turbulent drag for the portion of the plate beginning from 0 to the point where the turbulent boundary layer begins. For this problem:

$$\begin{aligned}\text{Total Drag} &= \text{laminar drag from } x = 0 \text{ to } x = 0.951 \text{ m}\\ &+ \text{turbulent drag from } x = 0 \text{ to } x = 5.0 \text{ m}\\ &- \text{turbulent drag from } x = 0 \text{ to } x = 0.951 \text{ m}\end{aligned}$$

or

$$\begin{aligned}\text{Total Drag} &= 0.664\sqrt{\mu\rho U^3 L} + 0.036\rho U^2 L\left(\frac{\nu}{UL}\right)^{1/5}\\ &- 0.036\rho U^2 L\left(\frac{\nu}{UL}\right)^{1/5}\\ &= 0.664\sqrt{18.22 \times 10^{-6} \times 1.197 \times 8^3 \times 0.951}\\ &+ 0.036 \times 1.197 \times 8^2 \times 5\left(\frac{18.22 \times 10^{-6}}{1.197 \times 8 \times 5}\right)^{1/5}\\ &- 0.036 \times 1.197 \times 8^2 \times 0.951\left(\frac{18.22 \times 10^{-6}}{1.197 \times 8 \times 0.951}\right)^{1/5}\\ &= 0.0684 + 0.717 - 0.190\\ \text{Total Drag} &= 0.595 \text{ N}\end{aligned}$$

The equation for turbulent drag used here is valid for $5 \times 10^5 \leqslant \mathbf{R}_L \leqslant 10^7$. For $\mathbf{R} = 10^7$, $x = 19$ m. Hence the use of equation for turbulent drag is valid. If the length of the plate had been greater than 19 m, we would have used the Schlichting relation for the portion of plate beyond 19 m.

Viscous Sublayer, Turbulent-Logarithmic Layer

Experimental results indicate that, actually, a turbulent boundary layer consists of three zones of flow with velocity distributions of their own. These three zones are: 1. a viscous sublayer, 2. a turbulent-logarithmic layer, and 3. a turbulent-velocity layer.

The *viscous sublayer* is the zone immediately adjacent to the wall, which, because of the damping effect of the wall, remains relatively smooth even though most of the flow in the boundary layer is turbulent. Newton's law of viscosity is applicable and the shear stress is approximately constant and equal to the shear stress at the wall. Let shear stress at the wall = τ_0. Then

$$\tau_0 = \mu \frac{du}{dy}$$

Integrating, we obtain

$$u = \frac{\tau_0 y}{\mu} \tag{7.30}$$

Dividing and multiplying the right-hand side by ρ, we obtain

$$u = \frac{\tau_0/\rho}{\mu/\rho} y$$

$$u = \frac{\sqrt{\tau_0/\rho} \cdot \sqrt{\tau_0/\rho}}{\nu} y \tag{7.31}$$

The term $\sqrt{\tau_0/\rho}$ has the dimension of velocity and is called *shear velocity*. Let shear velocity = u_*.

Then:

$$u_* = \sqrt{\tau_0/\rho} \tag{7.32}$$

Also, Eq. (7.31) becomes

$$u = \frac{u_*^2 y}{\nu}$$

or

$$\frac{u}{u_*} = \frac{y u_*}{\nu} \tag{7.33}$$

The term yu_*/ν is dimensionless and is called *shear number*, denoted by S. It is a measure of distance from the wall. It is found experimentally, that for the viscous sublayer, the shear number is approximately 5. Denoting the thickness of the sublayer by δ', we obtain

$$\delta' = \frac{5\nu}{u_*} \tag{7.34}$$

Next to the viscous sublayer is the *turbulent-logarithmic layer*. In this layer, the viscous forces become negligible. Instead, eddies and cross-currents found in the turbulent flow produce a shear effect on the fluid. These stresses in the turbulent flow are described in Section 7.3. From Eq. (7.8)

$$\tau = \rho \kappa^2 y^2 \left(\frac{du}{dy}\right)^2 \tag{7.35}$$

We will now assume that in the turbulent-log layer the shear stress is uniform and equal to the shear stress at the wall, τ_0.
Thus Eq. (7.35) becomes

$$\tau_0 = \rho\kappa^2 y^2 \left(\frac{du}{dy}\right)^2 \tag{7.36}$$

or

$$\frac{du}{dy} = \frac{1}{\kappa y}\sqrt{\frac{\tau_0}{\rho}} = \frac{u_*}{\kappa y}$$

Integrating with respect to y, we obtain

$$\frac{u}{u_*} = \frac{1}{\kappa}\ln(y) + C \tag{7.37}$$

The constant of integration C is experimentally found to be:

$$C = 5.56 - \frac{1}{\kappa}\ln\frac{\nu}{u_*} \tag{7.38}$$

Substituting in Eq. (7.37), and using $\kappa = 0.4$, we obtain

$$\frac{u}{u_*} = 2.50 \ln\frac{yu_*}{\nu} + 5.56 \tag{7.39}$$

$$\frac{u}{u_*} = 2.50 \ln(\mathbf{S}) + 5.56 \tag{7.40}$$

Eq. (7.40) when converted to $\log_{10}$ becomes:

$$\frac{u}{u_*} = 5.75 \log(\mathbf{S}) + 5.56 \tag{7.41}$$

The logarithmic velocity distribution given by Eq. (7.41) is valid for values of the shear number from about 30 to 500. To summarize, the turbulent boundary layer consists of a viscous sublayer which extends to a shear number of 5, and a turbulent-logarithmic layer which extends from a shear number of 30 to 500. From shear number 5 to 30, there is a buffer zone. Neither of the two velocity distributions apply in the buffer zone. The relationship between u/u_* and **S** up to shear number of 500 is known as the *law of the wall.* The actual thickness of these two zones is approximately 15% of the total thickness of the turbulent boundary layer. The third zone, which covers 85% of the turbulent boundary layer is the *turbulent-velocity layer.* The velocity distribution in this layer was developed earlier in the form of Eq. (7.20) to Eq. (7.29).

7.5 SEPARATION AND WAKE

The flat plate considered in Section 7.4 had zero pressure gradient along its length. The behavior of the boundary becomes different if there is a pressure gradient. Consider a curved surface placed in a stream of fluid of velocity U as shown in Fig. 7.11a. As the fluid reaches the curved surface, it is deflected. This narrows the area of flow causing the fluid to accelerate. The increase in velocity results in a corresponding decrease in pressure until it becomes minimum at the crest. It then rises again as the fluid decelerates as shown in Fig. 7.11b.

At points 1 and 2 the velocity gradient du/dy is positive. Also, since the pressure is decreasing in the direction of flow, it has a favorable effect on the boundary layer. Hence the growth of boundary layer is slower as compared with the flat plate with zero pressure gradient. This continues until point (3) is reached. Beyond point 3, the pressure increases along the direction of motion and, therefore, opposes the forward flow. This soon brings the slow moving fluid particles near the surface to a standstill. The velocity gradient du/dy, therefore, at some point 4 becomes zero. At a point downstream of point 4, (say point 5), the velocity gradient is actually reversed. This means that fluid is no longer able to follow the contour of the surface and separates from it. This phenomenon by which the fluid is separated from the surface before its end is reached is called *separation.* Separation actually starts at a point where the velocity gradient du/dy at the surface becomes zero. Separation can occur only if a positive pressure gradient along the surface is present.

A line joining the points of zero velocity along the direction of flow is known as the *separation streamline.* The reverse flow near the surface, after

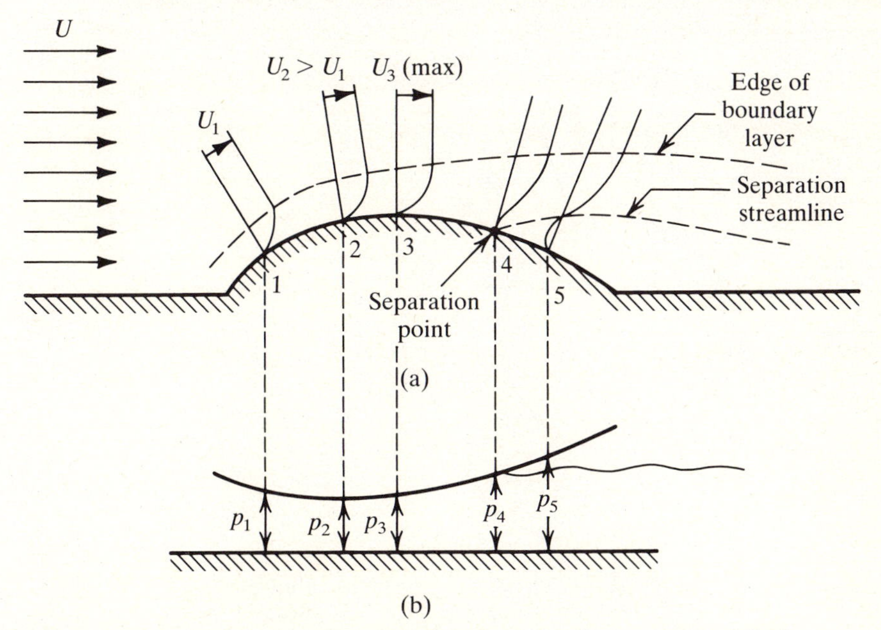

FIGURE 7.11 Separation over a curved surface placed in a stream of fluid.

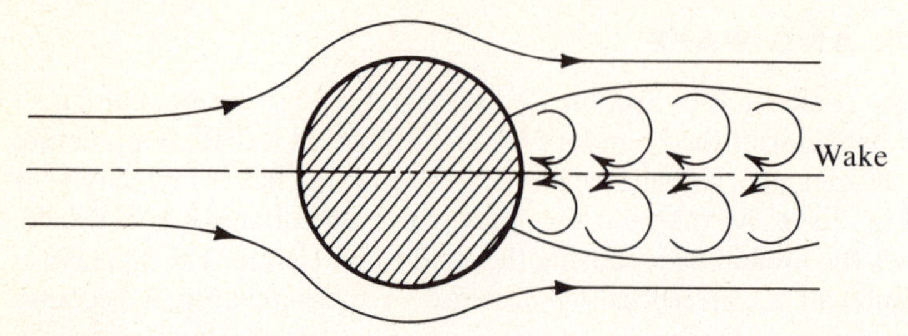

FIGURE 7.12 Formation of wake behind a sphere placed in a stream of fluid.

separation has occurred produces eddies, resulting in loss of energy. This causes the pressure to remain constant after separation and not let it rise which would have been the case in the absence of separation. A region downstream of the separation where lots of eddies are present and where pressure is constant is called a *wake* (Fig. 7.12).

Separation and formation of a wake can cause radical changes in the flow pattern by altering the effective boundary of flow. The solid surface no longer remains the effective boundary of flow and this may move the point of location of the minimum pressure and the point of separation upstream from where the pressure was originally a minimum.

Separation and wake occur with both laminar and turbulent boundary layers. Laminar layers, due to their slow velocities, are more prone to separation than turbulent layers.

A polar plot of pressure distribution on the surface of a cylinder immersed in the uniform flow of a fluid is shown in Fig. 7.13. The pressure distribution

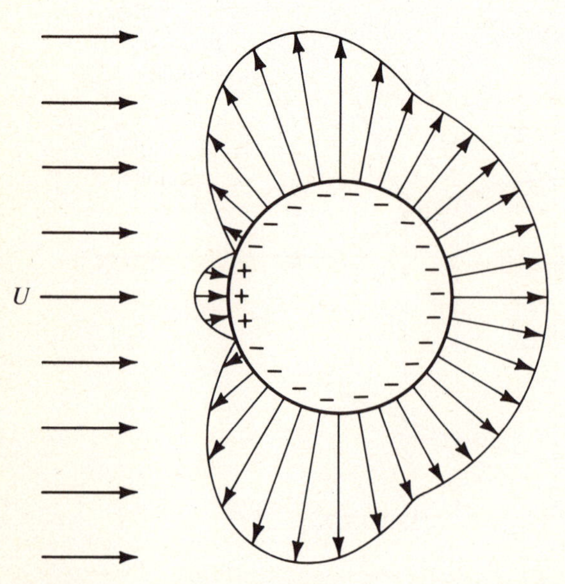

FIGURE 7.13 Pressure variation on the surface of a cylinder placed in a stream of fluid.

shows that the pressure at the front half is greater than the pressure at the rear half of the cylinder. This results in net force acting on the cylinder in the direction of flow. The drag resulting from this force is called the *pressure drag* or the *form drag*. Pressure drag combined with the skin friction drag gives the *total drag* on the cylinder.

7.6 DRAG ON IMMERSED BODIES

The *total drag* on a body immersed in a fluid is defined as that component of the total force exerted by the moving fluid on the body which is parallel to the approach velocity. It is immaterial whether the fluid is in motion and the body is at rest or the body is in motion and the fluid is at rest. The drag force is the same in both cases. It is customary to express the total drag in terms of a drag coefficient, the stagnation pressure $\rho U^2/2$ and the area of the body. Then

$$F_D = \tfrac{1}{2} C_D \rho U^2 A \tag{7.42}$$

where F_D = total drag; C_D = drag coefficient; U = approach velocity of the fluid; A = projected area of the body on a plane normal to approach velocity.

Separation has an important effect on the magnitude of the form drag. In the case of a sphere immersed in a flowing fluid, it is noticed that the point of separation is closer to the stagnation point in the case separation occurs in the laminar boundary layer when compared with the separation point in the case where the boundary layer becomes turbulent first and then separation occurs. It is clearly demonstrated by the photographs of Fig. 7.14. A plot of

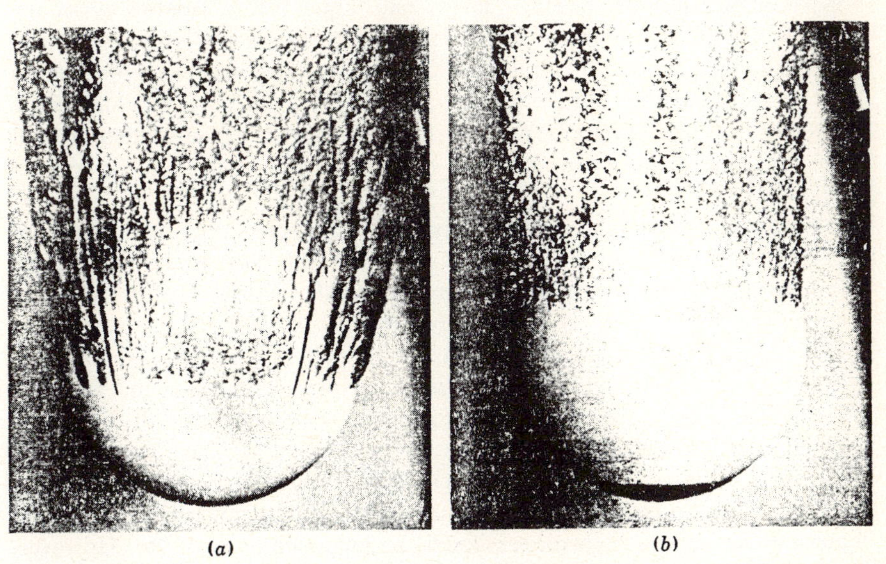

(a) (b)

FIGURE 7.14 Shift in separation point due to induced turbulence: (a) 8.5 in bowling ball, smooth surface, 25 ft/s entry velocity into water; (b) same except for 4-in diameter patch of sand on nose. (U.S. Navy Photograph).

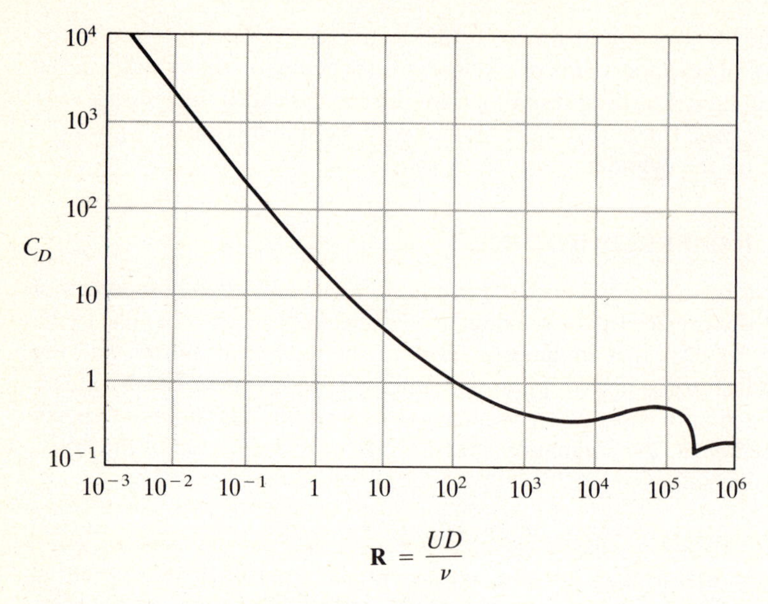

FIGURE 7.15 Drag coefficients for spheres.

drag coefficient against the Reynolds number for smooth spheres is shown in Fig. 7.15. It is evident from the diagram that as the Reynolds number increases, the C_D decreases. In other words, the drag becomes less and less as the Reynolds number becomes larger and larger before separation occurs. The shift to the turbulent boundary layer before separation is also evidenced by the sudden drop in the drag coefficient. Drag coefficients as a function of Reynolds number for infinitely long rough and smooth circular cylinders are given in Fig. 7.16. It may be noticed from the figure that the sudden drop in the drag coefficient occurs earlier in the rough cylinder than in the smooth cylinder.

From the foregoing discussion, it is evident that the drag on an immersed body due to the motion of the fluid around it can be reduced by delaying separation over the length of the body, or, in other words, by decreasing the size of the wake. It can be achieved by streamlining the body. A body is said to be streamlined if the form drag force is minimal. In a streamlined body, the thickness of the rear half is decreased gradually with length in such a way that the flow has no abrupt turn to make and that the separation occurs only over a small portion of the trailing edge as shown in Fig. 7.17. The power required to move an object through a fluid can be greatly reduced by streamlining the body. Examples of bodies in which streamlining can be important are automobiles, ships, airplanes, missiles, rockets, submarines, etc. However, when a body is streamlined by elongating, the skin friction drag on it increases due to increase in size. Therefore, an optimum condition must be sought when streamlining a body. The optimum condition would be obtained

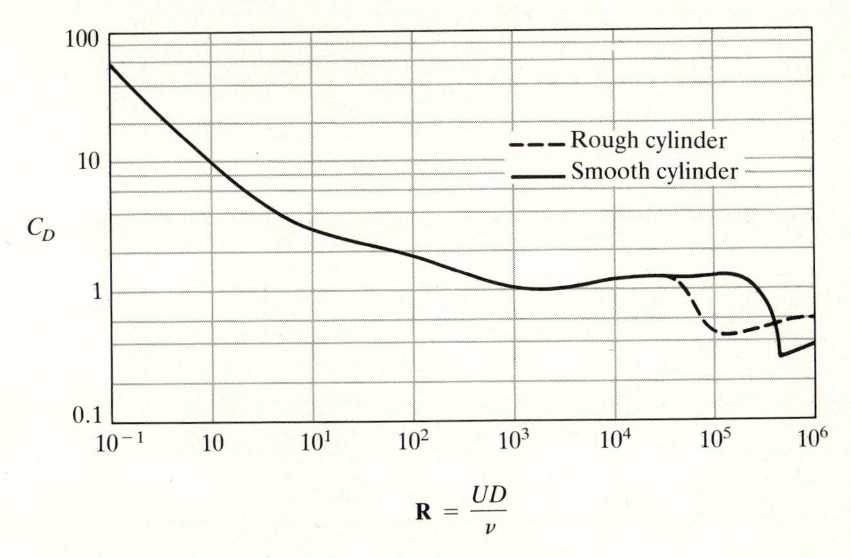

FIGURE 7.16 Drag coefficients as a function of Reynolds number for rough and smooth cylinders.

when the sum of the skin friction drag and the form drag is the minimum (Fig. 7.18).

Useful data for estimating automobile drag as a function of its body shape is given in Table 7.2. The drag coefficient for a particular body style is calculated using the equation

$$C_D = 0.16 + 0.0095 \sum_{i=1}^{8} C_{Di} \tag{7.43}$$

where the values of C_{Di} are read from Table 7.2 for each of the eight categories.

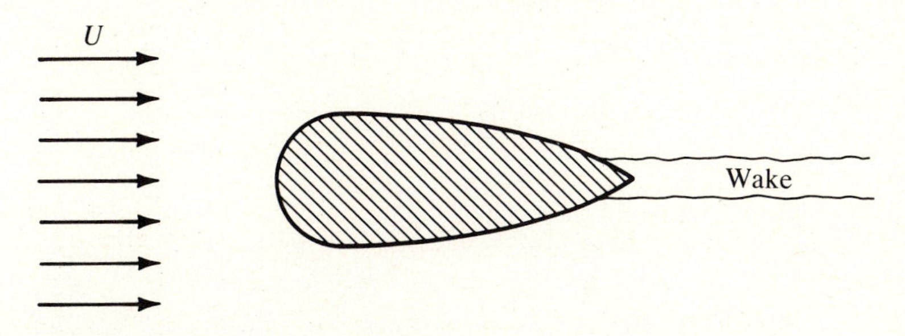

FIGURE 7.17 A streamlined body.

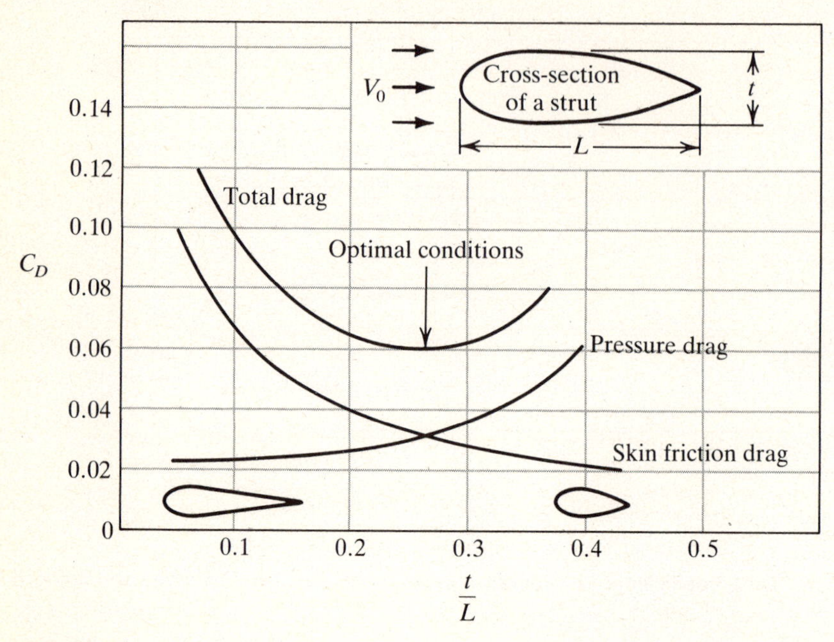

FIGURE 7.18 Drag coefficents for a family of struts. Source: Goldstein, S., *Modern developments in Fluid Dynamics*, Dover Publications, New York, 1965.

EXAMPLE 7.5

Determine the drag on a sphere 10 mm in diameter falling through glycerine at 5 cm/s.

Solution For glycerine, from Table A-4

$$S = 1.26; \qquad \mu = 0.862 \text{ N} \cdot \text{s/m}^2$$

Then,

$$\rho = 1.26 \times 1000 = 1260 \text{ kg/m}^3$$

$$\mathbf{R} = \frac{UD\rho}{\mu}$$

$$= \frac{0.05 \times 0.01 \times 1260}{0.862}$$

$$= 0.73$$

For $\mathbf{R} = 0.73$ $C_D = 35$ (from Fig. 7.15).
Hence:

$$F_D = \tfrac{1}{2} C_D \rho U^2 A$$

$$= \tfrac{1}{2} \times 35 \times 1260 \times 0.05^2 \times (\pi \times 0.01^2/4)$$

$$= 4.33 \times 10^{-3} \text{ N}$$

Table 7.2 ESTIMATES OF AUTOMOBILE DRAG COEFFICIENT

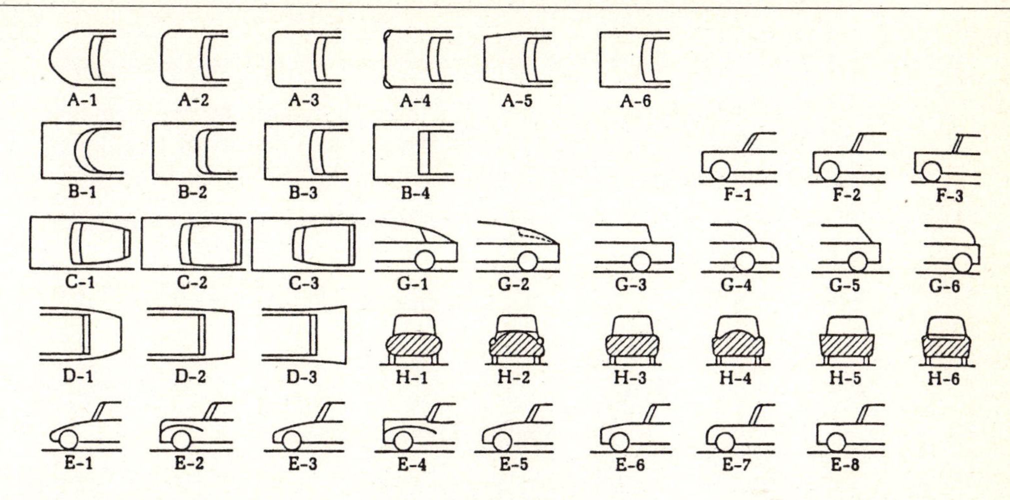

A. Plan view, front end

A-1 Approximately semicircular (1)

A-2 Well-rounded outer quarters (2)

A-3 Rounded corners without protuberances (3)

A-4 Rounded corners with protuberances (4)

A-5 Squared tapering-in corners (5)

A-6 Squared constant-width front (6)

B. Plan view, windshield

B-1 Full wraparound (approximately semicircular) (1)

B-2 Wraparound ends (2)

B-3 Bowed (3)

B-4 Flat (4)

C. Plan view, roof

C-1 Well- or medium-tapered to rear (1)

C-2 Tapering to front and rear (max. width at BC post) or approximately constant width (2)

C-3 Tapering to front (max. width at rear) (3)

D. Plan view, lower rear end

D-1 Well- or medium-tapered to rear (1)

D-2 Small taper to rear or constant width (2)

D-3 Outward taper (or flared-out fins) (3)

E. Side elevation, front end

E-1 Low, rounded front, sloping up (1)

E-2 High, tapered, rounded hood (1)

E-3 Low, squared front, sloping up (2)

E-4 High, tapered, squared hood (2)

E-5 Medium-height, rounded front, sloping up (3)

E-6 Medium-height, squared front, sloping up (4)

E-7 High, rounded front, with horizontal hood (4)

E-8 High, squared front, with horizontal hood (5)

F. Side elevation, windshield peak

F-1 Rounded (1)

F-2 Squared (including flanges or gutters) (2)

F-3 Forward-projecting peak (3)

G. Side elevation, rear roof/trunk

G-1 Fastback (roofline continuous to tail) (1)

G-2 Semifastback (with discontinuity in line to tail) (2)

G-3 Squared roof with trunk rear edge squared (3)

G-4 Rounded roof with rounded trunk (4)

G-5 Squared roof with short or no trunk (4)

G-6 Rounded roof with short or no trunk (5)

H. Front elevation, cowl and fender cross- section at windshield

H-1 Flush hood and fenders, well-rounded body sides (1)

H-2 High cowl, low fenders (2)

H-3 Hood flush with rounded-top fenders (3)

H-4 High cowl with rounded-top fenders (3)

H-5 Hood flush with square-edged fenders (4)

H-6 Depressed hood with high square-edged fenders (5)

Note: Drag rating values in parentheses are for use in Eq. (7.43).
Source: CRC Handbook of Tables for Applied Engineering Science, 2nd ed., (1973).

EXAMPLE 7.6

Determine the drag on the chimney of a mill 1 m in diameter and 50 m long when the wind is blowing at 40 km/h. The air has a density of 1.177 kg/m^3 and viscosity of 18.46×10^{-6} N·s/m^2. Assume the chimney to be smooth.

Solution Approach velocity of wind $= U = 40$ km/h or $U = 11.11$ m/s. The Reynolds number past the chimney is

$$\mathbf{R} = \frac{UD\rho}{\mu}$$

where

$$D = \text{diameter of the chimney}$$

Then:

$$\mathbf{R} = \frac{11.11 \times 1 \times 1.177}{18.46 \times 10^{-6}} = 7.08 \times 10^5$$

From Fig. 7.16 for a Reynolds number of 7.08×10^5 and for a smooth cylinder

$$C_D = 0.3$$

Hence:

$$\begin{aligned} F_D &= \tfrac{1}{2} C_D \rho U^2 A \\ &= \tfrac{1}{2} \times 0.3 \times 1.177 \times 11.11^2 \times (1 \times 50) \\ &= 1089 \text{ N} \end{aligned}$$

7.7 DRAG AND LIFT OF AIRFOILS

Let us first define some terms used in reference to airfoils. The geometry of an airfoil is described in Fig. 7.19.

Chord Line: A straight line joining the centers of curvatures of the front (i.e., leading) edge and the rear (i.e., trailing) edge. It is not necessarily an axis of symmetry.

Chord: Length of the chord line, denoted by c.

Span: The length of the airfoil in the direction normal to the chord line, denoted by b.

Plan Area: Plan area = Chord × Span, or

$$S = c \times b$$

where S is the plan area.

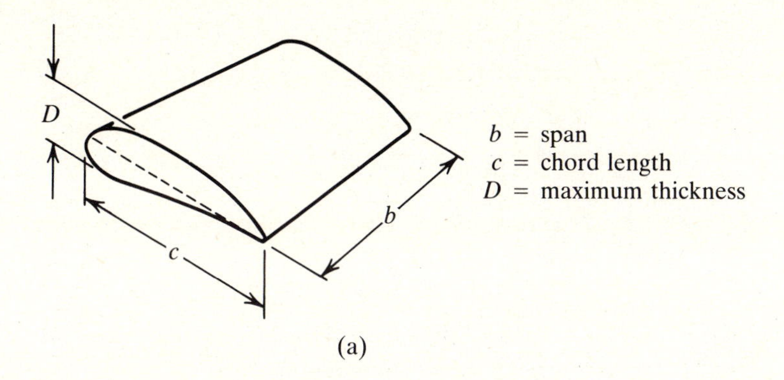

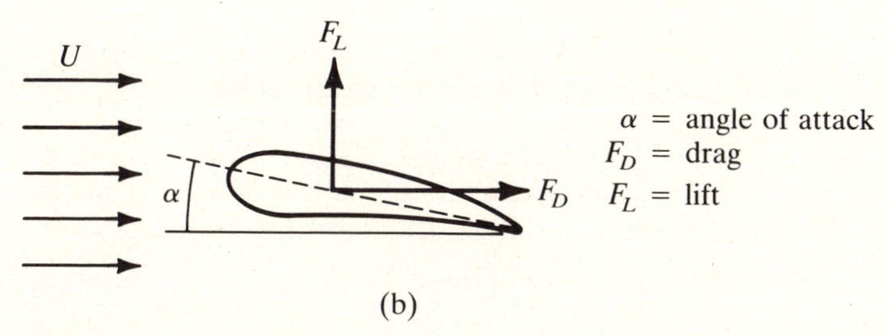

FIGURE 7.19 Geometry of an airfoil.

Mean Chord: Mean Chord = Plan area/Span, or

$$c' = S/b$$

where c' is the mean chord.

Aspect Ratio: Aspect Ratio = Span/mean chord, or

$$AR = b/c' = b^2/S$$

where AR is the aspect ratio.

Angle of Attack: The angle between the chord line and the direction of the fluid velocity, denoted by α.

Drag: That component of the aerodynamic force which is parallel to the direction of the oncoming air.

$$F_D = \tfrac{1}{2} C_D \rho U^2 S \tag{7.44}$$

where F_D = drag; C_D = drag coefficient; U = velocity of the oncoming air; S = plan area.

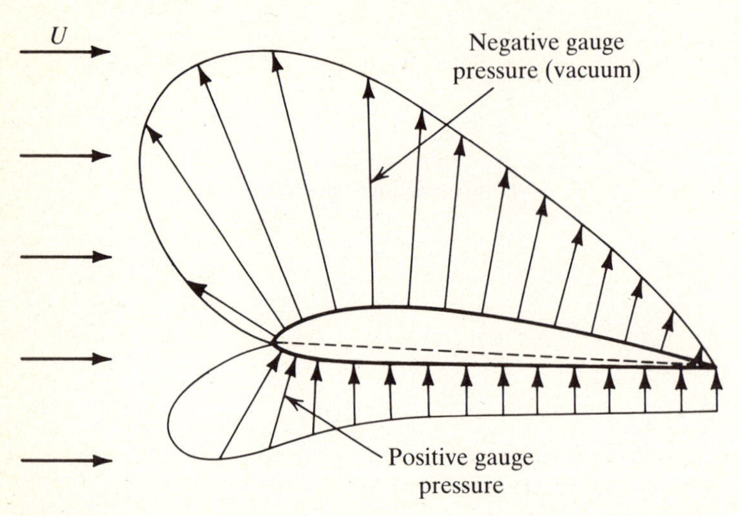

FIGURE 7.20 Pressure distribution over an airfoil placed in a stream of fluid.

Lift: That component of the aerodynamic force which is normal to the direction of the oncoming air. It is not necessarily vertical.

$$F_L = \tfrac{1}{2} C_L \rho U^2 S \tag{7.45}$$

where F_L = lift; C_L = lift coefficient; other variables are as defined before.

The pressure distribution over a typical airfoil in a stream of fluid is shown in Fig. 7.20. The pressure on the underside of the airfoil is positive while that on the upper side is negative. The net pressure is in the upward direction and produces the lift. The pressure distribution and the shape of streamlines around the airfoil depend on the angle of attack. The shape of streamlines and the size of the wake for various values of the angle of attack are shown in Fig. 7.21. As the angle of attack is increased, the point of separation moves upstream causing the drag force to increase. Also, as the angle of attack is increased, the fluid on top of the airfoil accelerates resulting in the lowering of the pressure. This causes the difference of pressure on both sides of the airfoil to increase in the upward direction resulting in increased lift. Hence, with the increasing angle of attack, the lift increases. When the separation point approaches the leading edge, the flow separates entirely from the upper side of the airfoil and the pressure on the upper side becomes almost equal to the upstream side and a decrease in the lift is noticed. This condition is known as *stall.* Typical lift and drag coefficients for an airfoil are shown in Fig. 7.22. It is evident from the figure that the coefficients of drag and lift increase with the angle of attack until the point of stall is reached. At the point of stall, a sudden decrease in the lift and sudden increase in the drag is noticed.

Airfoils of finite length also have an added drag and a reduced lift associated with the vortices generated at the tip of the airfoil as shown in Fig. 7.23. Since the pressure at the under side of the tip is higher than the pressure at the upper side, the wind circulates from the under side to the upper side.

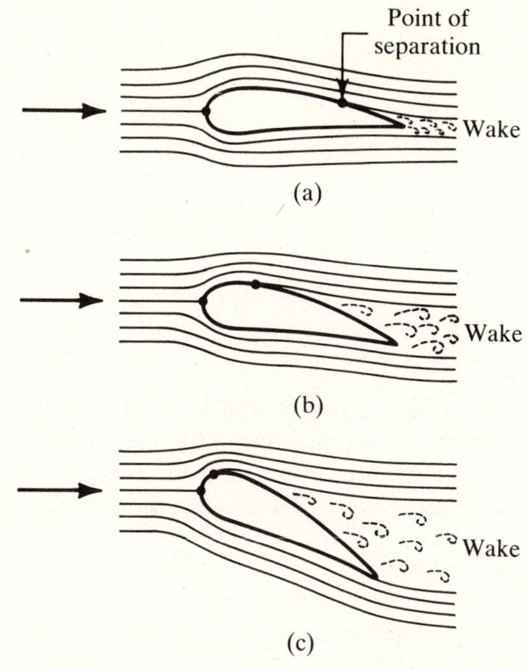

FIGURE 7.21 Streamlines around an airfoil as a function of angle of attack.

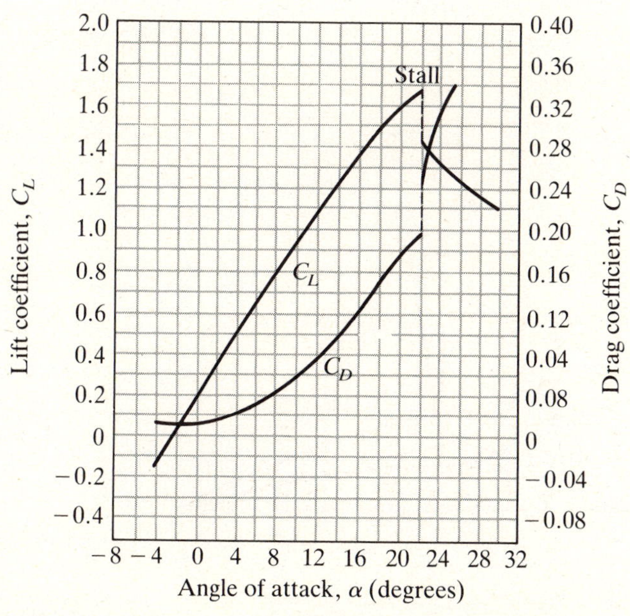

FIGURE 7.22 Lift and drag coefficients for an airfoil.

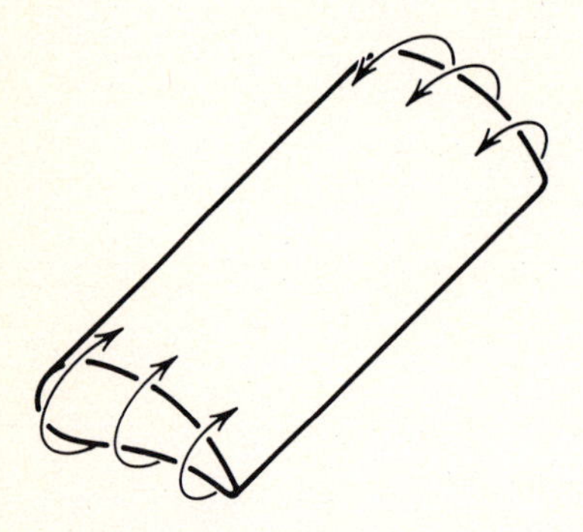

FIGURE 7.23 Vortices at the tip of a finite airfoil.

The net result of this circulation is to add a component u' to the free stream velocity U in the downward direction as shown in Fig. 7.24. The resulting effective velocity of approach (which is the vector sum of u' and U) acts at a reduced angle of attack, which in turn gives a lift smaller than the one associated with the actual angle of attack (Fig. 7.24). Also, the lift F_r so generated must be perpendicular to the effective velocity of approach. This force F_r can be resolved into two components, the true lift F_L normal to the actual approach velocity U, and a component F_{Di} parallel to U. This component (i.e., F_{Di}) is called *the induced drag*.

Prandtl showed that

$$u' = \frac{2F_L}{\pi \rho U b^2} \tag{7.46}$$

Therefore:

$$F_{Di} = F_L \phi = F_L \frac{u'}{U}$$

$$= \frac{2F_L^2}{\pi \rho U^2 b^2}$$

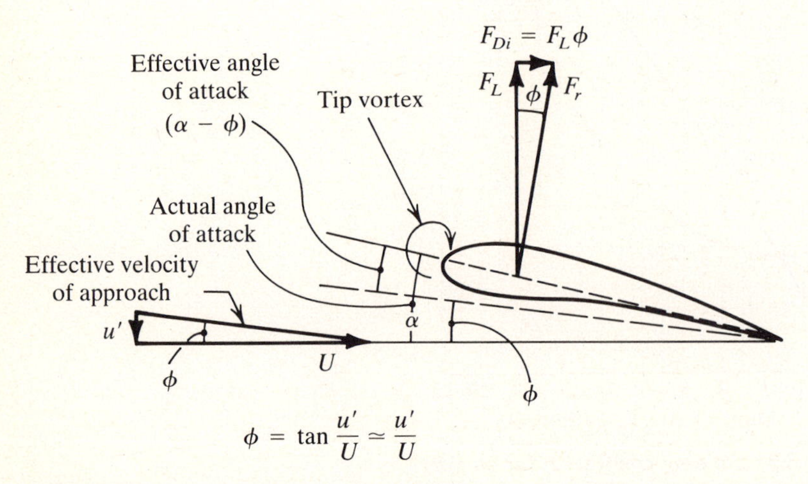

FIGURE 7.24 Induced drag over an airfoil.

For an airfoil of plan area S:

$$F_L = \frac{1}{2} C_L \rho U^2 S$$

Hence:

$$F_{Di} = \frac{2}{\pi \rho U^2 b^2} (\tfrac{1}{2} C_L \rho U^2 S)^2$$

$$= \frac{1}{2} \frac{C_L^2}{\pi (b^2/S)} \rho U^2 S$$

$$F_{Di} = \frac{1}{2} C_{Di} \rho U^2 S \tag{7.47}$$

where

$$C_{Di} = \frac{C_L^2}{\pi (b^2/S)} = \frac{C_L^2}{\pi AR} \tag{7.48}$$

Equation (7.47) is in the form of the drag equation and C_{Di} is the induced drag coefficient. It is evident from Eq. (7.48) that for a given wing section (i.e., constant C_L and constant c), the induced drag varies inversely with the span of the wing. Therefore, long slender wings will have less drags. Fig. 7.25 shows

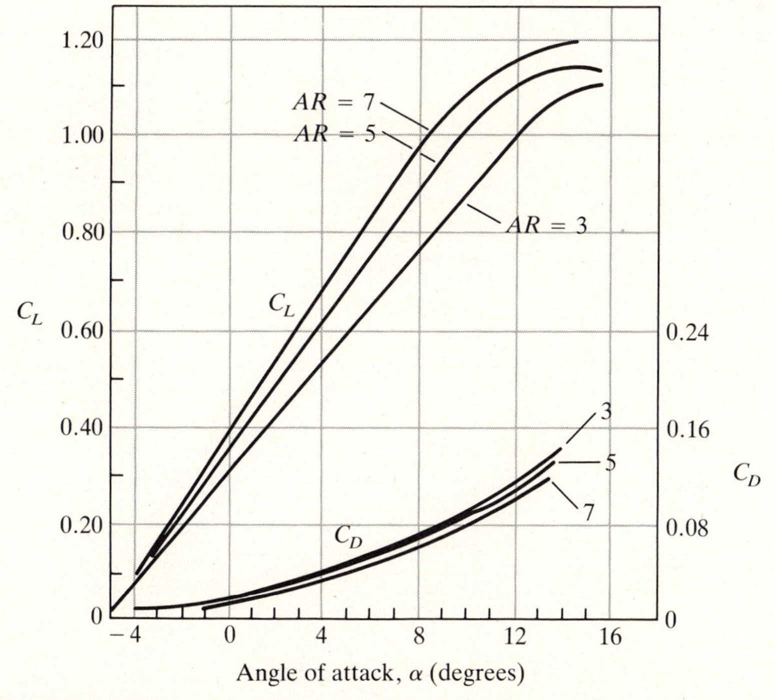

FIGURE 7.25 Lift and drag coefficients for various aspect ratios.

lift and drag characteristics of three wings of various aspect ratios. It is evident from the diagram that an airfoil with a larger aspect ratio has a smaller drag coefficient.

PROBLEMS

p.7.1 The Reynolds upper critical number is

a. of more importance than the Reynolds lower critical number.
b. 2,000.
c. 4,000.
d. 200.
e. none of these answers.

p.7.2 The Reynolds lower critical number is

a. 200.
b. 2,000.
c. 4,000.
d. 1,200.
e. none of these answers.

p.7.3 Water at 20°C is flowing through a round pipe 20 cm in diameter at the rate of 100 l/s. The Reynolds number is

a. 6.32×10^5.
b. 6.32×10^6.
c. 4.89×10^3.
d. 6.32×10^3.
e. none of these.

p.7.4 Crude oil at 30°C is being pumped through a 30 cm diameter pipeline at an average velocity of 5 m/s. The Reynolds number is

a. 2×10^6.
b. 2×10^3.
c. 2×10^5.
d. 1.4×10^5.
e. none of these.

p.7.5 Shear velocity is given by

a. τ_0/ρ.
b. $\sqrt{\tau_0/\rho}$.
c. $\sqrt{\rho/\tau_0}$.
d. $\sqrt{\tau_0/\gamma}$.
e. none of these answers.

p.7.6 The shear number is expressed by

a. $\dfrac{u_* y}{\rho}$.
b. $\dfrac{u_* \rho}{y}$.
c. $\dfrac{u_* y}{\nu}$.
d. $\dfrac{u_* \rho}{\mu}$.
e. none of these answers.

p.7.7 In a turbulent flow, the Von Kármán constant κ,

a. depends on the boundary configuration.
b. depends on the Reynolds number.
c. depends on the boundary configuration and the Reynolds number.
d. is independent of the boundary configuration and the Reynolds number.
e. none of these answers.

p.7.8 In a turbulent flow, the eddy viscosity η is

a. a property of the fluid.
b. a constant and does not change with the degree of turbulence.
c. a measure of the degree of turbulence.
d. depends on temperature.
e. none of these answers.

p.7.9 The Von Kármán constant κ of turbulent flow is taken to be equal to

a. 3.0.
b. 2.5.
c. 4.0.
d. 5.87.
e. none of these answers.

p.7.10 Calculate the Reynolds number if water at 15°C is flowing through a 50 mm diameter pipe at the rate of 5 l/s. Is the flow laminar or turbulent? What is the critical mean velocity?

p.7.11 A smooth flat plate 2 m long and 25 cm wide moves lengthwise through air at a speed of 5 m/s. Assuming the boundary layer to be entirely laminar, calculate halfway along the plate.

a. Thickness of the boundary layer
b. Shear stress.

Also determine the power required to move the plate. Assume air has a density of 1.21 kg/m^3 and kinematic viscosity of 14.9 mm^2/s.

p.7.12 Water at 15°C flows over a flat plate 2 m long and 25 cm wide at a velocity of 5 m/s along its length. Determine the drag on the plate.

p.7.13 Determine the drag on the plate in Problem 7.12 if the water flows along the width.

p.7.14 Glycerine flows past a thin flat plate at a free stream velocity of 7 m/s. What is the velocity 1 m down from the leading edge and 1.5 cm from the plate?

p.7.15 An aircraft pulls a sign 20 m long and 1.5 m wide through still air having a density of 1.2 kg/m^3 and viscosity of 1.8×10^{-5} N·s/m^2. What is the power required to pull the sign? Assume the sign to behave like a flat plate. The speed of the aircraft is 200 km/hr.

Chapter Eight

LAMINAR FLOW

In this chapter, we will derive the equations of laminar flow through pipes and between two parallel plates. We will also develop the Darcy-Weisbach equation for the head loss due to friction.

8.1 INTRODUCTION

In a laminar flow, the individual particles of a fluid move in a regular fashion, they do not cross each other's path and there is no mixing of the fluid. The fluid is thus considered to be moving in layers or laminae. However, it may be pointed out that, in more general terms, the velocity in a laminar flow may vary in all directions and not just in a direction perpendicular to the layers or laminae. Due to the low velocity, the viscous forces dominate over the inertia forces. Newton's law of viscosity given by

$$\tau = \mu \frac{\partial u}{\partial y} \tag{8.1}$$

is applicable. Whether the flow through a conduit is laminar or not, can be determined by calculating the Reynolds number. The flow is considered to be laminar if the Reynolds number is less than, or equal to, 2,000. A number of cases of laminar flow are now described.

8.2 LAMINAR FLOW THROUGH PIPES

In laminar flow, the flow takes place in layers with one layer gliding over the adjacent layer with a different velocity. Individual particles of fluid follow paths which do not cross those of neighboring particles. We recall from Section 1.5 that viscous forces are set up whenever there is a relative movement between fluid particles.

The experimental investigations of viscous flow of fluids in straight circular pipes were first made by a German engineer G. H. L. Hagen (1797–1884), and a French physician J. L. M. Poiseuille (1799–1869) working independently. Hagen experimented with the flow of water through small brass tubes, while Poiseuille studied the flow of water in fine glass capillary tubes, with the intention of drawing an analogy between the flow of water in glass capillary tubes and the flow of blood in veins. Their conclusions were similar in nature which suggested that the head loss through a given length of a tube was directly proportional to the rate of flow and inversely proportional to the fourth power of the diameter of the tube.

A theoretical development of the equations governing the flow of fluid in a round pipe is now presented.

Consider an elemental cylindrical volume of fluid having a radius r, cross-sectional area δA, and length δs inside a pipe of radius R as shown in Fig. 8.1. It is assumed that the flow is fully developed. The element is acted upon by the forces F_1 and F_2 at the end sections caused by the pressure of the flowing fluid and a shear force F_s on the curved surface caused by viscous forces. The weight of the element δw is another force which acts on the elemental volume. Let

p = pressure at the left end

τ = shear stress at a distance r from the center

γ = specific weight of the fluid

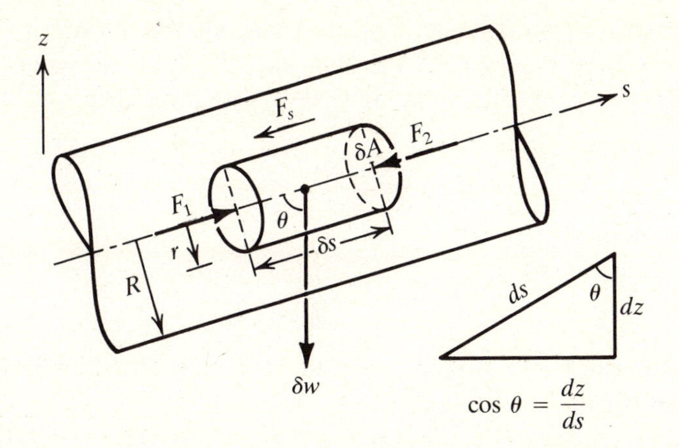

FIGURE 8.1 An elemental cylindrical volume in a case of laminar flow through a pipe.

Then

$$F_1 = p\delta A$$

$$F_2 = \left(p + \frac{dp}{ds}\,\delta s\right)\delta A$$

$$F_s = \tau 2\pi r\delta s$$

$$\delta w = \gamma\pi r^2\delta s$$

If the flow is steady and uniform, then according to Newton's second law of motion

$$\Sigma F_s = 0$$

$$F_1 - F_2 - F_s - \delta w \cos\theta = 0$$

and

$$p\delta A - \left(p + \frac{dp}{ds}\,\delta s\right)\delta A - \tau 2\pi r\delta s - \gamma\pi r^2\delta s\cos\theta = 0$$

$$-\frac{dp}{ds}\,\delta s\delta A - \tau\,\frac{2}{r}\,\delta A\delta s - \gamma\delta A\delta s\cos\theta = 0$$

Dividing by $\delta A\delta s$, and substituting dz/ds for $\cos\theta$, we obtain

$$-\frac{dp}{ds} - \frac{2\tau}{r} - \gamma\,\frac{dz}{ds} = 0$$

or

$$\tau = \frac{r}{2}\left(-\frac{dp}{ds} - \gamma\,\frac{dz}{ds}\right)$$

$$\tau = \frac{r}{2}\left[-\frac{d}{ds}(p + \gamma z)\right]$$

The term $d/ds\,(p + \gamma z)$ is the slope of $(p + \gamma z)$. Since the flow takes place in the direction of a decreasing $(p + \gamma z)$, the slope $d/ds\,(p + \gamma z)$ is always negative.
Let $-d/ds\,(p + \gamma z) = Sp$, so that Sp is always positive. Then

$$\tau = \frac{r}{2} Sp \tag{8.2}$$

From Newton's law of viscosity

$$\tau = -\mu \frac{dv}{dr} \tag{8.3}$$

where v is the velocity at a distance r from the centerline. Then from Eqs. (8.2) and (8.3), we obtain

$$-\mu \frac{dv}{dr} = \frac{r}{2} Sp$$

$$dv = -\frac{1}{2\mu} Spr\,dr$$

Integrating, we obtain

$$v = -\frac{r^2}{4\mu} Sp + A \tag{8.4}$$

where A is a constant of integration. It can be evaluated by using the known condition that $v = 0$ when $r = R$. This yields

$$A = \frac{R^2}{4\mu} Sp \tag{8.5}$$

Substituting in Eq. (8.4), we obtain

$$v = \frac{-r^2}{4\mu} Sp + \frac{R^2}{4\mu} Sp$$

$$v = \frac{R^2 - r^2}{4\mu} Sp \tag{8.6}$$

Maximum velocity occurs at the center where $r = 0$.
Hence:

$$v_{\max} = \frac{R^2}{4\mu} Sp \tag{8.7}$$

Also, maximum shear stress occurs at the walls where $r = R$.
Hence:

$$\tau_{\max} = \frac{R}{2} Sp \tag{8.8}$$

The velocity distribution and the shear distribution diagrams are given in Fig. 8.2.

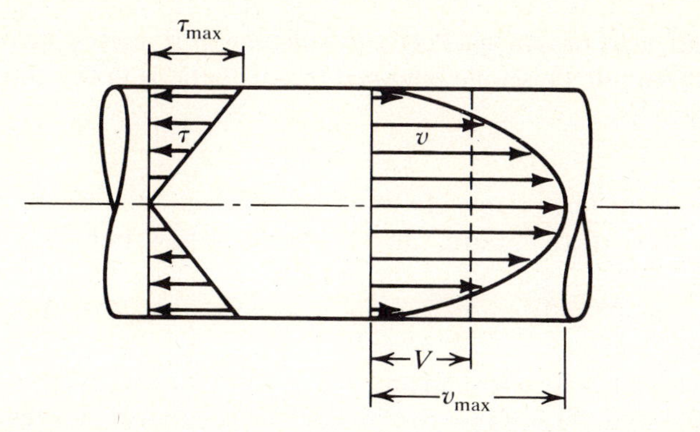

FIGURE 8.2 Velocity distribution and shear distribution over a cross-section of a pipe.

To calculate the discharge through the pipe, consider a concentric ring of thickness dr at a distance r from the center of the pipe. The discharge through the ring is given by $v2\pi r\,dr$. Then discharge through the entire pipe is given by

$$Q = \int_0^R v2\pi r\,dr$$

$$= \int_0^R \frac{R^2 - r^2}{4\mu} Sp2\pi r\,dr$$

$$= \frac{2\pi}{4\mu} Sp \int_0^R (R^2 r - r^3)\,dr$$

$$= \frac{2\pi}{4\mu} Sp \left[\frac{R^2 r^2}{2} - \frac{r^4}{4}\right]_0^R$$

$$Q = \frac{\pi R^4}{8\mu} Sp \tag{8.9}$$

The average velocity may be found by dividing the discharge by the area. Then

$$V = \frac{R^2}{8\mu} Sp \tag{8.10}$$

This indicates that average velocity is one-half the maximum velocity in a laminar flow.

For a horizontal pipe, z is constant; hence, $Sp = -dp/ds = \Delta p/L$ where Δp is the pressure drop over a length L of the pipe. Substituting in Eq. (8.9), we obtain

$$Q = \frac{\pi R^4 \Delta p}{8\mu L} \tag{8.11}$$

and substituting $R = D/2$, where D is the diameter of the pipe, we obtain

$$Q = \frac{\pi D^4 \Delta p}{128\mu L} \tag{8.12}$$

Eq. (8.12) is known as *the Hagen-Poiseuille equation.* It is interesting to note that neither Hagen nor Poiseuille obtained their equations in this form.

Comments

1. The numerical value of Sp will always be positive.
2. The Eqs. (8.2) to (8.12) are valid for laminar flow only—i.e., when $\mathbf{R} \leq 2000$.
3. Eqs. (8.11) and (8.12) are valid only if z is constant—i.e., the pipe is horizontal.
4. Eqs. (8.2) to (8.12) are not valid near the entrance to a pipe. The velocity distribution near the entrance to a pipe is almost uniform over the cross-section. A fluid must travel through a certain distance, S, in the pipe before a parabolic velocity distribution is attained. According to Langhaar:

$$S = \frac{0.058 V D^2 \rho}{\mu}$$

5. The kinetic energy correction factor, α, for laminar flow in a pipe of cross-sectional area A can be calculated using the equation

$$\alpha = \frac{1}{A} \int_A \left(\frac{v}{V}\right)^3 dA$$

where

$$v = \frac{R^2 - r^2}{4\mu} Sp \qquad \text{Eq. (8.6)}$$

$$V = \frac{R^2}{8\mu} Sp \qquad \text{Eq. (8.10)}$$

Hence

$$\alpha = \frac{2}{\pi R^2} \int_0^R \left[1 - \left(\frac{r}{R}\right)^2\right]^3 2\pi r\, dr = 2$$

6. The momentum correction factor, β, is given by

$$\beta = \frac{2}{\pi R^2} \int_0^R \left[1 - \left(\frac{r}{R}\right)^2\right]^2 2\pi r\, dr = 4/3$$

EXAMPLE 8.1

Determine the direction of flow of kerosene oil at 20°C through the tube shown in Fig. 8.3. Also determine the maximum velocity, the maximum shear stress and the discharge.

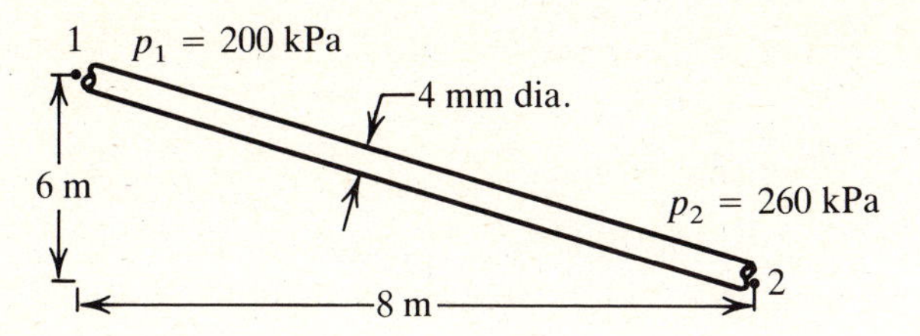

FIGURE 8.3 Example 8.1.

Solution For kerosene oil at 20°C

$$S = 0.81$$

$$\mu = 19.1 \times 10^{-4}\ \mathrm{N \cdot s/m^2} \qquad \text{(from Table A-4)}$$

The direction of flow can be determined by calculating $(p + \gamma z)$ at points 1 and 2. The direction of flow will be from the higher value to the lower value. Taking datum at point 2

$$(p + \gamma z)_1 = (200 + 9.81 \times 0.81 \times 6) = 247.68\ \mathrm{kN/m^2}$$

$$(p + \gamma z)_2 = (260 + 0) = 260\ \mathrm{kN/m^2}$$

This shows that the direction of flow is from point 2 to point 1

$$Sp = -\frac{d}{ds}(p + \gamma z) = -\frac{247.68 - 260}{\sqrt{6^2 + 8^2}} = 1.232\ \mathrm{kN/m^2/m} = 1232\ \mathrm{N/m^2/m}$$

From Eq. (8.8)

$$\tau_{\max} = \frac{R}{2} Sp$$

Hence:

$$\tau_{\max} = \frac{0.002}{2}(1232)$$

$$= 1.232\ \mathrm{N/m^2}$$

From Eq. (8.7)

$$v_{\max} = \frac{R^2}{4\mu} Sp$$

$$= \frac{(0.002)^2}{(4)(19.1 \times 10^{-4})}(1232)$$

$$v_{\max} = 0.645\ \mathrm{m/s}$$

From Eq. (8.9)

$$Q = \frac{\pi R^4}{8\mu} Sp$$

$$= \frac{\pi(0.002)^4}{8 \times 19.1 \times 10^{-4}}(1232)$$

$$= 4.05 \times 10^{-6} \text{m}^3/\text{s}$$

$$V = \frac{V_{\text{max}}}{2} = \frac{0.645}{2} = 0.3225 \text{ m/s}$$

$$\mathbf{R} = \frac{VD\rho}{\mu}$$

$$= \frac{0.3225 \times 0.004 \times 0.81 \times 1000}{19.1 \times 10^{-4}}$$

$$\mathbf{R} = 547$$

Since the Reynolds number is less than 2,000, the flow is laminar and the calculations are valid.

Head Loss in Laminar Flow

A part of the total head of a fluid is lost in overcoming the viscous resistance when the fluid moves from one point to another. Let us now develop an expression for this head loss in a laminar flow. Consider two points 1 and 2 in a pipe through which laminar flow is taking place as shown in Fig. 8.4. Applying the energy equation between these two points, we obtain

$$\frac{p_1}{\gamma} + z_1 + \frac{V_1^2}{2g} = \frac{p_2}{\gamma} + z_2 + \frac{V_2^2}{2g} + HL$$

The only head loss between these two points is due to friction. Let it be denoted by h_f. Thus, head loss $(HL) = h_f$. Also, $V_1 = V_2$.

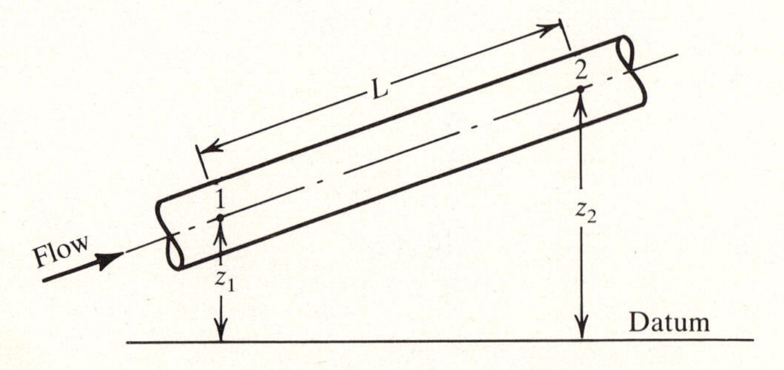

FIGURE 8.4 An inclined pipe containing laminar flow.

Then

$$\frac{p_1}{\gamma} + z_1 = \frac{p_2}{\gamma} + z_2 + h_f$$

$$\left(\frac{p_1}{\gamma} + z_1\right) - \left(\frac{p_2}{\gamma} + z_2\right) = h_f \tag{8.13}$$

Also, for the points 1 and 2

$$Sp = \frac{(p_1 + \gamma z_1) - (p_2 + \gamma z_2)}{L}$$

$$Sp = \frac{\gamma}{L}[(p_1/\gamma + z_1) - (p_2/\gamma + z_2)]$$

Substituting for $[(p_1/\gamma + z_1) - (p_2/\gamma + z_2)]$ from Eq. (8.13), we obtain

$$Sp = \frac{\gamma}{L} h_f$$

Substituting in Eq. (8.10) and replacing R by $D/2$, where D is the diameter of the pipe, we obtain

$$V = \frac{D^2}{32\mu}\left(\frac{\gamma}{L} h_f\right)$$

$$h_f = \frac{32\mu LV}{\gamma D^2} \tag{8.14}$$

By rearrangement, we obtain

$$h_f = 64\left(\frac{\mu}{VD\rho}\right)\left(\frac{L}{D}\right)\left(\frac{V^2}{2g}\right)$$

The term $(\mu/VD\rho)$ is the reciprocal of the Reynolds number.

Hence:

$$h_f = \frac{64}{\mathbf{R}}\left(\frac{L}{D}\right)\left(\frac{V^2}{2g}\right) \tag{8.15}$$

$$h_f = f\left(\frac{L}{D}\right)\left(\frac{V^2}{2g}\right) \tag{8.16}$$

where f is known as the *friction factor*. Eq. (8.16) is called the *Darcy-Weisbach equation.* It is evident from the above equation that, in a laminar flow, the friction factor is equal to $64/\mathbf{R}$.

EXAMPLE 8.2

Water flows in a pipe 2 cm in diameter and 30 m long; the pipe is running full. Find the head loss when the temperature is 5°C and the velocity is 10 cm/s.

Solution From Table A-1, for water at 5°C

$$\nu = 1.54 \times 10^{-6}\ \text{m}^2/\text{s}$$

$$\mathbf{R} = \frac{VD}{\nu} = \frac{0.1 \times 0.02}{1.54 \times 10^{-6}} = 1299$$

As **R** is less than 2,000, the flow is laminar.
Therefore

$$f = \frac{64}{\mathbf{R}} = \frac{64}{1299} = 0.0493$$

and

$$h_f = f\frac{L}{D}\frac{V^2}{2g}$$

$$= \frac{0.0493 \times 30 \times (0.1)^2}{0.02 \times 2 \times 9.81}$$

$$= 0.037 \text{ m of water}$$

or

$$h_f = 3.7 \text{ cm of water.}$$

EXAMPLE 8.3

Oil at 60°F has a specific gravity of 0.92 and kinematic viscosity of 0.0205 ft^2/s. Find the horsepower required to pump 50 tons of this oil per hour along a pipeline 9 inches in diameter and one mile long.

Solution For the oil $S = 0.92$
Hence:

$$\gamma = 0.92 \times 62.4 = 57.4 \text{ lbs/ft}^3$$

$$\nu = 0.0205 \text{ ft}^2/\text{s}$$

$$\text{Rate of flow} = \frac{50 \times 2000}{57.4}$$

$$= 1742.16 \text{ ft}^3/\text{hr}$$

or

$$Q = 0.484 \text{ ft}^3/\text{s}$$

$$A = \frac{\pi(9)^2}{4 \times 144} = 0.4418 \text{ ft}^2$$

$$V = \frac{0.484}{0.4418} = 1.095 \text{ ft/s}$$

$$\mathbf{R} = \frac{1.095 \times 0.75}{0.0205} = 40.06$$

R is < 2,000; therefore, flow is laminar

$$f = \frac{64}{\mathbf{R}} = \frac{64}{40.06} = 1.6$$

$$h_f = \frac{1.6 \times 5280 \times (1.095)^2}{2 \times 32.2 \times 0.75}$$

$$= 209.7 \text{ ft of oil}$$

$$\text{Horsepower required} = \frac{\gamma Q h_f}{550}$$

$$= \frac{57.4 \times 0.484 \times 209.7}{550}$$

$$= 10.6$$

EXAMPLE 8.4

The viscosity of a liquid is determined by timing the discharge of 50 cm^3 of the liquid from a vessel through a horizontal capillary tube. The vessel is an upright cylinder, open at the top, 5 cm in diameter, and the capillary tube is 1 mm bore and 10 cm long. The vessel is at first filled to a height of 5 cm above the axis of the tube, and it is found that 50 cm^3 are discharged in 20 minutes. Find the viscosity of the liquid in poise, given that its density is 0.88 g/cm^3. Neglect the end effects of the tube and the velocity head at the discharge.

Solution The setup of the apparatus is shown in Fig. 8.5. The velocity of flow in the tube depends on the height of fluid in the vessel. As the height decreases, the

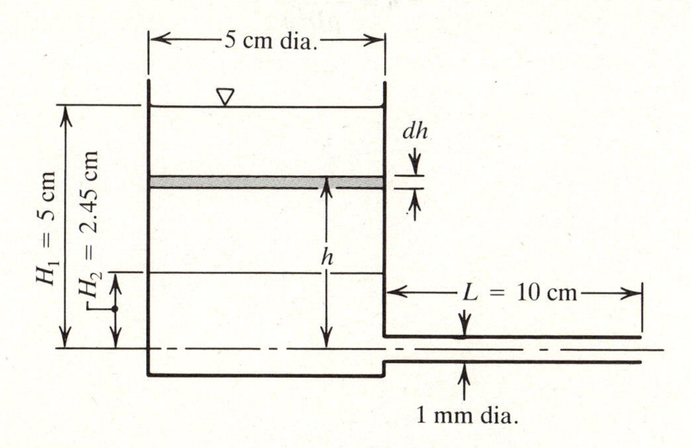

FIGURE 8.5 Example 8.4.

velocity also decreases. Consider the instant when the surface of the liquid is at a height h above the center of the tube. Let the surface fall by dh in time dt and let the corresponding quantity of flow out of the vessel be $d\forall$.

Then

$$d\forall = -A\,dh = aV\,dt \tag{1}$$

where

$$a = \text{area of the tube}$$

$$A = \text{surface area of vessel}$$

$$V = \text{velocity in the tube}$$

Rearranging equation (1)

$$dt = -\frac{A}{aV}\,dh \tag{2}$$

The time T taken by the fluid to fall from H_1 to H_2 can be calculated by integration of Eq. (2)

$$\int_0^T dt = -\frac{A}{a}\int_{H_1}^{H_2}\frac{dh}{V}$$

or

$$T = -\frac{A}{a}\int_{H_1}^{H_2}\frac{dh}{V} \tag{3}$$

The velocity in the tube can be calculated using Eq. (8.14). Since we are told to neglect the velocity head at the discharge, we can write

$$h_f = h = \frac{32\mu LV}{\rho D^2 g}$$

from which

$$V = \frac{\rho D^2 g}{32\mu L}\,h$$

Substituting in Eq. (3):

$$T = \frac{-A}{a}\int_{H_1}^{H_2}\frac{32\mu L}{\rho D^2 g}\frac{dh}{h}$$

or

$$T = \frac{32A\mu L}{\rho a D^2 g}\ln\left(\frac{H_1}{H_2}\right)$$

For this problem:

$$T = 20 \text{ min} = 20 \times 60 = 1200 \text{ s}$$

$$a = \frac{\pi}{4}(0.1)^2 = 0.00785 \text{ cm}^2$$

$$A = \frac{\pi}{4}(5)^2 = 19.63 \text{ cm}^2$$

$$D = 0.1 \text{ cm}$$

$$L = 10 \text{ cm}$$

$$\rho = 0.88 \text{ g/cm}^3$$

$$\mu = ?$$

$$H_1 = 5 \text{ cm}$$

H_2 can be calculated by using

$$\forall = (H_1 - H_2)A$$

or

$$50 = (5 - H_2)19.63$$

from which $H_2 = 2.45$ cm.

Therefore:

$$1200 = \frac{32 \times 19.63 \times \mu \times 10}{0.88 \times 0.00785 \times (0.1)^2 \times 981} \ln\left(\frac{5}{2.45}\right)$$

from which $\mu = 0.0182$ poise

8.3 LAMINAR FLOW BETWEEN FLAT PLATES

Another example of laminar flow may be encountered when a fluid flows between two flat surfaces with a low velocity. Consider two flat inclined parallel plates with a fluid in between (Fig. 8.6). The lower plate is fixed and

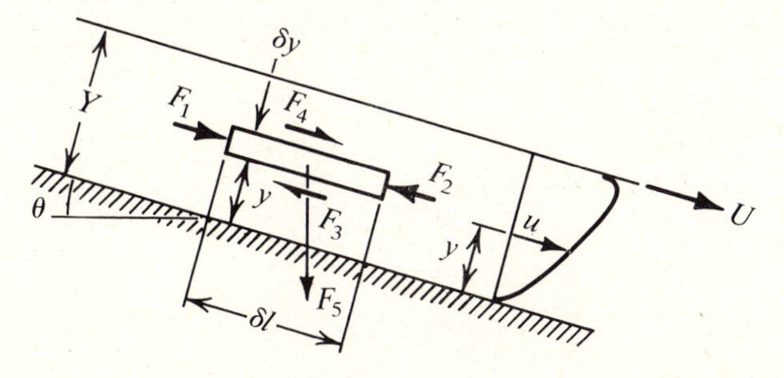

FIGURE 8.6 An elemental volume in a case of laminar flow between flat plates.

the upper plate is moving with a velocity U. Consider an element of fluid having a length of δl, a depth of δy, and a unit thickness located at a distance y from the lower fixed plate as shown in Fig. 8.6. The element is acted upon by the pressure forces F_1 and F_2, the viscous forces F_3 and F_4 and the gravity force F_5. Let p be the pressure at the left face and τ be the shear stress at the lower face of the fluid element. Then the magnitude of various forces will be given by

$$F_1 = p\delta y$$

$$F_2 = \left(p + \frac{\partial p}{\partial l}\delta l\right)\delta y$$

$$F_3 = \tau\delta l$$

$$F_4 = \left(\tau + \frac{\partial \tau}{\partial y}\delta y\right)\delta l$$

$$F_5 = \gamma\delta l\delta y$$

For steady flow, the equation of motion for the element of fluid in the l direction is given by

$$F_1 - F_2 - F_3 + F_4 + F_5 \sin\theta = 0$$

or

$$p\delta y - \left(p + \frac{\partial p}{\partial l}\delta l\right)\delta y - \tau\delta l + \left(\tau + \frac{\partial \tau}{\partial y}\delta y\right)\delta l + \gamma\delta l\,\delta y \sin\theta = 0$$

Dividing by $\delta l\delta y$, using $\sin\theta = -\partial z/\partial l$ and simplifying, we obtain

$$\frac{\partial \tau}{\partial y} = \frac{\partial}{\partial l}(p + \gamma z)$$

Since τ is function of y only and p and z are functions of l only, the partial derivatives in the above equation may be replaced by total derivatives.

Hence:

$$\frac{d\tau}{dy} = \frac{d}{dl}(p + \gamma z)$$

The term $d/dl\,(p + \gamma z)$ is the slope of $(p + \gamma z)$. Since the flow takes place in the direction of decreasing $(p + \gamma z)$, the slope $d/dl\,(p + \gamma z)$ is always negative. Let $-d/dl\,(p + \gamma z) = Sp$, so that Sp is always positive. Then

$$\frac{d\tau}{dy} = -Sp \tag{8.17}$$

By differentiating Eq. (8.1), we obtain

$$\frac{d\tau}{dy} = \mu\frac{d^2u}{dy^2}$$

$$\mu\frac{d^2u}{dy^2} = -Sp$$

Integrating the above equation twice with respect to y, we obtain

$$u = -\frac{Sp}{2\mu} y^2 + \frac{A}{\mu} y + B \tag{8.18}$$

in which A and B are constants of integration. These constants can be evaluated using the boundary conditions that $u = 0$ when $y = 0$, and $u = U$ when $y = Y$. These conditions when substituted in Eq. (8.18) yield

$$B = 0$$

$$A = \frac{\mu}{Y}\left(U + \frac{Sp}{2\mu} Y^2\right)$$

Substituting in Eq. (8.18), we obtain

$$u = \frac{Uy}{Y} + \frac{Sp}{2\mu}(Yy - y^2) \tag{8.19}$$

This equation gives the velocity distribution between the two plates when the lower plate is fixed and the upper plate is moving with a velocity U. If the upper plate is moving in the same direction as the fluid, the maximum velocity occurs at the moving plate (Fig. 8.7a). However, if the upper plate is moving in a direction opposite to that of the fluid, the maximum velocity may occur somewhere between the plates depending on the magnitude of U (Fig. 8.7b). The discharge per unit width past a cross-section, can be obtained using

$$q = \int_0^Y u \, dy$$

or

$$q = \int_0^Y \left[\frac{Uy}{Y} + \frac{Sp}{2\mu}(Yy - y^2)\right] dy$$

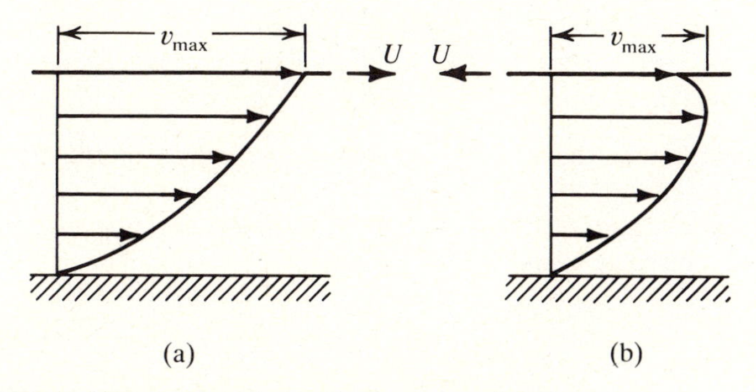

FIGURE 8.7 Velocity distribution between two flat plates.

Integrating, we obtain

$$q = \frac{UY}{2} + \frac{SpY^3}{12\mu} \tag{8.20}$$

The above equation shows that flow can take place between two plates if one of them is moving even if there is no difference of piezometric head between the two ends ($Sp = 0$).

If the upper plate is also fixed, $U = 0$ and Eq. (8.19) reduces to

$$u = \frac{Sp}{2\mu}(Yy - y^2) \tag{8.21}$$

This means that the velocity distribution between the fixed plates is parabolic, it being zero at the walls and maximum midway between the plates. The maximum velocity can be obtained by putting $y = Y/2$ in Eq. (8.21), which yields

$$v_{max} = \frac{Y^2}{8\mu} Sp \tag{8.22}$$

The discharge past a cross-section can be obtained by using the relation

$$q = \int_0^Y u\, dy$$

$$= \int_0^Y \left[\frac{Sp}{2\mu}(Yy - y^2)\right] dy$$

or

$$q = \frac{SpY^3}{12\mu} \tag{8.23}$$

The average velocity may be obtained by dividing the discharge with the cross-sectional area, which yields

$$V = \frac{Sp}{12\mu} Y^2 \tag{8.24}$$

This indicates that the average velocity is two-thirds of the maximum velocity.

For horizontal plates, $Sp = -dp/dl$. For two points having a pressure drop of Δp, and a distance L apart, $Sp = \Delta p/L$.

Hence:

$$q = \frac{\Delta p Y^3}{12\mu L} \tag{8.25}$$

It should be emphasized again that all the above equations are valid only if the flow is fully developed. They are not valid near entrances to and exits from the plates.

Eq. (8.25) is often used to determine leakage of lubricants from areas between a piston and a concentric cylinder.

EXAMPLE 8.5

For the two-plate system shown in Fig. 8.8, $U = 0$ ft/s, $p_1 = 6.5$ psi, $p_2 = 8$ psi, $l = 4$ ft, and $\theta = 30°$. The distance between the plates is 0.006 ft. The fluid is kerosene oil at 20°C. Determine the velocity distribution, the discharge and the shear exerted on the upper plate.

Solution For kerosene oil at 20°C

$$S = 0.81; \qquad \mu = 4.0 \times 10^{-5} \text{ lb} \cdot \text{s/ft}^2$$

$$\gamma = 62.4 \times 0.81 = 50.54 \text{ lbs/ft}^3$$

At point 1

$$(p + \gamma z)_1 = (6.5 \times 144) + 50.54 \times 4 \sin 30°$$
$$= 1037.08 \text{ psf}$$

At point 2

$$(p + \gamma z)_2 = (8 \times 144) = 1152 \text{ psf}$$

The flow is taking place from point 2 to point 1.

$$Sp = -\frac{d}{dl}(p + \gamma z) = \frac{1037.08 - 1152}{4}$$
$$= 28.73 \text{ lbs/ft}^3$$

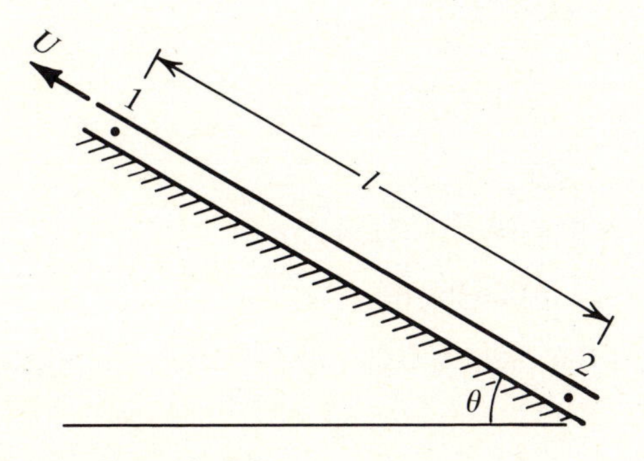

FIGURE 8.8 Example 8.5.

From Eq. (8.21)

$$u = \frac{Sp}{2\mu}(Yy - y^2)$$

$$= \frac{28.73}{2 \times 4 \times 10^{-5}}(0.006y - y^2)$$

$$= 2154.75\ y - 359125\ y^2$$

Hence, the velocity distribution is given by

$$u = 2154.75\ y - 359125\ y^2$$

Maximum velocity occurs where $du/dy = 0$

$$\frac{du}{dy} = 2154.75 - 718250\ y = 0$$

or

$$y = 0.003 \text{ ft}$$

The discharge per ft width $= q = \int_0^Y u\,dy$

$$= \int_0^{0.006} (2154.75\ y - 359125\ y^2)\,dy$$

$$= \left[\frac{2154.75\ y^2}{2}\right]_0^{0.006} - \left[\frac{359125\ y^3}{3}\right]_0^{0.006}$$

$$= 0.013 \text{ ft}^3\text{/s/ft}$$

Hence $q = 0.013$ ft^3/s per unit ft of plate.

To find shear force at the upper plate, find du/dy at the plate. At the plate, $y = 0.006$ ft.

Hence:

$$\frac{du}{dy} = 2154.75 - 718250 \times 0.006$$

$$= -2154.75 \text{ s}^{-1}$$

Therefore:

$$\tau = \mu \frac{du}{dy}$$

$$= 4 \times 10^{-5}(-2154.75)$$

$$= -0.086 \text{ lbs/ft}^2$$

Thus, the shear stress on the plate is 0.086 lbs/ft^2 and it is resisting the motion of the plate.

PROBLEMS

p.8.1 A fluid flows through a round pipe with **R** $< 2{,}000$. The shear stress is

a. zero at the wall and increases parabolically to the center.
b. zero at the wall and increases linearly to the center.
c. maximum at the wall and decreases linearly to zero at the center.
d. constant over the cross-section
e. none of these answers.

p.8.2 A fluid flows through a round pipe with **R** $< 2{,}000$. The velocity

a. is constant over the cross-section.
b. is zero at the wall and increases linearly toward the center.
c. varies parabolically across the section.
d. is maximum at the wall
e. is none of these answers.

p.8.3 A fluid flows through a round tube with **R** $< 2{,}000$. The discharge varies

a. linearly as the viscosity.
b. inversely as the pressure drops.
c. directly as the square of the diameter.
d. inversely as the square of the diameter.
e. none of these answers.

p.8.4 A fluid flows between two plates. The flow is laminar. The lower plate is fixed and the upper plate is moving. The shear stress

a. is constant over the distance between the plates.
b. is zero at the lower plate and increases linearly to a maximum at the upper plate.
c. is maximum at the lower plate and decreases linearly to zero at the upper plate.
d. varies parabolically across the distance between the plates.
e. none of these answers.

p.8.5 The shear stress in a fluid flowing between two fixed parallel plates with laminar flow

a. is constant over the distance between the plates.
b. varies parabolically over the distance between the plates.
c. is zero in the center and increases linearly towards the plates.
d. is zero at the plates and increases linearly towards the center.
e. none of these answers.

p.8.6 The velocity in a fluid flowing between two fixed parallel plates with laminar flow is

a. zero at the plates and increases linearly to a maximum at the center.
b. zero at the plates and increases parabolically to a maximum at the center.
c. maximum at the plates and decreases parabolically to zero at the center.
d. constant over the distance between the plates.
e. none of these answers.

p.8.7 Air of kinematic viscosity 15.5×10^{-5} ft^2/s flows through a pipe of 1.5 in in diameter. What is the maximum velocity for laminar flow?

p.8.8 Water at 20°C flows through a pipe of 25 cm in diameter and 6000 m long with a velocity of 3.75 m/s. Find the head loss.

p.8.9 Oil at a temperature of 20°C is pumped through a pipeline of 15 cm in diameter and 15,000 m long. Find the horsepower required to pump 100 liters of this oil per second. The oil has specific gravity of 0.9 and kinematic viscosity at 20°C is 1.67×10^{-5} m^2/s.

p.8.10 In Fig. 8.9 $\theta = 30°$, $p_1 = 10$ psi, $p_2 = 15$ psi, $Y = 0.005$ in, $l = 10$ ft, $\gamma = 50$ lb/ft^3, $U = 5$ ft/s. Determine the shear stress per sq ft on the upper plate and the rate of flow through the plates.

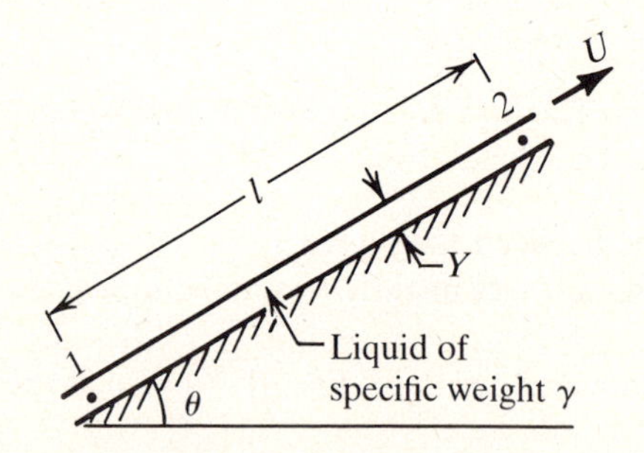

FIGURE 8.9 Problems 8.10, 8.11.

p.8.11 In Fig. 8.9 $\theta = 45°$, $p_1 = 50$ kPa, $p_2 = 30$ kPa, $Y = 0.009$ cm, $l = 100$ cm, $\gamma = 8500$ N/m^3, $U = 2$ m/s, determine the shear stress per sq meter on the upper plate and the rate of flow through the plates.

p.8.12 The conveyor belt system shown in Fig. 8.10 is used to transfer a fluid of specific weight γ from reservoir A to reservoir B. Derive an expression for velocity u in terms of U, θ, Y, y and γ. Also derive an expression for maximum rate at which the fluid can be transferred from A to B.

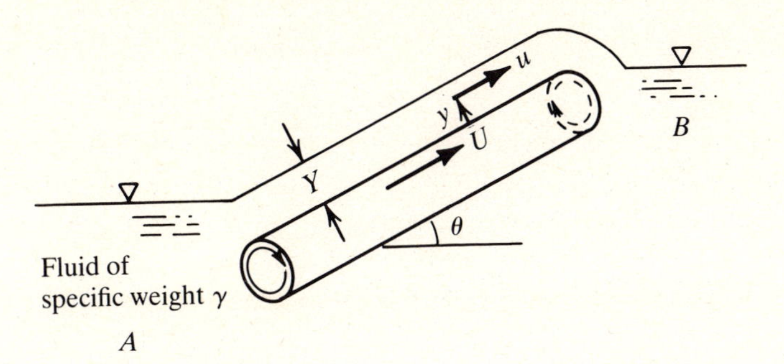

FIGURE 8.10 Problem 8.12.

Chapter Nine

TURBULENT FLOW IN PIPES

In Chapter Eight, we derived the Darcy–Weisbach equation for the head loss due to friction in laminar flow through pipes. In this chapter, the use of the Darcy–Weisbach equation will be extended to include turbulent flow in pipes. Three basic types of pipe problems will be introduced. Also, procedures for solving various types of pipe systems such as pipes connected in series, parallel, branches and networks will be developed.

9.1 INTRODUCTION

In turbulent flow, the motion of individual fluid particles is highly erratic producing eddies and cross-currents. Precise mathematical analysis of turbulent flow is difficult; therefore, experimentation is usually needed. Even the most advanced theories of turbulent flow utilize some experimental data. Most of the experimental work in turbulent flow was done using flow in round pipes; however, the findings can be generalized and extended to other shapes such as rectangular ducts and irregularly shaped sewers.

One of the pipe flow problems of interest to engineers is the estimation of energy required to force a fluid through a conduit with a certain velocity or pressure. Since some of the energy is lost in overcoming frictional resistance, estimation of such losses is important before a reliable estimate of total required energy can be made.

9.2 HEAD LOSS DUE TO FRICTION

An expression for head loss due to viscous resistance, known as the *Darcy-Weisbach equation*, was developed in Section 8.2. The form of the equation for laminar flow was

$$h_f = f\left(\frac{L}{D}\right)\left(\frac{V^2}{2g}\right) \tag{9.1}$$

where f, known as the *friction factor*, was shown to be a function of the Reynolds number and equal to $64/\mathbf{R}$. The other variables in Eq. (9.1) are: L = length of the pipe, D = diameter of the pipe, and V = average or mean velocity of flow. As the flow regime changes from laminar to turbulent, inertia forces start to become significant. The eddies and cross-currents appear which are aided by the roughness of the pipe. Experiments conducted by Henry Darcy (1803–58) on turbulent flow through long, unobstructed, straight pipes of uniform diameters showed that the head loss in a turbulent flow may also be calculated using Eq. (9.1). But in a case of turbulent flow the friction factor f is a function of the Reynolds number as well as of the pipe roughness.

All surfaces are rough to some extent. When viewed under a microscope, the surface of almost every material seems to have nail-like projections on it. The height, shape, size and spacing of these projections depend on the type of the material. In practice, usually the height of the roughness projections is taken to be the measure of the pipe roughness. Again, length of surface projections in absolute terms may not be a useful indicator since identical lengths of roughness projections may have significantly different effects in a small diameter pipe and in a large diameter pipe. Therefore, effects of pipe roughness are usually expressed in terms of *relative roughness*, which is defined as the ratio of the roughness projections to the diameter of the pipe.

We can then write:

$$\text{Relative roughness} = \frac{\varepsilon}{D} \tag{9.2}$$

where ε = height of roughness projections and D = diameter of the pipe.

In order to investigate the effect of the roughness of pipe walls on the resistance to flow, a German engineer, Johann Nikuradse, conducted a series of experiments on pipes having diameters of 2.5 cm, 5 cm, and 10 cm. The inner surfaces of these pipes were given different degrees of roughness by coating them with grains of sand of various coarseness. Six different values for ε/D ranging from 1/30 to 1/1014 were obtained. The resistance of each pipe was measured experimentally for various velocities of flow. The friction factors were thus obtained for various values of the Reynolds number and for various values of relative roughness.

The results of Nikuradse's experiments are shown in Fig. 9.1 where f and $\mathbf{R}$ have been plotted on a log-log scale. For small values of $\mathbf{R}$ the flow is laminar and follows the straight line AB. During this type of flow, the friction factor did not depend on the roughness of the pipe because all the results lie on the straight line AB regardless of the relative roughness ε/D. The turbulent flow follows another line (not straight) CE for smooth surfaces, while for rough surfaces the results seem to deviate from the line CE. The pipes with the roughest surface cause the points to break away from the line CE at smaller values of $\mathbf{R}$, while the smoother pipes cause the points to coincide

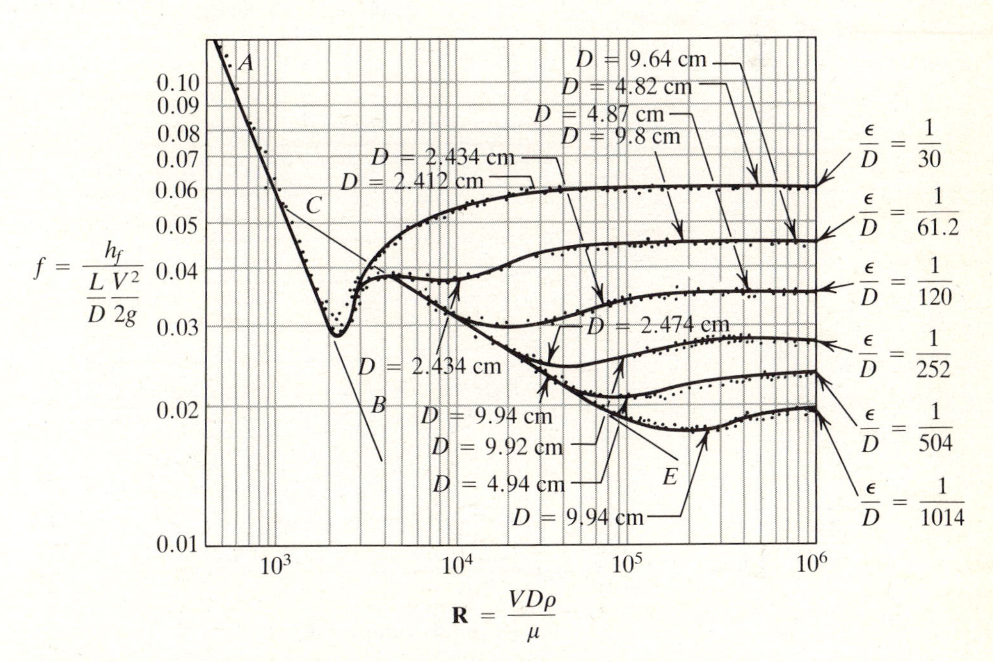

FIGURE 9.1 Results of Nikuradse's sand-roughened pipe experiments.

with line CE up to large values of $\mathbf{R}$. From this experiment, the following relationships between f, $\mathbf{R}$ and ε/D were obtained:

1. For laminar flow, when $\mathbf{R} \leqslant 2{,}000$

$$f = \frac{64}{\mathbf{R}} \tag{9.3}$$

2. For turbulent flow in smooth pipes (line CE)

$$\frac{1}{\sqrt{f}} = 0.86 \ln\left(\mathbf{R}\sqrt{f}\right) - 0.8 \tag{9.4}$$

3. For completely rough surfaces, when f becomes independent of $\mathbf{R}$

$$\frac{1}{\sqrt{f}} = 1.14 - 0.86 \ln \frac{\varepsilon}{D} \tag{9.5}$$

4. For the region between smooth and completely rough surfaces

$$\frac{1}{\sqrt{f}} = -0.86 \ln\left(\frac{\varepsilon/D}{3.7} + \frac{2.51}{\mathbf{R}\sqrt{f}}\right) \tag{9.6}$$

Eq. (9.6) is known as *the Colebrook equation* and is the basis for the Moody diagram given in Fig. 9.2. The values of ε for various pipe materials are given in Table 9.1.

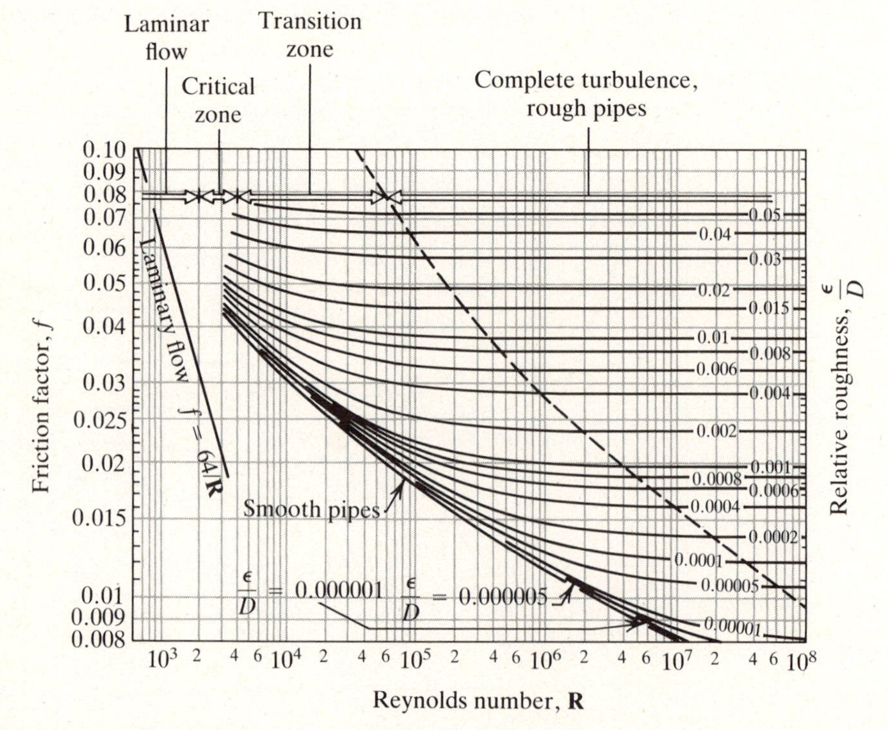

FIGURE 9.2 Moody diagram.

Table 9.1 VALUES OF ε FOR VARIOUS MATERIALS

	Feet	Millimeters
Drawn tubing, brass, lead, glass, centrifugally spun cement, bituminous lining, transite	0.000005	0.0015
Commercial steel or wrought iron	0.00015	0.046
Welded-steel pipe	0.00015	0.046
Asphalt-dipped cast iron	0.0004	0.12
Galvanized iron	0.0005	0.15
Cast iron, average	0.00085	0.25
Wood stave	0.0006 to 0.003	0.18 to 0.9
Concrete	0.001 to 0.01	0.3 to 3.0
Riveted steel	0.003 to 0.03	0.9 to 9.0

Swamee and Jain (1976) suggested an explicit formula which can be used in place of a Moody diagram with the restrictions placed on it. The formula and the restrictions are:

$$f = \frac{1.325}{\left[\ln\left(\frac{\varepsilon}{3.7D} + \frac{5.74}{\mathbf{R}^{0.9}}\right)\right]^2} \tag{9.7}$$

For

$$10^{-6} \leq \frac{\varepsilon}{D} \leq 10^{-2}$$

$$5000 \leq \mathbf{R} \leq 10^8$$

Results obtained from Eq. (9.7) match closely with those obtained from Eq. (9.6).

9.3 SIMPLE PIPE PROBLEMS

Generally, pipe flow problems can be divided into three types:

Type I. When Q, L, D, ν, and ε are given and the head loss h_f is required.
Type II. When h_f, L, D, ν, and ε are given and the discharge Q is required.
Type III. When h_f, Q, L, ν, ε, are given and the diameter, D, is required.

The fundamental relationships needed to solve pipe flow problems are:

1. Reynolds number: $\mathbf{R} = \dfrac{VD\rho}{\mu} = \dfrac{VD}{\nu}$

2. The Darcy–Weisbach equation: $h_f = \dfrac{fLV^2}{D2g}$

The solution procedure for each type is first described and then illustrated by numerical examples.

Problem Type I

The following steps may be followed to solve problem type I:

1. Determine ε of the pipe material from Table 9.1 and compute ε/D.
2. Determine Reynolds number using

$$\mathbf{R} = \frac{VD\rho}{\mu} = \frac{VD}{\nu}$$

3. Find friction factor f using an appropriate equation or Moody diagram.
4. Calculate head loss due to friction using

$$h_f = \frac{fLV^2}{2Dg}$$

EXAMPLE 9.1

Determine the head loss due to a flow of 100 1/s of glycerine at 20°C through 100.0 m of 20 cm diameter cast iron pipe. Rework the problem if the fluid is water.

Solution For glycerine, from Table A-4

$$S = 1.26$$

$$\mu = 0.862 \text{ N} \cdot \text{s/m}^2$$

Also, for water, from Table A-1

$$\rho_w \text{ at } 20^\circ\text{C} = 998 \text{ kg/m}^3$$

Hence

$$\rho_g = \text{density of glycerine}$$

$$= 1.26 \times 998 = 1257.5 \text{ kg/m}^3.$$

$$Q = 100 \text{ 1/s} = 0.1 \text{ m}^3\text{/s}$$

$$A = \frac{\pi(0.2)^2}{4} = 0.0314 \text{ m}^2$$

$$V = \frac{0.1}{0.0314} = 3.185 \text{ m/s}$$

$$\mathbf{R} = \frac{VD\rho}{\mu} = \frac{3.185 \times 0.2 \times 1257.5}{0.862} = 930 \text{ (laminar flow)}$$

Since the flow is laminar, it is independent of pipe roughness.

Also

$$f = \frac{64}{\mathbf{R}} = \frac{64}{930} = 0.0688$$

$$h_f = f\frac{L}{D}\frac{V^2}{2g}$$

$$= \frac{0.0688 \times 100 \times (3.185)^2}{0.2 \times 2 \times 9.81}$$

$$= 17.8 \text{ m of glycerine}$$

If the fluid flowing through the pipe is water, then from Table A-1,

$$\mu = 1.005 \times 10^{-3}\ \text{N} \cdot \text{s/m}^2$$

$$\rho = 998\ \text{kg/m}^3$$

Hence:

$$\mathbf{R} = \frac{3.185 \times 0.2 \times 998}{1.005 \times 10^{-3}} = 6.32 \times 10^5$$

The flow is turbulent; f will be found using the Moody diagram. From Table 9.1 ε for cast iron = 0.025 cm.

$$\varepsilon/D = \frac{0.025}{20} = 0.00125$$

Now

$$f(0.00125, 6.31 \times 10^5) = 0.021$$

Therefore,

$$h_f = \frac{0.021 \times 100 \times (3.185)^2}{0.2 \times 2 \times 9.81} = 5.43 \text{ m of water}$$

It may be noted that the head loss in the case of water is approximately one-third of that in the case of glycerine. In other words, three times as much power will be required to pump glycerine as water.

Computer Solution for Problem Type I

The solution to pipe problems can be handled easily and effectively by computers. A computer program written in BASIC is listed in Appendix B which can be used to solve the three types of pipe problems. The solution to Example 9.1 is shown below:

```
RUN
this program solves for three types of pipe problems
punch the variable to be determined equal to zero
title? Example 9.1
units(si/fps)? si
length of pipe(m)? 100
diameter of the pipe(m)? 0.2
epsilon(m)? 0.00025
k viscosity(sq.m/sec)? 1.007e-06
head loss(m)? 0
discharge(cu.m/sec)? 0.1

      **********
Example 9.1
head loss(m)=   5.487956
      **********
do you wish to solve another problem (y/n)? n
Break in 590
Ok
```

Problem Type II

This type of problem involves three unknowns—i.e., **R**, V and f and two equations, namely the head loss equation and the Reynolds number equation. Since the number of unknowns is greater than the number of useful equations, the unknowns cannot be obtained directly. However, an indirect iterative procedure may be devised as follows:

$$h_f = f \frac{L}{D} \frac{V^2}{2g}$$

from which

$$V = \sqrt{\frac{2gDh_f}{L}}\,(f)^{-1/2}$$

or

$$V = \frac{A}{\sqrt{f}} \tag{9.8}$$

where

$$A = \sqrt{\frac{2gDh_f}{L}} \qquad \text{(known constant for Problem Type II)} \tag{9.9}$$

Also

$$\mathbf{R} = \frac{VD}{\nu}$$

or

$$\mathbf{R} = BV \tag{9.10}$$

where

$$B = \frac{D}{\nu} \qquad \text{(known constant for Problem Type II)} \qquad (9.11)$$

After establishing Eqs. (9.8) and (9.10) by inserting numerical values for A and B, the following iterative procedure may be adopted:

1. Determine ε/D.
2. Assume a value for f. Rather than a mere guess, a reasonable value of f may be assumed using Moody diagrams. Locate ε/D on the diagram and go horizontally across to the f scale and read f.
3. Using the value of f, calculate V using Eq. (9.8).
4. Using the value of V obtained in Step 3, calculate **R** using Eq. (9.10).
5. Using ε/D and **R**, determine f.
6. Compare f obtained in Step 5 with the assumed value of f. If they are identical, stop. If not, repeat steps 3 through 6, until convergence is attained.
7. After convergence, calculate discharge using the equation

$$Q = \frac{V\pi D^2}{4}$$

EXAMPLE 9.2

Determine the discharge of water at 20°C through a 20 cm diameter cast iron pipe if the head loss in 100 m length of pipe is equal to 5.43 m.

Solution This is a Type II problem.

For water at 20°C, $\nu = 1.007 \times 10^{-6}$ m²/s (from Table A-1). ε for cast iron = 0.025 cm

$$L = 100 \text{ m}$$

$$D = 20 \text{ cm} = 0.2 \text{ m}$$

$$h_f = 5.43 \text{ m}$$

$$\varepsilon/D = \frac{0.025}{20} = 0.00125$$

Using Eq. (9.9)

$$A = \sqrt{\frac{2gDh_f}{L}}$$

$$= \sqrt{\frac{2 \times 9.81 \times 0.2 \times 5.43}{100}}$$

$$= 0.462$$

Hence, from Eq. (9.8)

$$V = \frac{0.462}{\sqrt{f}} \qquad (1)$$

Using Eq. (9.11)

$$B = \frac{D}{\nu} = \frac{0.2}{1.007 \times 10^{-6}} = 1.99 \times 10^5$$

Then, from Eq. (9.10)

$$\mathbf{R} = 1.99 \times 10^5 V \tag{2}$$

Read f against $\varepsilon/D = 0.00125$ from the Moody diagram

$$f = 0.021$$

Calculate V from Eq. (1)

$$V = \frac{0.462}{\sqrt{0.021}} = 3.19 \text{ m/s}$$

Calculate **R** from Eq. (2)

$$\mathbf{R} = 1.992 \times 10^5 \times 3.19 = 6.35 \times 10^5$$

From the Moody diagram:

$$f(0.00125, 6.35 \times 10^5) = 0.021$$

This is equal to the assumed value; hence, convergence is attained.

$$Q = AV = \frac{\pi(0.2)^2}{4} \times 3.19 = 0.1 \text{ m}^3\text{/s}$$

Computer Solution for Problem Type II

The solution to the Example 9.2 is shown below using the computer program listed in Appendix B.

```
RUN
this program solves for three types of pipe problems
punch the variable to be determined equal to zero
title? Example 9.2
units(si/fps)? si
length of pipe(m)? 100
diameter of the pipe(m)? 0.2
epsilon(m)? 0.00025
k viscosity(sq.m/sec)? 1.007e-06
head loss(m)? 5.49
discharge(cu.m/sec)? 0

          **********
Example 9.2
discharge(cu.m/sec)=            .1000506
          **********
do you wish to solve another problem (y/n)? n
Break in 590
Ok
```

Problem Type III

In this type of problem, there are four unknowns—i.e., **R** V, f, and D. Therefore, an iterative procedure will be employed. The number of unknowns can be reduced by expressing velocity in terms of discharge.

$$h_f = f \frac{L}{D} \frac{V^2}{2g}$$

Now

$$V = \frac{Q}{A} = \frac{4Q}{\pi D^2}$$

Hence:

$$h_f = f \frac{L}{D} \left(\frac{4Q}{\pi D^2} \right)^2 \frac{1}{2g}$$

from which

$$D^5 = \left(\frac{8LQ^2}{h_f g \pi^2} \right) f$$

$$D^5 = Af \tag{9.12}$$

where

$$A = \frac{8LQ^2}{h_f g \pi^2} \tag{9.13}$$

For Problem Type III, A is a known constant.
Also

$$\mathbf{R} = \frac{VD}{\nu}$$

Substituting $V = 4Q/\pi D^2$ in the above equation:

$$\mathbf{R} = \frac{4Q}{\pi \nu} \frac{1}{D}$$

$$\mathbf{R} = \frac{B}{D} \tag{9.14}$$

where

$$B = \frac{4Q}{\pi \nu} \tag{9.15}$$

For Problem Type III, B is a known constant.

After establishing Eqs. (9.12) and (9.14) by inserting numerical values for A and B, the following iterative procedure may be used:

1. Assume $f = 0.01$
2. Calculate D, using Eq. (9.12)
3. Calculate ε/D
4. Calculate **R** using Eq. (9.14)
5. Using ε/D and **R**, determine f.

6. Compare f obtained in Step 5 with the assumed value of f. If they are approximately equal, stop. If not, repeat steps 2 through 5 until convergence is attained.

EXAMPLE 9.3

Determine the size of a cast iron pipe required to convey 100 1/s. of water at 20°C with a head loss of 5.43 m in a 100 m length of the pipe.

Solution

$$Q = 100 \text{ 1/s} = 0.1 \text{ m}^3/\text{s}$$

$$h_f = 5.43 \text{ m}$$

$$L = 100 \text{ m}$$

$$\nu = 1.007 \times 10^{-6} \text{ m}^2/\text{s}$$

$$\varepsilon \text{ for cast iron} = 0.025 \text{ cm}$$

Using Eq. (9.13)

$$A = \frac{8LQ^2}{h_f g \pi^2}$$

$$= \frac{8 \times 100 \times (0.1)^2}{5.43 \times 9.81 \times \pi^2}$$

$$= 1.522 \times 10^{-2}$$

Hence, from Eq. (9.12)

$$D^5 = 1.522 \times 10^{-2} f \tag{1}$$

Using Eq. (9.15)

$$B = \frac{4Q}{\pi \nu}$$

$$= \frac{4 \times 0.1}{\pi \times 1.007 \times 10^{-6}}$$

$$= 1.26 \times 10^5$$

Hence, from Eq. (9.14)

$$\mathbf{R} = \frac{1.26 \times 10^5}{D} \tag{2}$$

Iteration #1
Assume $f = 0.01$.
From Eq. (1)

$$D = 0.172 \text{ m}$$

$$\frac{\varepsilon}{D} = \frac{0.00025}{0.172} = 1.453 \times 10^{-3}$$

From Eq. (2)

$$\mathbf{R} = 7.33 \times 10^5$$

$$f(1.453 \times 10^{-3}, 7.33 \times 10^5) = 0.022$$

This value of f is not the same as the starting value; therefore, a second iteration is required.

Iteration #2

$$f = 0.021$$

From Eq. (1)

$$D = 0.199 \text{ m}$$

From Eq. (2)

$$\mathbf{R} = 6.33 \times 10^5$$

and

$$\frac{\varepsilon}{D} = \frac{0.00025}{0.199} = 0.0013$$

$$f(0.0013, 6.33 \times 10^5) = 0.021$$

This is equal to the starting value; hence, convergence is attained.

$$D = 0.199 \simeq 0.2 \text{ m}$$

Computer Solution

The solution to the Example 9.3 is shown below using the computer program listed in Appendix B.

```
RUN
this program solves for three types of pipe problems
punch the variable to be determined equal to zero
title? Example 9.3
units(si/fps)? si
length of pipe(m)? 100
diameter of the pipe(m)? 0
epsilon(m)? 0.00025
k viscosity(sq.m/sec)? 1.007e-06
head loss(m)? 5.49
discharge(cu.m/sec)? 0.1

          **********
Example 9.3
diameter(m)=    .1999317
          **********
do you wish to solve another problem (y/n)? n
Break in 590
Ok
```

9.4 SOME OTHER PIPE FRICTION FORMULAS

The empirical pipe-friction formulas, also known as *industrial pipe-friction formulas*, take the form

$$\frac{h_f}{L} = \frac{RQ^n}{D^m} \tag{9.16}$$

or,

$$S = \frac{RQ^n}{D^m}$$

where h_f = head loss due to friction
L = length of pipe
D = diameter of pipe
Q = discharge
R = resistance coefficient
n, m = constants, experimentally obtained

Table 9.2 VALUES OF C FOR THE HAZEN–WILLIAMS FORMULA

Type of pipe	C
Asbestos cement	140
Brass	130–140
Brick sewer	100
Cast iron	
New, unlined	130
Old, unlined	40–120
Cement-lined	130–150
Bitumastic, enamel-lined	140–150
Tar-coated	115–135
Concrete or concrete-lined	
Steel forms	140
Wooden forms	120
Centrifugally spun	135
Copper	130–140
Fire hose (rubber-lined)	135
Galvanized iron	120
Glass	140
Lead	130–140
Plastic	140–150
Steel	
Coal-tar enamel-lined	145–150
New, unlined	140–150
Riveted	110
Tin	130
Vitrified clay	100–140

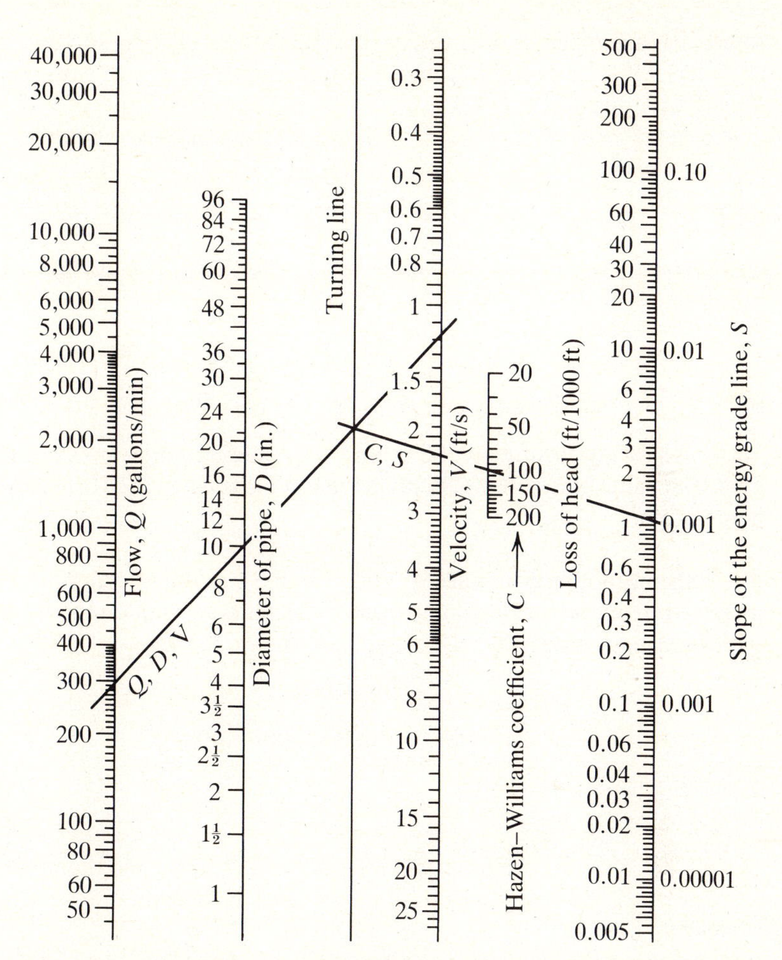

FIGURE 9.3 The Hazen–Williams nomograph.

The resistance coefficient R is a function of pipe roughness only. The value of R for a particular fluid is obtained experimentally and is valid for that fluid only. Since R is independent of viscosity, the changes of viscosity cannot be accounted for in the formula. Therefore, an empirical formula with specified values of n, m and R is valid only for the fluid viscosity for which it is developed, and it is normally limited to a range of Reynolds numbers and diameters.

Hazen and Williams developed an empirical formula in the form of Eq. (9.16) for water at ordinary temperatures. Their formula is given by:

$$S = \frac{4.727Q^{1.852}}{C^{1.852}D^{4.8704}} \qquad \text{in FPS units} \qquad (9.17a)$$

$$S = \frac{10.675Q^{1.852}}{C^{1.852}D^{4.8704}} \qquad \text{in SI units} \qquad (9.17b)$$

where C is a coefficient. The values of C for various types of surfaces is given in Table 9.2. The Hazen–Williams equation can be easily solved using a nomograph given in Fig. 9.3.

A general form of the Hazen–Williams formula may be written as

$$h_f = \frac{ALQ^n}{C^n D^m} \tag{9.18}$$

where $n = 1.852$, $m = 4.8704$, and $A = 4.727$ in the FPS units and 10.675 in the SI units.

EXAMPLE 9.4

Water is flowing through a 500 ft long, 9 in diameter pipe at the rate of 500 gpm. The value of C is 100. Calculate the head loss using the Hazen–Williams nomograph.

Solution In the nomograph, connect Q and D with a straight line and mark off the intersection on the turning line. Connect the marked point on the turning line with C with a straight line and read S on the slope line.

$$S = 0.004$$

$$h_f = SL = 0.004 \times 500 = 2.0 \text{ ft}$$

Therefore, head loss = 2.0 ft

Comment

It may be recalled that the head loss in a pipe is also calculated using the Darcy–Weisbach equation. The Hazen–Williams and the Darcy–Weisbach formulas can differ significantly in their results. The Darcy–Weisbach equation is more comprehensive and rationally based. It can account for pipe roughness as well as fluid viscosity and, therefore, has received wide acceptance. However, the empirical formulas developed from experimental results under specific conditions are generally more useful in the region over which the data were gathered.

Pipe Conveyance

For fully turbulent conditions at high Reynolds numbers, the friction factor is given by Eq. (9.5):

$$\frac{1}{\sqrt{f}} = 1.14 - 0.86 \ln \frac{\varepsilon}{D}$$

or,

$$\frac{1}{\sqrt{f}} = 1.14 + 2 \log \frac{D}{\varepsilon}$$

Substituting in the Darcy–Weisbach equation and converting V to Q, we obtain

$$Q = \left[\frac{\pi}{4}\sqrt{2g}\left(2\log\frac{D}{\varepsilon} + 1.14\right)D^{2.5}\right]\left(\frac{h_f}{L}\right)^{1/2} \tag{9.19}$$

or,

$$Q = \mathrm{K}\left(\frac{h_f}{L}\right)^{1/2} = \mathrm{K}\sqrt{S} \tag{9.20}$$

In Eq. (9.20), K is a constant for a particular pipe, and is called *conveyance.* Eq. (9.20) permits a direct solution of a pipe problem without any reference to the Moody diagram.

The formula, however, does not account for the viscosity of the fluid and, therefore, should be used with caution in problems where flows are not at high Reynolds numbers and viscosity may be important.

EXAMPLE 9.5

Determine the head loss when water flows at the rate of 100 1/s through a 30 cm diameter cast iron pipe. The pipe is 500 m long. Use the method of conveyance.

Solution For cast iron, $\varepsilon = 0.025$ cm

$$\mathrm{K} = \frac{\pi}{4}\sqrt{2g}\left(2\log\frac{D}{\varepsilon} + 1.14\right)D^{2.5}$$

$$= \frac{\pi}{4}\sqrt{2 \times 9.81}\left(2\log\frac{30}{0.025} + 1.14\right)(0.3)^{2.5}$$

$$= 1.25 \text{ m}^3/\text{s}$$

$$S = \left(\frac{Q}{\mathrm{K}}\right)^2 = \left(\frac{0.1}{1.25}\right)^2 = 0.0064$$

$$h_f = 0.0064 \times 500 = 3.2 \text{ m}$$

Therefore, head loss = 3.2 m.

EXAMPLE 9.6

Solve Example 9.2 using the conveyance method.

Solution For cast iron, $\varepsilon = 0.025$ cm

$$\mathrm{K} = \frac{\pi}{4}\sqrt{2 \times 9.81}\left(2\log\frac{20}{0.025} + 1.14\right)(0.2)^{2.5}$$

$$= 0.432 \text{ m}^3/\text{s}$$

$$Q = \mathrm{K}\sqrt{S} = 0.432\sqrt{\frac{5.43}{100}}$$

$$= 0.1 \text{ m}^3/\text{s}$$

Hence, the rate of flow is 0.1 m^3/sec. This is identical to what we obtained using the Darcy–Weisbach equation.

EXAMPLE 9.7

Solve Example 9.3 using the conveyance method.

Solution For cast iron, $\varepsilon = 0.025$ cm

$$K = \frac{Q}{\sqrt{S}} = \frac{0.1}{\sqrt{\frac{5.43}{100}}} = 0.432 \text{ m}^3/\text{s}$$

Also

$$0.432 = \frac{\pi}{4}\sqrt{2 \times 9.81}\left(2 \log \frac{D}{\varepsilon} + 1.14\right)D^{2.5}$$

$$0.432 = 3.479(2 \log D/\varepsilon + 1.14)D^{2.5}$$

A trial and error solution is required. Assume a value for D and evaluate the right-hand side. If it is equal to 0.432, stop. If not, assume another D and solve until convergence is attained. By trial and error, $D = 0.2$ m.

9.5 MINOR HEAD LOSSES

In addition to frictional head losses, a fluid flowing through a pipe is also subjected to head losses at changes in the cross-section, bends, and various other pipe fittings. The losses usually stem from the changes in the magnitude or direction of velocity which results in turbulence and eddies. This causes loss of energy in the form of heat. All such losses are termed *minor head losses* and are generally expressed as functions of velocity head, or

$$h_m = \frac{KV^2}{2g}$$

where h_m = minor head loss, K = an experimentally determined coefficient.

Although termed minor, such losses may exceed the frictional losses in small pipe systems. However, in long pipelines, minor head losses are small compared to the frictional head losses.

Head Loss Due To Contraction

A sudden contraction in a pipe is shown in Fig. 9.4. As the fluid approaches the contraction, the streamlines start to bend at a point A. This produces a

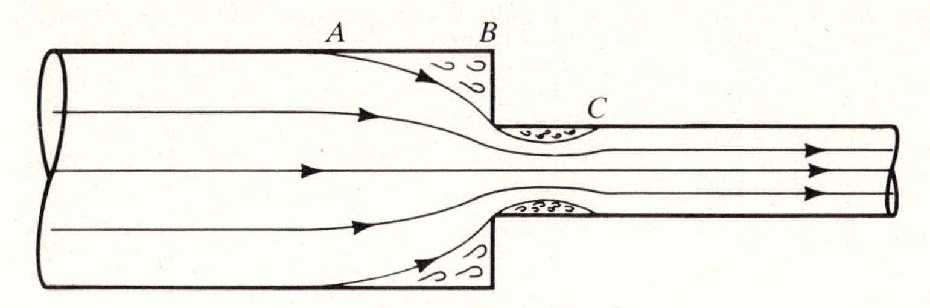

FIGURE 9.4 Streamlines in a sudden contraction.

zone of stagnant water with some eddies from A to B. At B, the fluid separates from the pipe and a *vena contracta* is formed. The stream then expands and reattaches itself to the wall at the point C. Eddies are formed between the *vena contracta* and the wall and it is here that most of the energy is dissipated as heat. The value of K can be found experimentally by using various diameters. Some typical results are shown in Table 9.3. For fully turbulent flow, K remains constant for a certain contraction ratio (i.e., d/D).

The head losses due to sudden contraction can be greatly reduced by providing a gradual transition in the form of a *confusor* (see diagram in Table 9.3). The head loss in a confusor is given by

$$h_m = \frac{K(V_1 - V_2)^2}{2g}$$

where K is a function of angle α. The values of K for various angles are given in Table 9.3.

A special case of sudden contraction is encountered when a fluid enters into a pipe from a reservoir. In this case, a contraction ratio of zero may be assumed. The value of K depends on the conditions at the entrance to the pipe. The entrance may be squared, rounded, reentrant or skewed. The values of K for these conditions are given in Table 9.3.

Head Loss Due to Expansion

A sudden expansion in a pipe is shown in Fig. 9.5. The streamlines emerging from the smaller diameter pipe cannot conform to the abrupt change in diameter at Section A and separate from the wall. The stream later expands and reattaches itself to the wall at B. In the corner, pockets of turbulent eddies are formed and energy is dissipated in the form of heat.

An expression for the head loss in sudden expansion was first obtained by J. C. Borda (1733–99) and L. Carnot (1753–1823) using momentum principles. The form of their expression is

$$h_m = \frac{(V_1 - V_2)^2}{2g}$$

Table 9.3 MINOR HEAD LOSS COEFFICIENTS

Diagram	Description	Value of K in $h_m = K(V^2/2g)$
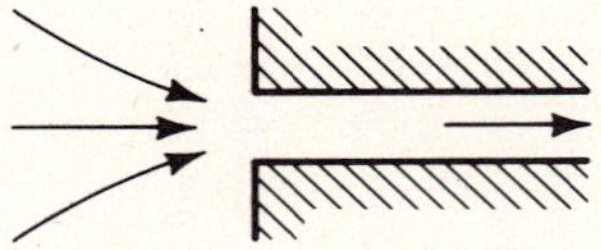	Perpendicular square entrance	0.50
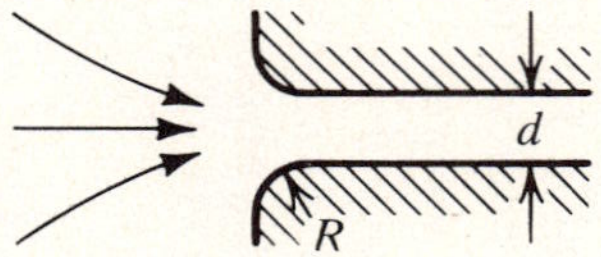	Perpendicular rounded entrance	$R/d =$ 0.05 | 0.1 | 0.2 | 0.3 | 0.4 $K =$ 0.25 | 0.17 | 0.08 | 0.05 | 0.04
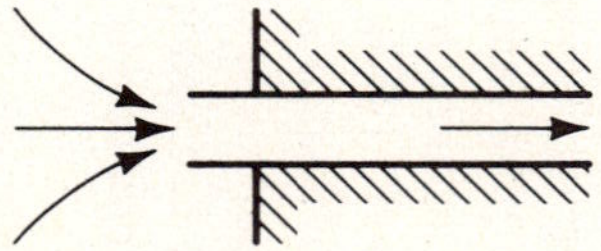	Perpendicular reentrant entrance	0.8
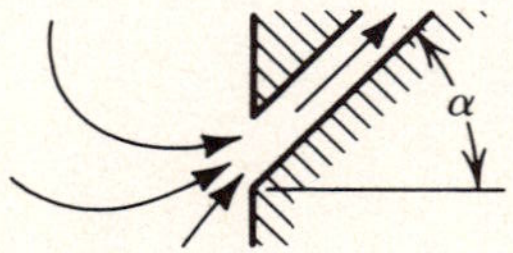	Additional loss due to skewed entrance	$K = 0.505 + 0.303 \sin \alpha + 0.226 \sin^2 \alpha$
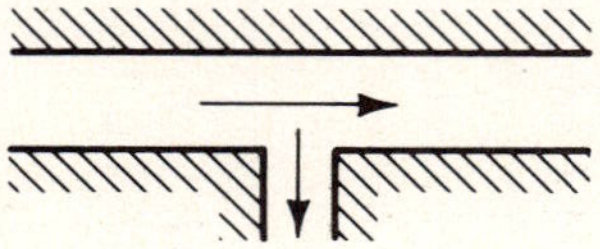	Standard tee, entrance to minor line	1.8

Sudden expansion

$$h_m = \frac{(V_1 - V_2)^2}{2g}$$

or

$$h_m = \left(\frac{V_1}{V_2} - 1\right)^2 \frac{V_2^2}{2g}$$

Sudden contraction

$(d/D)^2 =$	0.01	0.1	0.2	0.4	0.6	0.8
$K =$	0.5	0.5	0.45	0.38	0.28	0.14

Diffusor

$$h_m = \frac{K(V_1 - V_2)^2}{2g}$$

$\alpha^\circ =$	6	10	20	40	60	80	100	120	140
K for $D = 3d$	0.12	0.16	0.40	0.80	1.0	1.06	1.04	1.04	1.04
$D = 1.5d$	0.12	0.16	0.40	0.90	1.21	1.16	1.09	1.06	1.04
$D = 2d$	0.12	0.16	0.40	1.02	1.21	1.17	1.10		

Confusor

$$h_m = \frac{K(V_1 - V_2)^2}{2g}$$

$\alpha^\circ =$	20	40	60	80
$K =$	0.20	0.28	0.32	0.35

(Continued)

Table 9.3 *(Continued)*

<table>
<tr><th>Diagram</th><th>Description</th><th>Value of K in $h_m = K(V^2/2g)$</th></tr>
<tr><td>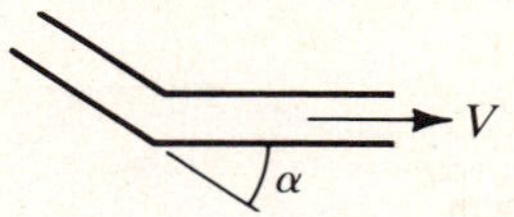</td><td>Sharp elbow</td><td>$K = 67.6 \times 10^{-6}(\alpha^\circ)^{2.17}$</td></tr>
<tr><td>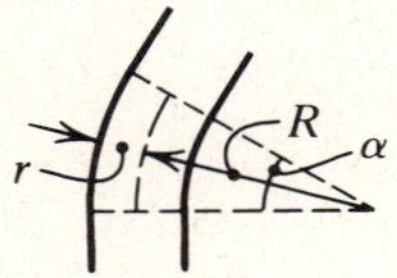</td><td>Bends</td><td>$[0.13 + 1.85(r/R)^{3.5}]\sqrt{\alpha^\circ/180^\circ}$</td></tr>
<tr><td>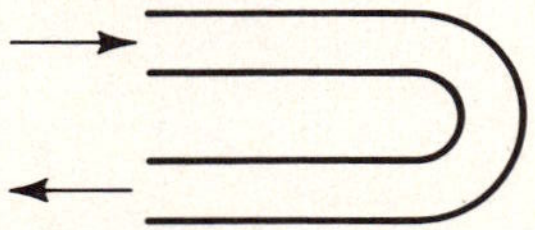</td><td>Close return bend</td><td>2.2</td></tr>
<tr><td>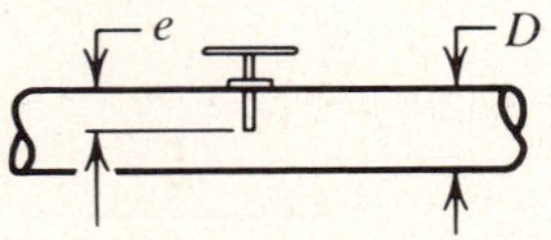</td><td>Gate valve</td><td><table><tr><td>$\varepsilon/D =$</td><td>0</td><td>1/4</td><td>3/8</td><td>1/2</td><td>5/8</td><td>3/4</td><td>7/8</td></tr><tr><td>$K =$</td><td>0.15</td><td>0.26</td><td>0.81</td><td>2.06</td><td>5.52</td><td>17.0</td><td>97.8</td></tr></table></td></tr>
<tr><td>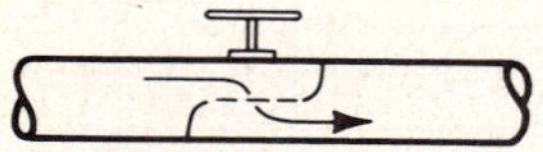</td><td>Globe valve</td><td>10 when fully open</td></tr>
<tr><td>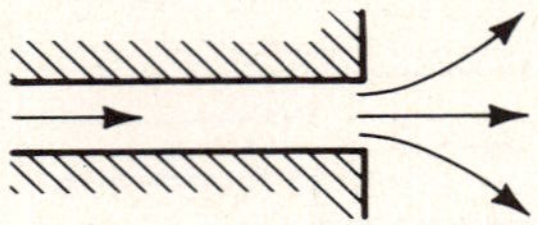</td><td>Exit from pipe
into reservoir</td><td>1.0</td></tr>
</table>

Note: Use the equation $h_m = K(V^2/2g)$ unless otherwise indicated.

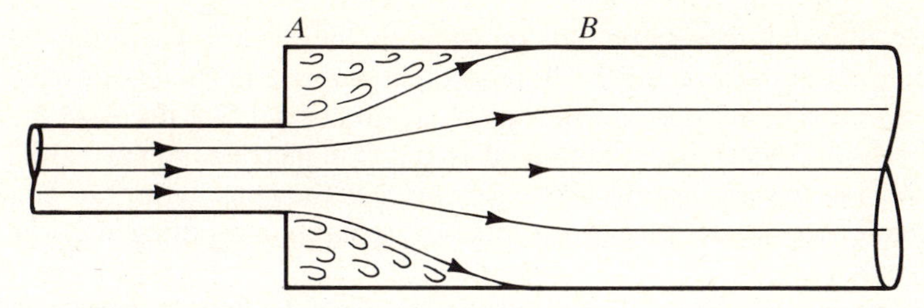

FIGURE 9.5 Streamlines in a sudden expansion.

The head loss due to sudden expansion can be greatly reduced by providing a gradual transition in the form of a *diffusor* (see diagram in Table 9.3). The head loss in a diffusor is given by

$$h_m = \frac{K(V_1 - V_2)^2}{2g}$$

where K depends on the expansion angle α and the expansion ratio (D/d). The values of K for various angles and expansion ratios are given in Table 9.3.

A special case of sudden expansion is encountered when a pipe discharges into a large reservoir. The head loss in this case is given by

$$h_m = \frac{V^2}{2g}$$

where V is the velocity in the pipe.

Head Loss In Bends

When a fluid flows through a bend it suffers a head loss due to change in direction. Consider a fluid flowing in a bend as shown in Fig. 9.6a. The

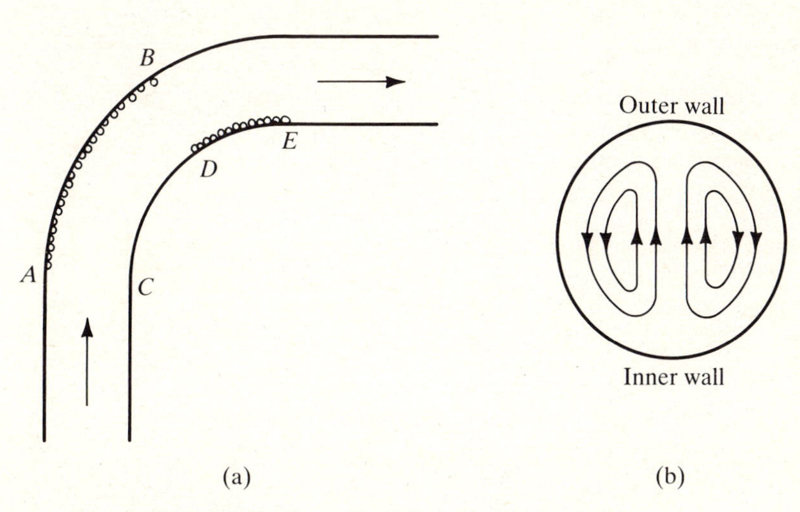

FIGURE 9.6 Causes of head loss when a fluid flows through a bend.

centrifugal forces cause the pressure to increase at the outer edge and decrease at the inner edge. The increase in pressure starts at point A, reaches a maximum at point B, and then decreases to normal. The decrease in pressure starts at C, reaches a minimum at D and then increases to normal. Thus, the fluid has to work against adverse pressure in the range from A to B and from D to E. This causes turbulent eddies and separation resulting in dissipation of energy as heat.

In addition, secondary currents, in the form of a double spiral as shown in Fig. 9.6b, are produced in the radial plane due to a pressure differential from outer-wall to inner-wall and from inside to outside. The double spiral motion may persist for a distance downstream as much as 50 to 75 times the pipe diameter until it is dissipated by viscous friction.

The head loss in a bend may be expressed as

$$h_m = \frac{KV^2}{2g}$$

where K depends on the radius of curvature and the angle of the bend. The values of K for various angles and radii of curvature may be determined using the relationship given in Table 9.3.

Head Loss In Pipe Fittings

Additional head is lost in other pipe fittings such as valves, couplings, elbows, etc. The head losses in such fittings are also expressed as

$$h_m = \frac{KV^2}{2g}$$

The values of K for various fittings are given in Table 9.3.

EXAMPLE 9.8

Determine the head loss through a 10 cm diameter cast iron pipe of 30 m length when water at 20°C flows through at the rate of 75 l/s. The system includes a sudden entrance and a globe valve.

Solution

$$Q = 75 \text{ l/s} = 0.075 \text{ m}^3\text{/s}$$

$$L = 30 \text{ m}$$

$$D = 10 \text{ cm} = 0.1 \text{ m}$$

$$A = \frac{\pi(0.1)^2}{4} = 0.007854 \text{ m}^2$$

$$V = \frac{Q}{A} = \frac{0.075}{0.007854} = 9.55 \text{ m/s}$$

For water at 20°C,

$$\nu = 1.007 \times 10^{-6}\ \text{m}^2/\text{s}$$

For cast iron,

$$\varepsilon = 0.025\ \text{cm}$$

$$\frac{\varepsilon}{D} = \frac{0.025}{10} = 0.0025$$

$$\mathbf{R} = \frac{VD}{\nu} = \frac{9.55 \times 0.1}{1.007 \times 10^{-6}} = 9.5 \times 10^5$$

$$f(0.0025, 9.5 \times 10^5) = 0.025 \text{ (from Moody diagram)}$$

Total head loss in the system = head loss due to friction + head loss due to sudden entrance + head loss due to globe valve

a. Head loss due to friction

$$= f\frac{L}{D}\frac{V^2}{2g} = \frac{0.025 \times 30 \times (9.55)^2}{0.1 \times 2 \times 9.81} = 34.9\ \text{m}$$

b. Head loss due to sudden contraction $= \dfrac{KV^2}{2g}$

where

$$K = 0.5$$

$$h_m = \frac{0.5(9.55)^2}{2 \times 9.81} = 2.3\ \text{m}$$

c. Head loss due to globe valve $= \dfrac{KV^2}{2g}$

where

$$K = 10$$

$$h_m = \frac{10(9.55)^2}{2 \times 9.81} = 46.5\ \text{m}$$

Hence, total head loss

$$= 34.9 + 2.3 + 46.5$$

$$= 83.7\ \text{m}$$

Comment

Although head losses due to changes of section, direction and the presence of obstructions, are termed minor head losses, they may be significant. Depending upon the nature of the flow system, they may even be larger than the frictional head loss as was evident from Example 9.8.

9.6 PIPES IN SERIES

Two or more pipes are said to be connected *in series* if the head of one is connected to the tail of the other and so on. Two important features of pipes connected in series are:

1. The head loss is cumulative—i.e., the total head loss is equal to the sum of the head losses in individual pipes.
2. The discharge through each pipe is the same.

All series pipe problems may fall into the following two categories:

1. Given the flow rate and pipe combinations, determine the total head loss.
2. Given the allowable head loss and the combination of the pipes, determine the flow rate.

The solution procedure is illustrated by the following example.

EXAMPLE 9.9

Find the discharge through the system shown in Fig. 9.7. All pipes are made of cast iron.

Solution For water at 20°C,

$$\nu = 1.007 \times 10^{-6}\ \text{m}^2/\text{s}$$

Let

$$V_1 = \text{velocity in 30 cm dia. pipe}$$

$$V_2 = \text{velocity in 15 cm dia. pipe}$$

$$V_3 = \text{velocity in 8 cm dia. nozzle}$$

Also let A be a point on the water surface in the reservoir and D be a point slightly outside the nozzle, such that the pressure at D is atmospheric. Taking

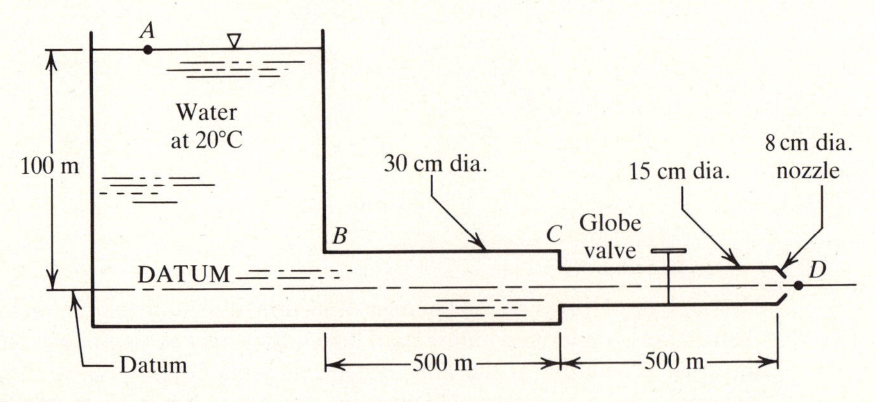

FIGURE 9.7 Example 9.9.

datum as shown and applying Bernoulli equation to points A and D, we obtain

$$\frac{p_A}{\gamma} + z_A + \frac{V_A^2}{2g} = \frac{p_D}{\gamma} + z_D + \frac{V_D^2}{2g} + HL_{(A-D)}$$

or

$$0 + 100 + 0 = 0 + 0 + \frac{V_D^2}{2g} + HL_{(A-D)}$$

Also

$$V_D = \text{velocity in the nozzle} = V_3.$$

Hence:

$$100 = \frac{V_3^2}{2g} + HL_{(A-D)} \tag{1}$$

Now tabulate all the head losses between points A and D.

(i) At section B, there will be head loss due to entrance and equal to $0.5(V_1^2/2g)$.
(ii) In pipe BC there will be a head loss due to friction and equal to $f_1(L_1/D_1)(V_1^2/2g)$.
(iii) At section C, there will be head loss due to sudden contraction and equal to $0.5(V_2^2/2g)$.
(iv) In pipe CD there will be a head loss due to friction and equal to $f_2(L_2/D_2)(V_2^2/2g)$.
(v) Between C and D there will be head loss due to presence of a valve and equal to $10(V_2^2/2g)$.
(vi) At the nozzle there will be a head loss due to reduction in area and equal to $3(V_3^2/2g)$.

Adding all the losses,

$$HL_{(A-D)} = 0.5\frac{V_1^2}{2g} + \frac{f_1L_1}{D_1}\frac{V_1^2}{2g} + 0.5\frac{V_2^2}{2g} + \frac{f_2L_2}{D_2}\frac{V_2^2}{2g} + 10\frac{V_2^2}{2g} + 3\frac{V_3^2}{2g} \tag{2}$$

In the above equation there are three different types of velocities. Express them in terms of only one velocity, say V_3. From continuity:

$$V_1A_1 = V_2A_2 = V_3A_3$$

$$V_1 = \frac{V_3A_3}{A_1} = V_3\left(\frac{D_3}{D_1}\right)^2 = V_3\left(\frac{8}{30}\right)^2 = 0.071V_3$$

$$V_2 = \frac{V_3A_3}{A_2} = V_3\left(\frac{D_3}{D_2}\right)^2 = V_3\left(\frac{8}{15}\right)^2 = 0.284V_3$$

Substituting in Eq. (2) and writing numerical values for other variables:

$$HL_{(A-D)} = 0.5\frac{(0.071V_3)^2}{2g} + f_1\frac{500}{0.3}\frac{(0.071V_3)^2}{2g} + 0.5\frac{(0.284V_3)^2}{2g}$$

$$+ f_2\frac{500}{0.15}\frac{(0.284V_3)^2}{2g} + \frac{10(0.284V_2)^2}{2g} + \frac{3V_3^2}{2g}$$

$$= \frac{V_3^2}{2g}(0.0025 + 8.4f_1 + 0.04 + 268.85f_2 + 0.8 + 3)$$

$$= \frac{V_3^2}{2g}(8.4f_1 + 268.85f_2 + 3.85)$$

Substituting in Eq. (1)

$$100 = \frac{V_3^2}{2g} + \frac{V_3^2}{2g}(8.4f_1 + 268.85f_2 + 3.85)$$

$$= \frac{V_3^2}{2g}(1 + 8.4f_1 + 268.85f_2 + 3.85)$$

or

$$100 = \frac{V_3^2}{2g}(8.4f_1 + 268.85f_2 + 4.85) \qquad (3)$$

$$V_3 = \sqrt{\frac{2g \times 100}{8.4f_1 + 268.85f_2 + 4.85}} \qquad (4)$$

Eq. (4) has three unknowns, namely V_3, f_1, and f_2; therefore, V_3 cannot be found directly. An iterative procedure, however, can be used to solve Eq. (4). For cast iron $\varepsilon = 0.025$ cm

$$\frac{\varepsilon}{D_1} = \frac{0.025}{30} = 0.00083$$

$$\frac{\varepsilon}{D_2} = \frac{0.025}{15} = 0.00167$$

Iteration #1

From the Moody diagram, for $\frac{\varepsilon}{D_1} = 0.00083$; $f_1 = 0.019$

and for $\frac{\varepsilon}{D_2} = 0.00167$; $f_2 = 0.022$

Now

$$V_3 = \sqrt{\frac{2 \times 9.81 \times 100}{8.4(0.019) + 268.85(0.022) + 4.85}}$$

$$= 13.4 \text{ m/s}$$

Also

$$V_1 = 0.071 \times 13.40 = 0.95 \text{ m/s}$$

$$V_2 = 0.284V_3 = 0.284 \times 13.40 = 3.81 \text{ m/s}$$

Let

$$\mathbf{R}_1 = \text{Reynolds number for pipe } BC$$

$$\mathbf{R}_2 = \text{Reynolds number for pipe } CD$$

Then

$$\mathbf{R}_1 = \frac{V_1 D_1}{\nu} = \frac{0.95 \times 0.3}{1.007 \times 10^{-6}} = 2.81 \times 10^5$$

$$\mathbf{R}_2 = \frac{V_2 D_2}{\nu} = \frac{3.81 \times 0.15}{1.007 \times 10^{-6}} = 5.7 \times 10^5$$

When

$$\mathbf{R}_1 = 2.81 \times 10^5 \quad \text{and} \quad \frac{\varepsilon}{D_1} = 0.00083, \; f_1 = 0.02$$

When

$$\mathbf{R}_2 = 5.7 \times 10^5 \quad \text{and} \quad \frac{\varepsilon}{D_2} = 0.00167, \; f_2 = 0.022$$

Both f_1 and f_2 are very close to the assumed values. Hence, there is no need for another iteration. Therefore:

$$V_3 = 13.40 \text{ m/s}$$

$$Q = V_3 A_3 = 13.40 \frac{\pi(0.08)^2}{4} = 0.067 \text{ m}^3/\text{s}$$

Computer Solution

A computer program in BASIC is listed in Appendix B which can be used to solve the iterative part of the series pipe problem. Since every problem may be unique in some feature of the system, the head loss part of the calculations has not been programmed. The user must first convert the problem to the form:

$$\frac{V^2}{2g}(V_1 f_1 + V_2 f_2 + \cdots + B) = \text{C}$$

which is a general form of Eq. (3) of Example 9.9. Starting from this equation, the discharge can be computed using the computer program. The problem is now solved using the program.

```
RUN
This program solves for a series pipe problem.
The problem must first be reduced to the form
   v**2/2g(a1*f1 + a2*f2 + ...... + b ) = c
   where v = velocity at exit(nozzle or last pipes)
title of the problem ? Example 9.9
units(si/fps) ? si
is there a nozzle at the end (y/n) ? y
diameter of nozzle (M)? 0.08
number of pipes in series ? 2
diameter of pipe  1 (M)? 0.3
epsilon for pipe  1 (M)? 0.00025
diameter of pipe  2 (M)? 0.15
epsilon for pipe  2 (M)? 0.00025
coefficient a 1 ? 8.4
coefficient a 2 ? 268.85
coefficient b ? 4.85
coefficient c ? 100
k viscosity (sq.m/sec)? 1.007e-06
-----------------------------------------

Example 9.9
discharge(cu.m/sec)   =    6.669638E-02
-----------------------------------------

do you wish to solve another problem(y/n)? n
Break in 740
Ok
```

Equivalent Pipes

Two pipe systems are said to be *equivalent* when the same head loss produces the same discharge in both the systems—i.e., for two pipes or pipe systems to be equivalent:

$$(h_f)_1 = (h_f)_2$$

and

$$Q_1 = Q_2$$

The concept of equivalent pipes is useful in the problems of pipes in series, pipes in parallel, branching pipes and pipe networks. Using the concept, a flow system containing many pipes, pipe fittings and other features can be represented by a single pipe of a given diameter and equivalent length. The procedure is illustrated by the following example.

EXAMPLE 9.10

Replace the flow system shown in Fig. 9.8 by a single cast iron pipe of 15 cm diameter, when 100 l/s of water at 20°C is flowing. All pipes in the given system are 20 cm in diameter and of cast iron.

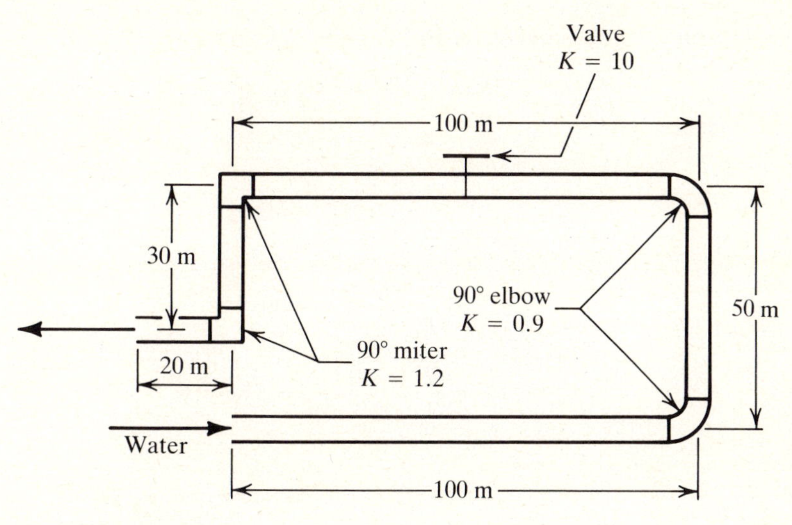

FIGURE 9.8 Example 9.10.

Solution

$$Q = 100 \text{ l/s} = 0.1 \text{ m}^3/\text{s}$$

For water at 20°C, $\nu = 1.007 \times 10^{-6}$ m²/s

$$\frac{\varepsilon}{D} = \frac{0.025}{20} = 0.00125$$

$$V = \frac{0.1}{\pi(0.2)^2/4} = 3.183 \text{ m/s}$$

$$\mathbf{R} = \frac{3.183 \times 0.2}{1.007 \times 10^{-6}} = 6.32 \times 10^5$$

$$f = 0.021 \text{ (from Moody diagram)}$$

Let

$(HL)_1$ = total head loss through the given system

$$(HL)_1 = \left[0.021 \frac{(100 + 50 + 30 + 20 + 100)}{(0.2)^5} \frac{16Q_1^2}{2\pi^2 g} + \frac{2(0.9)}{(0.2)^4} \frac{16Q_1^2}{2\pi^2 g} + \frac{10}{(0.2)^4} \frac{16Q_1^2}{2\pi^2 g} + \frac{2(1.2)}{(0.2)^4} \frac{16Q_1^2}{2\pi^2 g}\right]$$

$$(HL)_1 = \frac{16Q_1^2}{2\pi^2 g}(28562) \tag{1}$$

Let $(HL)_2$ = head loss through the equivalent pipe, D_2 = diameter of the equivalent pipe, Q_2 = discharge in the equivalent pipe, Then

$$(HL)_2 = f_2 \frac{L_2}{D_2^5} \frac{16Q_2^2}{2\pi^2 g} \tag{2}$$

For pipe 2 to be equivalent to the given system

$$(HL)_1 = (HL)_2 \quad \text{and} \quad Q_1 = Q_2$$

Hence,

$$28562 = f_2 \frac{L_2}{D_2^5} \tag{3}$$

For the equivalent pipe

$$\frac{\varepsilon}{D} = \frac{0.025}{0.15} = .0017$$

$$V = \frac{Q}{A} = \frac{0.1}{\frac{\pi(0.15)^2}{4}} = 5.66 \text{ m/s}$$

$$\mathbf{R} = \frac{VD}{\nu} = \frac{(5.66)(0.15)}{1.007 \times 10^{-6}} = 8.46 \times 10^5$$

$$f = 0.023$$

Substituting in Eq. (3)

$$L_2 = \frac{28562 \times (0.15)^5}{0.023} = 94.3 \text{ m}$$

9.7 PIPES IN PARALLEL

A *parallel pipe system* consists of a combination of two or more pipes connected in such a way that the flow is divided among the pipes and then is joined again. The junction at which the pipes branch out is called *the diverging node* and the junction at which the pipes meet again is called the *converging node*. A parallel pipe system is shown in Fig. 9.9. Two important features of pipes connected in parallel are:

1. The head loss through each pipe is the same, and is equal to the difference in the total energy between the diverging node and the converging node.
2. The discharge Q is cumulative—i.e., the discharge Q at point A will be distributed among the n parallel pipes in such a way that $Q = \sum_n Q_n$.

In a parallel pipe problem, it is always advisable to first express all branches in terms of equivalent pipes so that only head losses due to friction are considered. Otherwise, the problems become complicated and difficult to solve.

Generally, two cases occur: (1) given the head loss from A to B, find the discharge in each branch; (2) given discharge Q at A, find the distribution of

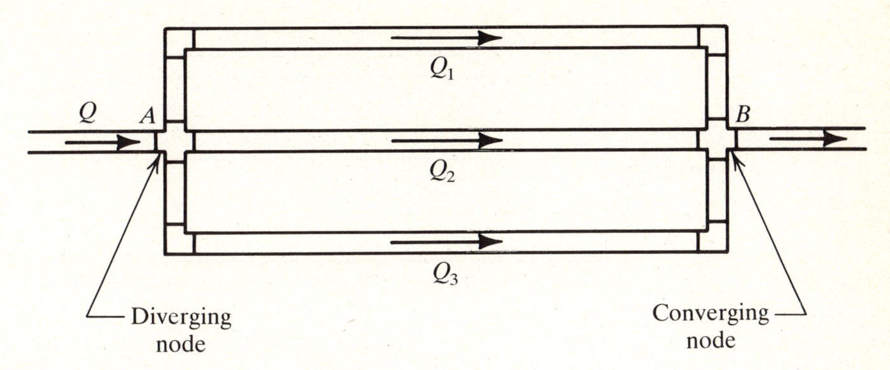

FIGURE 9.9 Pipes connected in parallel.

Q and head loss. In the first case, each branch is solved like problem type II knowing that the head loss is the same through each branch. In the second case, the following procedure is recommended for a problem consisting of N parallel pipes:

1. Assume discharge Q'_1 through pipe 1.
2. Determine the head loss $(h_f)_1$ in pipe 1. This is a problem type I.
3. Knowing that $(h_f)_1 = (h_f)_2 = \cdots = (h_f)_N$, determine $Q'_2, Q'_3, \ldots, Q'_N$. These are problem type II.
4. Determine actual discharges using

$$Q_1 = \frac{Q'_1}{\sum_{n=1}^{N} Q'_n} Q$$

$$Q_2 = \frac{Q'_2}{\sum_{n=1}^{N} Q'_n} Q$$

$$\vdots$$

$$Q_N = \frac{Q'_N}{\sum_{n=1}^{N} Q'_n} Q$$

5. Verify the correctness of the solution by computing $(h_f)_1, (h_f)_2, \ldots, (h_f)_N$ and making sure that they are equal.

The procedure is now illustrated by an example.

EXAMPLE 9.11

Given the parallel pipe system of Fig. 9.10, calculate the discharge through each branch. All pipes are made of cast iron. The fluid flowing is water at 20°C.

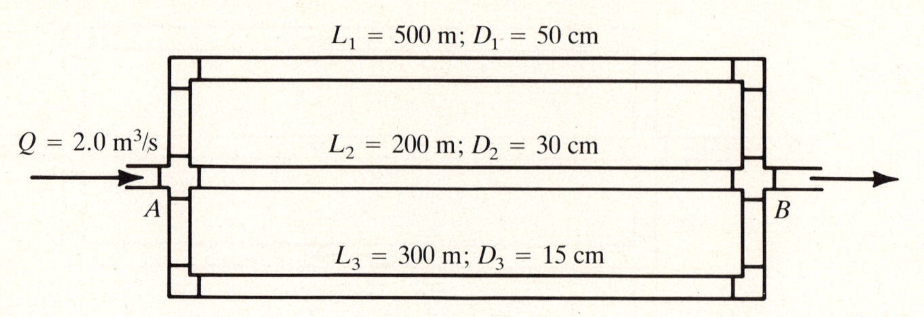

FIGURE 9.10 Example 9.11.

Solution For cast iron

$$\varepsilon = 0.025 \text{ cm}$$

For water at 20°C

$$\nu = 1.007 \times 10^{-6} \text{ m}^2/\text{s}$$

Assume

$$Q_1 = 0.5 \text{ m}^3/\text{s}$$

Pipe #1

$$\frac{\varepsilon_1}{D_1} = \frac{0.025}{50} = 0.0005$$

$$V_1 = \frac{Q_1}{A_1} = \frac{0.5}{\dfrac{\pi(0.5)^2}{4}} = 2.55 \text{ m/s}$$

$$\mathbf{R} = \frac{VD}{\nu} = \frac{(2.55)(0.5)}{1.007 \times 10^{-6}} = 1.27 \times 10^6$$

$$f(0.0005, 1.27 \times 10^6) = 0.018$$

$$h_f = \frac{f_1 L_1}{D_1}\frac{V_1^2}{2g} = \frac{(0.018)(500)}{0.5}\frac{(2.55)^2}{2(9.81)} = 5.96 \text{ m}$$

$$(h_f)_1 = 5.96 \text{ m}$$

Pipe #2

$$(h_f)_2 = (h_f)_1 = 5.96 \text{ m}$$

$$L_2 = 200 \text{ m}$$

$$D_2 = 0.3 \text{ m}$$

$$\frac{\varepsilon_2}{D_2} = 0.00083$$

$$V_2 = \frac{0.419}{\sqrt{f}}$$

Assume $$f = 0.02$$

$$V_2 = \frac{0.419}{\sqrt{0.02}} = 2.96 \text{ m/s}$$

$$\mathbf{R} = \frac{V_2 D_2}{\nu} = \frac{(2.96)(0.3)}{1.007 \times 10^{-6}} = 8.8 \times 10^5$$

This gives, $f = 0.02$

This is equal to the assumed value of f, hence, no further iteration is needed.

$$Q_2 = V_2 A_2 = 2.96\left[\frac{\pi(0.3)^2}{4}\right] = 0.209 \text{ m}^3/\text{s}$$

Pipe #3

$$(h_f)_3 = (h_f)_1 = 5.96 \text{ m}$$

$$L_3 = 300 \text{ m}$$

$$D_3 = 0.15 \text{ m}$$

$$\frac{\varepsilon_3}{D_3} = 0.00167$$

$$V_3 = \frac{0.242}{\sqrt{f}}$$

Assume $$f = 0.022$$

$$V_3 = \frac{0.242}{\sqrt{0.022}} = 1.63 \text{ m/sec}$$

$$\mathbf{R} = \frac{V_3 D_3}{\nu} = \frac{(1.63)(0.15)}{1.007 \times 10^{-6}} = 2.4 \times 10^5$$

This gives, $f = 0.022$: no further iteration needed.

$$Q_3 = V_3 A_3 = (1.63)\left[\frac{\pi(0.15)^2}{4}\right] = 0.028 \text{ m}^3/\text{s}$$

Now, find the actual discharge through each pipe,

$$Q_1 + Q_2 + Q_3 = 0.737 \text{ m}^3/\text{s}$$

$$Q_1 = 0.5\left(\frac{2.0}{0.737}\right) = 1.35 \text{ m}^3/\text{s}$$

$$Q_2 = 0.209\left(\frac{2.0}{0.737}\right) = 0.567 \text{ m}^3/\text{s}$$

$$Q_3 = 0.028\left(\frac{2.0}{0.737}\right) = 0.076 \text{ m}^3/\text{s}$$

Now, find the actual head loss.

$$Q_1 = 1.35 \text{ m}^3/\text{s}$$

$$V_1 = \frac{Q_1}{A_1} = \frac{1.35}{\frac{\pi(0.5)^2}{4}} = 6.88 \text{ m/s}$$

$$\mathbf{R} = \frac{V_1 D_1}{\nu_1} = \frac{(6.88)(0.5)}{1.007 \times 10^{-6}} = 2.4 \times 10^6$$

$$\frac{\varepsilon}{D} = 0.0005$$

$$f = 0.018$$

$$(h_f)_1 = \frac{0.018(500)}{0.5}\frac{(6.88)^2}{2g} = 43.3 \text{ m}$$

Similarly, it is found that $(h_f)_2 = (h_f)_3 = (h_f)_1 = 43.3$ m. Therefore, the distribution of discharge is correct.

Computer Solution

A computer program in BASIC is listed in Appendix B which can be used to solve a parallel-pipe problem. Example 9.11 is now solved using the program.

```
RUN
This program solves for the distribution of discharge in
pipes connected in parallel

title of the problem? Example 9.11
units(si/fps) ? si
number of pipes connected in parllel ? 3
diameter of pipe 1 (m)? 0.5
length of pipe 1 (m)? 500
epsilon of pipe 1 (m)? 0.00025
diameter of pipe 2 (m)? 0.3
length of pipe 2 (m)? 200
epsilon of pipe 2 (m)? 0.00025
diameter of pipe 3 (m)? 0.15
length of pipe 3 (m)? 300
epsilon of pipe 3 (m)? 0.00025
k. viscosity of the fluid (sq.m/sec)? 1.007e-06
total discharge (cu.m/sec)? 2
-_-_-_-_-_-_-_-_-_-_-_-_-_-_-_-_-_

Example 9.11
discharge in pipe 1 (cu.m/sec)  =   1.359805
discharge in pipe 2 (cu.m/sec)  =   .5657635
discharge in pipe 3 (cu.m/sec)  =   7.443155E-02
head loss in each pipe (m)         =    41.30691
-_-_-_-_-_-_-_-_-_-_-_-_-_-_-_-_-_

do you wish to solve another problem(y/n) ? n
Break in 680
Ok
```

9.8 BRANCHING PIPES

A pipe system in which fluid is brought to a junction where three or more pipes meet is termed a *branching pipe system* (Fig. 9.11). There are two important features of a branching pipe system: (1) the total amount of fluid entering into the junction must be equal to the total amount of fluid leaving the junction—i.e., the algebraic sum of all the flows at the junction must be zero, or $\Sigma Q = 0$; (2) all the pipes that meet at the junction have the same pressure at the junction.

The direction of flow in a branch is determined by the relative positions of the total energy line at the two ends of the branch. For example, for the branching pipe system shown in Fig. 9.11, the total energy at junction J is given by the elevation of the fluid in the manometer connected at junction J, and the total energy at the surface of each reservoir is given by the elevation of the fluid level in that reservoir. Therefore, fluid will flow from A to J; J to B and J to C. A typical branching pipe problem would be to determine the flow in each branch.

A branching pipe problem having N number of pipes can be solved as follows:

1. Assume height of the total energy line at J.
2. Calculate $h'_{f_1}, h'_{f_2}, h'_{f_3}, \ldots, h'_{f_N}$
3. Calculate $Q'_1, Q'_2, Q'_3, \ldots, Q'_N$. This is pipe problem type II.
4. Check for $\sum_{n=1}^{N} Q'_n = 0$. If satisfied, the procedure is terminated. If not, go to step 1.

Only three iterations are required. If convergence is not attained in three iterations, $\Sigma Q'$ may be plotted against the total energy (TEL). From the plot,

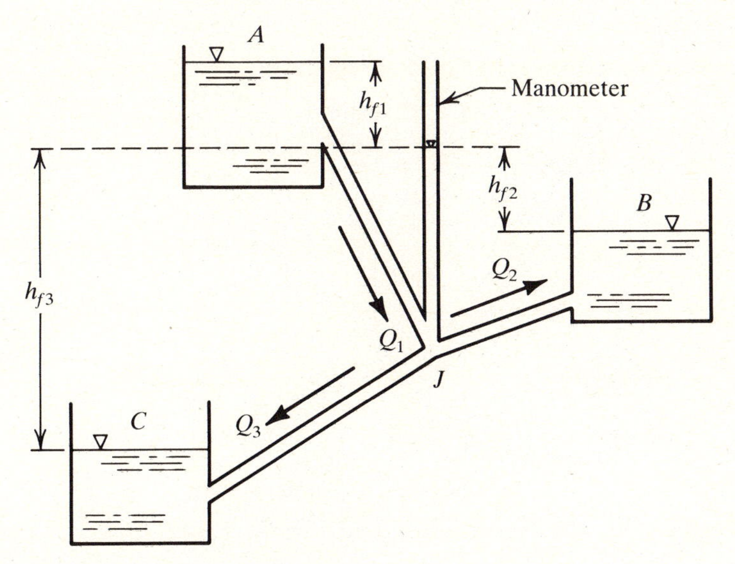

FIGURE 9.11 A branching pipe system.

TEL corresponding to $\Sigma Q' = 0$ may be read. The steps are repeated once again to determine correct flows. The procedure is illustrated by the following example.

EXAMPLE 9.12

For the branching pipe system shown in Figure 9.12, determine the flow in each branch. The fluid in each reservoir is water at 20°C. All pipes are made of cast iron.

Solution Let us assume that the elevation of the total energy line (TEL) at $J = 125$ m. It may be noticed that the elevation of TEL at J cannot be more than 135 m and less than 112 m, which are, respectively, the elevations of the highest and the lowest reservoirs. Let Q_1, Q_2, Q_3 be the flow rates and h_{f1}, h_{f2}, and h_{f3} be the head losses in the pipelines AJ, BJ and CJ, respectively. Now determine Q_1, Q_2, and Q_3 using the iterative procedure of problem type II. Here, the computer program for three types of pipe problems listed in Appendix B can be used to save time. For water at 20°C, $\nu = 1.007 \times 10^{-6}$ m^2/s and for cast iron, $\varepsilon = 0.025$ cm. Now, $h_{f1} = 135 - 125 = 10$ m, and water flows from A to J. $Q_1 = +3.654$ m^3/s. (Designate flows entering the junction with a + sign and those leaving the junction with a − sign.)

$$h_{f2} = 125 - 120 = 5 \text{ m: water flows form } J \text{ to } B$$

$$Q_2 = -0.369 \text{ m}^3/\text{s}$$

$$h_{f3} = 125 - 112 = 13 \text{ m: water flows from } J \text{ to } C$$

$$Q_3 = -1.258 \text{ m}^3/\text{s}$$

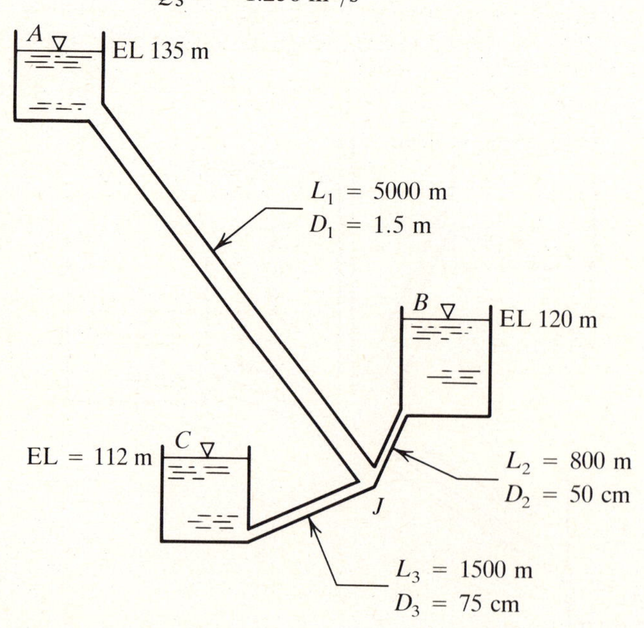

FIGURE 9.12 Example 9.12.

Now, find the algebraic sum of Q_1, Q_2 and Q_3, which is 3.654 – 0.369 – 1.258 = 2.027 m^3/s. Since the algebraic sum is not zero, the assumed elevation of TEL at point J is not correct. Guess another elevation. Here, we would like to decrease Q_1 and increase Q_2 and Q_3. To achieve this, the elevation at J must be raised. Let it be 130 m. Then:

$$h_{f1} = 135 - 130 = 5 \text{ m} \qquad Q_1 = 2.567 \text{ m}^3/\text{s}$$

$$h_{f2} = 130 - 120 = 10 \text{ m} \qquad Q_2 = -0.524 \text{ m}^3/\text{s}$$

$$h_{f3} = 130 - 112 = 18 \text{ m} \qquad Q_3 = -1.483 \text{ m}^3/\text{s}$$

$$\Sigma Q = 2.567 - 0.524 - 1.483 = 0.56 \text{ m}^3/\text{s}$$

The elevation at J must again be raised. Let it be 132 m. Then

$$h_{f1} = 135 - 132 = 3 \text{ m} \qquad Q_1 = 1.985 \text{ m}^3/\text{s}$$

$$h_{f2} = 132 - 120 = 12 \text{ m} \qquad Q_2 = -0.575 \text{ m}^3/\text{s}$$

$$h_{f3} = 132 - 112 = 20 \text{ m} \qquad Q_3 = -1.565 \text{ m}^3/\text{s}$$

$$\Sigma Q = 1.985 - 0.575 - 1.565 = -0.155$$

This time the elevation at J must be decreased. Let it be 131.5 m. Then

$$h_{f1} = 135 - 131.5 = 3.5 \text{ m} \qquad Q_1 = 2.148 \text{ m}^3/\text{s}$$

$$h_{f2} = 131.5 - 120.0 = 11.5 \text{ m} \qquad Q_2 = -0.563 \text{ m}^3/\text{s}$$

$$h_{f3} = 131.5 - 112.0 = 19.5 \text{ m} \qquad Q_3 = -1.545 \text{ m}^3/\text{s}$$

$$\Sigma Q = 2.148 - 0.563 - 1.545 = 0.04 \text{ m}^3/\text{s}$$

As a final iteration, the elevation of TEL at J must be increased slightly. Let it be 131.6 m. Then

$$h_{f1} = 135 - 131.6 = 3.4 \text{ m} \qquad Q_1 = 2.116 \text{ m}^3/\text{s}$$

$$h_{f2} = 131.6 - 120 = 11.6 \text{ m} \qquad Q_2 = -0.565 \text{ m}^3/\text{s}$$

$$h_{f3} = 131.6 - 112 = 19.6 \text{ m} \qquad Q_3 = -1.549 \text{ m}^3/\text{s}$$

$$\Sigma Q = 2.116 - 0.565 - 1.549 = 0.002 \text{ m} \simeq 0.0.$$

Therefore, the elevation of the total energy line at J is 131.6 m and rates of flow are:

$$Q_1 = 2.116 \text{ m}^3/\text{s} \qquad \text{from } A \text{ to } J$$

$$Q_2 = 0.565 \text{ m}^3/\text{s} \qquad \text{from } J \text{ to } B$$

$$Q_3 = 1.549 \text{ m}^3/\text{s} \qquad \text{from } J \text{ to } C$$

Computer Solution

A computer program in BASIC is listed in Appendix B, which can be used to solve branching pipe problems. The solution to Example 9.12 is now listed.

```
RUN
This program solves branching pipe problems

title of the problem? Example 9.12
units(si/fps)? si
number of pipes ? 3
diameter of pipe  1 (m)? 1.5
length of pipe  1 (m)? 5000
epsilon for pipe  1 (m)? 0.00025
diameter of pipe  2 (m)? 0.5
length of pipe  2 (m)? 800
epsilon for pipe  2 (m)? 0.00025
diameter of pipe  3 (m)? 0.75
length of pipe  3 (m)? 1500
epsilon for pipe  3 (m)? 0.00025
elevation of reservoir  1 (m)? 135
elevation of reservoir  2 (m)? 120
elevation of reservoir  3 (m)? 112
k. viscosity of the fluid (sq.m/sec)? 1.007e-06

_ _ _ _ _ _ _ _ _ _ _ _ _ _ _ _ _ _ _ _ _ _ _ _ _ _ _ _ _ _ _ _
Example 9.12
 Q for branch  1                2.103572 (cu.m/sec)
 Q for branch  2               -.5653691 (cu.m/sec)
 Q for branch  3               -1.548131 (cu.m/sec)
 total head at junction              131.6317 (m)

NOTE: negative Q means water flowing
      away from the junction
_ _ _ _ _ _ _ _ _ _ _ _ _ _ _ _ _ _ _ _ _ _ _ _ _ _ _ _ _ _ _ _
 do you wish to solve another problem (y/n) ? n
Break in 590
Ok
```

9.9 PIPE NETWORKS

Pipes connected in loops and branches form a *pipe network* (Fig. 9.13). A point at which more than one pipe meet is called *a node*; for example, in Fig. 9.13 points *a*, *b*, *c* etc., are nodes. A closed circuit consisting of more than one pipe is called a *loop*; for example, pipes *ac*, *cd*, *de* and *ea* form a loop. A loop starts from a node and terminates on the same node. Two important features of pipes connected to form a network are: (1) the algebraic sum of discharges entering and leaving a node is zero, or $\Sigma Q = 0$, and (2) the algebraic sum of the total head loss in a loop is zero, or $\Sigma h_f = 0$.

A typical problem may consist of determining the discharge in each pipe when the discharge entering and leaving the network is known. Two methods of solution can be used.

One method is to write a series of simultaneous equations based on the above-mentioned two features—i.e., $\Sigma Q = 0$ for a node, and $\Sigma h_f = 0$ for a loop. Usually, enough numbers of equations can be written to solve the

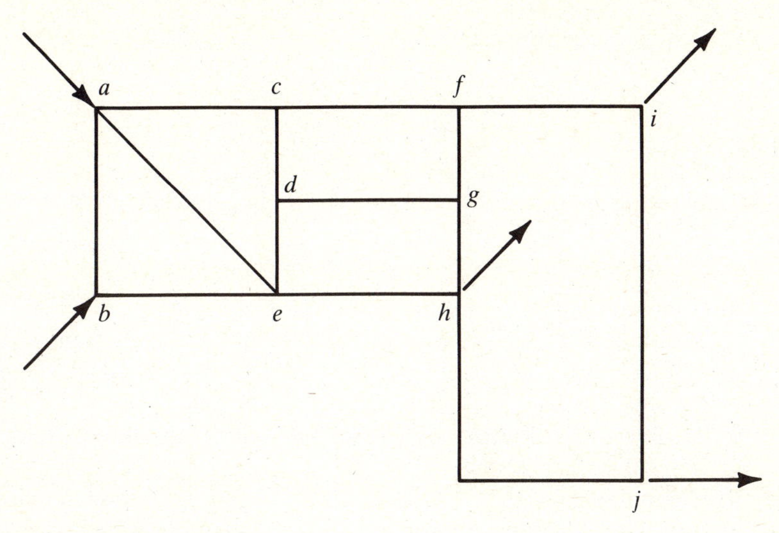

FIGURE 9.13 A pipe network.

problem. However, for large networks, the procedure may become tedious and difficult.

Another method, known as the *Hardy-Cross method*, uses successive approximations based on the two conditions stated above. The head loss in each pipe can be calculated using either the Darcy–Weisbach equation or the Hazen–Williams formula. The Hazen–Williams equation is good for water only. Minor losses should be included as equivalent lengths in each pipe. If the Hazen–Williams equation is used, a formula for calculating the correction can be obtained as follows: From Equation (9.18):

$$h_f = \frac{ALQ^n}{C^n D^m}$$

$$h_f = rQ^n$$

where $r = AL/C^n D^m$ and is constant for a given pipe. Let Q_0 be the assumed value of flow in a pipe and ΔQ be the correction, which, when applied to Q_0 will bring Q_0 closer to the actual Q in the pipe, or $Q = Q_0 + \Delta Q$. Then, for a pipe

$$h_f = r(Q_0 + \Delta Q)^n = r(Q_0^n + Q_0^{n-1}\Delta Q + \cdots)$$

If ΔQ is small, all terms in the Taylor's expansions involving powers of ΔQ can be eliminated. Now for a loop:

$$\Sigma h_f = \Sigma r(Q_0^n + nQ_0^{n-1}\Delta Q) = 0$$

Assuming ΔQ to be the same for each pipe in the loop and expressing Q_0^n as $Q_0|Q_0|^{n-1}$ to account for the direction of summation around the loop, the above equation becomes

$$\Sigma r Q_0 |Q_0|^{n-1} + \Delta Q \Sigma rn |Q_0|^{n-1} = 0$$

$$\Delta Q = -\frac{\Sigma r Q_0 |Q_0|^{n-1}}{\Sigma rn |Q_0|^{n-1}} \tag{9.21}$$

The following steps may be used to solve a pipe network problem;

1. Convert all minor head losses into equivalent lengths and add to the pipe lengths.
2. Compute r for each pipe, using $r = AL/C^n D^m$ where L = length of pipe, D = diameter of pipe, C = Hazen–Williams constant from Table 9.2, $n = 1.852$, $m = 4.8704$, $A = 4.727$ in the FPS system and 10.675 in the SI system.
3. Assume a distribution of discharge in the network so as to satisfy $\Sigma Q = 0$ at each node.
4. Compute $\Sigma r Q_0 |Q_0|^{n-1}$ and $\Sigma rn |Q_0|^{n-1}$ for each loop moving in a clockwise direction. Compute ΔQ using Equation (9.21).
5. Add ΔQ algebraically to each discharge in the loop. If ΔQ has a + sign, it will be added to the clockwise flows and subtracted from the counter-clockwise flows. If ΔQ has a − sign, it will be subtracted from the clockwise flows and added to the counter-clockwise flows.

EXAMPLE 9.13

Determine the discharge through each pipe of the network shown in Fig. 9.14. Use $C = 130$ in the Hazen–Williams equation.

Solution Follow the procedure step by step as outlined.

1. Let us assume that minor losses are already included as equivalent lengths in the given lengths of pipes.
2. Compute r for each pipe using $r = AL/C^n D^m$ where $A = 4.727$, $n = 1.852$, $m = 4.8704$, $C = 130$. Then $r = 0.0005749L/D^{4.8704}$. The values of r for the pipes in this network are shown in Fig. 9.15.

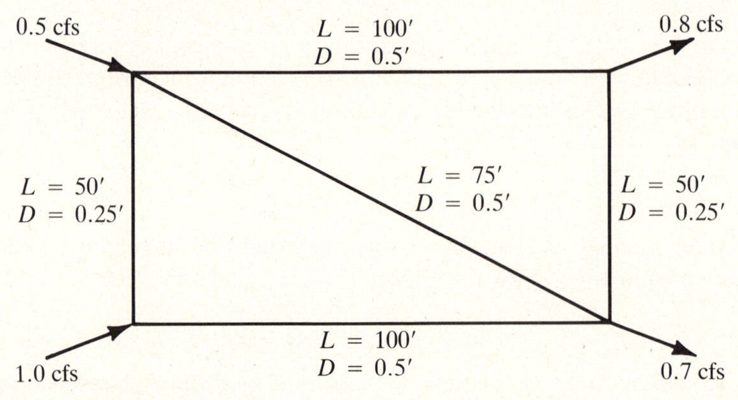

FIGURE 9.14 Example 9.13.

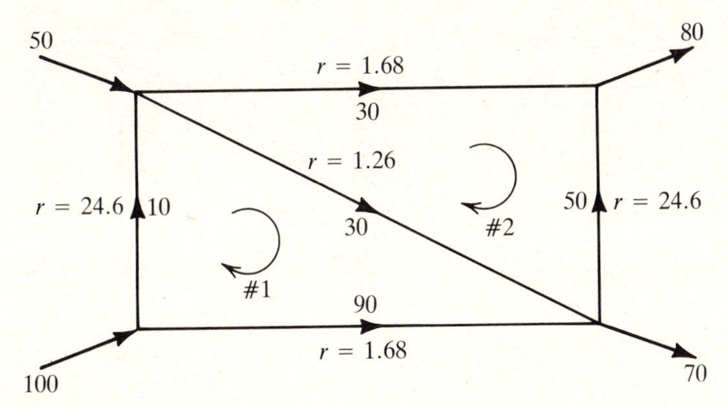

FIGURE 9.15 Example 9.13.

3. Since the discharges are small, a suitable multiple of the given values may be used for convenience in calculations. Let us multiply all discharges by 100. Later we will divide our answers by 100 to get the correct flows. The new values are shown in Fig. 9.15. The assumed distribution of discharge in the network is also shown in Fig. 9.15.
4. *Iteration #1*

   Loop #1

$rQ_0\|Q_0\|^{0.852}$		$r(1.852)\|Q_0\|^{0.852}$	
$24.6 \times 10\|10\|^{0.852}$	$= 1749.6$	$24.6 \times 1.852 \times \|10\|^{0.852}$	$= 324.0$
$1.26 \times 30\|30\|^{0.852}$	$= 685.5$	$1.26 \times 1.852 \times \|30\|^{0.852}$	$= 42.3$
$1.68 \times (-90)\|90\|^{0.852}$	$= -6991.4$	$1.68 \times 1.852 \times \|90\|^{0.852}$	$= 143.9$
	-4556.3		510.2

$$\Delta Q = -\frac{-4556.3}{510.2} = 8.93$$

Apply this correction to all the pipes in loop #1.

Loop #2

In this loop, for a pipe which is common with loop #1, use the corrected value of Q.

$1.68 \times 30\|30\|^{0.852}$	$= 914.0$	$1.68 \times 1.852 \times \|30\|^{0.852}$	$= 56.4$
$24.6 \times (-50)\|50\|^{0.852}$	$= -34468.9$	$24.6 \times 1.852 \times \|50\|^{0.852}$	$= 1276.7$
$1.26 \times (-38.93)\|38.93\|^{0.852}$	$= -1110.6$	$1.26 \times 1.852 \times \|38.93\|^{0.852}$	$= 52.8$
	-34665.5		1385.9

$$\Delta Q = -\frac{-34665.5}{1385.9} = 25.01$$

The corrected rates of flow after iteration #1 are shown in Fig. 9.16. Notice that the diagonal pipe is corrected twice; once in loop #1 and then again in loop #2. Since ΔQ is large, another iteration is needed.

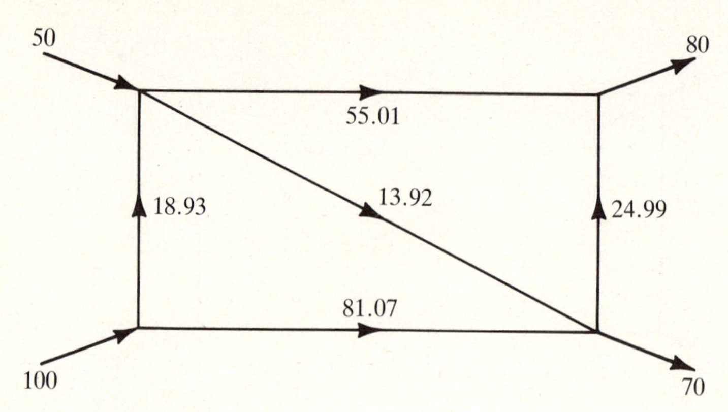

FIGURE 9.16 Example 9.13.

Iteration #2
Loop #1

$$24.6 \times 18.93 \times |18.93|^{0.852} = 5704.5 \qquad 24.6 \times 1.852 \times |18.93|^{0.852} = 558.1$$
$$1.26 \times 13.93 \times |13.93|^{0.852} = 165.6 \qquad 1.26 \times 1.852 \times |13.93|^{0.852} = 22.0$$
$$1.68 \times (-81.07) \times |81.07|^{0.852} = -5761.3 \qquad 1.68 \times 1.852 \times |81.07|^{0.852} = 131.6$$
$$\overline{108.8} \qquad \overline{711.7}$$

$$\Delta Q = -\frac{108.8}{711.7} = -0.15$$

Loop #2

$$1.68 \times 55.01 \times |55.01|^{0.852} = 2809.4 \qquad 1.68 \times 1.852 \times |55.01|^{0.852} = 94.6$$
$$24.6 \times (-24.99) \times |24.99|^{0.852} = -9541.1 \qquad 24.6 \times 1.852 \times |24.99|^{0.852} = 707.1$$
$$1.26 \times (-13.77) \times |13.77|^{0.852} = -162.1 \qquad 1.26 \times 1.852 \times |13.77|^{0.852} = 21.8$$
$$\overline{-6893.8} \qquad \overline{823.5}$$

$$\Delta Q = -\frac{-6893.8}{823.5} = 8.37$$

The corrected rates of flow after iteration #2 are shown in Fig. 9.17. Since ΔQ in loop #2 is large, another iteration is needed.

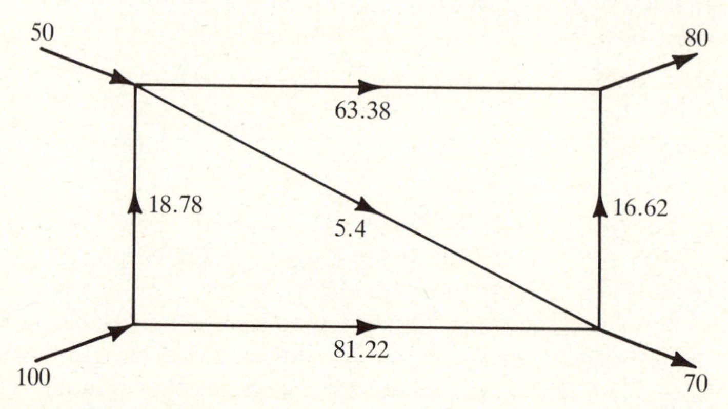

FIGURE 9.17 Example 9.13.

Iteration #3
Loop #1

$24.6 \times 18.78 \times \|18.78\|^{0.852} = 5621.1$		$24.6 \times 1.852 \times \|18.78\|^{0.852} = 554.3$
$1.26 \times 5.4 \times \|5.4\|^{0.852} = 28.6$		$1.26 \times 1.852 \times \|5.4\|^{0.852} = 9.8$
$1.68 \times (-81.22) \times \|81.22\|^{0.852} = -5781.0$		$1.68 \times 1.852 \times \|81.22\|^{0.852} = 131.8$
-131.3		695.9

$$\Delta Q = -\frac{-131.3}{695.9} = 0.189$$

Loop #2

$1.68 \times 63.38 \times \|63.38\|^{0.852} = 3651.9$		$1.68 \times 1.852 \times \|63.38\|^{0.852} = 106.7$
$24.6 \times (-16.62) \times \|16.62\|^{0.852} = -4482.7$		$24.6 \times 1.852 \times \|16.62\|^{0.852} = 499.5$
$1.26 \times (-5.59) \times \|5.59\|^{0.852} = -30.5$		$1.26 \times 1.852 \times \|5.59\|^{0.852} = 10.1$
-861.3		616.3

$$\Delta Q = -\frac{-861.3}{616.3} = +1.40$$

The corrected rates of flow after iteration #3 are given in Fig. 9.18. More iterations can be carried out if further refinement is desired. The actual flow rates will be 1/100 of the flow rates shown in Fig. 9.18.

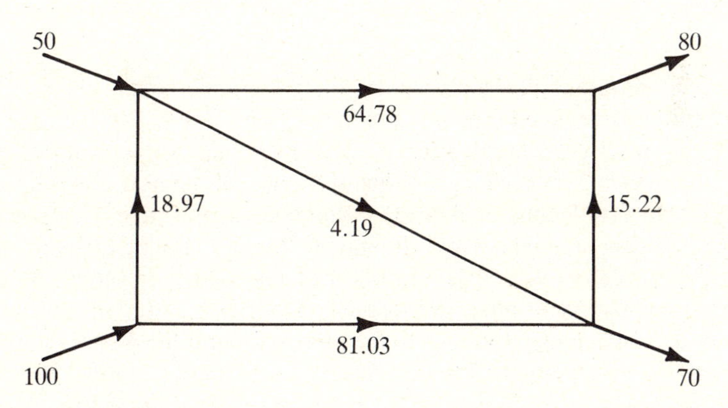

FIGURE 9.18 Example 9.13.

Computer Solution

It is clear from this example that the computational procedure is simple but lengthy. For a network consisting of a large number of pipes and many loops, the solution may take a considerable amount of time. However, solutions can be carried out easily on a digital computer. A program in BASIC is listed in Appendix B which uses the Hardy-Cross method to determine flow rates in a pipe network. The solution to Example 9.13 using this program is shown below.

```
RUN
this program solves for a pipe network problem

title of the problem? Example 9.13
system of units (si/fps)? fps
number of iterations allowed? 10
discharge tolerance? 0.001

****  Example 9.13  ****

          no. of pipe            discharge
                                (cu.ft/sec)
               1                  18.99632
               2                  81.00369
               3                  4.182796
               4                  64.81353
               5                  15.18648
largest delq =   4.895901E-04 no. of iterations =   5

NOTE:
Negative discharge means that the direction
of flow is opposite to that assumed initially.

Break in 1810
Ok
```

NOTE:
To use the program all the pipes must first be numbered and the initial flow distribution determined. The input is in the form of data cards placed at the beginning of the program. On the first line, punch N1 and N2, where N1 = number of pipes, N2 = number of loops. On the next line punch L, D, C, Q where L = length of pipe, D = diameter of the pipe, C = roughness coefficient, Q = assumed initial discharge. Prepare similar cards for all the pipes for a total of N1 pipes. On the next line punch N, $A(M)$, $M = 1, N$ where N = number of pipes in loop #1, A = pipe number in the clockwise direction. Assign a negative sign to the pipe number if the direction of flow in that pipe is anticlockwise. Prepare similar cards for each loop.

For Example 9.13, the data cards are arranged as given below.

```
10 DATA 5,2
20 DATA 50,0.25,130,10
30 DATA 100,0.5,130,90
40 DATA 75,0.5,130,30
50 DATA 100,0.5,130,30
60 DATA 50,0.25,130,50
70 DATA 3,1,3,-2
80 DATA 3,4,-5,-3
```

9.10 SIPHON FLOW

A closed conduit which lifts the liquid to an elevation higher than its free surface and then discharges it at a lower elevation is called *a siphon.* A typical siphon problem is illustrated by the following example.

EXAMPLE 9.14

Two reservoirs whose surface levels differ by 100 ft are connected by a pipe 2 ft in diameter and 10,000 ft long. The pipeline crosses a ridge whose summit is 30 ft above the level of, and 1,000 ft distant from, the higher reservoir. Find the minimum depth below the ridge at which the pipe must be laid if the absolute pressure in the pipe is not to fall below 10 ft of water, and calculate the discharge in cubic feet per second. The pipe roughness is 0.00015 ft and kinematic viscosity of water is 1.084×10^{-5} ft^2/s. Neglect head loss at entrance to the pipe.

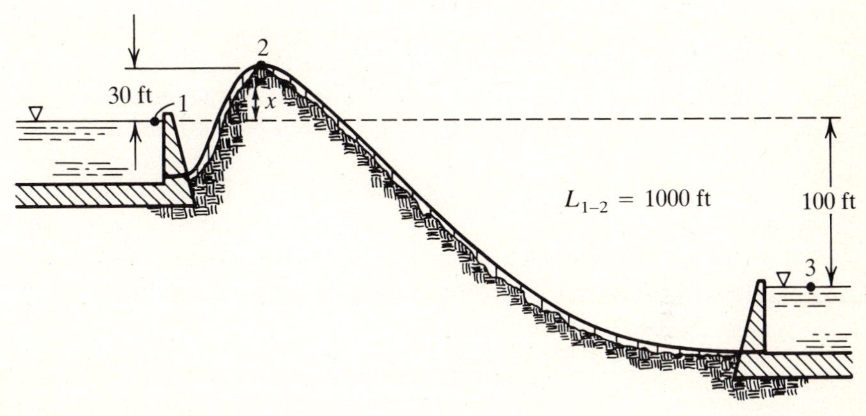

FIGURE 9.19 Example 9.14.

Solution First find the velocity through the siphon by applying Bernoulli equation between points 1 and 3. Or

$$\frac{p_1}{\gamma} + z_1 + \frac{V_1^2}{2g} = \frac{p_3}{\gamma} + z_3 + \frac{V_3^2}{2g} + HL_{(1-3)}$$

Then:

$$0 + 100 + 0 = 0 + 0 + 0 + \frac{f(10{,}000)}{2} \frac{V^2}{2 \times 32.2}$$

$$100 = 77.64\, fV^2 \quad \text{or} \quad V = \frac{1.1347}{\sqrt{f}} \tag{1}$$

$$\varepsilon/D = \frac{0.00015}{2} = 0.000075$$

Assume $f = 0.011$.

Then from Eq. (1)

$$V = 10.82 \text{ ft/s}$$

$$\mathbf{R} = \frac{10.82 \times 2}{1.084 \times 10^{-5}} = 2.0 \times 10^6$$

Now f from the Moody diagram = 0.0125. This is not close to the assumed value of f, hence, another iteration is needed. Using $f = 0.0125$,

$$V = 10.15 \text{ ft/s}$$

$$\mathbf{R} = 1.87 \times 10^6$$

$$f = 0.0125, \qquad \text{convergence attained}$$

Hence:

$$V = 10.15 \text{ ft/s}$$

$$Q = 10.15 \times \frac{\pi(2)^2}{4} = 31.9 \text{ ft}^3\text{/s}$$

Let x be the maximum height of the centerline of pipe above the level of the higher reservoir. Let there be a point 2 located at the centerline of the pipe at this level. Apply Bernoulli equation between points 1 and 2.

$$\frac{p_1}{\gamma} + z_1 + \frac{V_1^2}{2g} = \frac{p_2}{\gamma} + z_2 + \frac{V_2^2}{2g} + HL_{(1-2)}$$

Since pressure at point 2 is minus gauge, we will write all pressures in terms of absolute pressures. Therefore, in terms of absolute pressures,

$$34 + 0 + 0 = 10 + x + \frac{(10.15)^2}{2 \times 32.2} + \frac{0.0125 \times 1000 \times (10.15)^2}{2 \times 2 \times 32.2}$$

$$34 = 10 + x + 1.6 + 10.0$$

$$x = 34 - 10 - 1.6 - 10 = 12.4 \text{ ft}$$

Hence, minimum depth below the ridge = 30 − 12.4 = 17.6 ft.

PROBLEMS

p.9.1 Water flows through a 25 cm diameter pipe at the rate of 0.2 m³/s. The head loss in a 100 m length of this pipe was found to be 50 cm. What is the friction factor?

p.9.2 Determine the horsepower required to pump crude oil through a 6 in diameter wrought iron pipe at the rate of 5 cfs. The total length of the pipeline is 5000 ft.

p.9.3 Water at 20°C is being pumped through a 20 cm diameter cast iron pipe. A head loss of 6 m was recorded in a 150 m length of the pipe. Determine the discharge through the pipe.

p.9.4 A pipeline is to be designed to pump 0.5 m^3/s of water at 25°C through a concrete pipe. The head loss in a 120 m length of the pipe should not exceed 10 m. Determine the size of the concrete pipe.

p.9.5 Determine the flow rate in the system shown in Fig. 9.20. Assume the pipe to be smooth and neglect the losses at the entrance and at the exit to the pipe.

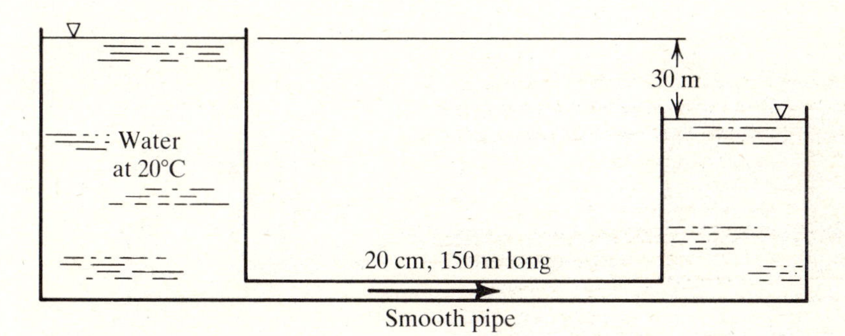

FIGURE 9.20 Problems 9.5, 9.6.

p.9.6 Determine the flow rate in the system of Fig. 9.20 considering the head losses at the entrace and at the exit to the pipe.

p.9.7 Determine the discharge through the system shown in Fig. 9.21. All pipes are made of cast iron.

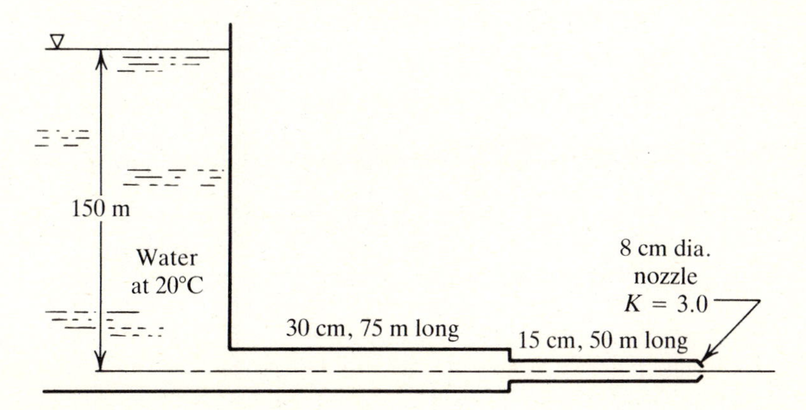

FIGURE 9.21 Problems 9.7, 9.8.

p.9.8 Determine the length of a 20 cm diameter cast iron pipe which is equivalent to the system shown in Fig. 9.21.

p.9.9 Determine the equivalent length of a 10 cm diameter cast iron pipe for (a) a sudden expansion from 20 cm to 30 cm diameter, (b) a globe valve and a 90° bend.

p.9.10 Water at 15°C is being pumped through the system shown in Fig. 9.22. Determine the head on the pump for a flow rate of 150 l/s. All pipes are commercial steel 20 cm in diameter.

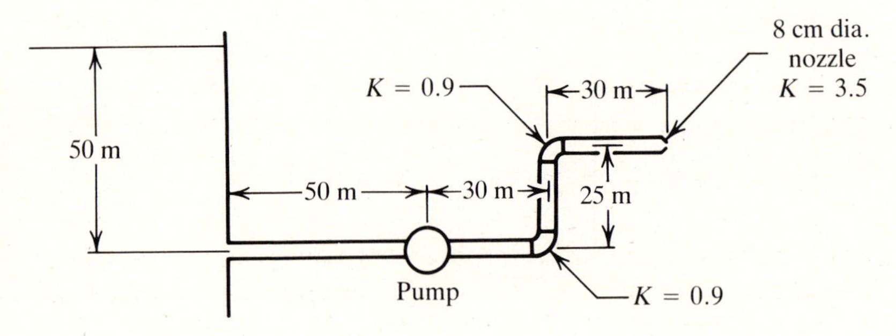

FIGURE 9.22 Problem 9.10.

p.9.11 Determine H in Fig. 9.23 such that the flow rate is 30 cfs. All pipes are commercial steel and the fluid is water at 20°C.

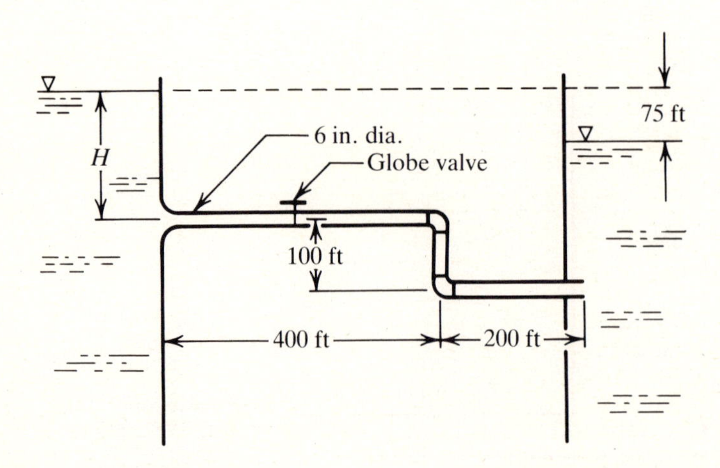

FIGURE 9.23 Problem 9.11.

p.9.12 Determine the rate of flow when H is 20 m in Fig. 9.24.

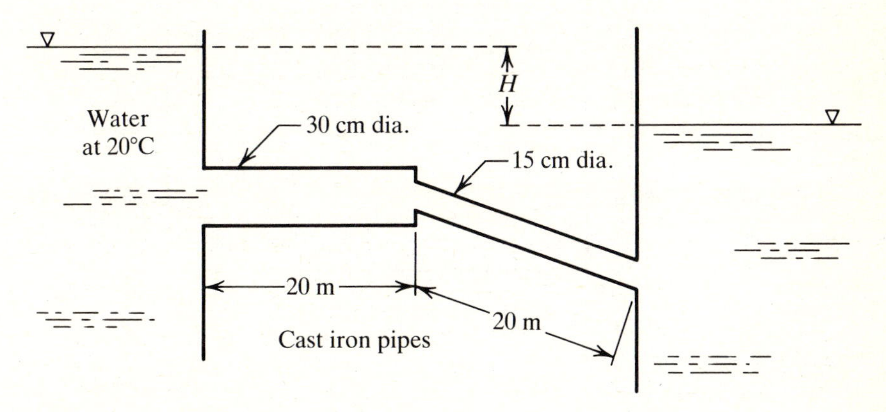

FIGURE 9.24 Problems 9.12, 9.13.

p.9.13 Determine H for a flow rate of 2 m^3/s, in Fig. 9.24.

p.9.14 The flow rate at point A in Fig. 9.25 is 1 m^3/s and the pressure is 300 KPa. Determine the pressure at point B. All pipes are commercial steel and the fluid is water at 20°C. Neglect all minor head losses.

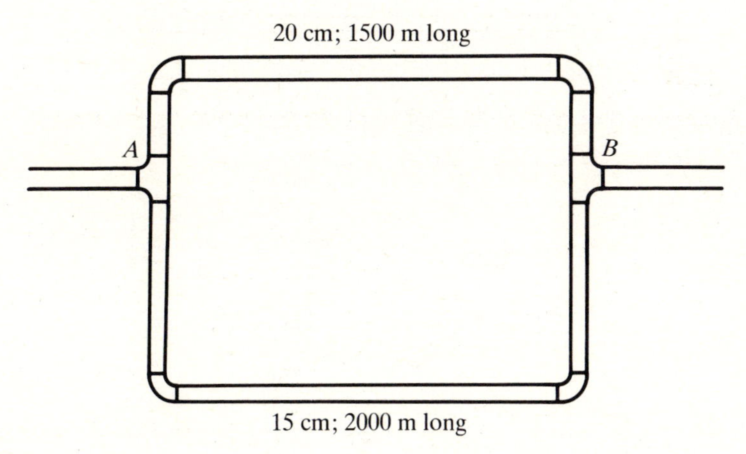

FIGURE 9.25 Problems 9.14, 9.15.

p.9.15 Determine the pressure at B in Fig. 9.25 considering the minor head losses.

p.9.16 The rate of flow at point A in Fig. 9.26 is 2 m^3/s and the pressure is 350 KPa. Determine the pressure at B. All pipes are cast iron 50 cm in diameter. The fluid is water at 20°C. Neglect all minor head losses.

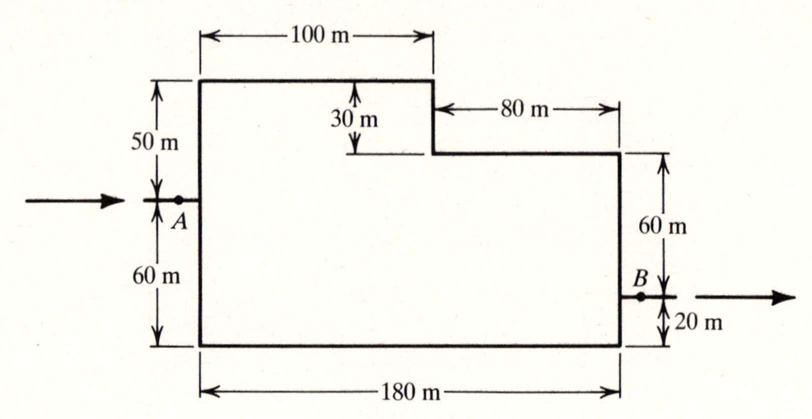

FIGURE 9.26 Problem 9.16.

p.9.17 Water at 68°F is flowing through the system shown in Fig. 9.27. Determine the distribution of flow in the three pipes. All pipes are of cast iron. Neglect minor head losses.

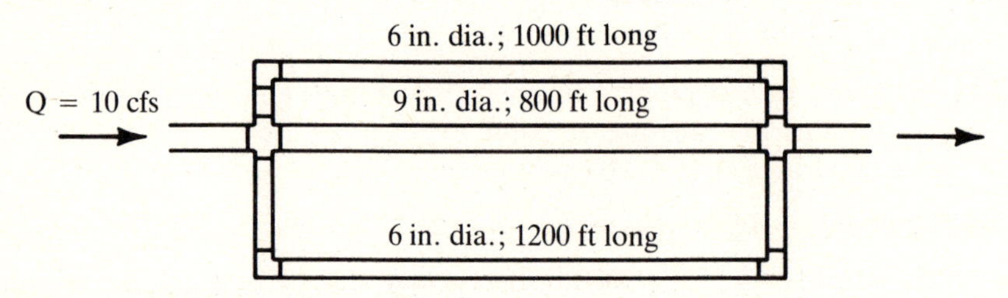

FIGURE 9.27 Problem 9.17.

p.9.18 Determine the flow rate through each branch of the system shown in Fig. 9.28. The fluid is water at 20°C.

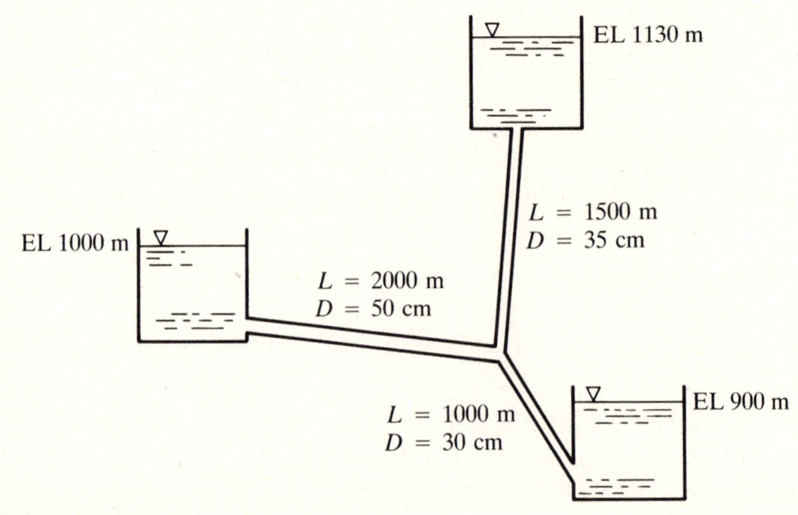

FIGURE 9.28 Problem 9.18.

p.9.19 Determine the flow rate with the pump removed in Fig. 9.29. The fluid is water at 20°C. All pipes are made of cast iron.

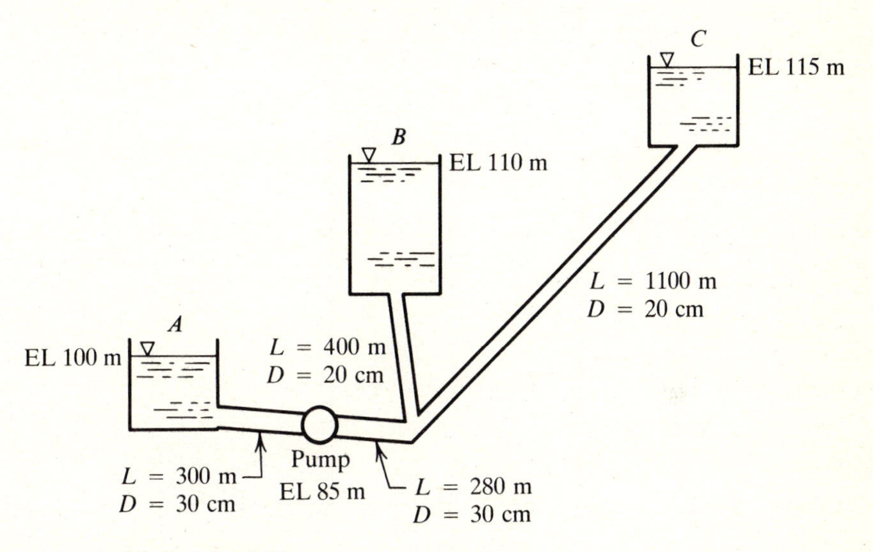

FIGURE 9.29 Problems 9.19, 9.20.

p.9.20 Water is being pumped from reservoir A to reservoirs B and C of Fig. 9.29. If the flow into reservoir B is to be maintained at 50 l/s, what power input would be required for the pump which is only 75% efficient?

p.9.21 Determine the flow through each of the pipes of the networks shown in Figs. 9.30a, b, c and d.

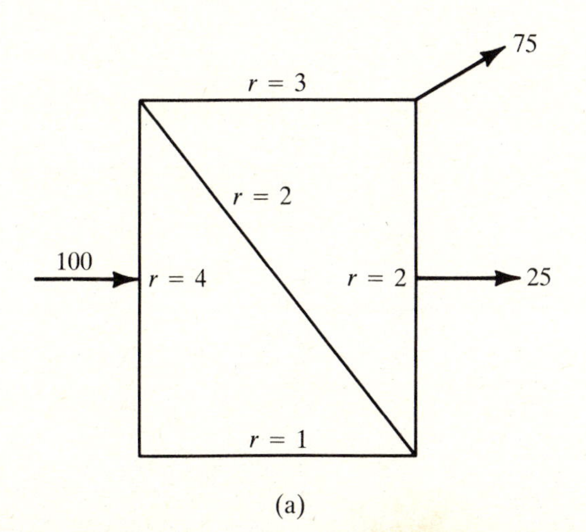

(a)

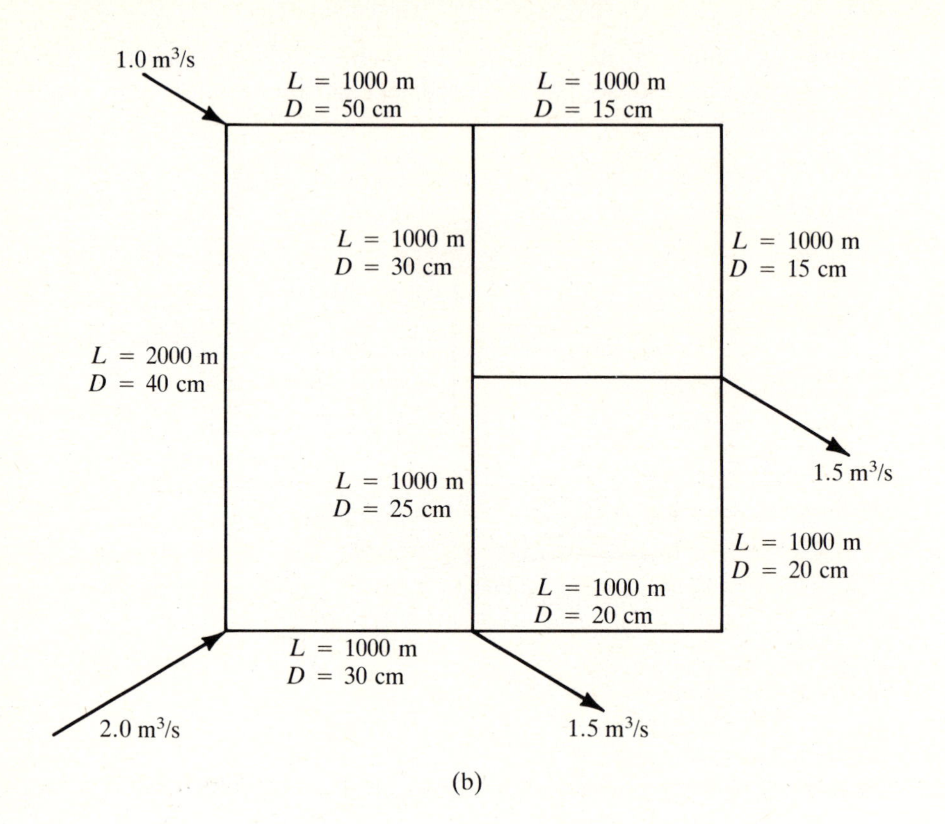
1.0 m³/s
L = 1000 m
D = 50 cm
L = 1000 m
D = 15 cm
L = 1000 m
D = 30 cm
L = 1000 m
D = 15 cm
L = 2000 m
D = 40 cm
1.5 m³/s
L = 1000 m
D = 25 cm
L = 1000 m
D = 20 cm
L = 1000 m
D = 20 cm
L = 1000 m
D = 30 cm
2.0 m³/s
1.5 m³/s

(b)

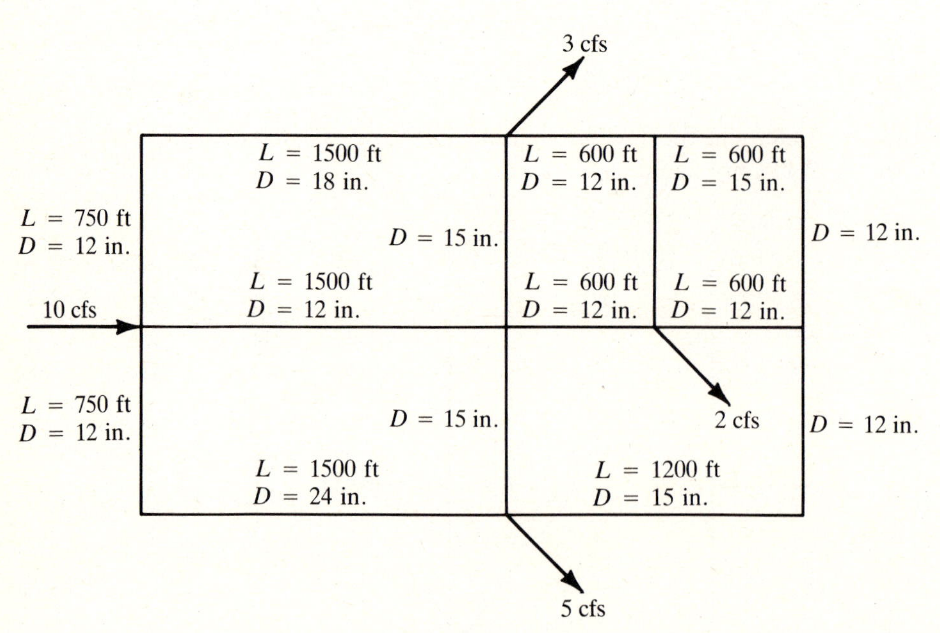
3 cfs
L = 1500 ft
D = 18 in.
L = 600 ft
D = 12 in.
L = 600 ft
D = 15 in.
L = 750 ft
D = 12 in.
D = 15 in.
D = 12 in.
L = 1500 ft
D = 12 in.
L = 600 ft
D = 12 in.
L = 600 ft
D = 12 in.
10 cfs
L = 750 ft
D = 12 in.
D = 15 in.
2 cfs
D = 12 in.
L = 1500 ft
D = 24 in.
L = 1200 ft
D = 15 in.
5 cfs

(c)

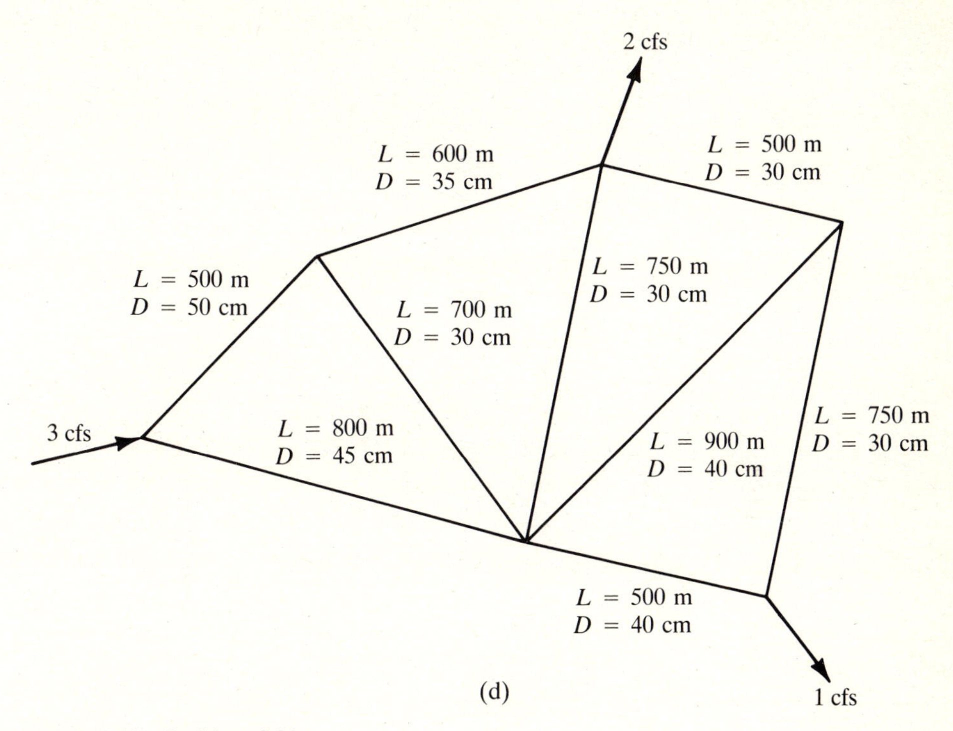

FIGURE 9.30 Problem 9.21.

Chapter Ten

OPEN CHANNEL FLOW

So far, we have considered the class of flow in which the fluid was under pressure. In this chapter, we will study another class of flow, known as the open channel flow, in which the pressure at the liquid surface is atmospheric. We will develop equations of flow in open channels and learn about hydraulic phenomena such as hydraulic jump and backwater curve. Open channel flow is frequently encountered in nature (flow in streams and rivers); therefore, knowledge pertaining to it is important to engineers.

10.1 INTRODUCTION

The term *open channel* applies to any passage through which a fluid is flowing when its free surface is under atmospheric pressure. The channel may be open or covered. If it is covered, it should not run full, otherwise pressure may not be atmospheric. A pipe which is not running full, a partially full sewer, an aqueduct, a stream or a river are examples of open channels.

Since the pressure everywhere on the fluid surface is atmospheric, the pressure gradient along the direction of flow is zero. The flow, therefore, in an open channel is not caused by the pressure gradient, rather it is caused by the force of gravity. Hence, the bed of the channel must slope for flow to take place.

A pipe flow and an open channel flow are compared in Fig. 10.1. Since water in the pipe of Fig. 10.1a is under pressure the hydraulic grade line (HGL) is located at a distance of p/γ from the centerline of the pipe. The

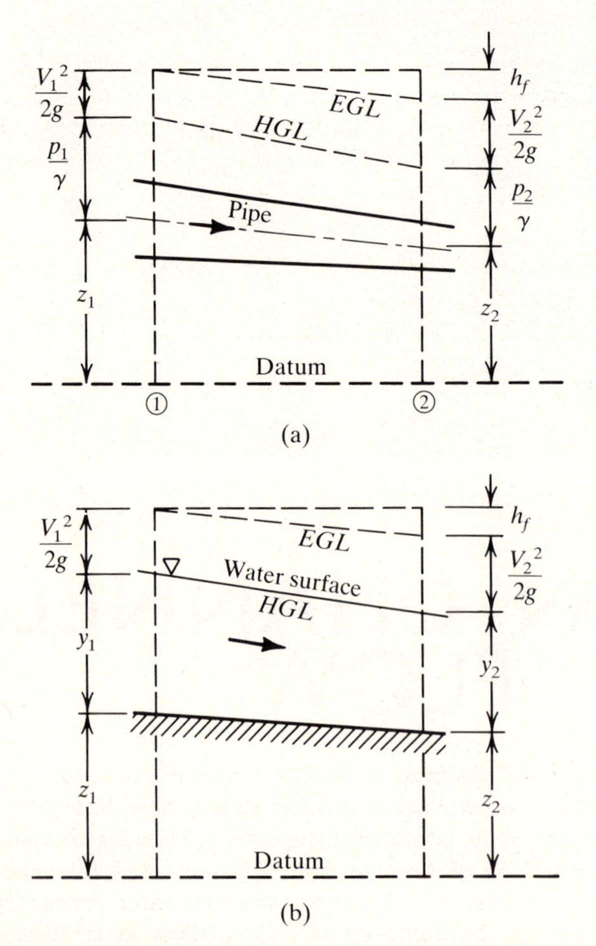

FIGURE 10.1 Open channel flow and pipe flow compared.

energy grade line (EGL) is at a distance of $V^2/2g$ above the HGL. The loss of energy when the water flows from section 1 to section 2 is shown as h_f in Fig. 10.1a. In the open channel flow shown in Fig. 10.1b, the water surface is free, hence, the HGL in general coincides with the water surface. The EGL is at a distance of $V^2/2g$ above the HGL. It may, however, be pointed out that in certain cases of open channel flow the HGL may not exactly coincide with the water surface.

Types of Flow

The flow in an open channel may be *steady, unsteady, uniform, nonuniform* or any combination thereof. *Steady* flow exists if the depth of flow at a section does not vary with time; otherwise, the flow is *unsteady*. *Uniform* flow exists if the depth of flow remains the same from section to section at any instant of time; otherwise, the flow is *nonuniform*. The flows may further be classified as *steady-uniform, steady-nonuniform, unsteady-uniform* or *unsteady-nonuniform* by combining the above definitions. *Unsteady-uniform* flow is rarely found in practice.

Nonuniform flow is also called *varied flow*. Varied flow may be gradually varied or rapidly varied depending on the distance over which the depth actually varies. A backwater curve behind a dam is an example of a gradually varied flow, while a hydraulic jump is an example of a rapidly varied flow. These definitions are further illustrated by Fig. 10.2.

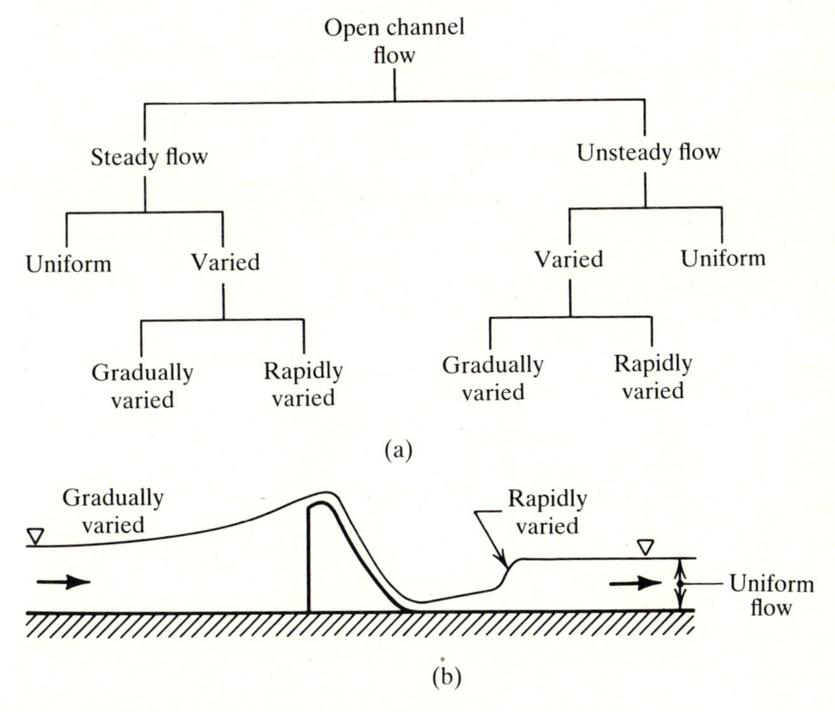

FIGURE 10.2 Various types of open channel flow.

Table 10.1 GEOMETRIC ELEMENTS OF CHANNEL SECTIONS

Section	Area A	Wetted perimeter P	Top width T	Hydraulic depth D
Rectangle	by	$b + 2y$	b	y
Trapezoid	$(b + zy)y$	$b + 2y\sqrt{1 + 2^2}$	$b + 2zy$	$\frac{(b + zy)y}{b + 2zy}$
Triangle	zy^2	$2y\sqrt{1 + 2^2}$	$2zy$	$\frac{1}{2y}$
Circle	$\frac{1}{8}(\theta - \sin\theta)d_0^2$	$\frac{1}{2}\theta d_0$	$\left(\sin\frac{1}{2}\theta\right)d_0$ or $2\sqrt{y(d_0 - y)}$	$\frac{1}{8}\left(\frac{\theta - \sin\theta}{\sin\frac{1}{2}\theta}\right)d_0$

Channel Geometry and Elements

A channel having a constant cross-section and constant bed slope is called a *prismatic channel*. Otherwise, the channel is nonprismatic. *A channel section* normally refers to the cross-section of a channel taken normal to the direction of the flow.

The *depth of flow* (y) is the vertical distance of the lowest point of a channel section from the free surface. The *stage* is the elevation or vertical distance of the free surface above a datum. If the lowest point of the channel section is chosen as the datum, the stage is identical with the depth of flow. The *top width* (T) is the width of the channel section at the free surface. The *area of flow* (A) is the cross-sectional area of the flow normal to the direction of flow. The *wetted perimeter* (P) is the length of the channel at a section wetted by the flowing liquid. The *hydraulic radius* (R) is the area of flow divided by the wetted perimeter, or $R = A/P$. The *hydraulic depth* (D) is the area of flow divided by the top width, or $D = A/T$. The *section factor* (Z) is the product of the area of flow and the square root of the hydraulic depth, or $Z = A\sqrt{D}$.

Table 10.1 provides a list of formulas for various types of channel sections.

EXAMPLE 10.1

A trapezoidal channel has a base of 3 m and side slopes of 2 vertical to 1 horizontal. The depth of water in the channel is 2 m. Compute the top width, area of flow, wetted perimeter, hydraulic radius, hydraulic depth and section factor.

Solution The cross-section of the channel is shown in Fig. 10.3. For the given section, $z = 0.5$. Using information contained in Table 10.1, for a trapezoidal channel:

$$\text{The top width} = T = b + 2zy$$

$$= 3 + 2 \times 0.5 \times 2 = 5 \text{ m}$$

$$\text{Area of flow} = A = (b + zy)y$$

$$= (3 + 0.5 \times 2)2$$

$$= 8.0 \text{ m}^2$$

$$\text{Wetted perimeter} \doteq P = b + 2y\sqrt{1 + z^2}$$

$$= 3 + 2 \times 2\sqrt{1 + (0.5)^2}$$

$$= 7.472 \text{ m}$$

$$\text{Hydraulic radius} = R = A/P = \frac{8}{7.472} = 1.07 \text{ m}$$

$$\text{Hydraulic depth} = D = A/T = \frac{8}{5} = 1.6 \text{ m}$$

$$\text{Section Factor} = Z = A\sqrt{D} = 8\sqrt{1.6} = 10.12 \text{ m}^{2.5}$$

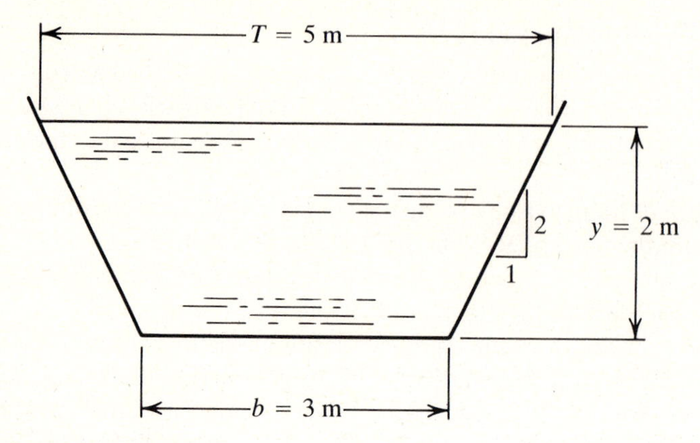

FIGURE 10.3 Example 10.1.

10.2 UNIFORM FLOW IN AN OPEN CHANNEL

The flow in an open channel is uniform if the depth of flow does not vary along the length of the channel. The average velocity also remains constant. Since the depth of flow does not vary and the velocity is constant, the channel bed, the water line, the hydraulic grade line, and the total energy line are parallel to each other. Let us now derive an expression for the velocity of flow.

Consider a channel of constant cross-section A carrying a uniform flow as shown in Fig. 10.4. Let L = length of the channel bounded between two sections ① and ②. The forces acting on the control volume, shown by dotted lines in Fig. 10.4 are: the hydrostatic forces F_1 and F_2 acting on sections ① and ② respectively, a friction force F_f acting on an area which is in contact with the fluid, and gravity force W. The equation of motion in the direction of

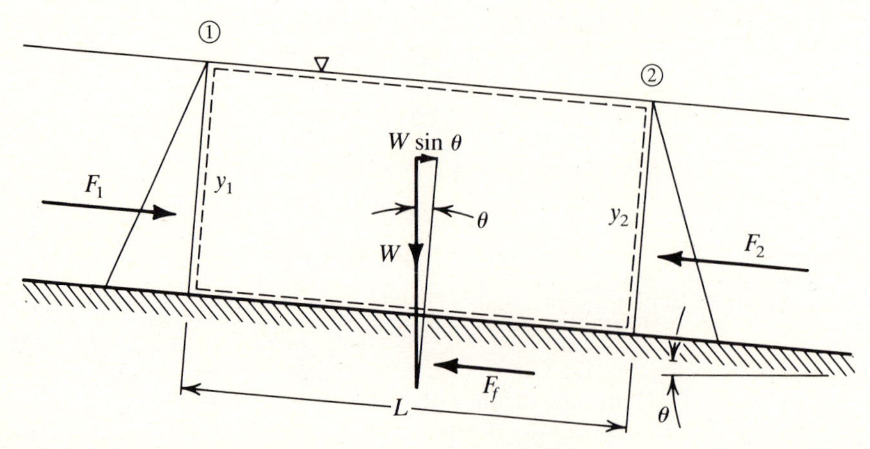

FIGURE 10.4 Forces in case of uniform flow in an open channel.

flow may be written as

$$F_1 + W \sin \theta - F_2 - F_f = \frac{\gamma}{g} a$$

Since the velocity is constant, there is no acceleration in the direction of flow and $a = 0$. Also, since the depth of flow is constant, the hydrostatic forces F_1 and F_2 are equal. Therefore, the above equation reduces to

$$W \sin \theta - F_f = 0 \tag{10.1}$$

The weight of the fluid contained within the control volume is γAL, where γ = specific weight of the fluid. Also, since θ is small, $\sin \theta = \tan \theta = S$, where S = slope of the channel bed. Hence, $W \sin \theta = \gamma ALS$.

In 1769, a French engineer, Antoine Chézy, through experimentation, determined that the shear stress on the channel bed was proportional to the square of the average velocity, or $\tau \alpha V^2$ or $\tau = KV^2$ where K is a constant whose value depends on the channel characteristics. Therefore, $F_f = \tau PL = KV^2PL$, where P = wetted perimeter. Substituting for $W \sin \theta$ and F_f in Eq. (10.1), we obtain

$$\gamma ALS - KV^2PL = 0$$

$$V = \sqrt{\left(\frac{\gamma}{K}\right)\left(\frac{A}{P}\right)S}$$

Denoting $A/P = R$ and $\sqrt{\dfrac{\gamma}{K}} = C$, we obtain

$$V = C\sqrt{RS} \tag{10.2}$$

This equation is known as the *Chézy Formula*, and C as the *Chézy constant*.

Over the past two centuries, many researchers have developed expressions for determining C. The most widely used expression was developed by an Irish engineer, Robert Manning, in 1891. According to him:

$$C = \frac{Cm}{n} R^{1/6}$$

where

$$Cm = 1.0 \text{ in SI units}$$

$$= 1.486 \text{ in FPS units}$$

$$n = \text{coefficient of friction}$$

Substituting for C in Eq. (10.2), we obtain

$$V = \frac{Cm}{n} R^{2/3} S^{1/2} \tag{10.3}$$

This equation is known as *Manning's formula*. Typical values of n for various channel conditions and material are given in Table 10.2.

Table 10.2 TYPICAL VALUES OF MANNING'S *n*

Type of channel	Minimum	Normal	Maximum
Riveted and spiral steel	0.013	0.016	0.017
Coated cast iron	0.01	0.013	0.014
Uncoated cast iron	0.011	0.014	0.016
Galvanized wrought iron	0.013	0.016	0.017
Black wrought iron	0.012	0.014	0.015
Corrugated metal drain	0.021	0.024	0.030
Glass	0.009	0.01	0.013
Cement mortar	0.011	0.013	0.015
Finished concrete channel	0.01	0.012	0.014
Concrete culvert, straight	0.01	0.011	0.013
Concrete culvert with bends, connections	0.011	0.013	0.014
Concrete sewer with manholes, inlets, etc.	0.013	0.015	0.017
Wood stave	0.01	0.012	0.014
Clay drainage tile	0.011	0.013	0.017
Brickwork	0.012	0.015	0.017
Earthen channel, straight, clean	0.017	0.02	0.025
Channel, straight with short grass, few weeds	0.022	0.027	0.033

Source: Open Channel Hydraulics by Chow, New York: McGraw-Hill.

EXAMPLE 10.2

Determine the discharge for a trapezoidal channel with $b = 3$ m and sides slope 1 on 1. The depth of flow is 1.5 m and the slope of the channel bed is 0.0009. The channel has a finished concrete lining.

Solution For the given flow conditions:

$$z = \text{side slopes} = 1.0$$

$$b = 3.0 \text{ m}$$

$$y = 1.5 \text{ m}$$

$$S = 0.0009$$

$$n = 0.012 \text{ (for concrete from Table 10.2)}$$

$$\begin{aligned} A &= (b + zy)y \\ &= (3 + 1.5)1.5 \\ &= 6.75 \text{ m}^2 \end{aligned}$$

$$P = b + 2y\sqrt{1 + z^2}$$
$$= 3 + 2 \times 1.5\sqrt{1 + 1}$$
$$= 7.24 \text{ m}$$
$$R = \frac{A}{P} = \frac{6.75}{7.24} = 0.93 \text{ m}$$

From Manning's equation:

$$Q = A \frac{1}{0.012} R^{2/3} S^{1/2}$$
$$= \frac{6.75}{0.012} (0.93)^{2/3} (0.0009)^{1/2}$$
$$= 16.1 \text{ m}^3/\text{s}$$

EXAMPLE 10.3

Determine the normal depth of flow in a trapezoidal channel carrying 16.1 m³/s. The channel has a width of 3.0 m and side slopes of 1 on 1. The slope of the channel bed is 0.0009. The channel has a finished concrete lining.

Solution For the channel, let y be the normal depth.

Also:

$$Q = 16.1 \text{ m}^3/\text{s}$$
$$b = 3.0 \text{ m}$$
$$z = 1$$
$$S = 0.0009$$
$$A = (b + zy)y = (3 + y)y$$
$$P = b + 2y\sqrt{1 + z^2} = 3 + 2.828y$$
$$R = \frac{A}{P} = \frac{(3 + y)y}{3 + 2.828y}$$

From Manning's equation:

$$Q = A \frac{Cm}{n} R^{2/3} S^{1/2}$$
$$16.1 = (3 + y)y \frac{1}{0.012} \left[\frac{(3 + y)y}{(3 + 2.828y)} \right]^{2/3} (0.0009)^{1/2}$$

By trial and error, $y = 1.5$ m.

Since y is obtained by trial and error, a computer solution may be helpful.

Computer Solution

A computer program written in BASIC is listed in Appendix B which may be used to determine y. In addition to calculating for the normal conditions, the program also calculates for the critical conditions, which are discussed in Section 10.4. The solution is given below:

```
RUN
open channel program

title? Example 10.3
system of unit(si/fps)? si
use the following code to describe the
channel type
 1 for triangular
 2 for trapezoidal
 3 for rectangular
 4 for circular

channel type? 2
bottom width (enter 0 for type 1) (m)? 3
left side slope (--:1) (enter 0 for type 3)? 1
right side slope (--:1) (enter 0 for type 3)? 1
bed slope? 0.0009
mannings n? 0.012
normal depth (m)? 0
discharge (cu.m/sec)? 16.1


Example 10.3
trapezoidal channel problem
***** normal flow calculations *****
rate of flow (cu.m/sec)        =    16.10083
normal velocity (m/sec)        =    2.385309
normal depth (m)               =    1.5
normal top width (m)           =    6
normal area of flow (sq.m)     =    6.75
froud number                   =    .7180155
***** critical flow calculations *****
rate of flow (cu.m/sec)        =    16.09956
critical velocity (m/sec)      =    3.066472
critical depth (m)             =    1.238648
critical top width (m)         =    5.477295
critical area of flow (sq.m)   =    5.25019
critical slope                 =    1.801343E-03

do you wish to solve another problem (y/n)? n
Break in 2140
Ok
```

Variation of Velocity over Cross-Section of a Channel

The velocity of flow varies at different points of the cross-section. The frictional resistance of the sides causes the water to slow down toward the sides of the channel, and the frictional resistance between the water surface and the atmosphere causes a slight reduction of velocity at the free surface.

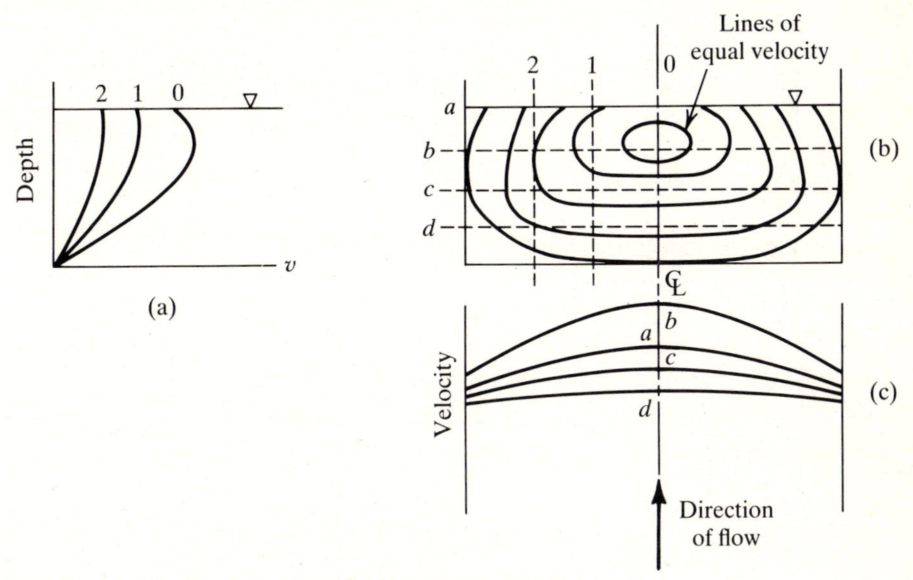

FIGURE 10.5 Velocity variation over a channel cross-section.

The maximum velocity will be on the vertical centerline of the channel at a point a little below the free surface.

The variation of velocity over the cross-section of a rectangular channel is shown in Fig. 10.5a. The curves shown are lines of equal velocity; they have the greatest value at the center, just below the water surface, and decrease toward the sides and base. Figure 10.5b and Fig. 10.5c show the velocity profiles at different points of the section lines drawn on Fig. 10.5a.

The mean velocity on any vertical section occurs at a depth of approximately 0.6 of the total depth; it varies with the type of channel and the friction of the sides. The discharge of the whole channel may be obtained by dividing the section into vertical rectangles and finding the mean velocity of each rectangle. Using this mean velocity, the discharge through each rectangle may be obtained. The sum of all these discharges will be the total discharge of the channel.

10.3 BEST HYDRAULIC SECTIONS

A channel with the best hydraulic section is one which can carry maximum discharge for a given amount of excavation.
From Chézy's equation

$$Q = AC\sqrt{\frac{AS}{P}}$$

If A, C and S are constant, the discharge Q will be maximum when the wetted perimeter P is minimum or when dP/dy is zero. The conditions for the most economical section for various types of channel sections are now derived.

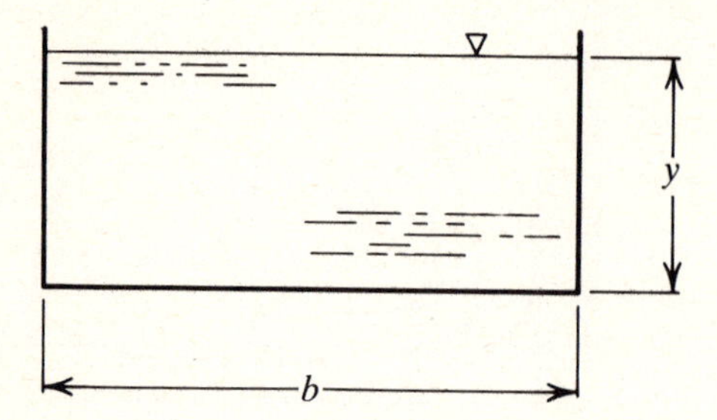

FIGURE 10.6 Cross-section of a rectangular channel.

Rectangular Channel: Depth for Maximum Discharge

Consider the rectangular section shown in Fig. 10.6. The wetted perimeter P is given by

$$P = b + 2y$$

Also

$$b = \frac{A}{y}$$

Therefore:

$$P = \frac{A}{y} + 2y$$

$$= Ay^{-1} + 2y$$

$$\frac{dP}{dy} = -Ay^{-2} + 2 = 0$$

$$A = 2y^2$$

$$by = 2y^2$$

$$y = \frac{b}{2} \tag{10.4}$$

Therefore, a rectangular channel will be most efficient when the depth of flow is half the width of the channel.

Trapezoidal Channel: Depth for Maximum Discharge

As shown earlier, for maximum Q, P should be minimum. For a trapezoidal channel (Fig. 10.7)

$$P = b + 2y\sqrt{1 + z^2} \tag{10.5}$$

Also, for a trapezoidal channel

$$A = (b + zy)y$$

$$A = by + zy^2 \tag{10.6}$$

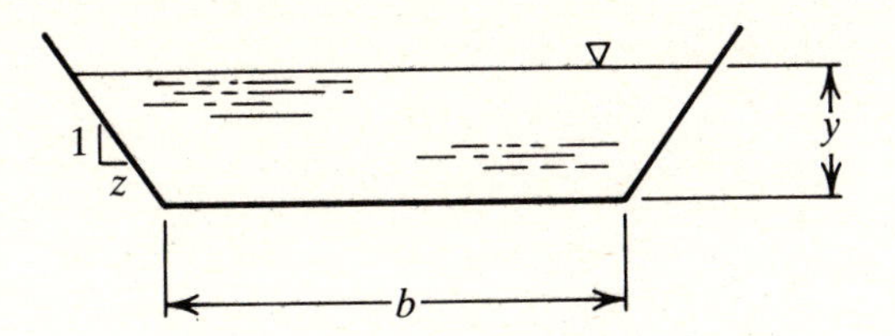

FIGURE 10.7 Cross-section of a trapezoidal channel.

from which

$$b = \frac{A - zy^2}{y} = \frac{A}{y} - zy \tag{10.7}$$

Substituting Eq. (10.7) in Eq. (10.5)

$$P = \frac{A}{y} - zy + 2y\sqrt{1 + z^2} \tag{10.8}$$

For a most economical section, $\dfrac{dP}{dy} = 0$

$$\frac{dP}{dy} = -\frac{A}{y^2} - z + 2\sqrt{1 + z^2} = 0$$

$$\frac{A}{y^2} = 2\sqrt{1 + z^2} - z$$

Substituting for A from Eq. (10.6)

$$\frac{by + zy^2}{y^2} = 2\sqrt{1 + z^2} - z$$

$$\frac{b}{y} + z = 2\sqrt{1 + z^2} - z$$

$$\frac{b}{y} = 2(\sqrt{1 + z^2} - z) \tag{10.9}$$

Equation (10.9) gives the condition for a most economical trapezoidal section.

Note also:

$$R = \frac{A}{P} = \frac{by + zy^2}{b + 2y\sqrt{1 + z^2}} \tag{10.10}$$

From Eq. (10.9):

$$b = 2y(\sqrt{1 + z^2} - z)$$

Substituting in Eq. (10.10)

$$R = \frac{2y^2(\sqrt{1+z^2} - z) + zy^2}{2y(\sqrt{1+z^2} - z) + 2y\sqrt{1+z^2}}$$

And simplifying, we obtain

$$R = \frac{y}{2}$$

which is another condition for the most economical trapezoidal section.

EXAMPLE 10.4

A trapezoidal channel is to be designed for conveying 10,000 ft^3 of water per minute. Determine the cross-sectional dimensions of the channel if the slope is 1 in 1600, the sides are inclined at 45°, and cross-section is to be a minimum. Use $C = 90$ in the Chézy formula.

Solution Using Eq. (10.9), and putting $z = 1$

$$b = 0.828y$$

$$A = by + zy^2 = by + y^2 \text{ when } z = 1$$

Hence:

$$A = 0.828y^2 + y^2 = 1.828y^2$$

Also, for most economical section

$$R = 0.5y$$

From the Chézy formula:

$$V = 90\sqrt{RS}$$

$$= 90\sqrt{0.5y\frac{1}{1600}}$$

$$= 2.25\sqrt{0.5y}$$

Now

$$Q = AV$$

Therefore:

$$\frac{10{,}000}{60} = 1.828y^2 \times 2.25\sqrt{0.5y}$$

Squaring both sides and simplifying

$$y = 5.04 \text{ ft}$$

$$b = 0.828y = 4.17 \text{ ft}$$

Therefore, a trapezoidal channel having sides inclined at 45°, a depth of flow equal to 5.04 ft and base width equal to 4.17 ft will be the most economical channel to convey a discharge of 10,000 ft^3 per minute.

Circular Channel: Depth for Maximum Velocity

The velocity of flow in a given circular channel will depend upon the depth of the water (Fig. 10.8).

Let y = depth of water for maximum velocity, θ = one-half the angle subtended at center by the water line, in radians, r = radius of the channel section.

Then:

$$A = \text{area of flow}$$

$$= \text{area of sector} - \text{area of triangle OAB}$$

$$= r^2\theta - \frac{r^2 \sin 2\theta}{2}$$

$$A = r^2\left(\theta - \frac{\sin 2\theta}{2}\right) \tag{10.11}$$

$$P = 2r\theta \tag{10.12}$$

Also, from Chézy's equation:

$$V = C\sqrt{RS}$$

For constant values of C and S, V is maximum, when R is maximum, or V is maximum when $dR/d\theta = 0$.

$$\frac{dR}{d\theta} = \frac{d(A/P)}{d\theta} = P\frac{dA}{d\theta} - A\frac{dP}{d\theta} = 0 \tag{10.13}$$

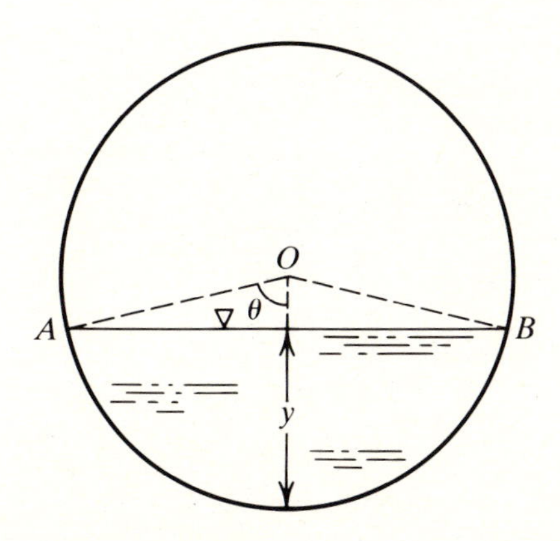

FIGURE 10.8 Cross-section of a circular channel.

Now

$$\frac{dA}{d\theta} = r^2(1 - \cos 2\theta)$$

$$\frac{dP}{d\theta} = 2r$$

Substituting in Eq. (10.13)

$$Pr^2(1 - \cos 2\theta) - A2r = 0$$

Substituting for A and P from Eqs. (10.11) and (10.12), and simplifying

$$2\theta = \tan 2\theta \tag{10.14}$$

A trial and error solution of Eq. (10.14) yields

$$2\theta = 257.5°$$

Now

$$\begin{aligned} y &= r - r\cos\theta \\ &= r - r\cos\frac{257.5}{2} \\ &= r(1 + 0.62) \\ &= 1.62r \\ &= 0.81D \end{aligned}$$

Hence:

$$y = 0.81D$$

Therefore, the depth for maximum velocity = 0.81 × diameter of channel.

Circular Channel: Depth for Maximum Discharge

From Chézy's equation:

$$V = C\sqrt{RS}$$

and

$$Q = AC\sqrt{RS} = AC\sqrt{\frac{A}{P}S}$$

or

$$Q = C\sqrt{\frac{A^3}{P}S}$$

Therefore, discharge is maximum when A^3/P is maximum, or $d(A^3/P)/d\theta$ should be zero for maximum discharge.

$$\frac{d(A^3/P)}{d\theta} = \frac{3PA^2 \dfrac{dA}{d\theta} - A^3 \dfrac{dP}{d\theta}}{P^2} = 0$$

$$3PA^2 \frac{dA}{d\theta} - A^3 \frac{dP}{d\theta} = 0$$

$$3P \frac{dA}{d\theta} - A \frac{dP}{d\theta} = 0$$

Substituting for A and P from Eqs. (10.11) and (10.12), and simplifying

$$4\theta - 6\theta \cos 2\theta = -\sin 2\theta$$

By trial and error, $\theta = 154°$
Thus, for maximum discharge:

$$y = r - r\cos\theta$$

$$= 1.9r$$

$$y = 0.95D$$

Therefore, depth for maximum discharge $= 0.95 \times$ diameter of channel.

10.4 SPECIFIC ENERGY

Consider a cross-section of a channel (Fig. 10.9a) through which a fluid is flowing at a constant rate Q. The average velocity at the cross-section is then given by

$$V = \frac{Q}{A}$$

where A is the area of the cross-section. The *specific energy* E of the channel at this cross-section is defined as

$$E = y + \frac{V^2}{2g} \tag{10.15}$$

This is not the total energy of the water as datum energy is not included; consequently, the slope of the channel is not included in the term 'specific energy.'

Figure 10.9b shows the kinetic energy, the static energy and the specific energy for this section for a fixed quantity of flow and for various depths of flow. It is noticed that the specific energy at first decreases with the increase in depth. At point b, the specific energy has its minimum value; beyond this point, there is an increase in the specific energy as the depth increases. The depth at point b, where E is minimum, is called *the critical depth* (y_c) and the

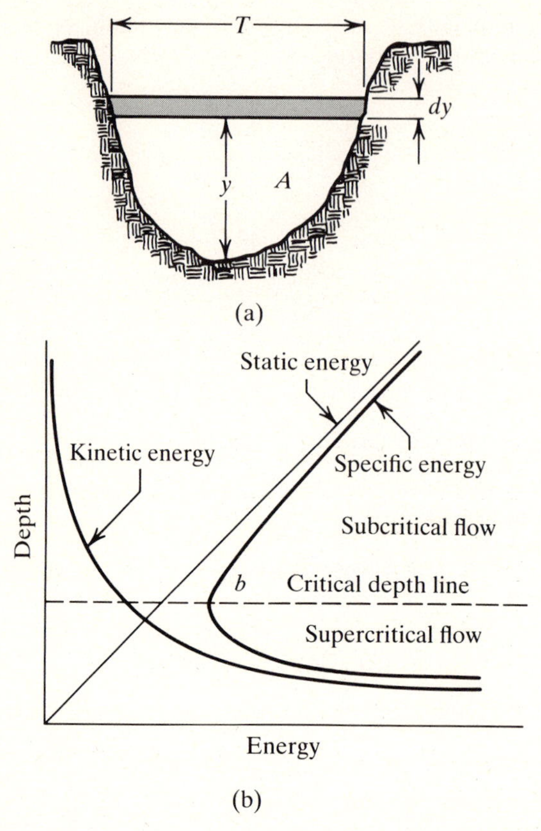

FIGURE 10.9 Specific energy diagram for an arbitrary cross-section of a channel.

corresponding velocity of flow is called the *critical velocity* (V_c). For each value of specific energy to the right of b, there are two depths which produce the same E for a given discharge. These depths are called *the conjugate depths* or *the alternative depths.* When the depth of flow is greater than the critical depth, the flow is *subcritical* or *tranquil*; when it is less, the flow is *supercritical* or *rapid.*

An expression for the critical depth can be obtained by differentiating Eq. (10.15) with respect to y and equating to zero. Then:

$$E = y + \frac{V^2}{2g}$$

$$E = y + \frac{Q^2}{2A^2 g}$$

$$\frac{dE}{dy} = 1 - \frac{Q^2}{gA^3}\frac{dA}{dy} \tag{10.16}$$

From Fig. 10.9a, near the water surface

$$dA = T\,dy \tag{10.17}$$

$$\frac{dA}{dy} = T$$

Substituting in Eq. (10.16), and equating to zero, we obtain

$$\frac{dE}{dy} = 1 - \frac{Q^2 T}{gA^3} = 0$$

$$\frac{Q^2 T}{gA^3} = 1 \tag{10.18}$$

At critical conditions, $Q/A = V_c$. Also, $A/T = D$ = hydraulic depth. Substituting in Eq. (10.18), we obtain

$$\frac{V_c^2}{gD} = 1$$

$$\frac{V_c}{\sqrt{gD}} = 1 \tag{10.19}$$

We can recognize the expression $V_c/\sqrt{gD}$ as the Froude number **F**. Therefore, at critical conditions, the Froude number is equal to unity.

For a rectangular channel of unit width, $V_c = Q/y_c$; $D = y_c$. Substituting in Eq. (10.19), we obtain

$$\frac{Q}{y_c\sqrt{gy_c}} = 1$$

or

$$y_c = \left(\frac{Q^2}{g}\right)^{1/3} \tag{10.20}$$

Depth for Maximum Flow at a Given Specific Energy

From the definition of specific energy

$$E = y + \frac{Q^2}{2A^2 g}$$

Now assume that Q varies and E is constant.

Hence:

$$Q = \sqrt{2A^2 gE - 2A^2 gy}$$

The discharge will be maximum, when $dQ/dy = 0$ or when

$$\frac{dQ}{dy} = \frac{4gEA\dfrac{dA}{dy} - 2A^2g - 4gyA\dfrac{dA}{dy}}{2(2A^2gE - 2A^2gy)^2} = 0$$

$$4gEA\frac{dA}{dy} - 2A^2g - 4gyA\frac{dA}{dy} = 0$$

Substituting $\dfrac{dA}{dy} = T$ and simplifying

$$2ET - TD - 2yT = 0$$

$$2E - D - 2y = 0$$

$$y = \frac{2E - D}{2}$$

$$y = E - \frac{D}{2} \tag{10.21}$$

Also, at critical conditions:

$$\frac{V_c^2}{gD} = 1$$

$$\frac{V_c^2}{2g} = \frac{D}{2}$$

$$y_c + \frac{V_c^2}{2g} = y_c + \frac{D}{2}$$

$$E_{\text{min}} = y_c + \frac{D}{2}$$

$$y_c = E_{\text{min}} - \frac{D}{2} \tag{10.22}$$

This equation is identical to Eq. (10.21). The comparison of the two equations suggests that the maximum discharge, for a given specific energy, occurs at the critical condition.

EXAMPLE 10.5

Water enters a rectangular channel of uniform width with a velocity of 8 m/s. The depth of water at entrance is 2.5 m. Calculate the critical depth. Also determine the critical depth graphically.

Solution Let b = width of channel. Then

$$Q = byV$$

$$= b \times 2.5 \times 8 = 20b \ \text{ft}^3/\text{s}$$

and

$$V_c = \frac{20b}{by_c} = \frac{20}{y_c}$$

For a rectangular channel:

$$D = \frac{A}{T} = \frac{yb}{b} = y$$

At critical state

$$\frac{V_c}{\sqrt{gD}} = 1$$

$$\frac{V_c^2}{gy_c} = 1$$

$$y_c = \frac{V_c^2}{g} = \frac{(20)^2}{(y_c)^2 9.81}$$

$$y_c = 3.44 \text{ m}$$

For a graphical determination of y_c, the static energy, kinetic energy and specific energy are calculated as shown in Table 10.3. The results are then plotted as shown in Fig. 10.10, and y_c is read against minimum specific energy, which is found to be 3.4 m.

Table 10.3 CALCULATIONS FOR EXAMPLE 10.5

y	V	$\frac{V^2}{2g}$	Static energy	Specific energy
0.5	40.00	81.55	0.5	82.05
1.0	20.00	20.39	1.0	21.39
1.5	13.33	9.06	1.5	10.56
2.0	10.00	5.10	2.0	7.1
2.5	8.00	3.26	2.5	5.76
3.0	6.67	2.27	3.6	5.27
3.5	5.71	1.66	3.5	5.16
4.0	5.00	1.27	4.0	5.27
4.5	4.40	0.98	4.5	5.49
5.0	4.00	0.82	5.0	5.82
5.5	3.64	0.68	5.5	6.18
6.0	3.33	0.57	6.0	6.57
6.5	3.08	0.48	6.5	6.98
7.0	2.86	0.42	7.0	7.42
7.5	2.67	0.36	7.5	7.86
8.0	2.50	0.32	8.0	8.32

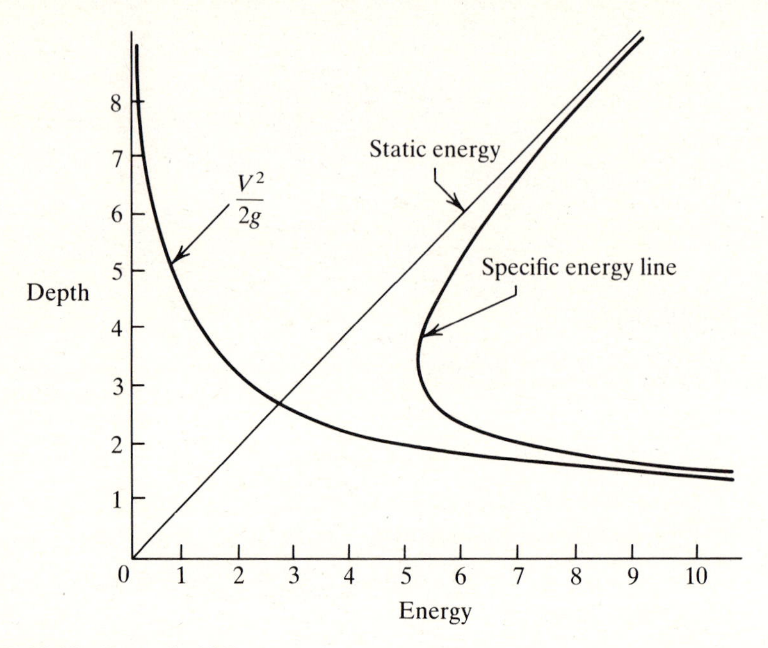

FIGURE 10.10 Example 10.5.

EXAMPLE 10.6

A concrete trapezoidal channel has a bottom width of 4.0 m and side slopes of 2 vertical to one horizontal. The bed slope is 0.0008, and the Manning n is 0.025. For a flow rate of 15 m^3/s, determine

(a) the normal depth
(b) the flow regime
(c) the critical depth.

Solution (a) The normal depth is calculated using the Manning equation

$$Q = A \frac{1.00}{0.025} R^{2/3} S^{1/2}$$

For this problem:

$$z = 0.5$$

$$S = 0.0008$$

$$n = 0.025$$

$$A = (b + zy)y$$

$$= (4 + 0.5y)y$$

$$P = b + 2y\sqrt{1 + z^2}$$
$$= 4 + 2y\sqrt{1 + 0.5^2}$$
$$= 4 + 2.236y$$

Hence

$$15 = (4 + 0.5y)y \frac{1}{0.025} \left[\frac{(4 + 0.5y)y}{4 + 2.236y}\right]^{2/3} (0.0008)^{1/2}$$

By trial and error, $y = 2.22$ m.

Therefore, the normal depth is equal to 2.22 m.

(b) To find the flow regime, determine the Froude number:

$$\mathbf{F} = \frac{V}{\sqrt{gD}}$$

$$A = (4 + 0.5 \times 2.22)2.22 = 11.3 \text{ m}^2$$

$$T = b + 2zy = 4 + 2 \times 0.5 \times 2.22 = 6.22 \text{ m}$$

$$D = \frac{A}{T} = \frac{11.3}{6.22} = 1.82 \text{ m}$$

$$V = \frac{Q}{A} = \frac{15}{11.3} = 1.32 \text{ m/s}$$

Hence:

$$\mathbf{F} = \frac{1.32}{\sqrt{9.81 \times 1.82}} = 0.31$$

Since the Froude number is less than 1, the flow is subcritical.

(c) At critical conditions:

$$\frac{V_c}{\sqrt{gD}} = 1$$

$$V_c = \sqrt{gD} = \sqrt{g\frac{A}{T}}$$

Also:

$$Q = AV_c$$
$$= (b + zy_c)y_c\sqrt{gA/T}$$
$$15 = (4 + 0.5y_c)y_c\left[\frac{9.81(4 + 0.5y_c)y_c}{4 + 2 \times 0.5 \times y_c}\right]^{1/2}$$

By trial and error, $y_c = 1.08$ m.

Hence, the critical depth is equal to 1.08 m.

Computer Solution

The following solution is obtained by using the computer program listed in Appendix B.

```
RUN
open channel program

title? Example 10.6
system of unit(si/fps)? si
use the following code to describe the
channel type
 1 for triangular
 2 for trapezoidal
 3 for rectangular
 4 for circular

channel type? 2
bottom width (enter 0 for type 1) (m)? 4
left side slope (--:1) (enter 0 for type 3)? 0.5
right side slope (--:1) (enter 0 for type 3)? 0.5
bed slope? 0.0008
mannings n? 0.025
normal depth (m)? 0
discharge (cu.m/sec)? 15


Example 10.6
trapezoidal channel problem
***** normal flow calculations *****
rate of flow (cu.m/sec)        =    14.99936
normal velocity (m/sec)        =    1.323289
normal depth (m)               =    2.218506
normal top width (m)           =    6.218506
normal area of flow (sq.m)     =    11.33491
froud number                   =    .3129351
***** critical flow calculations *****
rate of flow (cu.m/sec)        =    14.99967
critical velocity (m/sec)      =    3.071921
critical depth (m)             =    1.075989
critical top width (m)         =    5.075989
critical area of flow (sq.m)   =    4.882831
critical slope                 =    8.470692E-03

do you wish to solve another problem (y/n)? n
Break in 2140
Ok
```

EXAMPLE 10.7

A circular sewer 6.0 ft in diameter is laid on a slope of 0.0008. The Manning $n = 0.012$. Determine the parameters for the normal and the critical flow for the following flow rates:

(a) 100 cfs
(b) 400 cfs
(c) 600 cfs
(d) 1200 cfs.

Solution The problem is solved using the computer program listed in Appendix B. The solution is shown below. Notice that normal open channel flow is not possible for Q of 400, 600, and 1,200 cfs. For flows of 600 and 1200 cfs, open channel flow is not possible even under critical conditions.

```
RUN
open channel program
title? Example 10.7
system of unit(si/fps)? fps
use the following code to describe the
channel type
 1 for triangular
 2 for trapezoidal
 3 for rectangular
 4 for circular

channel type? 4
diameter (ft)? 6
bed slope? 0.0008
mannings n ? 0.012
normal depth (ft)? 0
discharge (cu.ft/sec)? 100

Example 10.7
circular channel problem
***** normal flow calculations *****
rate of flow (cu.ft/sec)       =     100.0006
normal velocity (ft/sec)       =     5.063622
normal depth (ft)              =     3.951477
normal top width (ft)          =     5.690234
normal area of flow (sq.ft)    =     19.74883
froud number                   =     .4789914

***** critical flow calculations *****
rate of flow (cu.ft/sec)         =   100.0005
critical velocity (ft/sec)       =   8.140979
critical depth (ft)              =   2.690521
critical top width (ft)          =   5.967989
critical area of flow (sq.ft)=       12.28359
critical slope                   =   2.772457E-03

do you wish to solve another problem (y/n)? y
title? part(b)
system of unit(si/fps)? fps
use the following code to describe the
channel type
 1 for triangular
 2 for trapezoidal
 3 for rectangular
 4 for circular
```

```
channel type? 4
diameter (ft)? 6
bed slope? 0.0008
mannings n ? 0.012
normal depth (ft)? 0
discharge (cu.ft/sec)? 400

normal open channel flow is not possible for the
given conditions. discharge is too large. pipe will
flow full.
***** critical flow calculations *****
rate of flow (cu.ft/sec)       =   400.0009
critical velocity (ft/sec)     =   15.06099
critical depth (ft)            =   5.333781
critical top width (ft)        =   3.770129
critical area of flow (sq.ft)= 26.55874
critical slope                 =   6.766719E-03

do you wish to solve another problem (y/n)? y
title? part(c)
system of unit(si/fps)? fps
use the following code to describe the
channel type
 1 for triangular
 2 for trapezoidal
 3 for rectangular
 4 for circular

channel type? 4
diameter (ft)? 6
bed slope? 0.0008
mannings n ? 0.012
normal depth (ft)? 0
discharge (cu.ft/sec)? 600
normal open channel flow is not possible for the
given conditions. discharge is too large. pipe will
flow full.
critical open channel flow is not possible for the
given conditions. discharge is too large. pipe will
flow full.

do you wish to solve another problem (y/n)? y
title? part(d)
system of unit(si/fps)? fps
use the following code to describe the
channel type
 1 for triangular
 2 for trapezoidal
 3 for rectangular
 4 for circular

channel type? 4
diameter (ft)? 6
bed slope? 0.0008
mannings n ? 0.012
normal depth (ft)? 0
```

```
discharge (cu.ft/sec)? 1200
normal open channel flow is not possible for the
given conditions. discharge is too large. pipe will
flow full.
critical open channel flow is not possible for the
given conditions. discharge is too large. pipe will
flow full.

do you wish to solve another problem (y/n)? n
Break in 1790
Ok
```

10.5 RAPIDLY VARIED FLOW

Rapidly varied flow is a type of nonuniform flow in which changes in depth occur over a short distance. Hence, a rapidly varying flow has a pronounced curvature of the streamlines. Such a sharp curvature may be seen when a mild slope changes suddenly to a steep slope causing the fluid surface to drop (known as a *hydraulic drop*). Sometimes, the change of curvature of streamlines may become so sharp as to cause the streamlines to break resulting in turbulence. Such a change is encountered in a hydraulic jump when the fluid passes from supercritical to subcritical conditions.

Since the curvature of the flow is highly pronounced, the pressure distribution in a rapidly varying flow is not hydrostatic. Also, since the depth changes over a short distance, the boundary friction does not play a significant role and may be neglected.

Hydraulic Jump

When, in a channel, the flow regime changes from supercritical to subcritical, an abrupt rise of water surface, known as *hydraulic jump*, occurs. A hydraulic jump frequently occurs in a canal below a regulating sluice, at the foot of a spillway or at a place where a steep channel slope suddenly becomes flat. The strength of the jump depends on the Froude number before the jump. The higher the Froude number, the stronger the jump.

The depth of flow before the jump is called *the initial depth* and that after the jump is called *the sequent depth.* A formula is now developed to determine the sequent depth in a rectangular channel when the initial depth is known. As shown in Fig. 10.11, consider two sections 1 and 2 and erect a control volume taking the sections, the bed of the channel and the water surface as control surfaces. Let V_1 = average velocity at section 1; V_2 = average velocity at section 2; y_1 = depth of flow at section 1; y_2 = depth of flow at section 2. For simplicity, let us assume that the floor of the channel is horizontal so that there is no component of the weight of the fluid in the direction of flow. Also, let us ignore the frictional forces between the fluid and the channel. Then the only forces acting on the control volume are the hydrostatic forces F_1 and F_2.

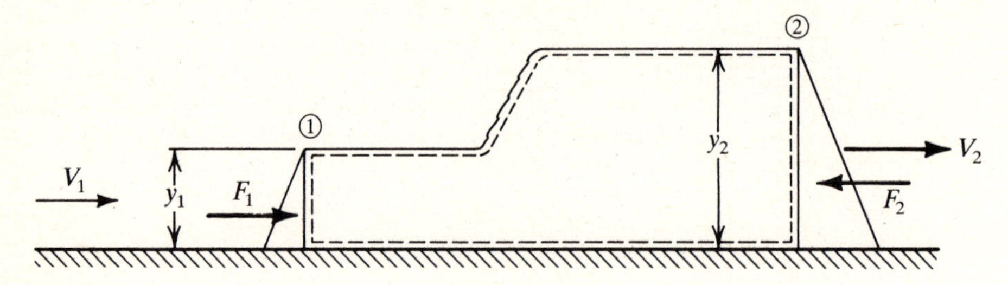

FIGURE 10.11 Hydraulic jump in a rectangular channel.

Here $F_1 = \gamma y_1^2/2$ and $F_2 = \gamma y_2^2/2$. Applying the momentum equation to the control volume, we obtain

$$\Sigma F = \dot{M}_O - \dot{M}_I$$

$$F_1 - F_2 = \rho_2 Q_2 V_2 - \rho_1 Q_1 V_1$$

Since ρ and Q are constants:

$$F_1 - F_2 = \rho Q(V_2 - V_1) \tag{10.23}$$

For a rectangular channel of unit width:

$$V_1 = \frac{Q}{y_1};\ V_2 = \frac{Q}{y_2}$$

Substituting for F_1, F_2, V_1 and V_2 in Eq. (10.23) and simplifying, we obtain

$$y_2^2 + y_1 y_2 - \frac{2Q^2}{g y_1} = 0$$

$$y_2 = \frac{-y_1}{2} \pm \frac{\sqrt{y_1^2 + \dfrac{8Q^2}{g y_1}}}{2}$$

$$y_2 = -\frac{y_1}{2} \pm \frac{\sqrt{y_1^2 + y_1^2 8\mathbf{F}_1^2}}{2}$$

where $\mathbf{F}_1$ is the Froude number at section 1.

Since a negative sign will give a negative depth, it can be discarded. Hence:

$$y_2 = \frac{1}{2} y_1 \left[\sqrt{1 + 8\mathbf{F}_1^2} - 1\right] \tag{10.24}$$

or

$$\frac{y_2}{y_1} = \frac{1}{2}\left[\sqrt{1 + 8\mathbf{F}_1^2} - 1\right] \tag{10.25}$$

From this equation, the depth of the channel on the downstream side of the jump can be calculated.

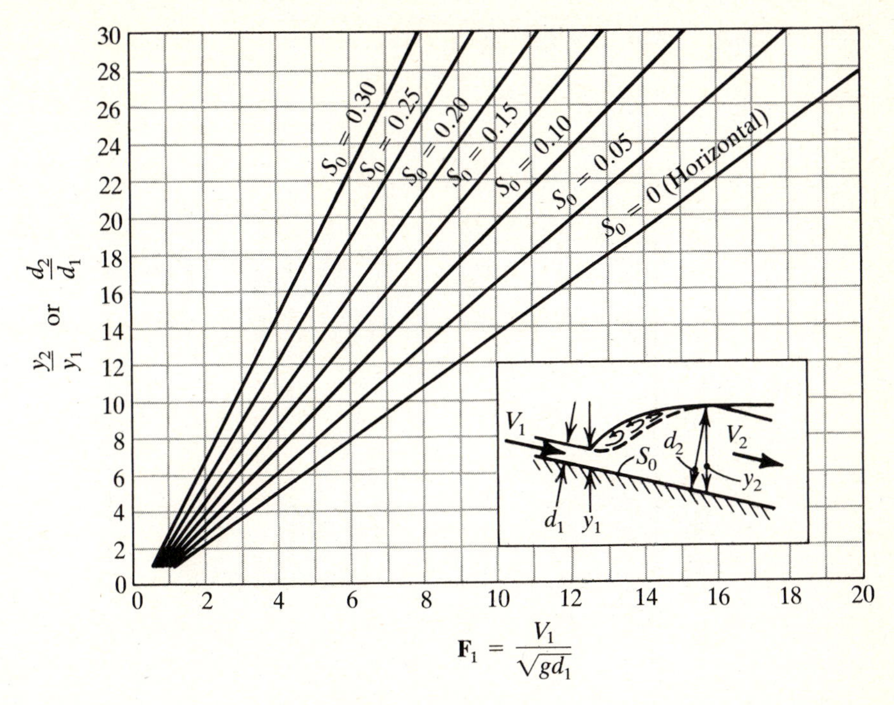

FIGURE 10.12 Experimental relations between $\mathbf{F}_1$ and y_2/y_1 or d_2/d_1 for jumps in horizontal and sloping channels. (Source: Open Channel Hydraulic by Chow, McGraw Hill, New York)

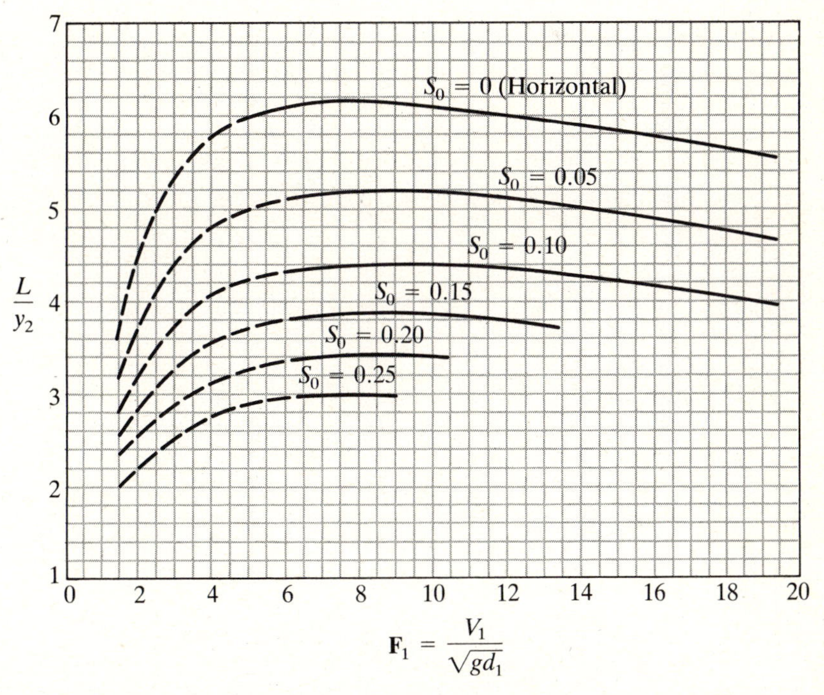

FIGURE 10.13 Length in terms of sequent depth of jumps in horizontal and sloping channels. (Source: Open Channel Hydraulics by Chow, McGraw Hill, New York)

Table 10.4 VARIOUS TYPES OF HYDRAULIC JUMPS

Froude number	Type of jump	Description	Diagram
1–1.7	Undular jump	The water surface shows only undulations.	F_1 = 1–1.7 Undular jump
1.7–2.5	Weak jump	Small rollers develop on the surface of the jump. The downstream water surface remains smooth, the velocity throughout is fairly uniform and the energy loss is low.	F_1 = 1.7–2.5 Weak jump
2.5–4.5	Oscillating jump	An oscillating jet enters the jump bottom to surface and back again with no periodicity. Each oscillation produces a large wave of irregular period traveling for miles.	Oscillating jet; Roller; F_1 = 2.5–4.5 Oscillating jump
4.5–9.0	Steady jump	The position and the action of the jump is least sensitive to variation in tailwater depth. The jump is well balanced with the best performance. The energy dissipation ranges from 45–70%.	F_1 = 4.5–9.0 Steady jump
9.0 and greater	Strong jump	Waves and rough surface is generated downstream. The jump action is rough but effective and the energy dissipation may reach 85%.	F_1 > 9.0 Strong jump

The energy per unit weight of the fluid before the jump is $(y_1 + V_1^2/2g)$ and after the jump is $(y_2 + V_2^2/2g)$. Hence, the loss of energy due to jump per unit weight of the fluid is $(y_1 + V_1^2/2g) - (y_2 + V_2^2/2g)$.

While deriving Eq. (10.25), the bed of the channel was assumed to be horizontal. This equation may be used for slightly sloping channels where the weight of the fluid in the jump has negligible effect on the jump. For a channel of large slope, the effect of the weight of the fluid may be appreciable and, therefore, should be included in the derivation. The results obtained by including the weight of the fluid for various slopes in a rectangular channel are given in Fig. 10.12.

The length of a jump can be defined as the distance between the front face of the jump and a point on the surface immediately downstream of the roller. A plot of $\mathbf{F}_1$ and L/y_2 for a horizontal or a sloping channel bed is given in Fig. 10.13. The curves were developed primarily for jumps occurring in rectangular channels, but may be applied approximately to jumps occurring in trapezoidal channels also.

Types of Jumps

The U.S. Bureau of Reclamation has classified the hydraulic jumps on a horizontal floor into various types depending on the Froude number of the incoming flow. The classification is described in Table 10.4.

Specific Force

The sum of the hydrostatic force F and the momentum ρQV at a section in a channel is known as *the specific force* F_s. It is expressed as:

$$F_s = F + \rho QV \tag{10.26}$$

For a given discharge, the specific force is a function of the depth of flow. A typical specific force diagram is shown in Fig. 10.14.

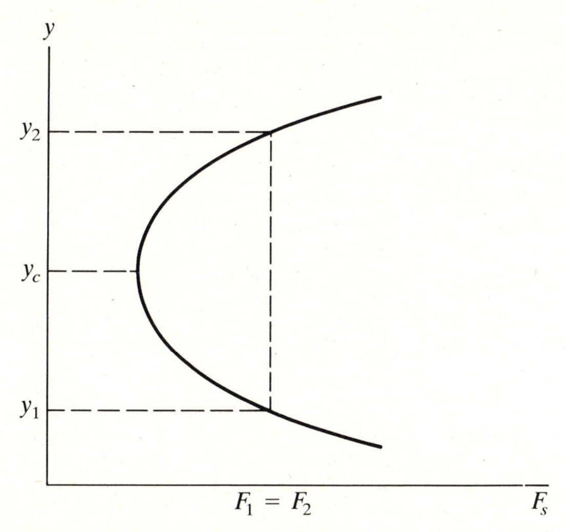

FIGURE 10.14 Specific force diagram.

For a rectangular channel of unit width

$$F_s = \frac{\gamma y^2}{2} + \frac{\gamma Q^2}{gy}$$

where y is the depth of flow and Q is the discharge. At minimum specific energy dF_s/dy must be zero.

Hence:

$$\frac{dF_s}{dy} = \gamma y - \frac{\gamma Q^2}{gy^2} = 0$$

$$y = \left(\frac{Q^2}{g}\right)^{1/3}$$

This equation is identical to Eq. (10.20). Therefore, y is the critical depth. It can be deduced that at a given channel cross-section, for a given rate of flow, the specific force is minimum at critical conditions.

It can be proved that the specific force at sections immediately before and after the hydraulic jump is the same. From Eq. (10.23),

$$F_1 + \rho_1 Q_1 V_1 = F_2 + \rho_2 Q_2 V_2$$

which shows that the specific force at section 1 is equal to the specific force at section 2 on the assumptions that the channel bed is horizontal and the frictional forces are negligible.

EXAMPLE 10.8

A sluice gate spans a channel of rectangular section 60 ft wide, has an opening 2.5 ft deep, and discharges 2500 ft^3/s of water. If a hydraulic jump is formed on the downstream side of the sluice, determine the height, the type and the length of the jump. Also find loss of energy.

Solution

$$b = 60 \text{ ft}$$

$$y_1 = 2.5 \text{ ft}$$

$$Q = 2500 \text{ ft}^3/\text{s}$$

$$V_1 = \frac{Q}{by_1} = \frac{2500}{(60)(2.5)} = 16.67 \text{ ft/s}$$

$$\mathbf{F}_1 = \frac{V_1}{\sqrt{gy_1}} = \frac{16.67}{\sqrt{32.2 \times 2.5}} = 1.858$$

From Eq. (10.25):

$$\frac{y_2}{y_1} = \frac{1}{2}\left[\sqrt{1 + 8\mathbf{F}_1^2} - 1\right]$$

$$= \frac{1}{2}\left[\sqrt{1 + 8(1.858)^2} - 1\right] = 2.175$$

$$y_2 = 2.5 \times 2.175 = 5.43 \text{ ft}$$

$$V_2 = \frac{Q}{y_2 b_2} = \frac{2500}{(5.43)(60)} = 7.67 \text{ ft/s}$$

Since $\mathbf{F}_1 = 1.858$, it is a weak jump.

The length of the jump may be determined using Fig. 10.13, which indicates that for $\mathbf{F}_1 = 1.858$ and $S_0 = 0$

$$\frac{L}{y_2} = 4.25$$

Hence:

$$L = 4.25 \times 5.43 = 23 \text{ ft}$$

$$\text{Energy/lb before jump} = y_1 + \frac{V_1^2}{2g}$$

$$= 2.5 + \frac{(16.67)^2}{64.4} = 6.815 \text{ ft} \cdot \text{lb/lb}$$

$$\text{Energy/lb after jump} = y_2 + \frac{V_2^2}{2g}$$

$$= 5.43 + \frac{(7.67)^2}{64.4} = 6.343 \text{ ft} \cdot \text{lbs/lb}$$

$$\text{Loss of energy} = 6.815 - 6.343 = 0.472 \text{ ft} \cdot \text{lb/lb}$$

$$\text{Total loss of energy} = (62.4)(2500)(0.472) \text{ ft} \cdot \text{lbs/s}$$

$$= 73{,}632 \text{ ft} \cdot \text{lbs/s}$$

$$= 133.8 \text{ HP}$$

10.6 GRADUALLY VARIED FLOW

Gradually varied flow is a steady nonuniform flow. It differs from uniform flow and rapidly varied flow in that the changes in depth of flow occur very gradually along the channel. The streamlines are practically parallel to one another and the pressure distribution over a section is almost hydrostatic. The flow of water behind a dam when the spillway is flowing at a steady rate is an example of a gradually varied flow.

In a uniform flow, the energy grade line, the water surface line and the channel bottom line are all parallel to each other. However, in a gradually varied flow the three lines are not parallel to each other.

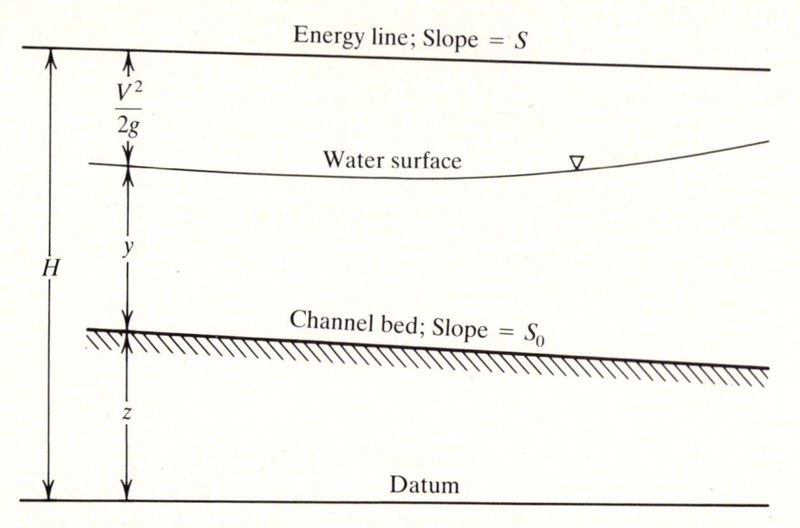

FIGURE 10.15 Gradually varied flow.

The development of the theory of the gradually varied flow dates back to the eighteenth century. All theories are based on the assumption that the uniform flow formula (i.e., Manning equation) can be used to evaluate the slope of the energy line of the gradually varied flow at a given section, and that the Manning's coefficients (i.e., n) developed for uniform flow may be used for gradually varied flow. According to Chow (1959)[1], these assumptions do not contribute to large errors in gradually varied flow computations.

Consider a gradually varied flow over an elementary length dx of a channel as shown in Fig. 10.15. The total energy head at any section is given by

$$H = \frac{V^2}{2g} + y + z = \frac{Q^2}{2gA^2} + y + z \tag{10.27}$$

where, for the section

$V =$ average velocity

$y =$ depth of flow

$z =$ elevation of the bed from a given datum

If x is the distance along the channel bed, then differentiation of Eq. (10.27) with respect to x yields

$$\frac{dH}{dx} = \frac{-Q^2}{gA^3}\frac{dA}{dx} + \frac{dy}{dx} + \frac{dz}{dx}$$

[1] Chow, V. T., 1959. *Open Channel Hydraulics*, New York: McGraw-Hill.

From Eq. (10.17), $dA = T\,dy$
Hence:

$$\frac{dH}{dx} = \frac{-Q^2 T}{gA^3}\frac{dy}{dx} + \frac{dy}{dx} + \frac{dz}{dx}$$

$$\frac{dy}{dx} = \frac{\dfrac{dH}{dx} - \dfrac{dz}{dx}}{1 - \dfrac{Q^2 T}{gA^3}} \tag{10.28}$$

In Eq. (10.28), the term dH/dx is the slope of the energy grade line. It is always negative, or $S = -dH/dx$. The term dz/dx is the slope of the channel bed. In general, $S_0 = -dz/dx$, assuming that the channel bed slopes in the direction of flow. As stated earlier, it is assumed that the slope of the energy grade line between two adjacent sections can be approximated by Manning's equation, or

$$S = -\frac{dH}{dx} = \left(\frac{nQ}{C_m R^{2/3} A}\right)^2 \tag{10.29}$$

The slope of the channel bed may also be obtained from Manning's equation if uniform flow is assumed to take place in the channel, or

$$S_0 = -\frac{dz}{dx} = \left(\frac{nQ}{C_m R_n^{2/3} A_n}\right)^2 \tag{10.30}$$

where R_n and A_n are for the normal flow.
From Eq. (10.18), for critical flow:

$$Q^2 = \frac{gA_c^3}{T_c} \tag{10.31}$$

where A_c and T_c are for the critical flow.

Substituting Eqs. (10.29), (10.30) and (10.31) in Eq. (10.28) and simplifying, we obtain

$$\frac{dy}{dx} = S_0 \frac{1 - \left(\dfrac{R_n}{R}\right)^{4/3}\left(\dfrac{A_n}{A}\right)^2}{1 - \left(\dfrac{A_c}{A}\right)^3\left(\dfrac{T}{T_c}\right)} \tag{10.32}$$

For simplicity, consider a wide rectangular channel, where $A = by$; $R \simeq y$. Substituting in Eq. (10.32), we obtain

$$\frac{dy}{dx} = S_0 \frac{1 - \left(\dfrac{y_n}{y}\right)^{10/3}}{1 - \left(\dfrac{y_c}{y}\right)^3} \tag{10.33}$$

Equation (10.33) is the general differential equation for gradually varied flow in a wide rectangular channel.

Classification of Gradually Varied Flow

The shape and the slope of the water surface line in a gradually varied flow depends on the slope of the channel bed and the relative position of the actual depth of flow, y, to the normal depth, y_n, and the critical depth y_c.

The channel slopes are classified into five categories, depending on the relative position of y_n to y_c. The criteria are as follows:

Mild slope $\quad : y_n > y_c \text{ or } \dfrac{y_n}{y_c} > 1$

Steep slope $\quad : y_n < y_c \text{ or } \dfrac{y_n}{y_c} < 1$

Critical slope $\quad : y_n = y_c \text{ or } \dfrac{y_n}{y_c} = 1$

Horizontal slope: $S_0 = 0 \text{ or } y_n = \infty$

Adverse slope $\quad : S_0 < 0 \text{ or } y_n < 0$

Various shapes of gradually varied flow on various types of slopes are given in Table 10.5.

Computations of Gradually Varied Flow

Computations of the gradually varied flow may be carried out either by the *standard step method* or by *numerical integration.*

Standard Step Method

Figure 10.16 represents a longitudinal section of a channel in which the velocity of water and depth of flow are not constant. Consider two vertical sections 1 and 2 at a distance Δx apart.

Let

V_1 = velocity of flow at section 1

y_1 = depth of flow at section 1

V_2 = velocity of flow at section 2

y_2 = depth of flow at section 2

S = slope of the total energy line

S_0 = slope of the channel bed

Applying the Bernoulli equation to the two vertical sections, and assuming the base of the channel at section 2 to be the datum line:

$$\frac{V_1^2}{2g} + y_1 + S_0\Delta x = \frac{V_2^2}{2g} + y_2 + S\Delta x$$

Table 10.5 GRADUALLY VARIED FLOW PROFILES

Slope	Regime	Sign of $1-(y_n/y)^{10/3}$	Sign of $1-(y_c/y)^3$	Sign of dy/dx	Depth	Type of surface profile	Figures
	$y > y_n$ $y > y_c$	+	+	+	Increases	M-1	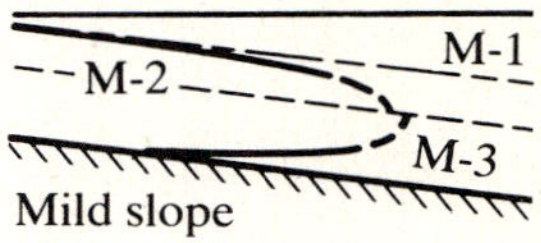
Mild $y_n > y_c$	$y < y_n$ $y > y_c$	−	+	−	Decreases	M-2	Mild slope
	$y < y_n$ $y < y_c$	−	−	+	Increases	M-3	
	$y > y_c$ $y > y_n$	+	+	+	Increases	S-1	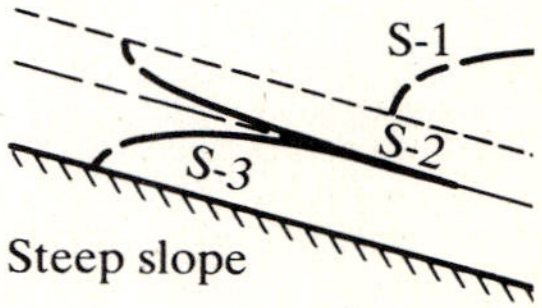 Steep slope
Steep $y_n < y_c$	$y < y_c$ $y > y_n$	+	−	−	Decreases	S-2	
	$y < y_c$ $y < y_n$	−	−	+	Increases	S-3	Critical slope
Critical $y_n = y_c$	$y > y_c$	+	+	+	Increases	C-1	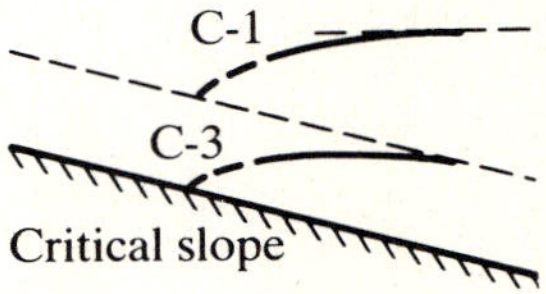
	$y < y_c$	−	−	+	Increases	C-3	Horizontal slope
Horizontal $y_n = \infty$	$y > y_c$	−	+	−	Decreases	H-2	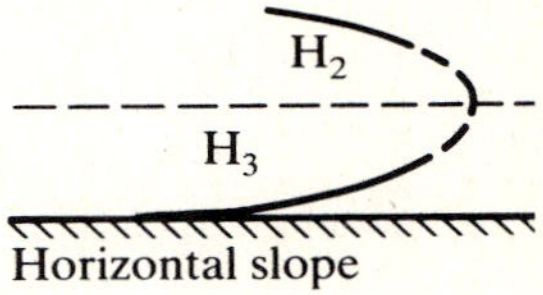
	$y < y_c$	−	−	+	Increases	H-3	
Adverse $y_n < 0$	$y > y_c$	−	+	−	Decreases	A-2	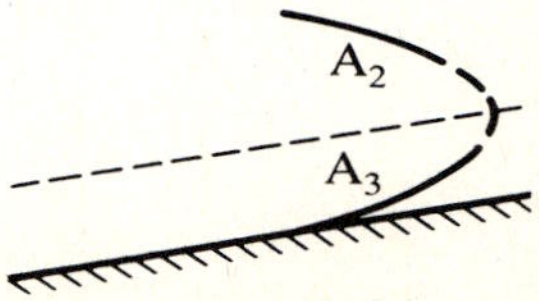
	$y < y_c$	−	−	+	Increases	A-3	

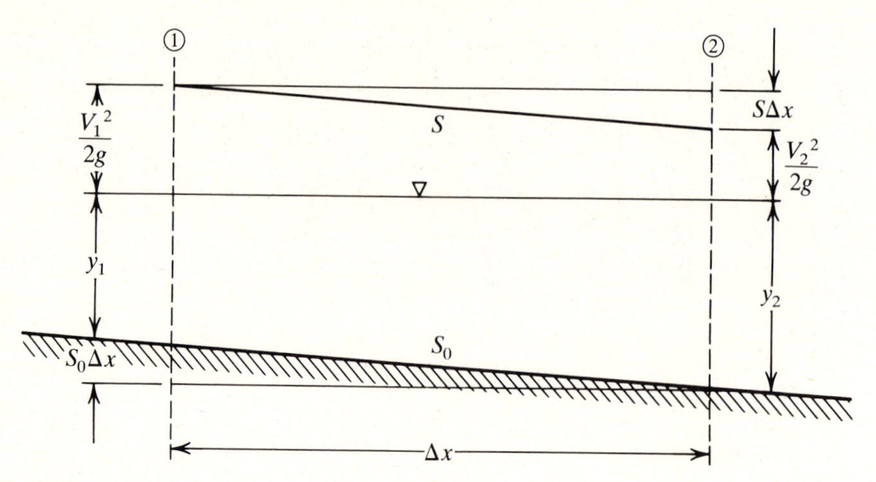

FIGURE 10.16 A channel section with nonuniform flow.

Let

$$y_1 + \frac{V_1^2}{2g} = E_1$$

$$y_2 + \frac{V_2^2}{2g} = E_2$$

$$E_1 + S_0\Delta x = E_2 + S\Delta x$$

$$S\Delta x - S_0\Delta x = E_1 - E_2$$

$$\Delta x = \frac{E_1 - E_2}{S - S_0} \qquad (10.34)$$

The problems involving gradually varied flow may be of two types. In a type I problem, conditions are known at two sections and it is required to calculate the distance Δx between the two sections. The solution to such a problem may be obtained directly, as follows:

1. Determine average conditions for the reach—i.e.:

$$\bar{A} = \frac{A_1 + A_2}{2}$$

$$\bar{R} = \frac{R_1 + R_2}{2}$$

2. Using average $\bar{A}$ and $\bar{R}$, calculate S using Eq. 10.29.
3. Compute Δx using Eq. (10.34)

In a type II problem, conditions are known at one section and it is required to calculate depth at another section Δx away. A trial solution is required. The following procedure may be used:

1. Assume a depth y_2; then compute A_2 and V_2.
2. For the assumed y_2, find average values for $\bar{A}$ and $\bar{R}$ for the reach. Using these average values, calculate S using Eq. (10.29).
3. Compute Δx from Eq. (10.34).
4. If the computed Δx is equal to the actual Δx, stop. If not, assume a new y_2 and repeat until convergence is attained.

EXAMPLE 10.9

A trapezoidal channel with $b = 10$ m, $z = 2$, $n = 0.012$, and $S_0 = 0.0002$ carries a discharge of 180 m^3/s. A dam built across the channel raises the height immediately upstream to 10 m. Determine the water surface profile up to a distance of 5000 m upstream of the dam.

Solution First of all, the type of the surface profile must be determined to find out whether the water surface would slope downward or upward from the dam. This is done by finding the normal and the critical depths.

$$A = y(b + zy) = y(10 + 2y)$$

$$P = b + 2y\sqrt{1 + z^2} = 10 + 4.472y$$

From Manning's equation

$$Q = A\frac{1}{n}\left(\frac{A}{P}\right)^{2/3}(S_0)^{1/2}$$

$$180 = y(10 + 2y)\frac{1}{0.012}\left[\frac{y(10 + 2y)}{10 + 4.472y}\right]^{2/3}(0.0002)^{1/2}$$

By trial and error, $y_n = 4.25$ m
To find y_c

$$\frac{Q^2T}{gA^3} = 1$$

$$\frac{(180)^2(10 + 4y_c)}{9.81[y(10 + 2y_c)]^3} = 1$$

By trial and error, $y_c = 2.67$ m
Since $y_n > y_c$, the slope is mild and $y > y_n > y_c$, the profile is M − 1 (see Table 10.5).

Table 10.6 SOLUTION FOR EXAMPLE 10.9

y (m)	A (m²)	P (m)	E (m)	$\bar{A}$ (m²)	$\bar{P}$ (m)	$\bar{R}$	$S = \left[\frac{nQ}{\bar{A}\bar{R}^{2/3}}\right]^2$	$\Delta x = \frac{E_1 - E_2}{S - S_0}$	Total distance
10.0	300.0	54.72	10.02						
9.8	290.1	53.83	9.82	295.05	54.28	5.436	5.6×10^{-6}	1028.8	1028.8
9.6	280.3	52.93	9.62	285.20	53.38	5.343	6.1×10^{-6}	1031.5	2060.3
9.4	270.7	52.04	9.42	275.50	52.49	5.249	6.73×10^{-6}	1034.8	3095.1
9.2	261.3	51.14	9.22	266.00	51.59	5.156	7.39×10^{-6}	1038.4	4133.5
9.0	252.0	50.25	9.02	256.65	50.70	5.062	8.14×10^{-6}	1042.4	5175.9

The water surface would slope down in the upstream direction. The control would be downstream and the calculations would start from $y = 10$ m and move upstream at intervals of, say, 0.2 m until a distance of 5,000 m is reached. The calculations are shown in Table 10.6.

Numerical Integration Method

Water surface profile in a prismatic channel of constant bed slope may be obtained using integration. From Fig. (10.15), for a reach dx, the rate of change of energy is given by

$$\frac{dH}{dx} = \frac{d}{dx}\left(y + \frac{V^2}{2g} - S_0 x\right)$$

where x is measured positive in the downstream direction. Differentiating

$$\frac{dH}{dx} = \frac{dy}{dx} + \frac{V}{g}\frac{dV}{dx} - S_0 \tag{10.35}$$

From the continuity equation

$$VA = Q$$

$$V\frac{dA}{dx} + A\frac{dV}{dx} = 0$$

$$\frac{dV}{dx} = -\frac{V}{A}\frac{dA}{dx}$$

From Eq. (10.17)

$$dA = Tdy$$

Hence:

$$\frac{dV}{dx} = -\frac{VT}{A}\frac{dy}{dx}$$

Substituting in Eq. (10.35)

$$\frac{dH}{dx} = \frac{dy}{dx} - \frac{V^2T}{Ag}\frac{dy}{dx} - S_0$$

$$\frac{dH}{dx} = \frac{dy}{dx}\left[1 - \frac{Q^2T}{gA^3}\right] - S_0 \tag{10.36}$$

Also from Eq. 10.29

$$\frac{dH}{dx} = -\left[\frac{nQ}{C_m AR^{2/3}}\right]^2$$

Substituting in Eq. (10.36), and solving for dx, yields

$$dx = \frac{1 - Q^2T/gA^3}{S_0 - (nQ/C_m AR^{2/3})^2}\,dy \tag{10.37}$$

$$dx = F(y)\,dy \tag{10.38}$$

For two sections, distance x apart and having depths of y_1 and y_2

$$x = \int_{y_1}^{y_2} F(y)\,dy \tag{10.39}$$

where

$$F(y) = \frac{1 - Q^2T/gA^3}{S_0 - (nQ/C_m AR^{2/3})^2}$$

Equation (10.39) may be solved numerically. If the depth is divided into small increments of Δy, then, distance Δx between two sections may be written

$$\Delta x = \left[\frac{F(y)_1 + F(y)_2}{2}\right]\Delta y \tag{10.40}$$

EXAMPLE 10.10

Solve Example 10.9 by the method of integration.

Solution From Example 10.9:

$$y_n = 4.25 \text{ m}$$

$$y_c = 2.67 \text{ m}$$

Since $y_n > y_c$, the slope is mild and $y > y_n > y$, the profile is M-1.

The water surface would slope down in the upstream direction. The control would be downstream and the calculations would start from $y = 10$ m and move upstream. Let the interval be 0.2 m. The calculations are given in Table 10.7.

TABLE 10.7 SOLUTION FOR EXAMPLE 10.10

y (m)	A (m^2)	P (m)	R (m)	T (m)	$F(y)$	Δx (m)	x (m)
10.0	300.0	54.72	5.482	50.0	5106	0	0
9.8	290.1	53.83	5.390	49.2	5116	1022	1022
9.6	280.3	52.93	5.296	48.4	5128	1024	2046
9.4	270.7	52.04	5.202	47.6	5142	1027	3073
9.2	261.3	51.14	5.109	46.8	5157	1030	4103
9.0	252.0	50.25	5.015	46.0	5228	1038	5141

Using Eq. (10.40), calculate Δx. For example, for the second entry under the column for Δx:

$$\left(\frac{5106 + 5116}{2}\right) \times 0.2 = 1022 \text{ m}$$

The depth of flow at 5,000 m can be interpolated from the last two results of Table 10.7

$$y \text{ at } 5{,}000 \text{ m} = 9.2 - \frac{(4103 - 5000)}{(4103 - 5141)} \times 0.2$$

$$= 9.03 \text{ m}$$

A computer program in BASIC is listed in Appendix B which can be used to calculate water surface profiles in a prismatic channel. The program is based on the method of integration. The solution to Example 10.10 with the computer program is as follows:

```
RUN
This program computes for water surface
profile in rectangular,trapezoidal or
triangular channel
title? Example 10.10
system of units (si/fps) ? si
length of reach (m)? 5000
channel width (m)? 10
left side slope(---:1) ? 2
right side slope(---:1) ? 2
mannings n ? 0.012
channel slope? 0.0002
discharge (cu.m/sec)? 180
depth of flow at control section (m)? 10
number of intervals ? 50
                 ************

Example 10.10
normal depth = 4.248657 (m)
critical depth = 2.665558 (m)
control is downstream, depth = 10 (m)

distance             depth            sp. energy          sp. force
  (m)                 (m)                (m)                 (N)
 0                   10                10.01835            1.1553E+07
 1173.769            9.770407          9.79023             1.089441E+07
 2350.529            9.540812          9.562264            1.026146E+07
 3530.714            9.311219          9.334469            9653704
 4714.834            9.081626          9.10687             9070682
 5000                9.026544          9.052299            8934441
**************
do you wish to continue (y/n) ? n
Break in 1380
Ok
```

EXAMPLE 10.11

For the dam-channel system of Fig. 10.17, (a) draw the backwater curve up to a distance of 10,000 ft upstream of the dam, (b) draw the water surface profile of the sheet of water flowing over the spillway, and (c) draw the water surface profile up to a distance of 5,000 ft downstream of the toe of the spillway. The following data are available:

Channel is rectangular and 50 ft wide
Manning's n for the channel = 0.025
Slope of the channel bed = 0.002
Depth of water immediately upstream of dam = 30.84 ft
Slope of downstream side of dam = 0.833
Manning's n for the dam = 0.012
Flow rate = 4,000 ft^3/s
Length of downstream slope of the dam = 39 ft

Solution The problem is solved using the program listed in Appendix B. The results are shown below in the form of a computer printout. The backwater curve upstream of the dam is first obtained by letting the control depth be equal to 30.84 ft. The first table in the printout lists the calculations for part (a). Then the profile of the sheet flowing over the dam is determined. The water surface elevations are computed starting from the critical depth at the top of the dam and moving toward the toe of the dam. It is found that the depth of flow at the toe of the dam is equal to 1.62 ft. The second table in the printout shows the calculations for part (b). Next, the profile downstream of the dam is computed starting from a depth of 1.62 ft. Since this is an M-3 curve (see Table 10.5), the program calculates the profile until critical depth is reached. Since the normal depth in the channel is 8.69 ft, a hydraulic jump is formed. The jump would form at a point where the specific forces after and before the jump are equal. Specific force after the jump can be computed using normal conditions. Hence:

$$F_s = \frac{1}{2}\gamma b y_n^2 + \rho Q V_n$$

$$= \frac{1}{2} \times 62.4 \times 50 \times (8.69)^2 + 1.94 \times 4000 \times \left(\frac{4000}{50 \times 8.69}\right)$$

$$= 189{,}300 \text{ lbs}$$

This value of F_s occurs somewhere between the depths of 3.589 ft and 3.87 ft (third table in the printout). By linear interpolation, the depth at which the specific force is 189,300 lbs is found to be 3.7 ft. Similarly, the distance downstream of the toe of the dam at which a depth of 3.7 ft occurs, is found to be 250 ft. After the jump, the depth of flow becomes normal and equal to 8.69 ft, and continues for the remainder of the distance. The water surface profile for the given system is shown in Fig. 10.18.

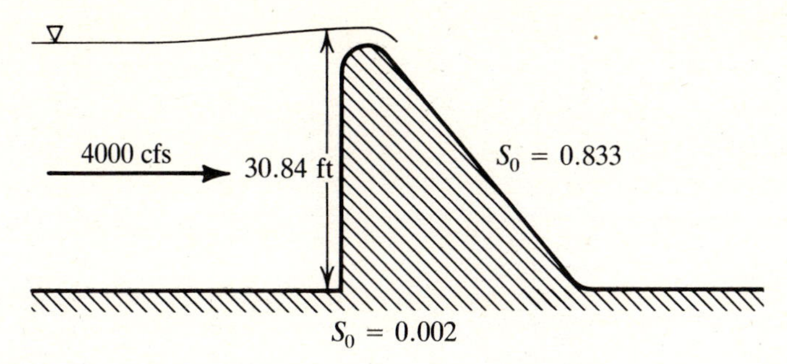

FIGURE 10.17 Example 10.11.

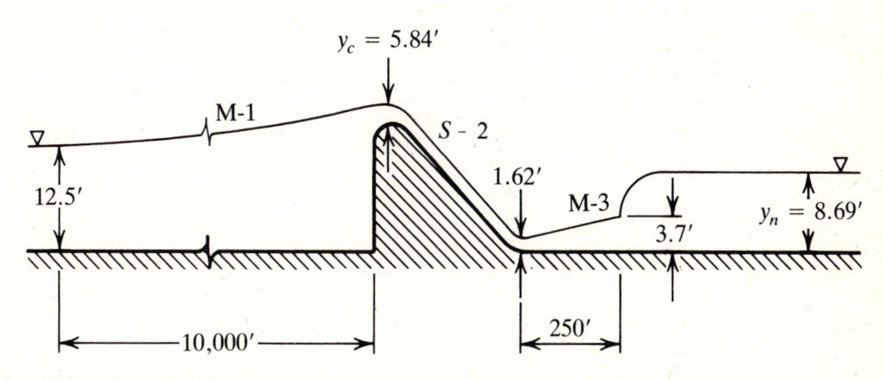

FIGURE 10.18 Example 10.11.

```
        RUN
This program computes for water surface
profile in rectangular,trapezoidal or
triangular channel
title? Example 10.11
system of units (si/fps) ? fps
length of reach (ft)? 10000
channel width (ft)? 50
left side slope(---:1) ? 0
right side slope(---:1) ? 0
mannings n ? 0.025
channel slope? 0.002
discharge cu.ft/sec)? 4000
depth of flow at control section (ft)? 30.84
number of intervals ? 50
                  ************

Example 10.11
normal depth = 8.689575 (ft)
critical depth = 5.836029 (ft)
control is downstream, depth = 30.84 (ft)
```

distance (ft)	depth (ft)	sp. energy (ft)	sp. force (lbs)
0	30.84	30.94449	1503833
452.516	29.95576	30.0665	1420563
905.8505	29.07151	29.1891	1339769
1360.103	28.18727	28.31235	1261454
1815.389	27.30302	27.43633	1185622
2271.844	26.41878	26.56116	1112278
2729.63	25.53453	25.68695	1041425
3188.938	24.65029	24.81384	973070
3649.998	23.76604	23.94199	907219.4
4113.089	22.8818	23.07161	843880.8
4578.553	21.99755	22.20293	783062.5
5046.81	21.11331	21.33624	724775
5518.39	20.22906	20.47191	669030.4
5993.965	19.34481	19.61038	615842.5
6474.4	18.46057	18.75218	565228.5
6960.839	17.57633	17.89801	517208.2
7454.816	16.69208	17.04876	471806.7
7958.457	15.80784	16.20553	429053.7
8474.796	14.92359	15.36981	388986.5
9008.356	14.03935	14.54354	351651.5
9566.251	13.1551	13.72936	317107.9
10000	12.50971	13.14475	293700.4

```
**************
do you wish to continue (y/n) ? y
title? flow over the spillway
system of units (si/fps) ? fps
length of reach (ft)? 39
channel width (ft)? 50
left side slope(---:1) ? 0
right side slope(---:1) ? 0
mannings n ? 0.012
channel slope? 0.8333
discharge cu.ft/sec)? 4000
depth of flow at control section (ft)? 0
number of intervals ? 30
            ************

flow over the spillway
normal depth = .8233643 (ft)
critical depth = 5.836029 (ft)
control is upstream, depth = 5.836029 (ft)
```

distance (ft)	depth (ft)	sp. energy (ft)	sp. force (lbs)
0	5.836029	8.753854	159390.3
8.919704E-03	5.669275	8.761272	159522.8
.0371714	5.50252	8.784763	159931.5
8.721782E-02	5.335765	8.82637	160634.1
.1619274	5.16901	8.888466	161650.8
.2646575	5.002256	8.973826	163004.1
.3993579	4.835501	9.085718	164720.1
.5707018	4.668747	9.227998	166828.2
.7842509	4.501992	9.405249	169362.4
1.046667	4.335237	9.622956	172361.8

```
 1.365985          4.168482          9.88772          175871.9
 1.751969          4.001728          10.20755         179945.7
 2.21658           3.834973          10.59222         184645.3
 2.774604          3.668218          11.05378         190044.3
 3.444496          3.501464          11.60724         196230.3
 4.249535          3.334709          12.27143         203308.1
 5.219447          3.167955          13.07026         211405.2
 6.392689          3.0012            14.03447         220676.7
 7.819767          2.834445          15.20411         231314.7
 9.568146          2.667691          16.63212         243559.2
 11.72973          2.500936          18.38966         257714.2
 14.4326           2.334181          20.57419         274170.5
 17.86012          2.167427          23.32205         293439.3
 22.28357          2.000672          26.82871         316202.2
 28.12093          1.833917          31.38237         343388.6
 36.0509           1.667162          37.42228         376299.8
 39                1.623277          39.33782         386130.6
**************
do you wish to continue (y/n) ? y
title? flow in the downstream channel
system of units (si/fps) ? fps
length of reach (ft)? 5000
channel width (ft)? 50
left side slope(---:1) ? 0
right side slope(---:1) ? 0
mannings n ? 0.025
channel slope? 0.002
discharge cu.ft/sec)? 4000
depth of flow at control section (ft)? 1.623277
number of intervals ? 30
              ************

flow in the downstream channel
normal depth = 8.689575 (ft)
critical depth = 5.836029 (ft)
control is upstream, depth = 1.623277 (ft)

distance           depth            sp. energy         sp. force
  (ft)             (ft)               (ft)              (lbs)
 0                 1.623277          39.33782         386130.6
 33.26071          1.904127          29.3137          331329.9
 67.31253          2.184977          23.00112         291260.3
 101.8212          2.465828          18.81022         260972.6
 136.4605          2.746678          15.9195          237541.5
 170.8992          3.027528          13.86974         219127.5
 204.7906          3.308378          12.38792         204515.3
 237.7626          3.589228          11.30345         192870.5
 269.4067          3.870078          10.50529         183600.5
 299.2651          4.150929          9.918641         176273.2
 326.8158          4.431779          9.491632         170566.1
 351.4517          4.712629          9.187366         166233.6
 372.4538          4.993479          8.979023         163085.1
 388.9537          5.274329          8.846726         160971
 399.8803          5.555179          8.775491         159771.6
 403.8816          5.836029          8.753854         159390.3
**************
```

```
do you wish to continue (y/n) ? n
Break in 1380
Ok
```

10.7 CHANNEL TRANSITIONS

A change in water surface occurs when the channel bottom is raised or lowered, when the width of the channel is increased or decreased, or when a combination of both exists. The flow in such channel transitions is rapidly varied flow.

Consider a channel shown in Fig. 10.19. The bed of the channel is raised by an amount Δz while the width of the channel remains constant. It is assumed that the transition is gradual, and that the head losses are negligible. Therefore,

$$y_1 + \frac{V_1^2}{2g} = y_2 + \frac{V_2^2}{2g} + \Delta z$$

$$E_1 = E_2 + \Delta z \tag{10.41}$$

In Eq. (10.41), for a constant rate of flow, E_1 is constant; therefore, E_2 must decrease when Δz increases. Two cases are now considered.

Case 1 (Fig. 10.20). The flow at section 1 is supercritical, or $y_1 < y_c$. For this case, a decrease in the specific energy E_2, at section 2 will result in an increase in the depth of flow y_2, as is evident from the specific energy diagram. As Δz is increased, E_2 decreases and y_2 increases. But E_2 cannot be decreased below E_{min} at which the depth approaches the critical depth y_c. Any further rise in Δz beyond this point would not change the depth over the hump, but would raise y_1. If $\Delta z > (E_1 - E_{min})$, a hydraulic jump is formed upstream of the hump and the flow becomes subcritical. Therefore, in the case of supercritical flow, creation of a hump in the channel bed would result in an increase in the

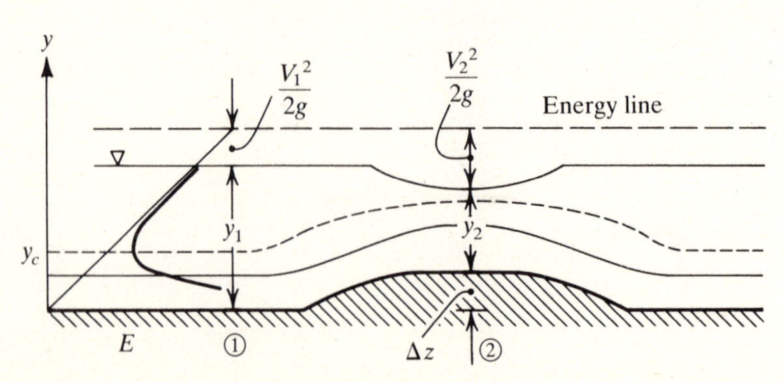

FIGURE 10.19 Channel with a streamlined hump.

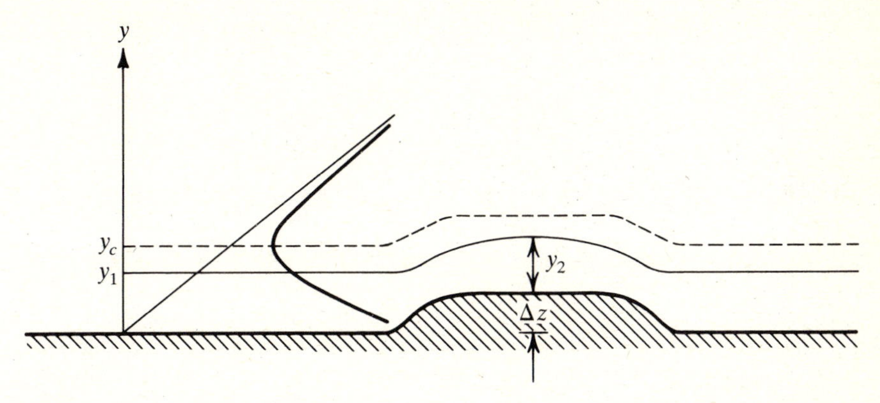

FIGURE 10.20 Water surface profile over a hump, when the flow before the hump is supercritical.

depth of flow over the hump, or $y_2 > y_1$. The limiting condition would be when $y_2 = y_c$.

Case 2 (Fig. 10.21). The flow at section 1 is subcritical, or $y_1 > y_c$. For this case, a decrease in E_2 will cause a decrease in y_2. As Δz is increased, both E_2 and y_2 decrease. The limiting condition would be when $y_2 = y_c$. Any further increase in Δz would not change y_2, but would raise y_1. Therefore, in the case of subcritical flow, creation of a hump in the channel bed would cause the water surface over the hump to fall. The limiting condition would be when $y_2 = y_c$.

Similar to raising the bed of the channel, the water surface profile in a channel is also affected if its sides are constricted. Consider a channel as shown in Fig. 10.22a. The top view of the channel with its constriction is shown in Fig. 10.22b. The sides of the channel have been constricted while no

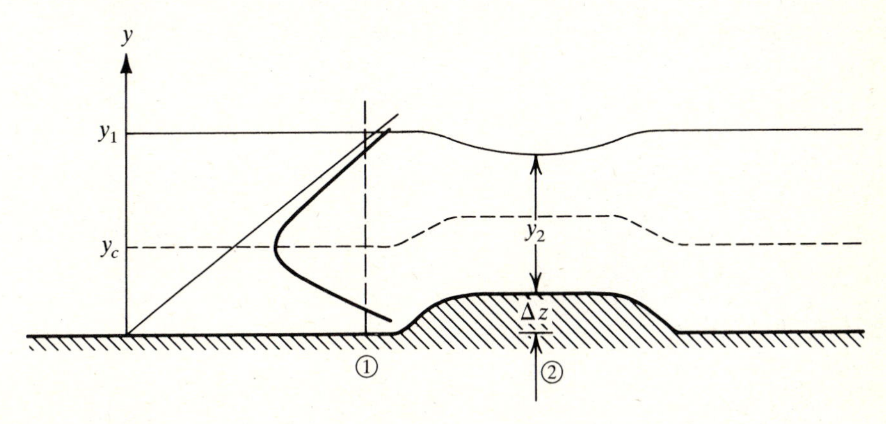

FIGURE 10.21 Water surface profile over a hump, when the flow before the hump is subcritical.

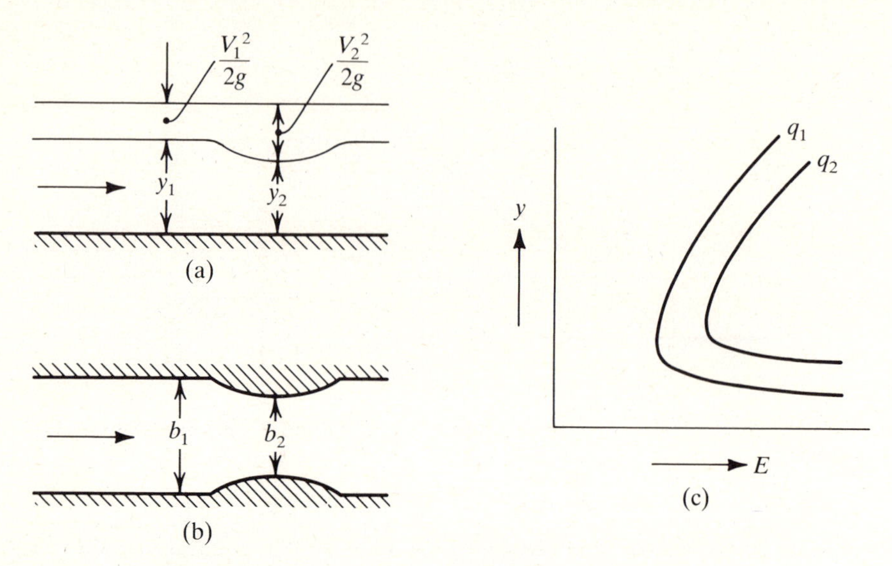

FIGURE 10.22 Channel with streamlined side constrictions.

change has been made in the channel bed. It is assumed that the transition is gradual and, hence, the head losses are negligible. Then

$$y_1 + \frac{V_1^2}{2g} = y_2 + \frac{V_2^2}{2g}$$

$$E_1 = E_2$$

This means that for a constant rate of flow, the specific energy before, and at, the transition remains constant. However, since the width of the channel at the constriction is smaller than that before it, the rate of flow per unit width of the channel would be larger at the constriction than before it. Two specific

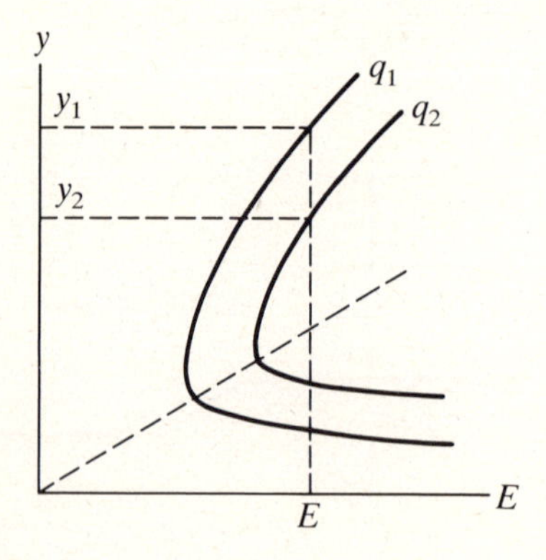

FIGURE 10.23 Specific energy diagrams when the flow before the constriction is subcritical.

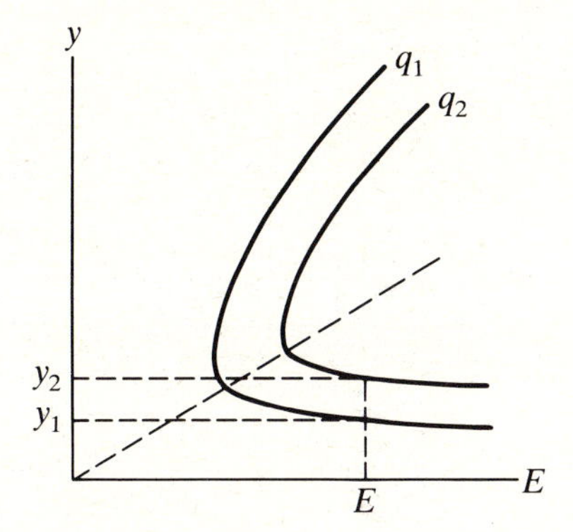

FIGURE 10.24 Specific energy diagrams when the flow before the constriction is supercritical.

energy curves have been drawn in Fig. 10.22c, one before the constriction where $q_1 = Q/b_1$ and the other at the constriction where $q_2 = Q/b_2$. Two cases are now considered.

Case No. 1. Let us consider the case when the depth of flow before the constriction is subcritical. Since E is the same before and at the constriction, the depth of flow decreases from y_1 to y_2 when we move from q_1 to q_2 as shown in Fig. 10.23. This means that, in this case, the water surface would fall at the constriction. The limiting condition would occur when the depth of flow at the constriction becomes critical. Beyond this, point y_1 would start to increase.

Case No. 2. Let us now consider the case when the depth of flow before the constriction is supercritical. Since E is the same before and at the constriction, the depth of flow increases from y_1 to y_2 when we move from q_1 to q_2 as shown in Fig. 10.24. This means that, in this case, the water surface would rise at the constriction. The limiting condition would be when the depth of flow at the constriction becomes critical.

EXAMPLE 10.12

Water flows at a depth of 6.0 ft in a rectangular channel 12 ft wide (Fig. 10.25). A streamlined hump 1.0 ft high in the channel bed produces a drop of 9 in in the water surface. Neglecting losses, determine the flow rate.

Solution

$$b = 12.0 \text{ ft}$$

$$y_1 = 6.0 \text{ ft}$$

$$y_2 = 6.0 - 1.0 - 0.75 = 4.25 \text{ ft}$$

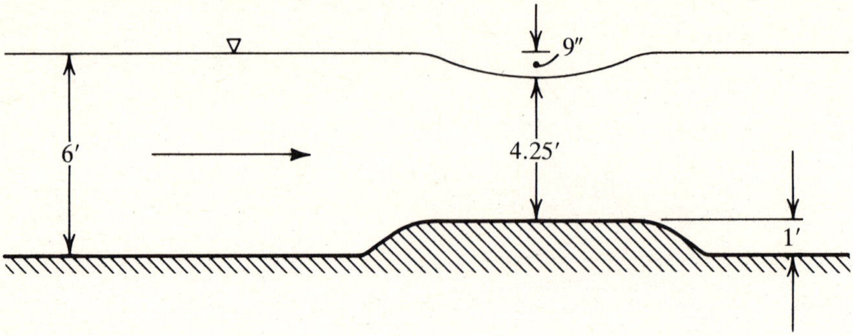

FIGURE 10.25 Example 10.12.

Also, neglecting the losses

$$E_1 = E_2 + \Delta z$$

$$y_1 + \frac{V_1^2}{2g} = y_2 + \frac{V_2^2}{2g} + \Delta z$$

$$6 + \frac{V_1^2}{2g} = 4.25 + \frac{V_2^2}{2g} + 1$$

$$V_2^2 - V_1^2 = 2g(0.75) \qquad (1)$$

From continuity

$$A_1 V_1 = A_2 V_2$$

$$(6 \times 12)V_1 = (4.25 \times 12)V_2$$

$$V_2 = 1.41 V_1$$

Substituting in (1)

$$(1.41 V_1)^2 - V_1^2 = 2g(0.75)$$

$$0.99 V_1^2 = 48.3$$

$$V_1 = 6.98 \text{ ft/s}$$

$$Q = A_1 V_1 = 6 \times 12 \times 6.98 = 503 \text{ cfs}$$

PROBLEMS

p.10.1 In an open channel flow, the pressure at the fluid surface is

a. always lower than the atmospheric pressure.
b. always equal to the atmospheric pressure.
c. always higher than the atmospheric pressure.

d. sometimes lower and sometimes higher than the atmospheric pressure.
e. none of these answers.

p.10.2 The correct expression for the hydraulic radius (R) is

a. A/P.
b. P/A.
c. A/T.
d. T/A.
e. none of these answers.

p.10.3 The correct expression for the hydraulic depth (D) is

a. A/P.
b. P/A.
c. A/T.
d. T/A.
e. none of these answers.

p.10.4 The expression for the Froude number (**F**) in an open channel flow is

a. V^2/gD.
b. $V/\sqrt{gD}$.
c. VR/ν.
d. $V/\sqrt{gR}$.
e. none of these answers.

p.10.5 Chézy's formula to calculate velocity in an open channel is given by

a. $V = \sqrt{CRS}$.
b. $V = CR^{2/3}S^{1/2}$.
c. $V = C\sqrt{RS}$.
d. $V = \dfrac{C_m}{n} R^{2/3}S^{1/2}$.
e. none of these answers.

p.10.6 Manning's formula to calculate velocity in an open channel is given by

a. $V = \sqrt{CRS}$.
b. $V = CR^{2/3}S^{1/2}$.
c. $V = C\sqrt{RS}$.
d. $V = \dfrac{C_m}{n} R^{2/3}S^{1/2}$.
e. none of these answers.

p.10.7 In an open channel flow, the specific energy is given by

a. $z + V^2/2g$.
b. $z + y + V^2/2g$.
c. $z + y$.
d. $y + V^2/2g$.
e. none of these answers.

p.10.8 In an open channel flow, critical flow occurs when

a. total energy is maximum.
b. specific energy is maximum.
c. specific energy is minimum.
d. specific energy is zero.
e. none of these answers.

p.10.9 In an open channel flow, let y = depth of flow; y_c = critical depth; V = velocity; V_c = critical velocity. For flow to be supercritical, the conditions to be satisfied are

a. $y > y_c$; $V > V_c$.
b. $y > y_c$; $V < V_c$.
c. $y < y_c$; $V > V_c$.
d. $y < y_c$; $V < V_c$.
e. none of these.

p.10.10 Using the notation of p.10.9, for flow to be subcritical, the conditions to be satisfied are

a. $y > y_c$; $V > V_c$.
b. $y > y_c$; $V < V_c$.
c. $y < y_c$; $V < V_c$.
d. $y < y_c$; $V > V_c$.
e. none of these conditions.

p.10.11 A hydraulic jump is formed whenever

a. flow regime changes from subcritical to supercritical.
b. flow regime changes from supercritical to subcritical.
c. flow occurs over a weir.
d. flow occurs through an orifice.
e. none of these answers.

p.10.12 In a gradually varied flow, the conditions are

a. steady and uniform.
b. steady and nonuniform.
c. unsteady and uniform.
d. unsteady and nonuniform.
e. none of these answers.

p.10.13 The slope of a channel is considered mild when

a. it is less than 0.001.
b. it is less than 0.002.
c. the normal depth is less than the critical depth.
d. the normal depth is greater than the critical depth.
e. none of these answers.

p.10.14 The slope of a channel is steep when

a. it is greater than 0.001.
b. the normal depth is less than the critical depth.
c. the normal depth is greater than the critical depth.
d. the actual depth of flow is greater than the critical depth.
e. the actual depth of flow is less than the critical depth.

p.10.15 A trapezoidal channel has a base of 20 ft and side slopes of 2 horizontal to 1 vertical. The depth of water in the channel is 4.0 ft. Compute the hydraulic radius, hydraulic depth and section factor.

p.10.16 A circular channel has a diameter of 4.0 ft. The depth of water in the channel is 2.75 ft. Compute the hydraulic radius, hydraulic depth and section factor.

p.10.17 Determine the rate of flow in a rectangular straight, clean, earthen channel 10 m wide and laid on a slope of 0.0016. The depth of flow is 2.5 m. Check your answer with the computer solution.

p.10.18 Solve p.10.17 if the channel is trapezoidal with a left side slope of 2:1 and a right side slope of 3:1. Check your answer with the computer solution.

p.10.19 Determine the rate of flow in a straight concrete culvert 72 in in diameter and laid on a 0.2% slope. The depth of flow is 4 ft. Check your answer with the computer solution.

p.10.20 Determine the normal depth of flow in a trapezoidal channel carrying 20 m^3/s. The channel is 5 m wide and has side slopes of 2:1. The channel is earthen with some grass and is laid on a 0.3% slope. Check your answer with the computer solution.

p.10.21 Determine the normal depth of flow in a concrete sewer carrying 40 ft^3/s. The sewer is 72 in in diameter and is laid on a slope of 0.16%. Check your answer with the computer solution.

p.10.22 A trapezoidal channel has a bottom width of 10 ft and side slopes of 2 horizontal to 1 vertical. The slope of the channel bed is 0.0005, and Manning's n is 0.025. For a flow rate of 350 ft^3/s, determine

a. The normal depth
b. The flow regime
c. The critical depth.

p.10.23 A circular channel has a diameter of 4.75 ft and is laid on a 0.5% slope. For a flow rate of 40 ft^3/s, determine the normal depth and the critical depth. Manning's n is 0.025.

p.10.24 For the channel of p.10.23 determine the normal and the critical depths if the flow rate is 75 ft^3/s. Is normal open channel flow possible for the given conditions?

p.10.25 Water is flowing in a rectangular channel 5 m wide at the rate of 50 m^3/s. A sluice gate decreases the depth to 1.5 m and a hydraulic jump is formed downstream. Determine the height, the type and the length of the jump. Also determine the loss of energy in the jump.

p.10.26 A hydraulic jump occurs at an initial depth of 1.45 m in a rectangular channel. If the sequent depth is 2.7 m, determine the length of the jump, the discharge, the type of the jump and the energy loss. The channel is 4 m wide.

p.10.27 Water flows under a sluice gate into a channel. After traversing a short length of the channel, a hydraulic jump occurs. The depth after the jump becomes 2.5 m and the velocity is 3 m/s. Calculate the depth of water on the upstream side of the hydraulic jump. Also calculate the length of the jump and the loss of energy per Newton of water.

p.10.28 A 3 m wide rectangular channel discharges 6 m^3/s. The slope is 0.0008 and Manning's coefficient is 0.012. At a certain point, the water depth is 0.8 m. How far downstream will the depth be 0.75 m? Use the standard step method.

p.10.29 At a section in a trapezoidal channel, $b = 5$ m, $z = 2$, $y = 3$ m. At another section 200 m downstream, $b = 4$ m, $z = 3$. The channel discharges 9 m^3/s, Manning's n is 0.012 and the slope is 0.0009. Determine the depth of flow at the second section.

p.10.30 A rectangular channel 15 ft wide is laid on a slope of 0.0002. A dam built across the channel raises the height of water immediately upstream to 12 ft. Determine the water surface profile to a distance of 5,000 ft upstream of the dam. The rate of flow is 300 cfs and Manning's n is 0.035.

p.10.31 Draw the water surface profile and locate the hydraulic jump for the condition given in Fig. 10.26. The channel is trapezoidal with $b = 10$ m, $z = 2$. It carries a discharge of 45 m^3/s. Use computer program listed in Appendix B.

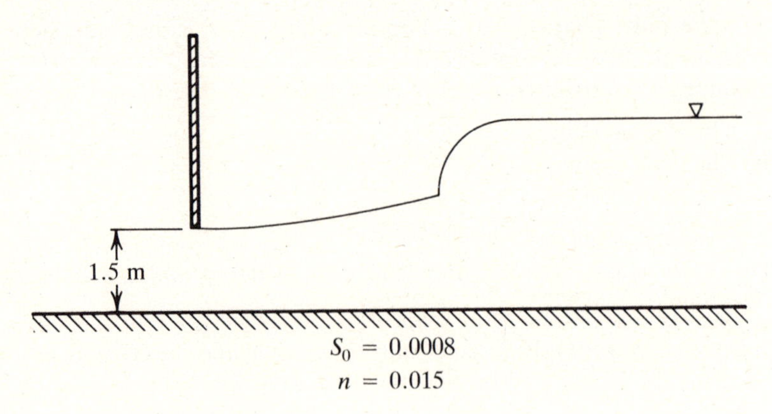

FIGURE 10.26 Problem 10.31.

Chapter Eleven

COMPRESSIBLE FLOW

In this chapter, we will consider fluids in which change in density is an important consideration. Mostly, gases fall into this class. The equation of state which relates pressure and density of a gas will be discussed in some detail. We will also learn about the speed of a sound wave, Mach number, shock waves and other relevent phenomena. The subject matter discussed in this chapter is particularly important to engineers who deal with gases.

11.1 THERMODYNAMIC CONCEPTS

Process

When the physical properties such as pressure, density, temperature, of a fluid are changed, it is said to undergo *a process.* If the process takes place in such a way that the fluid and its surroundings can be completely restored to their original conditions, it is said to be *reversible.* Otherwise, it is *irreversible.* All processes in the real world are irreversible due to viscous effects which cause mechanical energy to be converted to thermal energy. A *reversible process* is an idealized concept which is of great assistance in problem analysis. However, real processes in which velocity and temperature gradients are small may be assumed to be reversible without sacrificing much accuracy.

A process is said to be *adiabatic* if there is no transfer of heat between the fluid and its surroundings. There may be a change in the internal energy of a fluid during an adiabatic process, although there is no transfer of heat. Therefore, an adiabatic process is not always reversible.

Specific Heat, *c*

Specific heat is the property of a fluid which defines its ability to store thermal energy. Two types of specific heat are defined: specific heat at constant volume (c_v) and specific heat at constant pressure (c_p). The specific heat c_v of a fluid is the amount of heat required to raise the temperature of a unit mass of the fluid by one degree when the volume is held constant. The specific heat c_p of a fluid is the amount of heat required to raise the temperature of a unit mass of the fluid by one degree when the pressure is held constant. c_v and c_p have the units of kilocalories per kilogram per Kelvin (kcal/kg · K) in the SI System and Btu per slug per Rankine (Btu/slug · °R) in the FPS System. The numerical values of c_v and c_p are the same in both systems of units.

The specific heat ratio, k, is defined as the ratio of c_p to c_v, or $k = c_p/c_v$. The values of c_p, c_v, and k for various gases are given in Appendix A, Table A-3.

Internal Energy, *u*

The *internal energy* u, of a fluid is the energy per unit mass of the fluid due to its molecular spacing and molecular activity. It depends on the pressure, density, and temperature of the fluid. Internal energy has the units of kilocalories per kilogram (kcal/kg) or Joules per kilogram (J/kg) in the SI System and Btu per slug (Btu/slug) or ft · lb per slug (ft · lb/slug) in the FPS System. The conversion factor is 1 kcal = 4187 J and 1 Btu = 778 ft · lb.

Enthalpy, *h*

Enthalpy is an important property of a gas. Enthalpy per unit mass is given by $h = u + p/\rho^2$. Enthalpy has the same units as internal energy.

The specific heats c_v and c_p may also be defined in terms of internal energy and enthalpy. Then

$$c_v = \left(\frac{\partial u}{\partial T}\right) \text{ at constant volume}$$

$$c_p = \left(\frac{\partial h}{\partial T}\right) \text{ at constant pressure}$$

For a perfect gas, u and h are functions of temperature only. Hence:

$$c_v = \frac{du}{dT} \tag{11.1a}$$

or,

$$u = c_v T \tag{11.1b}$$

and,

$$c_p = \frac{dh}{dT} \tag{11.2a}$$

or,

$$h = c_p T \tag{11.2b}$$

The Perfect Gas

A *perfect gas* is defined as a fluid which has constant specific heats and obeys the equation of state given by

$$p = \rho R T \tag{11.3}$$

in which p = absolute pressure; ρ = density; R = gas constant and T = absolute temperature. The constant R of a gas is independent of temperature.

For a perfect gas, from Eqs. (11.1) and (11.2), we obtain

$$du = c_v dT \tag{11.4}$$

$$dh = c_p dT \tag{11.5}$$

But:

$$h = u + \frac{p}{\rho} = u + RT$$

Differentiating, we obtain

$$dh = du + RdT$$

$$\frac{dh}{dT} = \frac{du}{dT} + R$$

Substituting for dh/dT and du/dT from Eqs. (11.4) and (11.5), we obtain

$$c_p = c_v + R \tag{11.6}$$

Also

$$\frac{c_p}{c_v} = k$$

Hence

$$c_p = \frac{k}{k-1} R \tag{11.7}$$

$$c_v = \frac{R}{k-1} \tag{11.8}$$

Entropy

Entropy is a property of a gas which is used in the context of the second law of thermodynamics. If an amount of heat dq is supplied to a unit mass of gas at temperature T, then according to the second law of thermodynamics, the change in entropy, ds, of the gas is given by

$$ds = \frac{dq}{T} \tag{11.9}$$

From the first law of thermodynamics, for a closed system the heat supplied, dq, is equal to the change in internal energy, du, and the work done, dw. Then

$$dq = du + dw$$

$$dq = du + pd\forall s$$

where $\forall s$ is the specific volume of the gas. Substituting in Eq. (11.9), we obtain

$$ds = \frac{du + pd\forall s}{T}$$

$$Tds = du + pd\forall s \tag{11.10}$$

A process in which entropy remains constant, is called *isentropic*. Therefore, in an isentropic process, $ds = 0$. If $ds = 0$, then from Eq. (11.9), $dq = 0$. This means that a reversible adiabatic process is isentropic.

For a perfect gas, the change in internal energy and the change in enthalpy between any two states 1 and 2 may be written using Eqs. (11.4) and (11.5). Then

$$u_2 - u_1 = c_v(T_2 - T_1) \tag{11.11}$$

$$h_2 - h_1 = c_p(T_2 - T_1) \tag{11.12}$$

Substituting $c_v dT$ for du and ρRT for p, in Eq. (11.10), we obtain

$$Tds = c_v dT + \rho RT d\forall s$$

$$ds = c_v \frac{dT}{T} + \rho R d\forall s$$

Integrating between any two states 1 and 2 and replacing $\forall s$ with $1/\rho$, we obtain

$$s_2 - s_1 = c_v \ln\left(\frac{T_2}{T_1}\right) + R \ln\left(\frac{\rho_1}{\rho_2}\right) \tag{11.13}$$

From Eq. (11.8), $R = c_v(k - 1)$.

Substituting in Eq. (11.13) and simplifying, we obtain

$$s_2 - s_1 = c_v \ln\left[\left(\frac{T_2}{T_1}\right)\left(\frac{\rho_1}{\rho_2}\right)^{k-1}\right] \tag{11.14}$$

From Eq. (11.3)

$$T_2 = \frac{p_2}{\rho_2 R}; \; T_1 = \frac{p_1}{\rho_1 R}$$

Substituting in Eq. (11.14), we obtain

$$s_2 - s_1 = c_v \ln\left[\left(\frac{p_2}{p_1}\right)\left(\frac{\rho_1}{\rho_2}\right)^{k}\right] \tag{11.15}$$

Also, substituting for ρ_1 and ρ_2 in Eq. (11.14), we obtain

$$s_2 - s_1 = c_v \ln\left[\left(\frac{T_2}{T_1}\right)^{k}\left(\frac{p_2}{p_1}\right)^{1-k}\right] \tag{11.16}$$

For an adiabatic process, there is no transfer of heat; therefore, $ds = 0$ or $s =$ constant. Therefore, in a reversible adiabatic process, there is no change in entropy, or the process is isentropic.

Substituting $S_1 = S_2$ in Eq. (11.15), we obtain

$$\frac{p_2}{p_1} = \left(\frac{\rho_2}{\rho_1}\right)^k$$

$$\frac{p_1}{\rho_1^k} = \frac{p_2}{\rho_2^k}$$

Hence, for an isentropic process

$$\frac{p}{\rho^k} = \text{Constant} \tag{11.17}$$

There are some actual processes which are not isentropic—i.e., p/ρ^k is not constant, but they can be expressed approximately by a relation of the type $p/\rho^n =$ constant, where n is a positive constant. Such processes are called *polytropic*.

EXAMPLE 11.1

The initial conditions of a 10 kg mass of air are: $p_1 = 40$ kPa and $T_1 = 15°C$. The air is then compressed in such a way that $p_2 = 360$ kPa and $T_2 = 90°C$. Compute the change in the enthalpy of the air.

Solution The change in the enthalpy per unit mass of a gas is a function of temperature only and is given by Eq. (11.12). From this equation:

$$h_2 - h_1 = c_p(T_2 - T_1)$$

where, for the given problem

$$T_1 = 15 + 273 = 288 \text{ K}$$

$$T_2 = 90 + 273 = 363 \text{ K}$$

$$c_p \text{ for air} = 0.24 \text{ kcal/kg}\cdot\text{K (from Table A.3)}$$

Hence, change in enthalpy per kg of air

$$= 0.24(363 - 288) = 18 \text{ kcal}$$

The total change in enthalpy of 10 kg of air

$$= 10 \times 18 = 180 \text{ kcal}$$

Notice that the information on pressure is not used because enthalpy is a function of temperature only.

EXAMPLE 11.2

Determine the change in entropy of a 10 kg of air where the initial conditions are: $p_1 = 40$ kPa, $T_1 = 15°C$ and the final conditions are $p_2 = 360$ kPa, $T_2 = 90°C$. Assume atmospheric pressure to be 101.3 kPa.

Solution The change in entropy per unit mass of a gas is a function of temperature and pressure and is given by Eq. (11.16). From this equation

$$s_2 - s_1 = c_v \ln\left[\left(\frac{T_2}{T_1}\right)^k \left(\frac{p_2}{p_1}\right)^{1-k}\right]$$

where

$$T_1 = 15 + 273 = 288 \text{ K}$$

$$T_2 = 90 + 273 = 363 \text{ K}$$

$$p_1 = 40 + 101.3 = 141.3 \text{ kPa abs}$$

$$p_2 = 360 + 101.3 = 461.3 \text{ kPa abs}$$

From Table A-3

$$c_v = 0.171$$

$$k = 1.4$$

Hence, change in entropy per unit mass

$$= 0.171 \ln\left[\left(\frac{363}{288}\right)^{1.4}\left(\frac{461.3}{141.3}\right)^{-0.4}\right]$$

$$= -0.0255 \text{ kcal/kg}\cdot\text{K}$$

The total change in entropy of 10 kg of air

$$= 10(-0.0255) = -0.255 \text{ kcal/K}$$

EXAMPLE 11.3

A 10 slug mass of oxygen of initial conditions $p_1 = 10$ psia, $T_1 = 110°F$ is compressed isentropically to a pressure of 30 psia. Determine the final temperature of the gas and also the work done on it.

Solution In an isentropic process, according to Eq. (11.17),

$$\frac{p}{\rho^k} = \text{constant}$$

$$\frac{p_1}{\rho_1^k} = \frac{p_2}{\rho_2^k}$$

Substituting $\rho_1 = p_1 R T_1$; $\rho_2 = p_2 R T_2$ and simplifying, we obtain

$$T_2 = T_1 \left(\frac{p_2}{p_1}\right)^{(k-1)/k}$$

where for the given problem

$$p_1 = 10 \text{ psia}$$

$$p_2 = 30 \text{ psia}$$

$$T_1 = 110 + 460 = 570°\text{R}$$

$$k = 1.4 \text{ (from Table A-3)}$$

Hence:

$$(k-1)k = (1.4 - 1)/1.4 = 0.2857$$

$$T_2 = 570\left(\frac{30}{10}\right)^{0.2857}$$

$$= 780.2°\text{R}$$

$$T_2 = 320.2°\text{F}.$$

The work done on a gas is equal to the increase in its internal energy, which, for an isentropic process, is given by Eq. (11.11).
Hence:

$$\text{work done} = u_2 - u_1 = c_v(T_2 - T_1)$$

where, for the given problem

$$T_1 = 570°\text{R}$$

$$T_2 = 780.2°\text{R}$$

$$c_v = 0.157 \text{ (from Table A-3)}$$

Hence, work done per unit mass of oxygen

$$= 0.157(780.2 - 570)$$

$$= 33 \text{ Btu/slug}$$

Total work done on 10 slugs

$$= 10 \times 33 = 330 \text{ Btu}$$

11.2 ENERGY EQUATION WITH VARIABLE DENSITY: STATIC AND STAGNATION TEMPERATURE

The energy equation for the steady flow of a fluid between two points 1 and 2 was developed in Chapter Four and is given by Eq. (4.17),

$$q = \left(\frac{p_2}{\rho_2} + \frac{V_2^2}{2} + gz_2\right) - \left(\frac{p_1}{\rho_1} + \frac{V_1^2}{2} + gz_1\right) + (u_2 - u_1) + W'$$

If there is no transfer of heat between the fluid and its surroundings and no mechanical work is done, then $q = 0$, $W' = 0$ and the above equation reduces to

$$\left(u_2 + \frac{p_2}{\rho_2} + \frac{V_2^2}{2} + gz_2\right) - \left(u_1 + \frac{p_1}{\rho_1} + \frac{V_1^2}{2} + gz_1\right) = 0 \qquad (11.18)$$

But $(u + p/\rho)$ is the enthalpy h; hence:

$$\left(h_2 + \frac{V_2^2}{2} + gz_2\right) - \left(h_1 + \frac{V_1^2}{2} + gz_1\right) = 0$$

$$h + \frac{V^2}{2} + gz = \text{constant} \qquad (11.19)$$

This equation represents a steady adiabatic flow of a fluid along a streamline in which neither the fluid does work on its surroundings nor is work done on the fluid. If the fluid is a perfect gas, then from Eq. (11.2b), $h = c_p T$.

Substituting in Eq. (11.19), we obtain

$$c_p T + \frac{V^2}{2} + gz = \text{constant} \tag{11.20}$$

Also

$$c_p = \left(\frac{k}{k-1}\right) R \text{ from Eq. (11.7)}$$

Hence:

$$\left(\frac{k}{k-1}\right) RT + \frac{V^2}{2} + gz = \text{constant}$$

Also $RT = p/\rho$, hence,

$$\left(\frac{k}{k-1}\right)\frac{p}{\rho} + \frac{V^2}{2} + gz = \text{constant} \tag{11.21}$$

In a gas flow, the term gz is usually small in comparison with the other terms, and can be neglected. Eqs. (11.19) and (11.21) may then be written as

$$h + \frac{V^2}{2} = \text{constant} \tag{11.22}$$

$$\left(\frac{k}{k-1}\right)\frac{p}{\rho} + \frac{V^2}{2} = \text{constant} \tag{11.23}$$

From Eq. (11.22), it follows that for a given streamline, enthalpy h will be maximum when velocity V is zero (the stagnation point). This maximum value of enthalpy is called *the stagnation enthalpy*, h_0. The temperature corresponding to the stagnation enthalpy is called *the stagnation temperature*, T_0. For a perfect gas

$$T_0 = \frac{h_0}{c_p}$$

Eq. (11.22), when written between the stagnation point and any other point, will have the form

$$h + \frac{V^2}{2} = h_0$$

$$c_p T + \frac{V^2}{2} = c_p T_0 \tag{11.24}$$

It is clear from the above equation that the stagnation temperature, T_0 is greater than the ordinary static temperature, T by an amount equal to $V^2/2c_p$. A thermometer inserted in a flowing stream of gas measures the stagnation temperature and not the static temperature.

11.3 PROPAGATION OF SOUND WAVES

When a small pressure change is applied at any point in a fluid, it is transmitted equally in all directions. The speed with which this pressure change is transmitted depends upon the elasticity of the fluid. For a strictly incompressible fluid, the transmission of pressure change is instantaneous, but for a compressible fluid the transmission takes place progressively although it is very rapid. The speed with which this pressure adjustment occurs in fluids is of great practical importance and may be determined using the elastic properties of the fluid.

Consider a duct of constant cross-sectional area A containing a fluid of density ρ under a pressure p, moving with a velocity V. Let a pressure difference dp be applied at a cross-section by moving a piston inside the duct. Not all the fluid has yet adjusted to the pressure change; therefore, there exists a cross-section where there is more or less abrupt discontinuity of pressure. This discontinuity is known as *the pressure wave*. On the upstream side of the pressure wave are the new pressure $p + dp$, the new velocity $V + dV$, the new density $\rho + d\rho$, and on the downstream side of the pressure wave are the original pressure p, the original velocity V and the original density ρ (Fig. 11.1a). The figure also shows that the pressure wave, W is being propagated toward the right at a speed C.

Since the velocity of the fluid at point changes as the wave passes that point, the flow is unsteady and cannot be analyzed using the steady flow equations. However, if the coordinates also move with the wave, then the flow with respect to the moving coordinates will become steady. The new situation is shown in Fig. 11.1b. Let us now apply the continuity equation and the momentum equation to the control volume consisting of the portions immediately upstream and downstream of the pressure wave.

From continuity, the flow across the wave must be constant. Hence:

$$(\rho + d\rho)(V - C + dV)(A) = \rho(V - C)(A)$$

or

$$(C - V)\, d\rho = (\rho + d\rho)\, dV \tag{11.25}$$

From the momentum equation, we obtain

$$(p + dp)A - pA = \rho(V - C)A(-dV)$$

$$dp = \rho(C - V)\, dV \tag{11.26}$$

Eliminating dV from Eqs. (11.25) and (11.26), we obtain

$$(C - V)^2 = \left(\frac{\rho + d\rho}{\rho}\right)\frac{dp}{d\rho} \tag{11.27}$$

For a weak pressure wave, the change in pressure tends to become zero and, change in density also tends to become zero. This condition when applied to Eq. (11.27), yields

$$(C - V)^2 = \frac{dp}{d\rho} \tag{11.28}$$

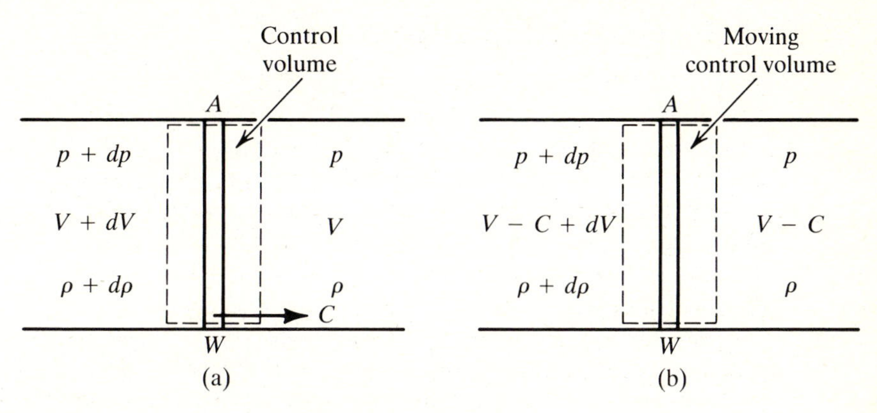

FIGURE 11.1 A pressure wave traveling through a duct.

where $(C - V)$ is the speed of the pressure wave relative to the speed of the fluid. Let it be denoted by c.

Then:

$$c = \sqrt{\frac{dp}{d\rho}} \tag{11.29}$$

For a weak pressure wave, the changes in pressure, density, and temperature are exceedingly small. Since the change in velocity is small, the resulting loss due to friction is negligible. Also, since the temperature change is very small and the wave propagates very rapidly, there is a negligible transfer of heat across the wave. Consequently, the passage of a wave is a process which is both frictionless and adiabatic and, therefore, is isentropic. For an isentropic process, from Eq. (1.24), the bulk modulus of elasticity is given by

$$\beta = kP = \left(\frac{dp}{d\rho/\rho}\right)$$

$$\frac{dp}{d\rho} = \frac{kP}{\rho} = \frac{k\rho RT}{\rho} = kRT$$

Hence:

$$c = \sqrt{kRT} \tag{11.30}$$

This equation is valid only for gases. Another expression for C applicable to both gases and liquids may be obtained by using the general expression for the bulk modulus of elasticity. Then:

$$\beta = \frac{dp}{(d\rho/\rho)}$$

$$\frac{dp}{d\rho} = \frac{\beta}{\rho}$$

Substituting in Eq. (11.29), we obtain

$$c = \sqrt{\beta/\rho} \tag{11.31}$$

The velocity given by Eq. (11.29) is called *the velocity of sound* or *sonic velocity*. In the problems involving propagation of pressure waves, it is useful to express the velocity of the fluid in terms of the sonic velocity. The ratio of fluid velocity to the sonic velocity is known as the *Mach number* (**M**). Fluid velocities less than the sonic velocity ($\mathbf{M} < 1$) are known as *subsonic*, those greater then the sonic velocity ($\mathbf{M} > 1$) are known as *supersonic*.

Eq. (11.30) and its variations are valid only if the pressure change is very small compared with the pressure itself.

EXAMPLE 11.4

Determine the speed of sound in water, ethyl alcohol and air at 20°C and 101.3 kPa.

Solution Since water and ethyl alcohol are liquids, Eq. (11.31) will be used to determine the speed of sound.

Hence:

$$c = \sqrt{\frac{\beta}{\rho}}$$

For water, from Table A-4

$$S = 1$$

$$\rho = 1 \times 1000 = 1000 \text{ kg/m}^3$$

$$\beta = 2.2 \times 10^9 \text{ N/m}^2$$

Therefore,

$$c = \sqrt{\frac{2.2 \times 10^9}{1000}} = 1483 \text{ m/s}$$

For ethyl alcohol, from Table A-4

$$S = 0.79$$

$$\rho = 0.79 \times 1000 = 790 \text{ kg/m}^3$$

$$\beta = 1.21 \times 10^9 \text{ N/m}^2$$

Therefore;

$$c = \sqrt{\frac{1.21 \times 10^9}{790}} = 1237 \text{ m/s}$$

The speed of sound in air is determined using Eq. (11.30).

Hence:

$$c = \sqrt{kRT}$$

where, for air, from Table A-3

$$k = 1.4$$

$$R = 287 \text{ m} \cdot \text{N/kg} \cdot \text{K}$$

$$T = 20 + 273 = 293\text{K}$$

Therefore:

$$c = \sqrt{1.4 \times 287 \times 293}$$

$$= 343 \text{ m/s}$$

The Mach Cone

Consider a source capable of emitting a disturbance so as to produce sound waves. These disturbances propagate in all directions at a speed c. For the purpose of this discussion, it is immaterial whether the origin is stationary and the fluid is moving or the fluid is stationary and the origin is moving.

Let us first consider the case when the velocity, V, of the moving source is less than the velocity of sound, c. If the body is at point A (Fig. 11.2) at time $t = 0$, then at time $t = t_1$, the body would have reached point B which is at a distance of Vt_1 from A. But the pressure wave at time $t = t_1$, would have traveled a distance of ct_1 in all directions from A. Since $c > V$, the pressure wave, in the direction of motion of the source, is able to travel ahead of the source. The waves traveling ahead of the source inform the fluid of the approaching source and prepare for its arrival. The sound emitted at times $3/4t_1$, $1/2t_1$, and $1/4t_1$ are also shown in Fig. 11.2 by the nonintersecting eccentric circles. To an observer stationed at B, the frequency of the sound pulses would appear higher than the emitted frequency but to an observer stationed at A, the frequency of the sound pulses would appear lower than the emitted frequency. This is known as *the Doppler effect.*

Let us now consider the case, when the velocity of the moving source, V, is greater than the velocity of sound c. After a time $t = t_1$ (Fig. 11.3), the body has traveled a distance of Vt_1. Since $V > c$, the body at time t_1 is outside the sphere formed by the pressure waves. In this case, the body moves faster than the message and arrives unannounced, at point B. The unprepared fluid has to adjust sharply to the change in pressure producing what is known as *a shock wave.* The circles showing the sound emitted at times $3/4t_1$, $1/2t_1$, and $1/4t_1$ are shown in Fig. 11.3. In this case, the circles intersect each other. The

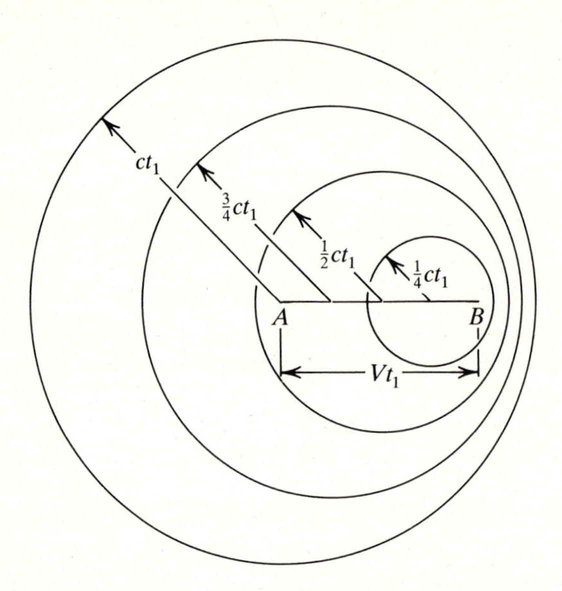

FIGURE 11.2 Propagation of soundwaves when the source is traveling with subsonic speed.

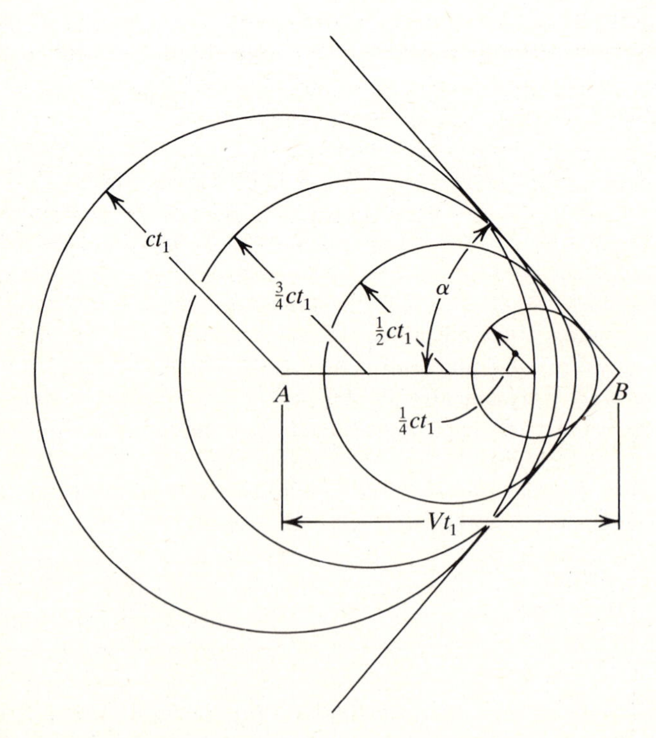

FIGURE 11.3 Propagation of sound waves when the source is traveling with supersonic speed.

locus of the leading surfaces of the sound waves is a cone. No sound is heard in front of this cone. From the geometry of the cone, it is seen that

$$\sin \alpha = \frac{ct_1}{Vt_1} = \frac{c}{V} = \frac{1}{\mathbf{M}}$$

$$\alpha = \sin^{-1}\left(\frac{1}{\mathbf{M}}\right)$$

The angle α is known as the *Mach angle* and the cone as the *Mach cone*. The region outside the Mach cone is completely unaffected by the sound waves and is called *the zone of silence*. The region inside the cone is termed *the zone of action* or the *region of influence*.

Remember that this discussion is limited to the source of disturbance in the form of a point. For large bodies, such as aircraft, the Mach cone will have a rounded apex.

11.4 NORMAL SHOCK WAVES

The discussion in Section 11.3 was valid for a pressure change of infinitesimal size. An abrupt finite pressure change, known as *a shock*, is now considered. A *shock wave* may be *normal* or *oblique*. A *normal* shock wave is one which is perpendicular to the direction of flow. Such shocks may occur in the diverging section of a convergent-divergent nozzle or in the front of a blunt-nosed body. In normal shock, the flow upstream of the shock is supersonic while that downstream is subsonic.

To determine the relationship between the quantities upstream and downstream of a shock wave, let us draw a control surface around the shock wave (Fig. 11.4). Let the quantities upstream be denoted by subscript '1' and those downstream by subscript '2.' Let us now write the continuity, momentum and energy equations for the control volume of cross-sectional area A.

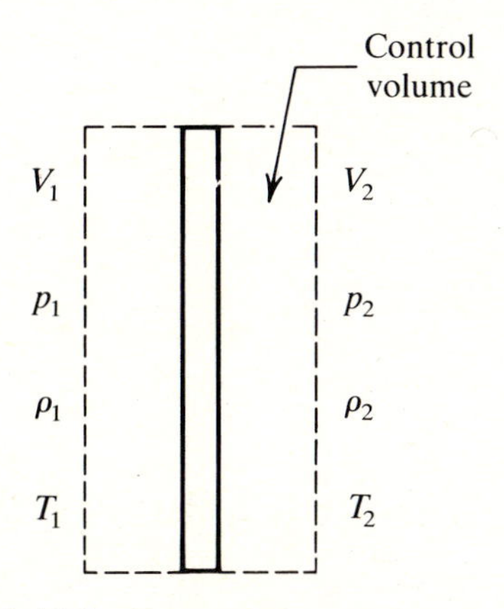

FIGURE 11.4 A normal shock wave.

From the continuity equation:

$$\rho_1 V_1 A = \rho_2 V_2 A$$

$$\rho_1 V_1 = \rho_2 V_2 \tag{11.32}$$

From the perfect gas law,

$$\rho = \frac{p}{RT}$$

From Eq. (11.30),

$$c = \frac{V}{\mathbf{M}} = \sqrt{kRT}$$

$$V = \mathbf{M}\sqrt{kRT}$$

Substituting in Eq. (11.32), we obtain

$$\frac{p_1}{RT_1}\mathbf{M}_1\sqrt{kRT_1} = \frac{p_2}{RT_2}\mathbf{M}_2\sqrt{kRT_2}$$

$$\frac{T_2}{T_1} = \left(\frac{p_2\mathbf{M}_2}{p_1\mathbf{M}_1}\right)^2 \tag{11.33}$$

From the momentum equation:

$$(p_1 - p_2)A = \rho_2 V_2^2 A - \rho_1 V_1^2 A$$

$$p_1 - p_2 = \rho_2 V_2^2 - \rho_1 V_1^2$$

$$p_1 - p_2 = \frac{p_2}{RT_2}\mathbf{M}_2^2 (kRT_2) - \frac{p_1}{RT_1}\mathbf{M}_1^2 (kRT_1)$$

$$= kp_2\mathbf{M}_2^2 - kp_1\mathbf{M}_1^2$$

$$\frac{p_2}{p_1} = \frac{1 + k\mathbf{M}_1^2}{1 + k\mathbf{M}_2^2} \tag{11.34}$$

From the energy equation, assuming adiabatic conditions:

$$c_p T_1 + \frac{1}{2}V_1^2 = c_p T_2 + \frac{1}{2}V_2^2$$

$$c_p T_1 + \frac{1}{2}\mathbf{M}_1^2 (kRT_1) = c_p T_2 + \frac{1}{2}\mathbf{M}_2^2 (kRT_2)$$

$$\frac{T_2}{T_1} = \frac{c_p + \frac{1}{2}\mathbf{M}_1^2 kR}{c_p + \frac{1}{2}\mathbf{M}_2^2 kR}$$

From Eq. (11.7), $c_p = [k/(k-1)]R$. Substituting and simplifying we obtain:

$$\frac{T_2}{T_1} = \frac{1 + \frac{1}{2}(k-1)\mathbf{M}_1^2}{1 + \frac{1}{2}(k-1)\mathbf{M}_2^2} \tag{11.35}$$

Substituting in Eq. (11.33) for T_2/T_1 from Eq. (11.35) and for p_2/p_1 from Eq. (11.34), we obtain:

$$\frac{1 + \frac{1}{2}(k-1)\mathbf{M}_1^2}{1 + \frac{1}{2}(k-1)\mathbf{M}_2^2} = \left(\frac{1 + k\mathbf{M}_1^2}{1 + k\mathbf{M}_2^2}\right)^2 \left(\frac{\mathbf{M}_2}{\mathbf{M}_1}\right)^2$$

Solving this equation for $\mathbf{M}_2$ as a function of $\mathbf{M}_1$, we obtain two solutions. One solution is $\mathbf{M}_1 = \mathbf{M}_2$, which is trivial, and corresponds to no shock wave. The other solution is

$$\mathbf{M}_2^2 = \frac{2 + (k-1)\mathbf{M}_1^2}{2k\mathbf{M}_1^2 - (k-1)} \tag{11.36}$$

From this equation, the Mach number downstream of the shock wave may be calculated. The other quantities may then be calculated using Eqs. (11.34) and (11.35). It is also evident from Eq. (11.36) that if $\mathbf{M}_1 = 1$, then $\mathbf{M}_2 = 1$ and that if $\mathbf{M}_1 > 1$, then $\mathbf{M}_2 > 1$. Also, from the above equations, we may obtain

$$\frac{\rho_2}{\rho_1} = \frac{(k-1) + (k+1)(p_2/p_1)}{(k+1) + (k-1)(p_2/p_1)} \tag{11.37}$$

This equation is known as the *Rankine–Hugoniot relation.* The fact that this equation differs from the isentropic equation (which is $p/\rho^k = \text{constant}$), suggests that the production of shock waves is not an isentropic process. From Eq. (11.13), the change in entropy is given by

$$s_2 - s_1 = c_v \ln\left(\frac{T_2}{T_1}\right) + R \ln\left(\frac{\rho_1}{\rho_2}\right)$$

Substituting for T_2/T_1 from Eq. (11.35), for ρ_1/ρ_2 from Eq. (11.37) and then for $\mathbf{M}_2$ from Eq. (11.36), we obtain:

$$s_2 - s_1 = c_v \ln\left\{\frac{2k\mathbf{M}_1^2 - k + 1}{(k+1)}\left[\frac{2 + (k-1)\mathbf{M}_1^2}{(k+1)\mathbf{M}_1^2}\right]^k\right\} \tag{11.38}$$

From the above expression:

when $\mathbf{M}_1 > 1$, $s_2 - s_1 > 0$.

when $\mathbf{M}_1 < 1$, $s_2 - s_1 < 0$.

Since $(s_2 - s_1)$ cannot be negative in an adiabatic process, a shock wave cannot exist if the upstream flow is subsonic.

EXAMPLE 11.5

Air flows at initial conditions of $p_1 = 1.5$ psia, $T_1 = 40°\text{F}$ and $V_1 = 1500$ ft/s. If a normal shock wave occurs in the flow, determine p_2, V_2 and T_2 downstream of the shock wave.

Solution For air, from Table A-3:

$$k = 1.4, \quad R = 1716 \text{ ft}\cdot\text{lb/slug}\cdot{}^\circ\text{R}$$

Let us first find the Mach number upstream of the shock wave.

$$c_1 = \sqrt{kRT} = \sqrt{1.4 \times 1716 \times (40 + 460)}$$
$$= 1096 \text{ ft/s}$$
$$\mathbf{M}_1 = \frac{V_1}{c_1} = \frac{1500}{1096} = 1.369$$

Since $\mathbf{M}_1 > 1$, a shock wave does exist. From Eq. (11.36):

$$\mathbf{M}_2^2 = \frac{2 + (k-1)\mathbf{M}_1^2}{2k\mathbf{M}_1^2 - (k-1)}$$
$$= \frac{2 + (1.4-1)(1.369)^2}{2 \times 1.4(1.369)^2 - (1.4-1)} = 0.567$$
$$\mathbf{M}_2 = 0.753$$

Therefore, the Mach number downstream of the shockwave is 0.753, which is subsonic.

The pressure downstream of the shock wave may be calculated using Eq. (11.34).

Hence:

$$\frac{p_2}{p_1} = \frac{1 + k\mathbf{M}_1^2}{1 + k\mathbf{M}_2^2}$$
$$= \frac{1 + 1.4(1.369)^2}{1 + 1.4(0.753)^2} = 2.02$$
$$p_2 = 1.5 \times 2.02 = 3.03 \text{ psia}$$

The temperature downstream of the shock wave may be determined using Eq. (11.35).

Hence:

$$\frac{T_2}{T_1} = \frac{1 + 0.5(k-1)\mathbf{M}_1^2}{1 + 0.5(k-1)\mathbf{M}_2^2}$$
$$= \frac{1 + 0.5(1.4-1)(1.369)^2}{1 + 0.5(1.4-1)(0.753)^2} = 1.2348$$
$$T_2 = 500 \times 1.2348 = 617.4^\circ\text{R}$$
$$T_2 = 617.4 - 460 = 157.4^\circ\text{F}$$

11.5 ONE-DIMENSIONAL ISENTROPIC FLOW

In one-dimensional flow, all the flow parameters are expressed as a function of distance measured along the centerline of some conduit in which the fluid is flowing. Any variation of the parameters over the cross-section is neglected.

Consider a fluid of density ρ flowing in a conduit. Let the velocity at a point where the cross-section is A, be V. Then from continuity:

$$\rho AV = \text{constant} \tag{11.39}$$

Differentiating and then dividing by ρAV, we obtain

$$\frac{d\rho}{\rho} + \frac{dA}{A} + \frac{dV}{V} = 0 \tag{11.40}$$

From Euler's equation, for a steady flow of a frictionless fluid, if we neglect the dz term in Eq. 4.3 and multiply by ds we obtain:

$$\frac{dp}{\rho} + V\,dV = 0 \tag{11.41}$$

The expression dp/ρ can be written as $(dp/d\rho)(d\rho/\rho)$. But for an isentropic flow

$$\frac{dp}{d\rho} = c^2$$

Therefore:

$$\frac{dp}{\rho} = c^2 \frac{d\rho}{\rho}$$

Substituting in Eq. (11.41), we obtain:

$$c^2 \frac{d\rho}{\rho} + V\,dV = 0$$

Substituting for $d\rho/\rho$ from Eq. (11.40) and rearranging, we obtain:

$$\frac{dA}{A} = \frac{dV}{V}\left(\frac{V^2}{c^2} - 1\right)$$

or

$$\frac{dA}{A} = \frac{dV}{V}(\mathbf{M}^2 - 1) \tag{11.42}$$

From the above equation, the following conclusions may be drawn regarding the subsonic, sonic, and supersonic flows:

1. If the flow is subsonic, $\mathbf{M} < 1$. Then, dA has a sign opposite to that of dV. This means that if the area decreases, the velocity increases and if the area increases, the velocity decreases. Also, $dA = 0$ where $dV = 0$. This means that when the cross-sectional area is minimum, the velocity is maximum.
2. If the flow is sonic, $\mathbf{M} = 1$ and $dA = 0$. This means that A must be minimum. In other words, if the velocity of flow is sonic, then the area of flow must be minimum. The flow is said to be critical when the velocity of flow is equal to the sonic velocity.

3. If the flow is supersonic, $\mathbf{M} > 1$ and dA has the same sign as dV. This means that if the area decreases, the velocity decreases and if the area increases, the velocity increases. Also $dA = 0$ when $dV = 0$. This means that when the cross-sectional area is minimum, the velocity is also minimum.

If the conduit is being supplied from a reservoir, the conditions in the reservoir are stagnant. Let us apply Eq. (11.23) between the stagnation point and any other point in the conduit. Then:

$$\left(\frac{k}{k-1}\right)\frac{p}{\rho} + \frac{V^2}{2} = \left(\frac{k}{k-1}\right)\frac{p_0}{\rho_0}$$

where the quantities with the zero subscript refer to the stagnation point. Replacing p/ρ by RT and V by $\mathbf{M}\sqrt{kRT}$, we obtain:

$$\left(\frac{k}{k-1}\right)RT + \frac{\mathbf{M}^2(kRT)}{2} = \left(\frac{k}{k-1}\right)RT_0$$

$$\frac{T}{T_0} = \frac{1}{1 + \left[\left(\frac{k-1}{2}\right)\right]\mathbf{M}^2} \tag{11.42}$$

From the perfect gas law, $T = p/\rho R$. Also, for an isentropic flow $\rho/\rho_0 = (p/p_0)^{1/k}$. Substituting in Eq. (11.42) and simplifying, we obtain:

$$\frac{p}{p_0} = \frac{1}{\{1 + [(k-1)/2]\mathbf{M}^2\}^{k/(k-1)}} \tag{11.43}$$

Substituting $p/p_0 = (\rho/\rho_0)^k$ in Eq. (11.43), we obtain:

$$\frac{\rho}{\rho_0} = \frac{1}{\{1 + [(k-1)/2]\mathbf{M}^2\}^{1/(k-1)}} \tag{11.44}$$

Flow Through a Nozzle

If the conduit is a converging-diverging nozzle, such as a de Laval nozzle shown in Fig. 11.5 (named after its inventor Carl Gustaf Patrik de Laval [1845–1913], a Swedish engineer), and the conditions at the throat are sonic, then $\mathbf{M} = 1$. Using an asterisk to denote the conditions at the throat, Eqs. (11.42) to (11.44) when applied between the stagnation point and the throat, will yield:

$$\frac{T_*}{T_0} = \frac{2}{k+1} \tag{11.45}$$

$$\frac{p_*}{p_0} = \frac{1}{(k+1)^{k/(k-1)}} \tag{11.46}$$

$$\frac{\rho_*}{\rho_0} = \frac{1}{(k+1)^{1/(k-1)}} \tag{11.47}$$

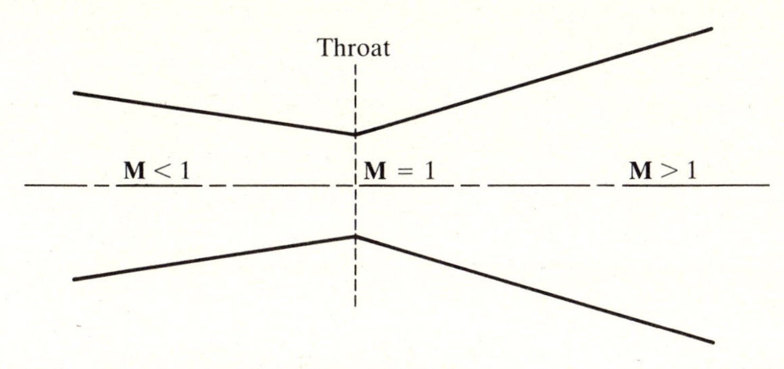

FIGURE 11.5 A converging-diverging nozzle.

The ratios of the area A/A_*, where A_* is the throat area, may be found using continuity. From continuity, mass rate of flow is the same throughout the nozzle.

Hence:

$$\rho A V = \text{constant}$$

Applying this equation to the throat and any other section, we obtain:

$$\rho A V = \rho_* A_* V_* \tag{11.48}$$

where the asterisk signifies quantities at the throat which are sonic (corresponding to $\mathbf{M} = 1$). From Eq. (11.48)

$$\frac{A}{A_*} = \frac{\rho_* V_*}{\rho V}$$

We had shown previously that for a perfect gas under isentropic conditions $V = \mathbf{M}\sqrt{kRT}$.

Hence:

$$\frac{A}{A_*} = \frac{\rho_* \mathbf{M}_* \sqrt{kRT_*}}{\rho \mathbf{M} \sqrt{kRT}}$$

$$\frac{A}{A_*} = \frac{\rho_*}{\rho \mathbf{M}} \left(\frac{T_*}{T}\right)^{1/2} \tag{11.49}$$

Applying Eq. (11.23) between the throat and any other section, we obtain:

$$\left(\frac{k}{k-1}\right)\frac{p}{\rho} + \frac{V^2}{2} = \left(\frac{k}{k-1}\right)\frac{p_*}{\rho_*} + \frac{V_*^2}{2}$$

Substituting $p/\rho = RT$; $V = \mathbf{M}\sqrt{kRT}$, and simplifying, we obtain:

$$\frac{T}{T_*} = \frac{\dfrac{(k+1)}{2}}{1 + \left[\dfrac{(k-1)}{2}\right]\mathbf{M}^2} \tag{11.50}$$

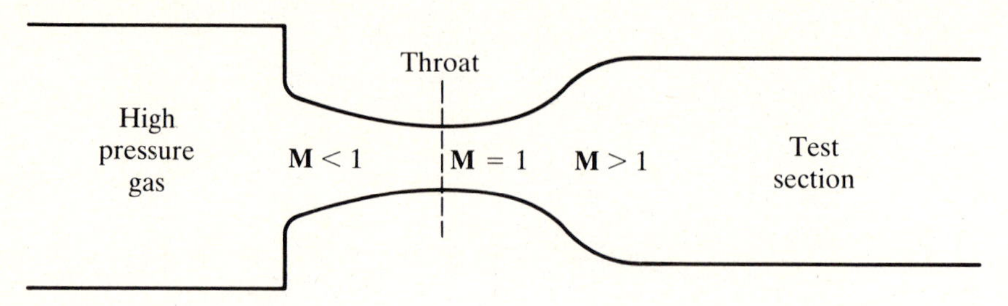

FIGURE 11.6 A supersonic wind tunnel.

And expressing T/T_* in terms of ρ/ρ_*, we obtain:

$$\frac{\rho}{\rho_*} = \left\{ \frac{\dfrac{(k+1)}{2}}{1 + \left[\dfrac{(k-1)}{2}\right]\mathbf{M}^2} \right\}^{1/(k-1)} \tag{11.51}$$

Substituting for T_*/T and ρ_*/ρ in Eq. (11.49) from Eqs. (11.50) and (11.51), we obtain:

$$\frac{A}{A_*} = \frac{1}{\mathbf{M}} \left\{ \frac{1 + \dfrac{(k-1)}{2}\mathbf{M}^2}{\dfrac{(k+1)}{2}} \right\}^{[(k+1)/2](k-1)} \tag{11.52}$$

The above equation is valid for all Mach numbers and for all sections of the nozzle. It is also used to design the test section of a supersonic wind tunnel. A supersonic wind tunnel may be constructed by passing a high pressure gas through a de Laval nozzle and then into a test section as shown in Fig. 11.6.

PROBLEMS

p.11.1 A process in which no transfer of heat takes place between the fluid and its surroundings is

a. isentropic.
b. reversible.
c. irreversible.
d. adiabatic.
e. none of these answers.

p.11.2 The SI units of specific heat is

a. kg/kcal · K.
b. kg/kcal · °R.
c. kcal/kg · K.
d. kcal/kg · R.
e. none of these answers.

p.11.3 Enthalpy per unit mass of a gas is given by

a. $h = c_v + c_p$.
b. $h = v + c_v$.
c. $h = v + 2/p$.
d. $h = v + p/2$.
e. none of these.

p.11.4 One expression for specific heat at Constant Pressure is

a. $dv/dT = c_p$.
b. $[k/(k-1)]R = c_p$.
c. $R/(k-1) = c_p$.
d. $v + RT = c_p$.
e. none of these.

p.11.5 An isentropic process is given by the expression

a. $\frac{p}{n} = \text{constant.}$

b. $\frac{p}{\rho^n} = \text{constant.}$

c. $\frac{p}{k} = \text{constant.}$

d. $\frac{p}{\rho^k} = \text{constant.}$

e. not listed above.

p.11.6 The internal energy of a fluid depends on

a. pressure only.
b. density only.
c. temperature only.
d. pressure, density and temperature.
e. none of these.

p.11.7 The change in the enthalpy per unit mass of a gas is a function of

a. pressure only.
b. temperature only.
c. pressure and temperature.
d. pressure, density and temperature.
e. none of these.

p.11.8 The change in entropy per unit mass of a gas is a function of

a. pressure only.
b. temperature only.
c. pressure and temperature.
d. pressure, density and temperature.
e. none of these.

p.11.9 For a given streamline enthalpy h is maximum when

a. velocity V is maximum.
b. velocity V is zero.
c. pressure p is maximum.
d. pressure p is zero.
e. none of these.

p.11.10 Select the correct statement in case of a perfect gas.

a. The relationship between stagnation enthalpy h_0 and stagnation temperature T_0 is given by $h_0 = T_0/c_p$.
b. The stagnation temperature T_0 is always less than the static temperature T.
c. A thermometer inserted in a flowing stream of gas measures the stagnation temperature and not the static temperature.
d. A thermometer inserted in a flowing stream of gas measures the static temperature and not the stagnation temperature.

p.11.11 One expression for the speed of sound is

a. $c = kRT$.
b. $c = RT$.
c. $c = \sqrt{kRT}$.
d. $c = \sqrt{RT}$.
e. none of these.

p.11.12 The propagation of sound wave in a fluid is approximately

a. frictionless and adiabatic.
b. frictionless and isothermal.
c. irreversible and isothermal.
d. adiabatic and isothermal.
e. none of these.

p.11.13 The velocity of air is supersonic when

a. $\mathbf{M} = 1$.
b. $\mathbf{M} < 1$.
c. $\mathbf{M} > 1$.
d. $\mathbf{R} > 4{,}000$.
e. none of these.

p.11.14 The speed of sound in water is usually calculated using

a. $c = kRT$.
b. $c = (k/k + 1)RT$.
c. $c = \sqrt{\frac{\beta}{\rho}}$.
d. $d = \sqrt{\frac{\rho}{\beta}}$.
e. none of these.

p.11.15 Define the following:

- **a.** Doppler effect.
- **b.** Mach Cone.
- **c.** The zone of silence.
- **d.** The zone of action.
- **e.** A normal shock wave.

p.11.16 A 10 kg mass of air has initial temperature and pressure of 20°C and 50 kPa respectively. The air is then compressed to a final temperature and pressure of 80°C and 320 kPa. Compute the change in the enthalpy of the air.

p.11.17 Determine the change in the entropy of a 2 slug of air when the initial conditions are: $p_1 = 10$ psi, $T_1 = 60$°F and the final conditions are: $P_2 =$ 100 psi, $T_2 = 200$°F. Use standard atmospheric conditions.

p.11.18 A 15 slug mass of air has initial conditions of $p_1 = 20$ psi, $T_1 = 100$°F. It is compressed isentropically to a pressure of 60 psi. Determine the final temperature of the gas and the work done on it. Assume standard atmospheric conditions.

p.11.19 Determine the speed of sound in ocean water at 70°F at a depth of one mile below the surface.

p.11.20 Determine the speed of sound at a distance of 30,000 ft. above the sea level. Use U.S. standard atmosphere.

p.11.21 Air flows in a tunnel at initial conditions of $p_1 = 7$ kPa absolute, $T_1 = 9$°C. If a normal shock wave occurs in the flow, determine p_2, v_2 and T_2 downstream of the shock wave.

p.11.22 In a de-Laval nozzle, air is flowing. The throat diameter is 50 mm. Determine the temperature ratio T/T^* and the pressure ratio P/p^* at a diverging section where the diameter is 55 mm.

Chapter Twelve

TURBOMACHINES

In this chapter, we will learn about turbomachines, such as pumps and turbines. Pumps are used to increase the pressure in a flow system by converting mechanical/electrical energy into pressure energy. Turbines use the pressure energy to produce mechanical/electrical energy. We will also learn the basic theory of pumps and turbines and the methods of selecting these machines to perform specific jobs.

12.1 INTRODUCTION

A rotodynamic machine or *turbomachine* is one which adds energy to, or extracts energy from, a fluid by virtue of a rotating system of blades within the machine. Pumps and compressors; fans and blowers; and water, steam, or gas turbines are examples of turbomachines. The rotating element is called a *rotor* in a compressor and a gas or steam turbine, an *impeller* in a pump, and a *runner* in a hydraulic turbine.

12.2 PUMPS

Pumps add energy to water. They are classified on the basis of the ways they move liquid: centrifugal, rotary and reciprocating (Fig. 12.1). The *centrifugal pumps* are further classified as axial-flow, radial-flow or mixed-flow. The classification of centrifugal pumps is based on the way the impeller imparts energy to the water. In a radial-flow pump, the impeller is shaped to force water outward at right angles to its axis. In an axial-flow pump, the impeller forces water in the axial direction only, while in a mixed-flow pump the impeller forces water in a radial as well as in an axial direction.

A pump can be single stage or multistage depending upon the number of impellers. Another distinguishing characteristic is the position of the shaft, which can be vertical or horizontal.

Power and Efficiency

The impeller of a pump is rotated by a force applied from an external source such as an electric motor. The power input to the pump (P_i) is given by:

$$P_i = T\omega \tag{12.1}$$

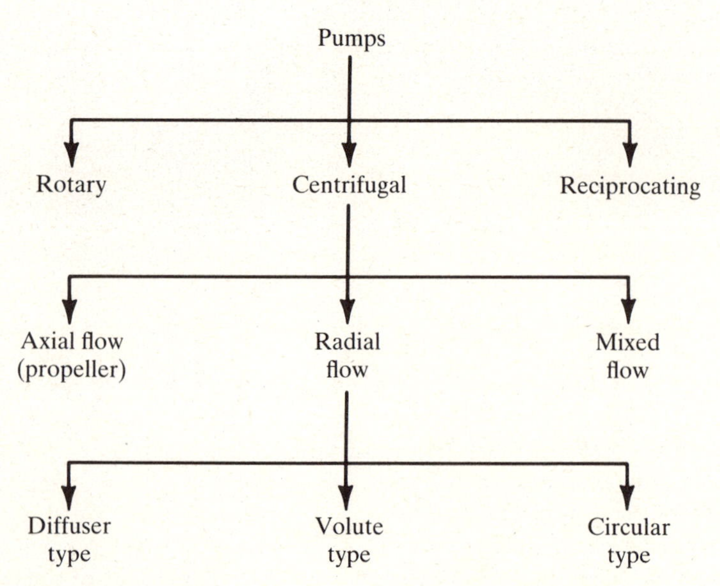

FIGURE 12.1 Types of pump.

where T is the torque applied to the pump by the motor and ω is the angular velocity.

The power output of a pump (P_0) is measured in terms of head produced at a certain discharge and is expressed as:

$$P_0 = \gamma Q H \tag{12.2}$$

where γ is the specific weight of the fluid, Q is the rate of flow and H is the head produced. Pumps are never 100% efficient. Therefore, the power output is always less than the power input. The *efficiency* of a pump (η) is defined as:

$$\eta = \frac{P_0}{P_i} = \frac{\gamma Q H}{T\omega} \tag{12.3}$$

EXAMPLE 12.1

A centrifugal pump delivers 0.25 m^3/s of oil of specific gravity 0.8 against an elevation head of 25 m when rotating at 1800 rpm. If the torque measured at the drive shaft is 350 N·m, determine the efficiency of the pump.

Solution Given:

$$Q = 0.25 \text{ m}^3/\text{s}$$

$$\gamma = 0.8 \times 9810 = 7848 \text{ N/m}^3$$

$$H = 25 \text{ m}$$

$$T = 350 \text{ N}\cdot\text{m}$$

$$N = 1800 \text{ rpm}$$

$$\omega = \frac{2\pi \times 1800}{60} = 188.5 \text{ rad/s}$$

Using Eq. (12.3)

$$\eta = \frac{\gamma Q H}{T\omega}$$

$$= \frac{7848 \times 0.25 \times 25}{350 \times 188.5}$$

$$= 0.74$$

$$= 0.74 \times 100 = 74\%$$

Hence, efficiency of the pump is 74%.

Comment

Care must be exercised to use consistent units. When using the SI System, express all variables in N, and *m* units and when using the English system, express all variables in lbs and ft.

12.3 CENTRIFUGAL PUMPS

In a centrifugal pump, the fluid enters into the propeller along a blade and is rotated vigorously. The fluid then leaves the propeller with a very high velocity. Since the propeller is housed in a casing, the kinetic energy is converted into pressure energy and the fluid rises into the pipe attached to the outlet of the casing. The two most important elements of a centrifugal pump are: (1) the rotating element called the impeller, and (2) the *casing* or *housing* which encloses the impeller and does the important job of converting the kinetic energy into the pressure energy.

The power needed to rotate the impeller is supplied by an external source such as a motor. The housing is designed in the shape of an expanding spiral such that the liquid leaving the propeller is led toward the discharge pipe with minimum loss (Fig. 12.2).

Theory

The velocity vector diagrams at entrance and exit to a single blade of a pump impeller are shown in Fig. 12.3.

Let

r_1 = radius at entrance to the impeller

r_2 = radius at exit to the impeller

ω = angular velocity of the impeller

V_1 = absolute velocity of fluid entering the impeller

V_2 = absolute velocity of fluid leaving the impeller

u_1 = peripheral velocity of the impeller at entrance = $r_1\omega$

u_2 = peripheral velocity of the impeller at exit = $r_2\omega$

v_1 = fluid velocity relative to the impeller at entrance

v_2 = fluid velocity relative to the impeller at exit

V_{1r} = radial component of V_1 at entrance

V_{2r} = radial component of V_2 at exit

V_{1t} = tangential component of V_1 at entrance, also called the whirl velocity

V_{2t} = tangential component of V_2 at exit

α_1 = angle which the absolute velocity V_1 makes with the peripheral velocity u_1 at entrance

α_2 = angle which the absolute velocity V_2 makes with the peripheral velocity u_2 at exit

β_1 = blade angle at entrance

β_2 = blade angle at exit

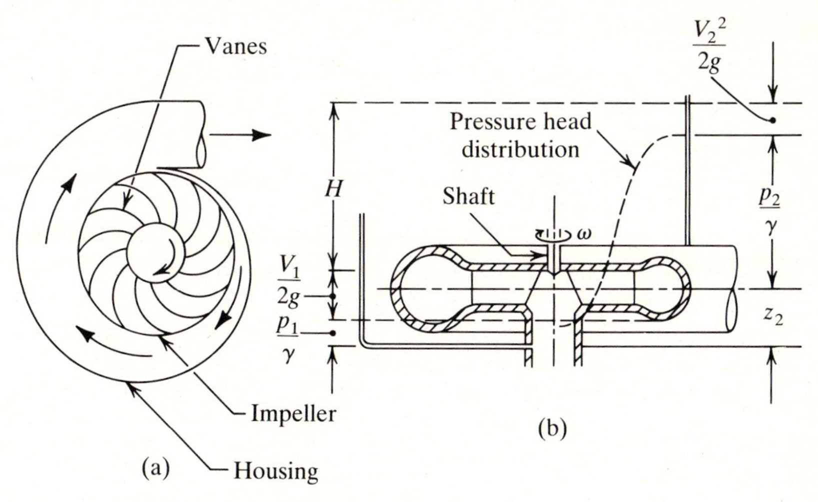

FIGURE 12.2 Cross-section of a centrifugal pump.

The basic principle employed is that the torque required to drive an impeller is equal to the change of moment of momentum of the fluid passing through the impeller.

Assume that the frictional losses are negligible and that the fluid has perfect guidance through the casing so that the relative velocity v is always tangent to the vane.

$$\text{moment of momentum at entrance} = T_1 = (\rho_1 Q_1 V_{1t})r_1$$

$$\text{moment of momentum at exit} = T_2 = (\rho_2 Q_2 V_{2t})r_2$$

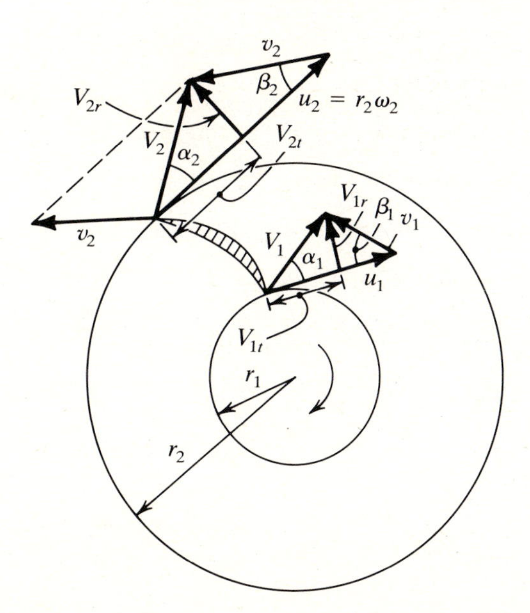

FIGURE 12.3 Velocity vector diagrams at entrance and exit to a single blade of a pump impeller.

Let

$$T = \text{torque acting on the fluid}$$

Then

$$T = T_2 - T_1$$
$$= \rho_2 Q_2 V_{2t} r_2 - \rho_1 Q_1 V_{1t} r_1$$

When $\rho_1 = \rho_2$ and Q_1 and Q_2:

$$T = \rho Q[V_{2t} r_2 - V_{1t} r_1] \tag{12.4}$$

The pumps are generally so designed that V_{1t} is zero. Then

$$T = \rho Q r_2 V_{2t} \tag{12.5}$$

From Eq. (12.3), for $\eta = 1$

$$T = \frac{\gamma Q H}{\omega}$$

Therefore

$$\frac{\gamma Q H}{\omega} = \rho Q r_2 V_{2t} \tag{12.6}$$

$$H = \frac{\omega r_2 V_{2t}}{g}$$

Noting that $\omega r_2 = u_2$

$$H = \frac{u_2 V_{2t}}{g} \tag{12.7}$$

The discharge passing through the pump is given by

$$Q = 2\pi r_2 b_2 V_{2r} \tag{12.8}$$

where b_2 = width of the impeller at exit.

From the velocity diagram at the exit in Fig. 12.3, we have:

$$V_{2t} = u_2 - V_{2r} \cot \beta_2$$

Substituting in Eq. (12.7) we obtain:

$$H = \frac{u_2}{g}[u_2 - V_{2r} \cot \beta_2]$$

$$H = \frac{u_2^2}{g} - \frac{u_2 V_{2r}}{g} \cot \beta_2 \tag{12.9}$$

From Eq. (12.8)

$$V_{2r} = \frac{Q}{2\pi r_2 b_2}$$

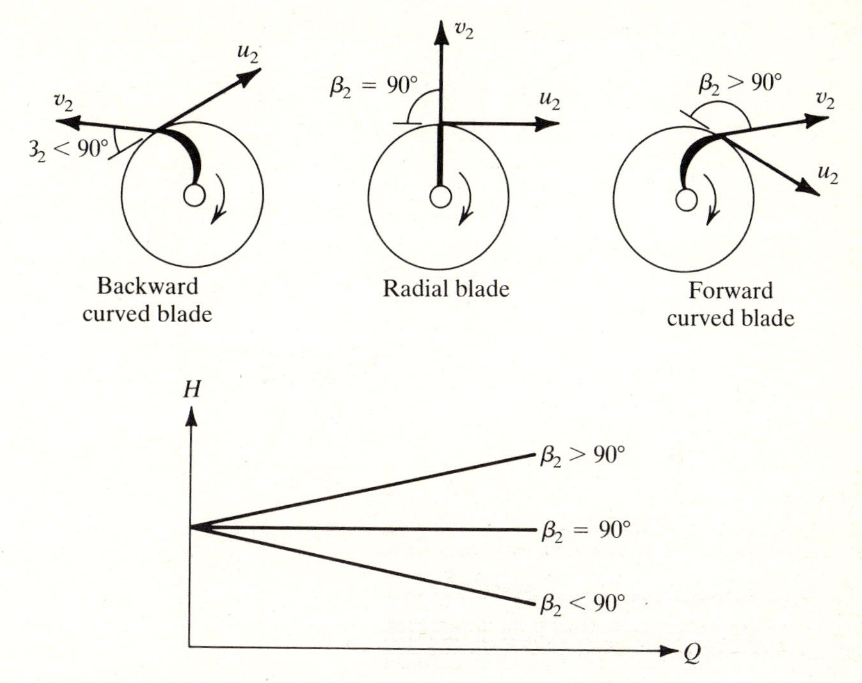

FIGURE 12.4 The relationship between H and Q for various blade shapes.

Substituting in Eq. (12.9), we obtain:

$$H = \frac{u_2^2}{g} - \frac{u_2 Q}{2\pi r_2 b_2 g} \cot \beta_2 \tag{12.10}$$

Equation (12.10) indicates that for a given impeller at a prescribed peripheral speed, the pump head H is a function of Q only. There are three possible shapes of impellers: *backward-curved, radial,* and *forward-curved.* The relationship between H and Q for various shapes is given in Fig. 12.4. For a backward-curved impeller $\beta_2 < 90°$. Hence, H decreases with an increase in Q. For a radial blade $\beta = 90°$. Hence, H is constant. For a forward-curved blade, $\beta_2 > 90°$. Hence, H increases as Q increases.

EXAMPLE 12.2

A centrifugal pump impeller has a diameter of 40 cm, a width of 3 cm, a blade angle of 60° and rotates at 1,500 rpm. The flow rate is 24 m^3 per minute. Calculate:

(a) The radial, whirl, relative and actual fluid velocities at the exit.
(b) For no inlet whirl, what theoretical head is added to the water by the pump?
(c) What power is required to rotate the impeller?

Solution The following information is given:

$$D_2 = 40 \text{ cm}; r_2 = 20 \text{ cm} = 0.2 \text{ m}$$

$$b = 3 \text{ cm} = 0.03 \text{ m}$$

$$\beta_2 = 60°$$

$$N = 1500 \text{ rpm}$$

$$Q = 24 \text{ m}^3/\text{min} = 0.4 \text{ m}^3/\text{s}$$

(a) The velocity vector diagram at the exit is drawn in Fig. 12.5.

$$\omega = \frac{2\pi N}{60}$$

$$= \frac{2\pi \times 1500}{60}$$

$$\omega = 157.1 \text{ rad/s}$$

$$u_2 = r_2 \omega$$

$$= 0.2 \times 157.1$$

$$= 31.4 \text{ m/s}$$

From Eq. (12.8)

$$V_{2r} = \frac{Q}{2\pi r_2 b_2}$$

$$= \frac{0.4}{2\pi \times 0.2 \times 0.03}$$

$$V_{2r} = 10.6 \text{ m/s}$$

Hence, the radial velocity = 10.6 m/s

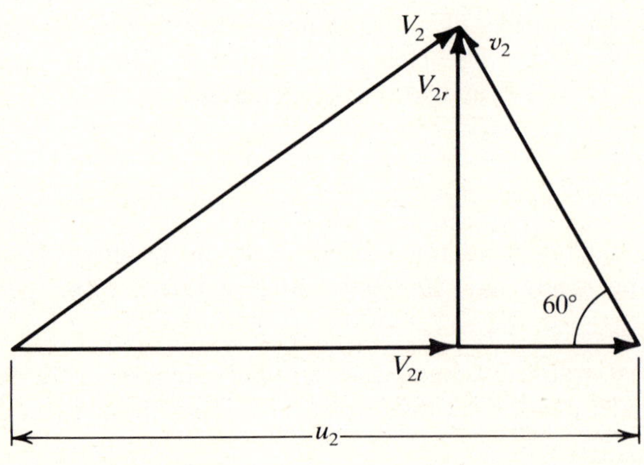

FIGURE 12.5 Example 12.2

From Fig. (12.5)

$$V_{2t} = u_2 - V_{2r} \cot 60°$$

$$= 31.4 - (10.6 \times 0.577)$$

$$V_{2t} = 25.3 \text{ m/s}$$

Hence, the whirl velocity = 25.3 m/s

$$v_2 = \frac{V_{2r}}{\sin 60°} = \frac{10.6}{0.866} = 12.25 \text{ m/s}$$

Hence, the relative velocity is 12.25 m/s

$$V_2 = \sqrt{(V_{2t})^2 + (V_{2r})^2}$$

$$= \sqrt{(25.3)^2 + (10.6)^2}$$

$$= 27.43 \text{ m/s}$$

Hence, the actual velocity = 27.4 m/s

(b) For no inlet whirl:

$$H = \frac{u_2 V_{2t}}{g}$$

$$= \frac{31.4 \times 25.3}{9.81}$$

$$= 81 \text{ m}$$

Hence, the theoretical head added to water = 81 m.

(c) Power required = γQH

$$= 9.81 \times 0.4 \times 81$$

$$= 317.8 \text{ kW}$$

Hence, the power required = 317.8 kW

Pump Characteristics

In a centrifugal pump, the relationships between (1) discharge and head, (2) discharge and efficiency, and (3) discharge and power input, are called *the pump characteristics*. The characteristic curves of a typical pump are shown in Fig. 12.6.

The efficiency of the pump largely depends on the design of the housing and blades. It also depends on the operating conditions. A pump is usually designed for maximum efficiency.

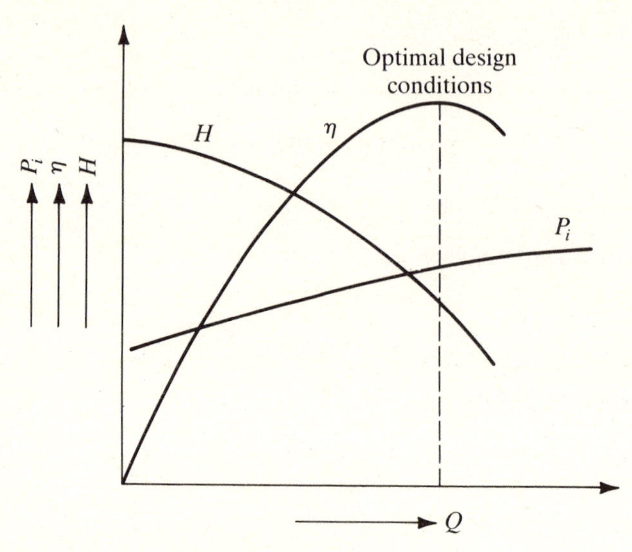

FIGURE 12.6 Pump characteristics.

12.4 HOMOLOGOUS PUMPS

Two pumps which are geometrically similar and which have similar velocity vector diagrams at inlet and exit to the impeller, are called *homologous pumps.* Homologous pumps will always have geometrically similar streamlines. For homologous pumps, the following dimensionless parameters remain constant:

$$\frac{Q}{ND^3} = \text{constant} \tag{12.11}$$

$$\frac{H}{N^2D^2} = \text{constant} \tag{12.12}$$

where

Q = discharge of the pump

D = diameter of the impeller

H = dynamic head on the pump

N = speed of the pump

These parameters may be used to relate the characteristics of different sized, but similar pumps, operating under dynamically similar conditions.

EXAMPLE 12.3

A 36 cm diameter impeller in a centrifugal pump has the following characteristics at 1,500 rpm:

H(m)	Q(l/s)	η(%)
25	35	46
22.5	55	65
20	69	75
17.5	82	80
15	93	78
12.5	101	66
10.0	108	54

Determine the impeller size and the speed of a homologous pump to deliver 100 l/s at 20 m head at best efficiency.

Solution For best efficiency, $Q_1 = 82$ l/s, $H_1 = 17.5$ m at $\eta = 80\%$. From Eqs. (12.11) and (12.12)

$$\frac{Q_1}{N_1 D_1^3} = \frac{Q_2}{N_2 D_2^3}$$

and

$$\frac{H_1}{N_1^2 D_1^2} = \frac{H_2}{N_2^2 D_2^2}$$

where

$$Q_1 = 82 \text{ l/s} \qquad Q_2 = 100 \text{ l/s}$$

$$H_1 = 17.5 \text{ m} \qquad H_2 = 20 \text{ m}$$

$$D_1 = 36 \text{ cm} \qquad D_2 = ?$$

$$N_1 = 1500 \text{ rpm} \qquad N_2 = ?$$

Hence:

$$\frac{82}{1500 \times (36)^3} = \frac{100}{N_2 D_2^3}$$

$$N_2 D_2^3 = 8.53 \times 10^7 \qquad (1)$$

Also

$$\frac{17.5}{(1500)^2 (36)^2} = \frac{20}{N_2^2 D_2^2}$$

$$N_2^2 D_2^2 = 3.33 \times 10^9 \qquad (2)$$

Simultaneous solution of Eqs. (1) and (2) yields:

$$N_2 = 1500 \text{ rpm}$$

$$D_2 = 38.45 \text{ cm}$$

Comments

1. For most pumps, standard electric motors are used. Standard speed of AC synchronous induction motors at 60 cycles and 220 to 440 volts is given by:

$$\text{synchronous speed} = \frac{3600}{\text{No. of pair of poles}}$$

For a 50-cycle operation:

$$\text{synchronous speed} = \frac{3000}{\text{No. of pair of poles}}$$

2. Since speed of motor is required to calculate specific speed, the availability or selection of motor type will determine the size and type of the pump.
3. The power delivered by the pump is called *the water power*. Since pumps never perform at 100% efficiency, the power needed by the pump, called *brake power*, will be larger than the water power, or

$$\text{brake power} = \frac{\text{water power}}{\text{efficiency}}$$

12.5 SPECIFIC SPEED

The *specific speed* N_s of a pump is a parameter widely used in selecting a pump and in preliminary design. The specific speed of a homologous series of pumps is constant and may be interpreted as the speed at which a pump delivers a unit discharge at unit head. Eliminating D in Eqs. (12.11) and (12.12), and rearranging, we obtain:

$$\frac{N\sqrt{Q}}{H^{3/4}} = \text{constant}$$

By definition of specific speed, when $H = 1$ and $Q = 1$, the constant is N_s. Therefore:

$$N_s = \frac{N\sqrt{Q}}{H^{3/4}} \tag{12.13}$$

where, at maximum efficiency, N = speed of pump impeller in rpm; Q = discharge in gpm; H = dynamic pump head in ft. The equivalent formula in SI units is

$$N_s = \frac{284N\sqrt{Q}}{H^{3/4}}$$

where, at maximum efficiency, N = speed in rpm; Q = discharge in m^3/s; H = dynamic head in m.

Typical values of specific speeds for various types of pump impellers, together with corresponding pump characteristic curves are shown in

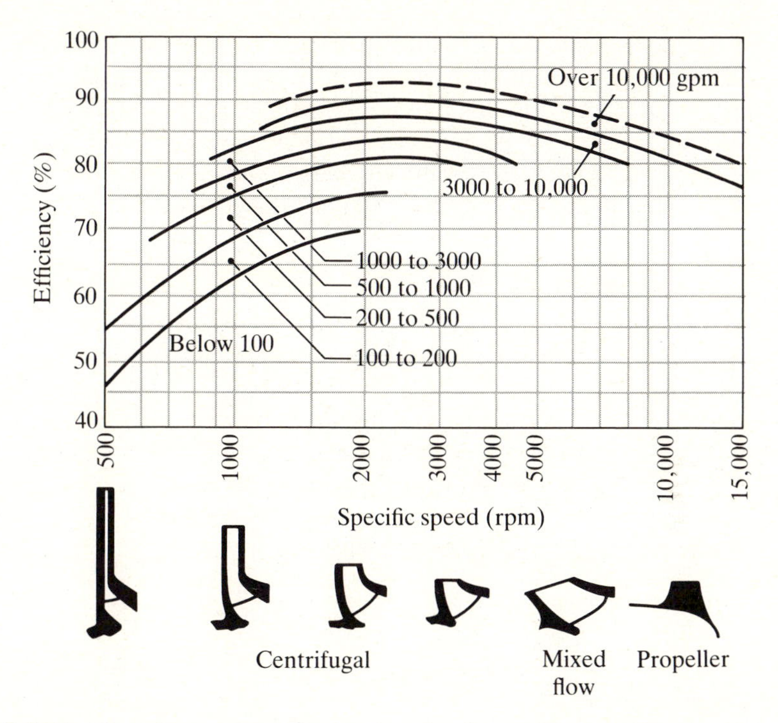

FIGURE 12.7 The relation of specific speed to the shape of the impeller, the discharge, and the efficiency of the pump.

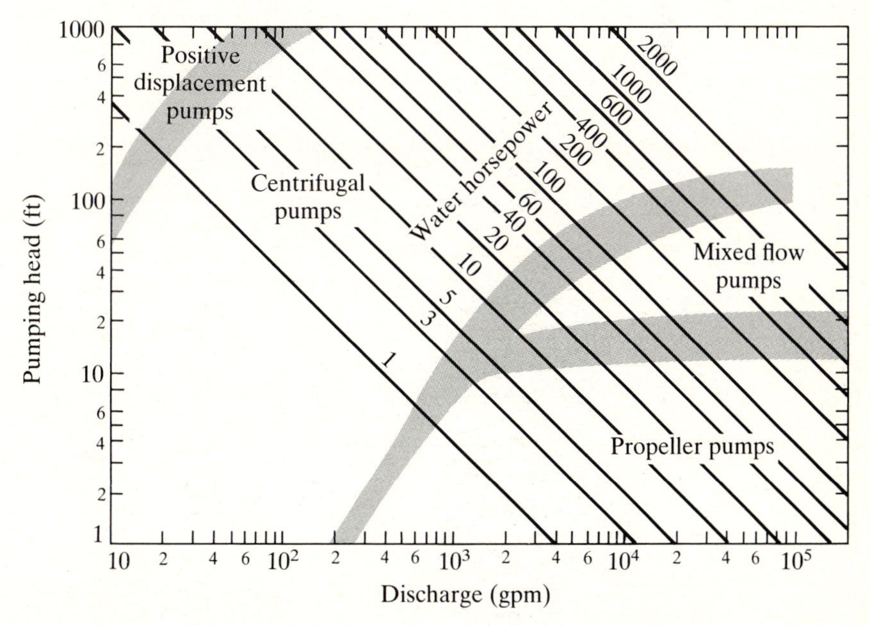

FIGURE 12.8 Power requirement for various discharges and heads of different types of pumps at optimum efficiency.

Fig. 12.7. Generally, pumps with low specific speeds (500 to 2,000 rpm) are made to deliver small discharges at high pressures (centrifugal pumps). Pumps characterized by high specific speeds (5,000 to 15,000 rpm) deliver large discharges at low pressures (mixed-flow, propeller-type pumps).

Power requirements for various discharges and heads of different types of pumps at optimum efficiency are given in Fig. 12.8.

EXAMPLE 12.4

Determine the pump size (Q, H, N_s) to delivei 500 gpm for the system shown in Fig. 12.9. The pressure required at the outlet is 30 psi. The pipe diameter is 8 in and the motor is 1,800 rpm at 60 cycles.

Solution

$$Q = 500 \text{ gpm}$$

$$= \frac{500}{449} = 1.11 \text{ ft}^3/\text{s}$$

$$D = 8 \text{ in} = 0.667 \text{ ft}$$

$$\text{Area of pipe} = \frac{\pi(0.667)^2}{4}$$

$$= 0.349 \text{ ft}^2$$

$$V = \frac{1.11}{0.349} = 3.18 \text{ ft/s}$$

Applying the energy equation between points 1 and 2, we obtain:

$$\frac{p_1}{\gamma} + z_1 + \frac{V_1^2}{2g} = \frac{p_2}{\gamma} + z_2 + \frac{V_2^2}{2g} + HL_{(1-2)} - h_s$$

Taking a line passing through point 1 as datum, and substituting for various terms, we obtain:

$$0 + 0 + 0 = \frac{30 \times 144}{62.4} + 50 + \frac{(3.18)^2}{64.4} + HL_{(1-2)} - h_s$$

$$h_s = 119.4 + HL_{(1-2)}$$

$$HL_{(1-2)} = 0.5\frac{V^2}{2g} + 2 \times 0.8\frac{V^2}{2g} + 10\frac{V^2}{2g} + \frac{f \times 150 \times V^2}{0.667 \times 2g}$$

$$HL_{(1-2)} = \frac{V^2}{2g}(12.1 + 224.9f)$$

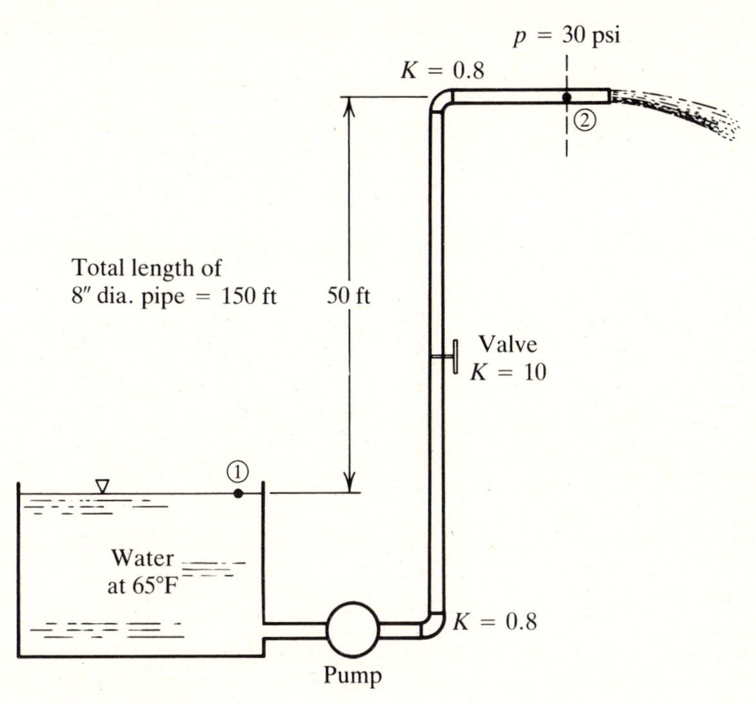

FIGURE 12.9 Example 12.4.

For water at 65°F:

$$\nu = 1.134 \times 10^{-5} \text{ ft}^2/\text{s}$$

$$\epsilon = 0.00085 \text{ ft}$$

$$\frac{\epsilon}{D} = 0.001275$$

$$\mathbf{R} = \frac{3.18 \times 0.667}{1.134 \times 10^{-5}} = 1.87 \times 10^5$$

$$f \text{ (from Moody's diagram)} = 0.022$$

Hence:

$$HL_{(1-2)} = \frac{(3.18)^2}{64.4}[12.1 + 224.9 \times 0.022]$$

$$= 2.68 \text{ ft}$$

Therefore:

$$h_s = 119.4 + 2.68 = 122.08 \text{ ft}$$

$$= 122.1 \text{ ft}$$

Therefore, the pump must be capable of delivering a dynamic head of 122.1 ft.

$$\text{Specific speed} = N_s = \frac{N\sqrt{Q}}{H^{3/4}}$$

$$= \frac{1800\sqrt{500}}{(122.1)^{3/4}}$$

$$= 1096$$

Hence the specific speed = 1096.

Consulting Fig. 12.7, it is found that a centrifugal pump of approximately 75% efficiency will meet the requirements. It is evident from the graph that the best efficiency of about 80% for the same discharge occurs when $N_s = 2{,}500$. Since Q and H cannot be changed, the value of N in the specific speed equation may be changed if maximum efficiency is to be attained. The new N is 4,100 rpm. This can be accomplished by either rewiring or buying a new motor.

12.6 CAVITATION

When the absolute pressure at some point in a flow system reaches the vapor pressure, the liquid starts to evaporate. In hydraulic machines, this phenomenon is called *cavitation*. Cavitation, if severe enough, will affect pump performance, cause noisy operation, enhance vibration and erode metal from the impeller.

Negative gauge pressures can develop on the suction side of the pump due to lowering of pressures caused by the lift head. Therefore, one of the important considerations in pump installation is the relative elevation between the pump and the water surface in the supply reservoir.

Consider two points 1 and 2 as shown in Fig. 12.10. Point 1 is located at the free surface of the lower reservoir and point 2 at the entrance side of the pump. Applying Bernoulli's equation to points 1 and 2

$$\frac{p_1}{\gamma} + z_1 + \frac{V_1^2}{2g} = \frac{p_2}{\gamma} + z_2 + \frac{V_2^2}{2g} + HL_{(1-2)}$$

Assuming that the pressure at point 1 is atmospheric (p_a) and taking the datum passing through point 1, we get:

$$\frac{p_a}{\gamma} + 0 + 0 = \frac{p_2}{\gamma} + H_s + \frac{V_2^2}{2g} + HL_{(1-2)}$$

$$\frac{p_2}{\gamma} + \frac{V_2^2}{2g} = \frac{p_a}{\gamma} - H_s - HL_{(1-2)}$$

The term H_s is called the *static suction lift*. Subtracting the vapor pressure head p_v/γ, from both sides:

$$\frac{p_2}{\gamma} + \frac{V_2^2}{2g} - \frac{p_v}{\gamma} = \frac{p_a}{\gamma} - H_s - HL_{(1-2)} - \frac{p_v}{\gamma} \qquad (12.14)$$

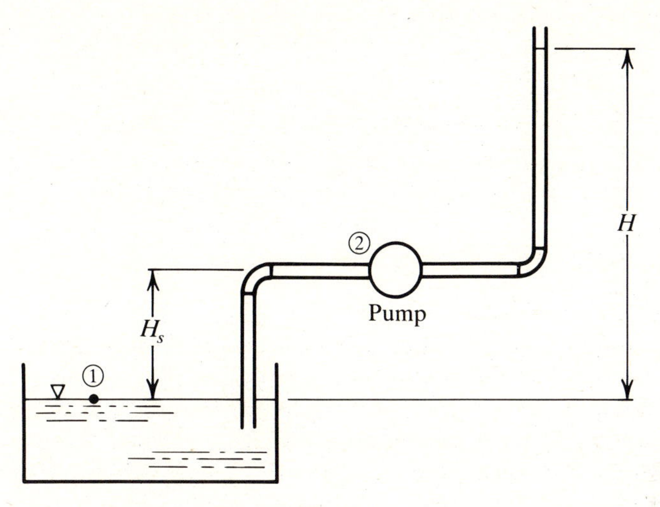

FIGURE 12.10 A pump installed in a line.

The term $(p_2/\gamma + V_2^2/2g - p_v/\gamma)$ is called the *net positive suction head* (NPSH).
Hence, from Eq. (12.14)

$$\text{NPSH} = \frac{p_a}{\gamma} - H_s - HL_{(1-2)} - \frac{p_v}{\gamma}$$

$$H_s = \frac{p_a}{\gamma} - \frac{p_v}{\gamma} - HPSH - HL_{(1-2)}$$

The value of NPSH corresponding to maximum allowable H_s is called the *minimum net positive suction head.*
Hence:

$$(H_s)_{\max} = \frac{p_a}{\gamma} - \frac{p_v}{\gamma} - (\text{NPSH})_{\min} - HL_{(1-2)} \tag{12.15}$$

The *cavitation number* σ for a pump is defined as the net positive suction head divided by the total head against which the pump operates, or

$$\sigma = \frac{\text{NPSH}}{H} \tag{12.16}$$

The cavitation number corresponding to the minimum net positive suction head is called the *critical cavitation number* σ_c, and

$$\sigma_c = \frac{(\text{NPSH})_{\min}}{H} \tag{12.17}$$

EXAMPLE 12.5

A pump with a critical cavitation number equal to 0.08 pumps a total head of 60 m. Water temp = 15°C, barometric pressure is 98 kPa, head loss in a suction pipe is 1 m. Determine the maximum value of the static suction lift.

Solution

$$H = 60 \text{ m}$$

$$\text{Temp} = 15°\text{C}$$

$$\text{Barometric pressure} = 98 \text{ kPa} = 98{,}000 \text{ N/m}^2$$

$$HL = 1 \text{ m}$$

From Table A-1,

$$\text{Vapor pressure} = 1707 \text{ N/m}^2 = p_v$$

$$\sigma \text{ critical} = \frac{(\text{NPSH})_{\text{min}}}{H}$$

Hence,

$$0.08 = \frac{(\text{NPSH})_{\text{min}}}{60}$$

$$(\text{NPSH})\text{min} = 4.8 \text{ m}$$

$$(H_s)_{\text{max}} = \frac{p_a}{\gamma} - \frac{p_v}{\gamma} - (\text{NPSH})_{\text{min}} - HL$$

$$= \frac{98000}{9810} - \frac{1707}{9810} - 4.8 - 1$$

$$= 4.02 \text{ m}$$

12.7 HYDRAULIC TURBINES

Hydraulic turbines extract energy from flowing water as it passes through the turbine runner. In order to produce power from flowing water, there must be a drop in the elevation so that the potential energy is converted into kinetic energy which, in turn, is used to drive the turbine. In most cases, the required drop in the elevation may be obtained by building a dam. A dam raises the water to a desirable level to create a head. The available head on the turbine is the difference in elevation between the water surface in the reservoir and the water surface in the river below the dam at a point where the turbine is located. The power input to the turbine is given by

$$P_1 = \gamma QH \tag{12.18}$$

where γ = the specific weight of water; Q = the rate of flow; H = the net effective head on the turbine. The net effective head on the turbine is determined by deducting losses from the gross head (gross head at any particular site is the distance from the reservoir elevation to the tailwater elevation) as the water flows to the turbine and as it leaves the turbine. These losses include losses due to friction in the penstock and all other minor head losses. The specific speed of a hydraulic turbine is defined as

$$N_s = \frac{N\sqrt{HP}}{(H)^{5/4}} \tag{12.19}$$

There are two basic classes of hydraulic turbines—*impulse turbines* and *reaction turbines.* They differ in that impulse turbines are driven by kinetic energy (jet action), while reaction turbines utilize both kinetic and pressure energy.

12.8 IMPULSE TURBINES

In the impulse turbine configuration, the available head is converted into a jet of high velocity. This high velocity jet then strikes a series of vanes, or cups, arranged on the periphery of a wheel. After impact, the momentum of the jet is changed which results in a force being applied on the wheel. The force produces a torque which then turns the wheel (Fig. 12.11).

A typical double-cupped bucket is shown in Fig. 12.12. Let a jet of velocity V strike the cup and be deflected through an angle θ. Let the velocity of the bucket be u. Then, from the momentum equation, the force F which is applied on the bucket in the x-direction is given by

$$F = \rho Q[(V - u) - (V - u)\cos\theta]$$

$$F = \rho Q(V - u)[1 - \cos\theta]$$

The power P exerted on the cup is then given by

$$P = Fu = \rho Qu(V - u)[1 - \cos\theta] \tag{12.20}$$

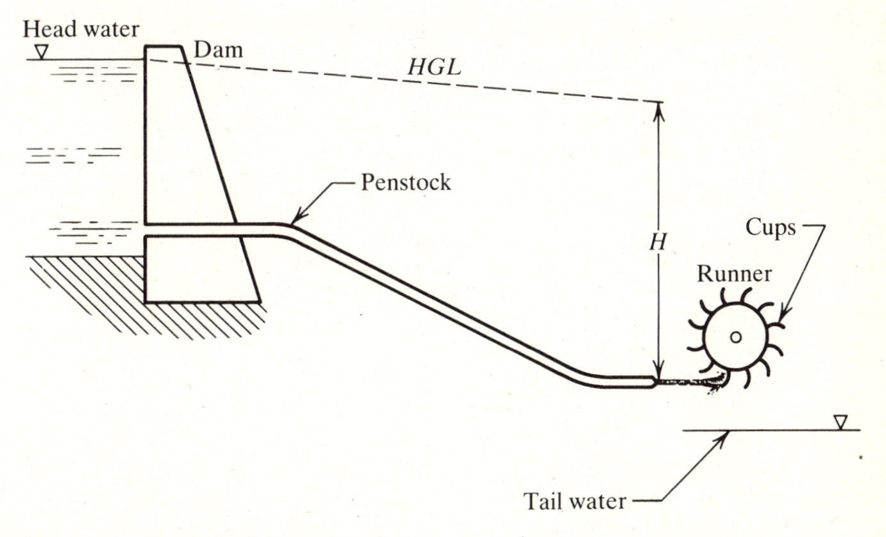

FIGURE 12.11 Impulse turbine configuration.

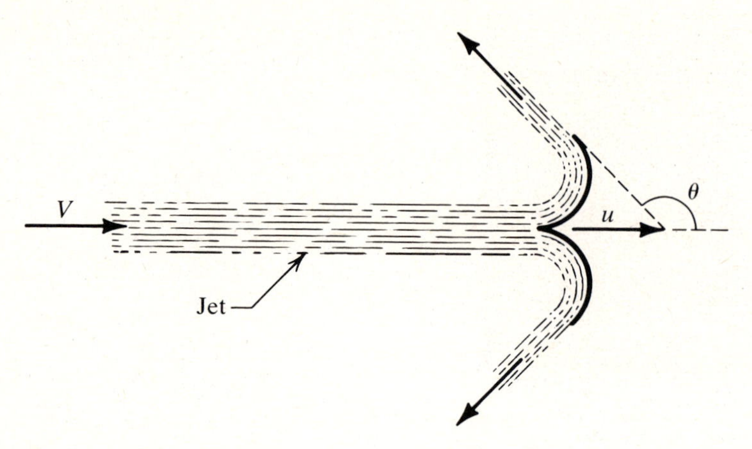

FIGURE 12.12 A double-cupped bucket.

To obtain maximum power, θ must be 180° ($\cos\theta = -1$) and $(V - u)$ must be maximum. To determine the condition when $(V - u)$ is maximum, differentiate Eq. (12.20) with respect to u and equate to zero. Hence (assuming ρ and Q to be constants, and $\cos\theta = -1$):

$$\frac{dP}{du} = 2(\rho QV - 2\rho Qu) = 0$$

$$(V - 2u) = 0$$

$$u = \frac{V}{2} \tag{12.21}$$

Therefore, maximum power is obtained when $\theta = 180°$ and $u = V/2$. Substituting these conditions in Eq. (12.20), we obtain

$$P = \frac{\rho QV^2}{2}$$
$$P = \frac{\gamma QV^2}{2g} \tag{12.22}$$

Thus, for idealized conditions, the entire kinetic energy of the jet will be converted into power and the exiting jet speed will be zero. However, practically, the exiting jet should maintain enough speed to get out of the next incoming bucket. To achieve this, the angle θ is usually kept somewhat below 180°.

The most widely used impulse turbine is the Pelton wheel (Fig. 12.13). In a Pelton wheel, the jet strikes a bucket tangentially and is split in half and turned in a horizontal plane through an angle of 173° to 176°. Also, in a Pelton wheel, there are losses due to the 'splitter ridges' and to friction

FIGURE 12.13 Pelton wheel runner for the 150,000 H.P., 2,500 ft head Kemano plant (source: Allis Chalmers Mfg. Co.).

between jet and bucket surface. This results in u being somewhat less than $V/2$. The ratio of u to V is known as the *speed factor*, ϕ. Thus:

$$\phi = \frac{u}{V} = \frac{u}{\sqrt{2gH}} \tag{12.23}$$

where H is the available head.

A Pelton wheel with a single jet will operate from 3 to 10 rpm based on the runner size necessary to obtain 1 HP under a 1 ft head. For horizontal shaft installations a maximum of two jets per runner will increase the speed approximately 40% as a result of the smaller physical size required for the same horsepower and head relationship. If the turbine shaft is arranged vertically, a maximum of six jets may be used and since the runner would be even smaller, the speed would be increased almost $2\frac{1}{2}$ times.

Two other types of impulse turbines are the Turgo or diagonal-type impulse runner and the Mitchell or cross-flow-type impulse runner.

The impulse runners must operate in air to be efficient. For this reason they must be located above the maximum tailwater elevation. This reduces the effective head on the nozzle by a distance from the tailwater elevation to the centerline of the nozzle. The distance above tailwater for low head applications may be a very appreciable loss which can make impulse turbines undesirable for low heads. Pelton wheels are usually employed for high heads, from about 200 m to as high as pressure pipelines can be built.

12.9 REACTION TURBINES

There are two basic types of reaction turbines—*Francis* and *propeller* (see Figs 12.14 and 12.15). In a Francis turbine, the water enters the runner in a radial direction—i.e., direction normal to the axis of rotation, while in a propeller turbine, the water enters the runner along the axis of rotation.

FIGURE 12.14 Francis runner for 62,000 H.P., 190 ft head Bull Shoals project (source: Allis Chalmers Mfg. Co.).

FIGURE 12.15 A Kaplan-type propeller runner for 44,000 H.P., 48 ft head Kentucky plant (source: Allis Chalmers Mfg. Co.).

Francis Turbine

The Francis turbine is, in fact, a counterpart of the centrifugal pump. Fig. 12.16 shows a schematic cross-section through the runner of a Francis turbine. Application of the momentum equation yields exactly the same equation as obtained for the pump, except for interchange of subscripts. Thus, for no shock conditions at exit (from Eqs. 12.5 and 12.7)

$$T = \rho Q r_1 V_1 \tag{12.24}$$

$$H = \frac{u_1 v_1 \cos \alpha_1}{g} \tag{12.25}$$

where T and H are, respectively, the torque and the head imparted to the turbine runner by the flowing water.

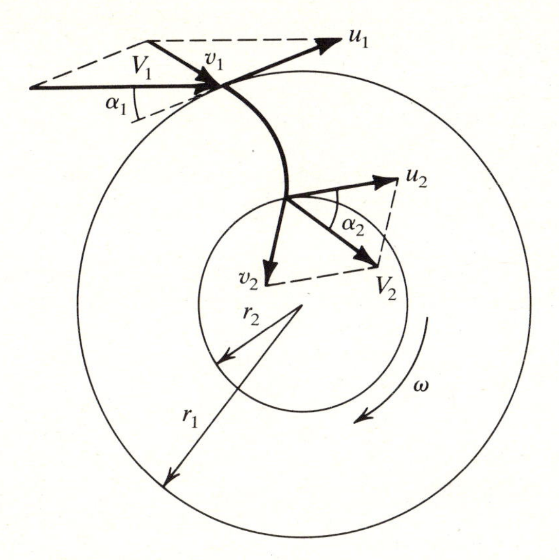

FIGURE 12.16 Velocity vector diagrams at the entrance and exit to a Francis turbine runner.

Equations (12.11) and (12.12), applicable to the homologous impellers of centrifugal pumps, can also be applied to homologous Francis turbine runners.

The normal specific speed range for Francis runners is from 10 to 100. High head Francis runners have large inlet diameter with low entrance height and a small discharge diameter, the specific speed being on the order of 20 rpm. Such runner designs have been used for heads in excess of 1,500 ft. However, for such high heads, the runner is generally located well below tailwater elevations in order to minimize cavitation damage.

For medium heads and a specific speed of approximately 50 rpm, the runner inlet and the discharge diameters are almost the same. For low head applications, at a specific speed of about 100 rpm, the inlet diameter of the runner is kept substantially smaller than the discharge diameter and the entrance height is made substantially large to provide a greater entrance area.

Propeller Turbine

In a propeller-type turbine, the water enters the runner along the axis of rotation, and the flow has no radial component. A schematic diagram of a propeller turbine setting is shown in Fig. 12.17. Also, in Fig. 12.18 is shown a single element of thickness dr of a single blade, at a radial distance r. The axial flow velocity as the water enters the region of the propeller at section 1-1 of Fig. 12.17 is given by

$$V_a = \frac{Q}{\pi(r_0^2 - r_1^2)} \tag{12.26}$$

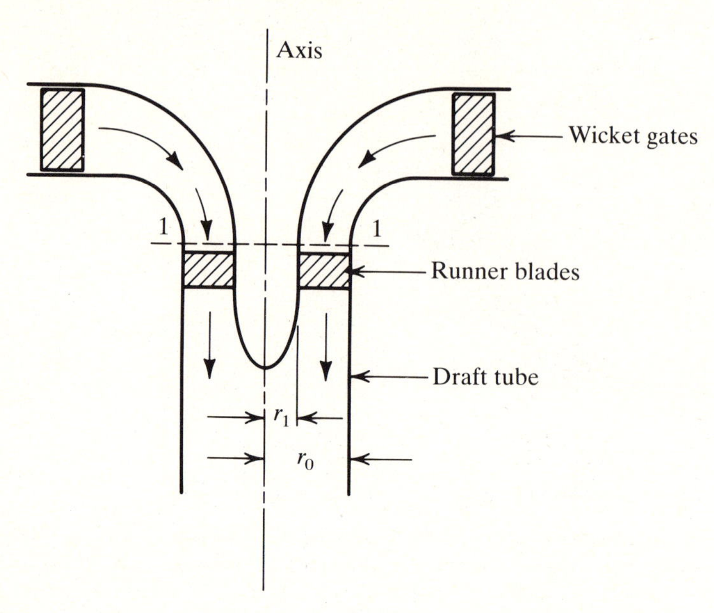

FIGURE 12.17 Schematic diagram of a propeller turbine setting.

where V_a is the axial component of the absolute velocity and Q is the total discharge. Hydraulic force and torque are imparted to the runner through the mechanism of lift and drag forces on the propeller blade. Each element of the blade may be considered as an airfoil section, upon which flow impinges at some angle of attack θ as shown in Fig. 12.19. The theory of an airfoil was developed in Chapter 7. The lift and drag forces are given by (Eqs. 7.44 and 7.45).

$$F_L = \frac{1}{2} C_L \rho u^2 S \tag{12.27}$$

$$F_D = \frac{1}{2} C_D \rho u^2 S \tag{12.28}$$

where C_L and C_D are, respectively, the lift coefficient and the drag coefficient. In the case of a turbine runner, u is the effective relative flow velocity at an angle β, such that

$$\cot \beta = \frac{1}{2} (\cot \beta_1 + \cot \beta_2) \tag{12.29}$$

Then

$$u = \frac{V_a}{\sin \beta} \tag{12.30}$$

The total force F on the span dr can then be resolved into axial and tangential forces as shown in Fig. 12.20. The tangential component F_T produces a

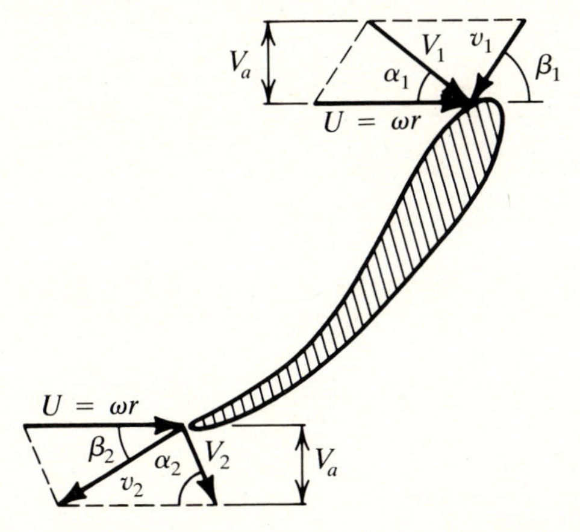

FIGURE 12.18 A single blade of a propeller turbine runner.

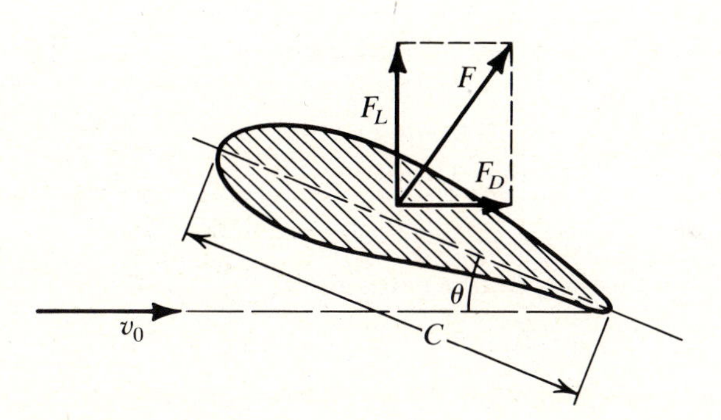

FIGURE 12.19 An element of a propeller turbine blade.

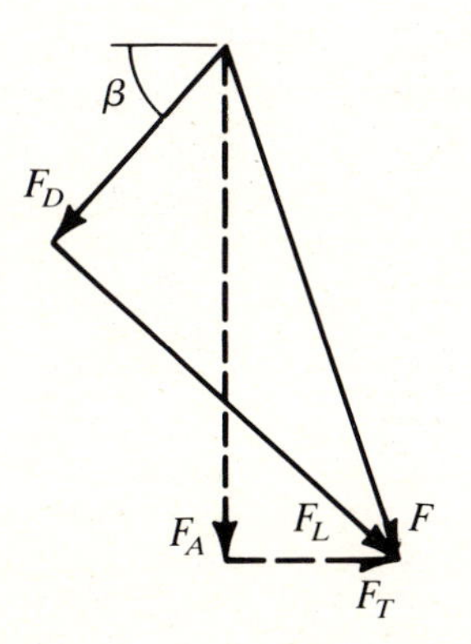

FIGURE 12.20 Vector diagrams for the element of the blade.

torque on the shaft, and the axial component causes a thrust on the bearing.

From Eqs. 12.27 and 12.28 and using the vector diagram of Fig. 12.20, expressions for thrust and torque for a single element of thickness dr and radial distance r of a single blade, may be developed:

$$\delta F_A = \frac{\rho u^2}{2} c(C_L \cos \beta + C_D \sin \beta)dr \tag{12.31}$$

$$\delta T = \delta F_T r = \rho \frac{u^2}{2} c(C_L \sin \beta + C_D \cos \beta)rdr \tag{12.32}$$

where c is the chord length. Total thrust and torque on a single blade can then be obtained by integration.

$$F_A = \int_{r_1}^{r_0} \delta F_A$$

$$T = \int_{r_1}^{r_0} \delta T$$

The blade angles β_1 and β_2 (Fig. 12.18) are designed to achieve certain directions for the relative velocities v_1 and v_2. Since v_1 and v_2 depend on u, which, in turn, depends on the radial distance r, each element of the blade must have a different shape corresponding to the requirements of β_1 and β_2. This means that in Eqs. (12.31) and (12.32), u, c, C_L, C_D, and β are functions of r.

The commonly used propeller turbines include: turbines with fixed or adjustable blades, fixed or adjustable gates and with horizontal, inclined or vertical axes. Arrangements typically used include bulb units, tube turbines and various vertical designs (Fig. 12.21).

The Kaplan turbine, named after Viktor Kaplan of Czechoslovakia, is a special kind of propeller turbine with adjustable runner blades which are automatically coordinated with operation of adjustable wicket gates to obtain the most efficient operation under varying head and load conditions. A Kaplan turbine can be used in any design arrangement—bulb or tubular, vertical or horizontal. With synchronized operation of blades and gates, it is especially applicable in low head plants where operating conditions often involve factors of varying load, flow, and head.

Generally, propeller-type hydraulic turbines can be applied for net heads up to 60 to 70 m with power output up to 200 MW. Within this range, Fig. 12.22 shows various types suited for various heads and output. For each of these arrangements, options exist for various combinations of fixed and adjustable blades with fixed and adjustable gates. See Fig. 12.23 for comparative performance of these options.

The type of turbine suited for a certain specific speed at maximum efficiency can be selected using Fig. 12.24. It is seen that an impulse turbine is most suited for specific speeds up to 10. Francis turbines are most effective at specific speeds up to 100 and propeller turbines for specific speeds greater than 100.

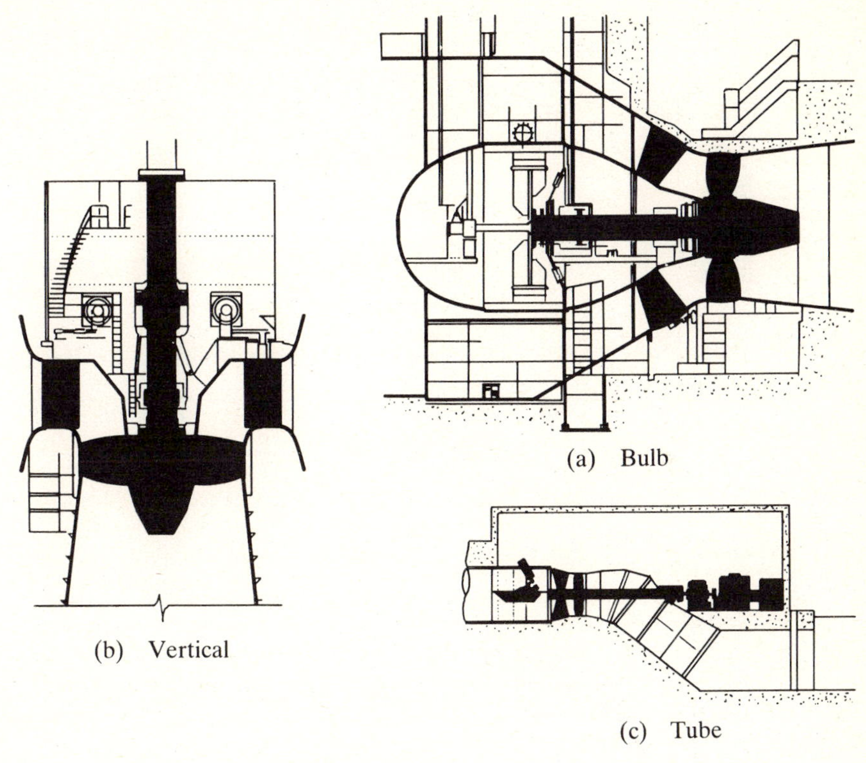

FIGURE 12.21 The three basic types of axial flow turbines (source: Allis Chalmers Mfg. Co.).

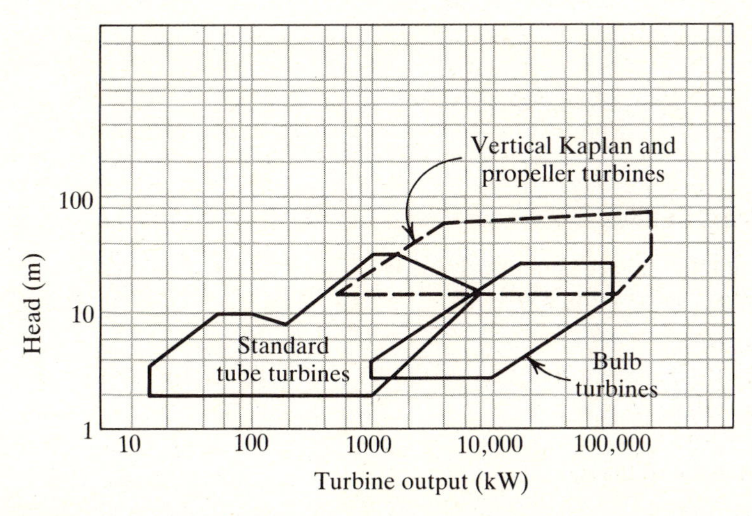

FIGURE 12.22 Common application ranges for axial flow hydraulic turbines (source: Allis Chalmers Mfg. Co.).

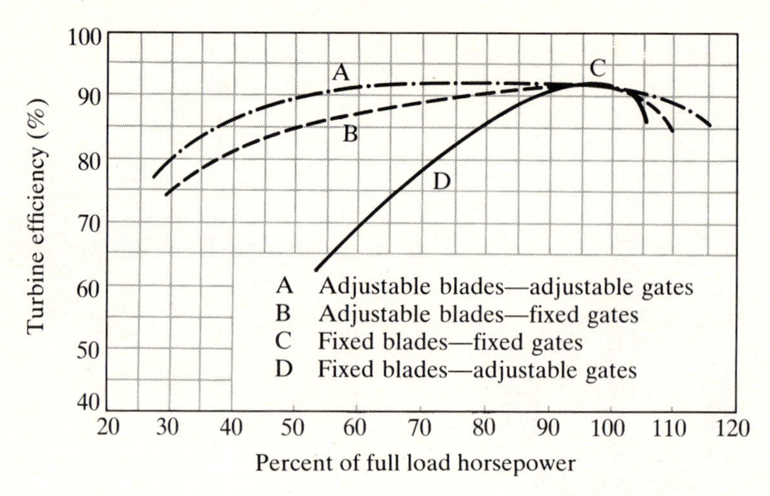

FIGURE 12.23 Comparison of propeller runner performance (source: Allis Chalmers Mfg. Co.).

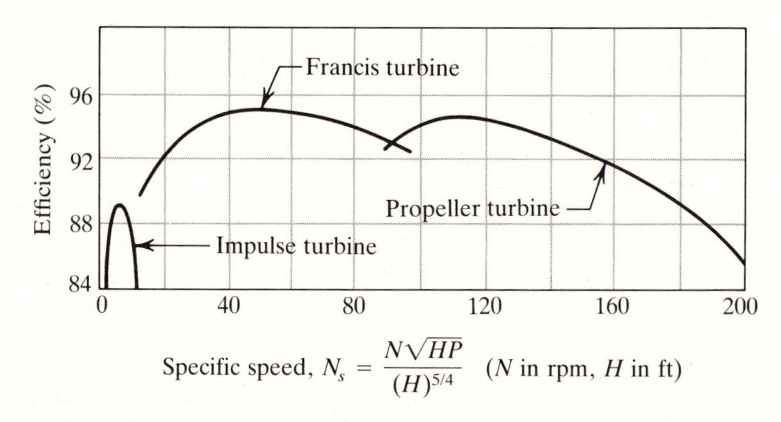

FIGURE 12.24 Efficiency vs specific speed for turbines (source: Morris, Wiggert, 1972).

PROBLEMS

p.12.1 An 80% efficient pump delivers 0.05 m^3/s at a dynamic head of 205 m. The power required in kW is

a. 100.6.
b. 125.7.
c. 639.6.
d. 799.5.
e. none of these.

p.12.2 An 80% efficient pump delivers 0.05 m^3/s at a dynamic head of 205 m. The electric motor driving the pump is operating at 3,600 rpm. The torque acting on the drive shaft in N.m is

a. 266.8.
b. 333.4.
c. 19846.2.
d. 33.3.
e. none of these.

p.12.3 The shaft of an electric motor is transmitting 200 kW at 1,800 rpm. The torque in N.m is

a. 111.1.
b. 1061.
c. 200.
d. 4.
e. none of these.

p.12.4 A centrifugal pump with a 10 ft diameter impeller discharges 100 cfs with a tangential velocity of 10 ft/s. The torque at exit in lb · ft is

a. 1,939.
b. 312,000.
c. 9,689.
d. 49,050.
e. none of these.

p.12.5 A centrifugal pump with a 10 ft diameter impeller has a blade angle of 60° at exit. At 1,800 rpm the tangential velocity of water at exit is 5 m/s. The theoretical head produced by the pump in m is

a. 240.2.
b. 14412.
c. 57.6.
d. 565.
e. none of these.

p.12.6 For a backward-curved blade, the head increases when the discharge

a. increases.
b. decreases.
c. remains constant.
d. is zero.
e. none of these.

p.12.7 For a radial blade, the head increases when the discharge

a. increases.
b. decreases.
c. remains constant.
d. is zero.
e. none of these.

p.12.8 A pump delivers 2 ft^3/s at 1800 rpm against a dynamic head of 150 ft. The specific speed is

a. 59.4.
b. 594.
c. 1,469.
d. 1,800.
e. none of these.

p.12.9 In terms of the symbols used in the text, the cavitation number is given by

a. $p_a/\gamma + p_v/\gamma$.
b. $p_a/\gamma - p_v/\gamma - H$.
c. H/NPSH.
d. NPSH/H.
e. none of these.

p.12.10 In a pump

a. $H \propto Q$.
b. $H \propto 1/Q$.
c. $H \propto Q^2$.
d. $H \propto Q^3$.
e. none of these.

p.12.11 When two identical pumps are connected in series

a. the head remains the same as for a single pump but the discharge nearly doubles.
b. the discharge remains the same as for a single pump but the head nearly doubles.
c. both the head and the discharge nearly double.
d. both the head and the discharge remain the same as for a single pump.
e. none of these.

p.12.12 When two identical pumps are connected in parallel

a. the head remains the same as for a single pump but the discharge nearly doubles.
b. the discharge remains the same as for a single pump but the head nearly doubles.
c. both the head and the discharge nearly double.
d. both the head and the discharge remain the same as for a single pump.
e. none of these.

p.12.13 The two basic classes of hydraulic turbines are

a. centrifugal and impulse.
b. rotary and reaction.
c. impulse and reaction.
d. rotary and impulse.
e. none of these.

p.12.14 The class of turbines which uses kinetic energy of a jet as its driving power is called

a. a reaction turbine.
b. a Francis turbine.
c. an impulse turbine.
d. a propeller turbine.
e. none of these.

p.12.15 In an impulse turbine, theoretical maximum power is attained when

a. $\theta = 90°$ and $u = V$.
b. $\theta = 0°$ and $u = V$.
c. $\theta = 0°$ and $u = V/2$.
d. $\theta = 90°$ and $u = V/2$.
e. none of these.

p.12.16 For an impulse turbine, the speed factor is defined as

a. $2gH$.
b. $u/2gH$.
c. $H/2gH$.
d. $H/2gu$.
e. none of these.

p.12.17 The impulse turbines are usually located

a. below the maximum tailwater elevation.
b. above the maximum tailwater elevation.
c. above the maximum elevation in the reservoir.
d. inside the draft tube.
e. none of these.

p.12.18 Generally, an impulse turbine is most suited for specific speeds up to

a. 10.
b. 50.
c. 100.
d. 500.
e. none of these.

p.12.19 Generally, a Francis turbine is most suited for specific speeds up to

a. 10.
b. 50.
c. 100.
d. 500.
e. none of these.

p.12.20 A centrifugal pump has $b_2 = 10$ cm, $r_1 = 30$ cm, $r_2 = 70$ cm, $\beta_2 = 45°$. It delivers a discharge of 2 m^3/s against a head of 12 m at 1,800 rpm. Calculate

a. the radial, whirl, relative and actual fluid velocities at the exit.

b. for no inlet whirl, what theoretical head is added to the water by the pump.
c. what power is required to rotate the impeller.

p.12.21 A centrifugal pump has $b_2 = 10$ cm, $r_1 = 30$ cm, $r_2 = 60$ cm, $\beta_2 = 45°$ and $\beta_1 = 60°$. It delivers 2.1 m^3/s against a head of 15 m. For no inlet whirl, what is the rotational speed? How efficient is the pump?

p.12.22 A centrifugal pump has $b_2 = 15$ cm, $r_2 = 60$ cm, $r_1 = 30$ cm and $\beta_2 = 45°$. If the absolute velocity at the outlet is 45 m/s, determine the torque on the impeller by the outflowing water.

p.12.23 A 36 cm diameter impeller in a centrifugal pump has the following characteristics at 1,500 rpm:

H(m)	25	22.5	20	17.5	15	12.5	10
Q(litre/s)	35	55	69	82	93	101	108
η(%)	46	65	75	80	78	66	54

Determine the impeller size and the speed of a homologous pump to deliver 200 l/s at 20 m head at best efficiency.

p.12.24 A centrifugal pump delivers 40,000 gpm against a head of 65 ft at its best efficiency of 80%. The pump has a 15 in impeller diameter and runs at 3000 rpm. Determine the impeller size and the speed of a homologous pump to deliver 100,000 gpm at 50 ft head at best efficiency. Also, determine the specific speed of the series of homologous pumps.

APPENDIX A

Physical Properties of Fluids

Table A-1 PHYSICAL PROPERTIES OF WATER IN SI UNITS

Temp °C	Specific weight γ N/m^3	Density ρ kg/m^3	Viscosity μ $N \cdot s/m^2$ $10^3\ \mu =$	Kinematic viscosity ν m^2/s $10^6\ \nu =$	Surface tension σ N/m $100\ \sigma =$	Vapor-pressure head p_v/γ m
0	9805	999.9	1.792	1.792	7.62	0.06
5	9806	1000.0	1.519	1.519	7.54	0.09
10	9803	999.7	1.308	1.308	7.48	0.12
15	9798	999.1	1.140	1.141	7.41	0.17
20	9789	998.2	1.005	1.007	7.36	0.25
25	9779	997.1	0.894	0.897	7.26	0.33
30	9767	995.7	0.801	0.804	7.18	0.44
35	9752	994.1	0.723	0.727	7.10	0.58
40	9737	992.2	0.656	0.661	7.01	0.76
45	9720	990.2	0.599	0.605	6.92	0.98
50	9697	988.1	0.549	0.556	6.82	1.26
55	9679	985.7	0.506	0.513	6.74	1.61
60	9658	983.2	0.469	0.477	6.68	2.03
65	9635	980.6	0.436	0.444	6.58	2.56
70	9600	977.8	0.406	0.415	6.50	3.20
75	9589	974.9	0.380	0.390	6.40	3.96
80	9557	971.8	0.357	0.367	6.30	4.86
85	9529	968.6	0.336	0.347	6.20	5.93
90	9499	965.3	0.317	0.328	6.12	7.18
95	9469	961.9	0.299	0.311	6.02	8.62
100	9438	958.4	0.284	0.296	4.94	10.33

Table A-2 PHYSICAL PROPERTIES OF WATER IN ENGLISH UNITS

Temp °F	Specific weight γ lb/ft^3	Density ρ slugs/ft^3	Viscosity μ lb·s/ft^2 $10^5\ \mu =$	Kinematic viscosity ν ft^2/s $10^5\ \nu =$	Surface tension σ lb/ft $100\ \sigma =$	Vapor-pressure head p_v/γ ft
32	62.42	1.940	3.746	1.931	0.518	0.20
40	62.43	1.940	3.229	1.664	0.514	0.28
50	62.41	1.940	2.735	1.410	0.509	0.41
60	62.37	1.938	2.359	1.217	0.504	0.59
70	62.30	1.936	2.050	1.059	0.500	0.84
80	62.22	1.934	1.799	0.930	0.492	1.17
90	62.11	1.931	1.595	0.826	0.486	1.61
100	62.00	1.927	1.424	0.739	0.480	2.19
110	61.86	1.923	1.284	0.667	0.473	2.95
120	61.71	1.918	1.168	0.609	0.465	3.91
130	61.55	1.913	1.069	0.558	0.460	5.13
140	61.38	1.908	0.981	0.514	0.454	6.67
150	61.20	1.902	0.905	0.476	0.447	8.58
160	61.00	1.896	0.838	0.442	0.441	10.95
170	60.80	1.890	0.780	0.413	0.433	13.83
180	60.58	1.883	0.726	0.385	0.426	17.33
190	60.36	1.876	0.678	0.362	0.419	21.55
200	60.12	1.868	0.627	0.341	0.412	26.59
212	59.83	1.860	0.593	0.319	0.404	33.90

Table A-3 PROPERTIES OF GASES AT LOW PRESSURES AND 80°F (26.67°C)

Gas	Chemical formula	Molecular weight	Gas constant R		Specific heat Btu/lb$_m$·°R or kcal/kg·K		Specific heat ratio k
			m·N/kg·K	ft·lb/slugs·°R	c_p	c_v	
Air	—	29.0	287	1716	0.240	0.171	1.40
Carbon monoxide	CO	28.0	297	1777	0.249	0.178	1.40
Helium	He	4.0	2077	12429	1.25	0.753	1.66
Hydrogen	H_2	2.0	4121	24665	3.43	2.44	1.40
Nitrogen	N_2	28.0	297	1777	0.248	0.177	1.40
Oxygen	O_2	32.0	260	1555	0.219	0.157	1.40
Water vapor	H_2O	18.0	462	2763	0.445	0.335	1.33

Table A-4 PHYSICAL PROPERTIES OF COMMON LIQUIDS AT 68°F (20°C) AND 14.7 psia (101,345 N/m²)

Liquid	Specific Gravity S	Dynamic viscosity μ		Surface tension σ		Vapor pressure* P_v		Bulk modulus β	
		$\frac{lb \cdot s}{ft^2}$ $\mu \times 10^5$	$\frac{N \cdot s}{m^2}$ $\mu \times 10^4$	$\frac{lb}{ft}$ $\sigma \times 10^3$	$\frac{N}{m}$ $\sigma \times 10^2$	$\frac{lb}{in^2}$	$\frac{N}{m^2}$ $p_v \times 10^{-3}$	$\frac{lb}{in^2}$ $\beta \times 10^{-5}$	$\frac{N}{m^2}$ $\beta \times 10^{-9}$
Benzene	0.90	1.4	6.7	2.0	2.89	1.50	10.3	1.5	1.0
Carbon tetrachloride	1.59	2.04	9.77	1.8	2.67	1.74	12.0	1.6	1.1
Castor oil	0.96	2060.00	9860.0	2.7	3.9			2.1	1.4
Ethyl alcohol	0.79	2.51	12.0	1.5	2.23	0.85	5.86	1.75	1.21
Gasoline	0.68	0.62	2.97						
Glycerine	1.26	1800.00	8620.0	4.3	6.3	2×10^{-6}	1.4×10^{-5}	6.3	4.3
Kerosene	0.81	4.0	19.1	1.7	2.5				
Linseed oil	0.94	92.0	440.0	2.3	3.4				
Mercury	13.55	3.3	15.8	32.0	51.0	2.33×10^{-5}	1.6×10^{-4}	38.0	26.0
Olive oil	0.91	176.0	842.0	1.8	2.6			2.3	1.6
SAE 10 oil	0.92	170.0	814.0	2.5	3.6				
SAE 30 oil	0.92	920.0	4400.0	2.4	3.5				
Turpentine	0.86	2.1	10.1	5.0	7.3	0.0077	0.053		
Water	1.00	2.1	10.1	5.0	7.36	0.34	2.3	3.2	2.2

Note: * in contact with air

Table A-5 BULK MODULUS OF ELASTICITY OF WATER, psi

Pressure		Temperature									
		32°F (0°C)		68°F (20°C)		120°F (48.9°C)		200°F (93.3°C)		300°F (148.9°C)	
psi	Pa $\times 10^{-5}$	psi $\beta \times 10^{-5}$	Pa $\beta \times 10^{-9}$	psi $\beta \times 10^{-5}$	Pa $\beta \times 10^{-9}$	psi $\beta \times 10^{-5}$	Pa $\beta \times 10^{-9}$	psi $\beta \times 10^{-5}$	Pa $\beta \times 10^{-9}$	psi $\beta \times 10^{-5}$	Pa $\beta \times 10^{-9}$
15	1.034	2.92	2.01	3.20	2.21	3.32	2.29	3.08	2.12		
1500	103.4	3.00	2.07	3.30	2.28	3.42	2.36	3.19	2.20	2.48	1.71
4500	310.2	3.17	2.19	3.48	2.40	3.62	2.50	3.38	2.33	2.71	1.87
15000	1034.0	3.80	2.62	4.10	2.83	4.26	2.94	4.05	2.79	3.50	2.41

APPENDIX B

```
10 PRINT"lock gate design program"
20 PRINT"title";
30 INPUT T$
40 PRINT"system of units (si/fps) ";
50 INPUT U$
60 IF U$ = "si" THEN GOSUB 1000
70 IF U$ = "fps" THEN GOSUB 2000
80 REM *****
90 REM ***** input data *****
100 REM *****
110 PRINT"height of the gate ";L$;
120 INPUT H
130 PRINT"fluid depth on the higher side ";L$;
140 INPUT H1
150 PRINT"fluid depth on the lower side ";L$;
160 INPUT H2
170 PRINT"width of the waterway ";L$;
180 INPUT B
190 PRINT"sp. wt. of the fluid ";G$;
200 INPUT G1
210 PRINT"minimum gate angle (degrees) ";
220 INPUT T1
230 PRINT"maximum gate angle (degrees) ";
240 INPUT T2
250 PRINT"angle to be incremented by (degrees) ";
260 INPUT T3
270 PRINT"allowable maximum thrust on each bearing ";F$;
280 INPUT TO
290 REM *****
300 REM ***** end of data. start of calculations. *****
310 REM *****
320 PRINT
330 PRINT T$
340 PRINT"abbreviations used in the table of printout"
350 PRINT" ttb = thrust at the top bearing"
360 PRINT" tbb = thrust at the bottom bearing"
370 PRINT" safety factor = maximum thrust/allowable thrust"
380 PRINT TAB(1);"angle";TAB(13);"width";TAB(25);"ttb";TAB(38);"tbb";
390 PRINT TAB(51);"safety"
400 PRINT TAB(1);"(deg)";TAB(13);L$;TAB(25);F$;TAB(38);F$;
410 PRINT TAB(51);"factor"
420 REM *****
430 FOR T4 = T1 TO T2 STEP T3
440 T = 3.1416*T4/180
450 W = B/2/COS(T)
460 P1 = G1*H1*H1*W*.5
470 P2 = G1*H2*H2*W*.5
480 R1 = (P1*H1 + P2*H2)/(6*SIN(T)*H)
490 R2 = ((P1-P2)/2 - R1*SIN(T))/SIN(T)
500 IF R1>R2 THEN GOTO 530
510 R3 = R2
520 GOTO 540
530 R3 = R1
540 S = TO/R3
550 PRINT TAB(2);T4;TAB(10);W;TAB(23);R1;TAB(36);R2;TAB(49);S
560 NEXT T4
570 PRINT"*****"
580 PRINT"do you wish to try another bearing (y/n) ";
590 INPUT P$
600 IF P$ = "y" THEN GOTO 270
610 PRINT"*****"
620 PRINT"do you wish to solve a new problem (y/n) ";
630 INPUT N$
640 IF N$ = "y" THEN GOTO 20
650 STOP
1000 L$ = "(m)"
1010 G$ = "(n/cu.m)"
1020 F$ = "(N)"
1030 RETURN
2000 L$ = "(ft)"
```

```
2010 G$ = "(lbs/cu.ft)"
2020 F$ = "(lbs)"
2030 RETURN
2040 END
```

FIGURE B.1 Lock gate design program.

```
10 PRINT"radial sluice gate design program "
20 PRINT"title ";
30 INPUT T$
40 PRINT"system of units (si/fps) ";
50 INPUT U$
60 IF U$ = "si" THEN GOSUB 1000
70 IF U$ = "fps" THEN GOSUB 2000
80 REM *****
90 REM ***** input data *****
100 REM *****
110 PRINT"maximum depth of water to be impounded";L$;
120 INPUT H
130 PRINT"width of the gate ";L$;
140 INPUT W
150 PRINT"sp. wt. of the fluid ";G$;
160 INPUT G1
170 PRINT"minimum angle theta (degrees) ";
180 INPUT T1
190 PRINT"maximum angle theta (degrees) ";
200 INPUT T2
210 PRINT"angle to be incremented by (degrees) ";
220 INPUT T3
230 PRINT"allowable maximum thrust on each bearing ";F$;
240 INPUT TO
250 REM *****
260 REM ***** end of data. start of calculations. *****
270 REM *****
280 PRINT
290 PRINT TAB(5);"angle";TAB(19);"radius";TAB(34);"thrust";
300 PRINT TAB(50);"safety"
310 PRINT TAB(5);"(deg)";TAB(20);L$;TAB(35);F$;TAB(50);"factor"
320 REM *****
330 FOR T4 = T1 TO T2 STEP T3
340 T = 3.1416*T4/180
350 R = H/2/SIN(T)
360 F1 = G1*H*H*W*.5
370 F2 = G1*W*R*R*(T - SIN(2*T)/2)
380 F = SQR(F1*F1 + F2*F2)
390 F3 = F/2
400 S = TO/F3
410 PRINT TAB(6);T4;TAB(17);R;TAB(32);F3;TAB(48);S
420 NEXT T4
430 PRINT"*****"
440 PRINT"do you wish to try another bearing (y/n) ";
450 INPUT P$
460 IF P$ = "y" THEN GOTO 230
470 PRINT"*****"
480 PRINT"do you wish to solve a new problem (y/n) ";
490 INPUT N$
500 IF N$ = "y" THEN GOTO 20
510 STOP
1000 L$ = "(M)"
1010 G$ = "(n/cu.m)"
1020 F$ = "(n)"
1030 RETURN
2000 L$ = "(ft)"
2010 G$ = "(lbs/cu.ft)"
2020 F$ = "(lbs)"
2030 RETURN
2040 END
```

FIGURE B.2 Radial sluice gate design program.

```
10 PRINT"this program solves for three types of pipe problems"
20 PRINT"punch the variable to be determined equal to zero"
30 PRINT"title";
40 INPUT T$
50 PRINT"units(si/fps)";
60 INPUT U$
70 IF U$ = "si" THEN GOTO 130
80 G = 32.2
90 L$ = "(ft)"
100 K$ = "(sq.ft/sec)"
110 D$ = "(cu.ft/sec)"
120 GOTO 170
130 G = 9.810001
140 L$ = "(m)"
150 K$ = "(sq.m/sec)"
160 D$ = "(cu.m/sec)"
170 PRINT"length of pipe";L$;
180 INPUT L
190 PRINT"diameter of the pipe";L$;
200 INPUT D
210 PRINT"epsilon";L$;
220 INPUT E
230 PRINT"k viscosity";K$;
240 INPUT N
250 PRINT "head loss";L$;
260 INPUT H
270 PRINT "discharge";D$;
280 INPUT Q
290 IF Q=0 THEN GOTO 410
300 IF D=0 THEN GOTO 560
310 A=3.1416*D*D/4
320 V=Q/A
330 GOSUB 740
340 H=P*L/D*V*V/(2*G)
350 PRINT
360 PRINT"           **********"
370 PRINT T$
380 PRINT"head loss";L$;"=",H
390 PRINT"           **********"
400 GOTO 700
410 A=SQR(2*G*D*H/L)
420 B=D/N
430 F=.02
440 V=A/SQR(F)
450 GOSUB 740
460 IF ABS(P-F)<=.0001 GOTO 490
470 F=P
480 GOTO 440
490 Q=3.1416*V*D*D/4
500 PRINT
510 PRINT"           **********"
520 PRINT T$
530 PRINT"discharge";D$;"=",Q
540 PRINT"           **********"
550 GOTO 700
560 A=(8*L*Q*Q)/(H*G*9.8696)
570 B=4*Q/3.1416/N
580 F=.01
590 D=(A*F)^.2
600 V=4*Q/(3.1416*D*D)
610 GOSUB 740
620 IF ABS(P-F)<=.0001 GOTO 650
630 F=P
640 GOTO 590
650 PRINT
660 PRINT"           **********"
670 PRINT T$
680 PRINT"diameter";L$;"=",D
690 PRINT"           **********"
700 PRINT"do you wish to solve another problem (y/n)";
```

```
710 INPUT P$
720 IF P$ = "y" THEN GOTO 30
730 STOP
740 R=V*D/N
750 C=LOG(E/(3.7*D)+5.74/R^.9)
760 P=1.325/(C*C)
770 RETURN
780 END
```

FIGURE B.3 Program for solving three types of pipe problems.

```
10 PRINT"This program solves for a series pipe problem."
20 PRINT"The problem must first be reduced to the form"
30 PRINT"   v**2/2g(a1*f1 + a2*f2 + ...... + b ) = c "
40 PRINT"   where v = velocity at exit(nozzle or last pipes)"
50 DIM A(20),D(20),E(20),V(20),F(20),P(20)
60 PRINT"title of the problem ";
70 INPUT T$
80 PRINT"units(si/fps) ";
90 INPUT U$
100 IF U$ = "si" THEN GOSUB 790
110 IF U$ = "fps" THEN GOSUB 840
120 PRINT"is there a nozzle at the end (y/n) ";
130 INPUT N$
140 IF N$ = "n" THEN GOTO 180
150 PRINT"diameter of nozzle ";L$;
160 INPUT D1
170 GOTO 190
180 D1 = 0!
190 PRINT"number of pipes in series ";
200 INPUT N
210 M=N
220 FOR I = 1 TO N
230 PRINT"diameter of pipe ";I;L$ ;
240 INPUT D(I)
250 PRINT"epsilon for pipe ";I;L$ ;
260 INPUT E(I)
270 NEXT I
280 FOR I = 1 TO N
290 PRINT"coefficient a";I;
300 INPUT A(I)
310 NEXT I
320 PRINT"coefficient b ";
330 INPUT B
340 PRINT"coefficient c ";
350 INPUT C
360 PRINT"k viscosity ";K$;
370 INPUT N1
380 IF D1 = 0! GOTO 410
390 M = N+1
400 D(M) = D1
410 FOR I = 1 TO N
420 F(I) = .02
430 NEXT I
440 A1 = 0!
450 FOR I = 1 TO N
460 A1 = A1+ F(I)*A(I)
470 NEXT I
480 A1 = A1+B
490 V(M) = SQR(2*G*C/A1)
500 FOR I = 1 TO N
510 V(I) = V(M)*(D(M)/D(I))^2
520 NEXT I
530 FOR I = 1 TO N
540 GOSUB 750
550 NEXT I
560 FOR I = 1 TO N
570 IF ABS(P(I)-F(I))>.0001 GOTO 600
```

```
580 NEXT I
590 GOTO 640
600 FOR I = 1 TO N
610 F(I) = P(I)
620 NEXT I
630 GOTO 440
640 Q = V(M)*3.1416*D(M)^2/4
650 PRINT"_ _ _ _ _ _ _ _ _ _ _ _ _ _ _ _ _ _ _ _ _"
660 PRINT
670 PRINT T$
680 PRINT"discharge";Q$;"  =  ";Q
690 PRINT"_ _ _ _ _ _ _ _ _ _ _ _ _ _ _ _ _ _ _ _ _"
700 PRINT
710 PRINT"do you wish to solve another problem(y/n)";
720 INPUT P$
730 IF P$ = "y" THEN GOTO 60
740 STOP
750 R = V(I)*D(I)/N1
760 H =LOG(E(I)/(3.7*D(I))+5.74/R^.9)
770 P(I) = 1.325/(H*H)
780 RETURN
790 G = 9.810001
800 L$ = "(M)"
810 K$ = "(sq.m/sec)"
820 Q$ = "(cu.m/sec)"
830 RETURN
840 G = 32.2
850 L$ = "(ft)"
860 K$ = "(sq.ft/sec)"
870 Q$ = "(cu.ft/sec)"
880 RETURN
890 END
```

FIGURE B.4 Series pipe program.

```
10 PRINT"This program solves for the distribution of discharge in"
20 PRINT"pipes connected in parallel"
30 PRINT
40 DIM D(20),L(20),E(20),Q(20),V(20),A(20)
50 PRINT"title of the problem";
60 INPUT T$
70 PRINT"units(si/fps) ";
80 INPUT U$
90 IF U$ = "si" THEN GOSUB 720
100 IF U$ ="fps" THEN GOSUB 770
110 PRINT"number of pipes connected in parllel ";
120 INPUT N
130 FOR I = 1 TO N
140 PRINT"diameter of pipe";I;L$;
150 INPUT D(I)
160 PRINT"length of pipe";I;L$;
170 INPUT L(I)
180 PRINT"epsilon of pipe";I;L$;
190 INPUT E(I)
200 NEXT I
210 PRINT"k. viscosity of the fluid ";K$;
220 INPUT N1
230 PRINT"total discharge ";Q$;
240 INPUT Q1
250 FOR I = 1 TO N
260 A(I) = 3.1416*D(I)^2/4
270 NEXT I
280 I = 1
290 Q(I) = Q1/3
300 V(I) = Q(I)/A(I)
310 GOSUB 680
320 H1 = P*L(I)/D(I)*V(I)^2/(2*G)
330 FOR I = 2 TO N
```

```
340 A1 = SQR(2*G*D(I)*H1/L(I))
350 B1 = D(I)/N1
360 F = .02
370 V(I) = A1/SQR(F)
380 GOSUB 680
390 IF ABS(P-F) <= .0001 GOTO 420
400 F = P
410 GOTO 370
420 Q(I) = V(I)*A(I)
430 NEXT I
440 Q2 = 0!
450 FOR I = 1 TO N
460 Q2 = Q2 + Q(I)
470 NEXT I
480 FOR I = 1 TO N
490 Q(I) = Q(I)/Q2*Q1
500 NEXT I
510 I = 1
520 V(I) = Q(I)/A(I)
530 GOSUB 680
540 H1 = P*L(I)/D(I)*V(I)^2/(2*G)
550 PRINT"_-_-_-_-_-_-_-_-_-_-_-_-_-_-_-_-_-_"
560 PRINT
570 PRINT T$
580 FOR I = 1 TO N
590 PRINT"discharge in pipe";I;Q$;"  =  ";Q(I)
600 NEXT I
610 PRINT"head loss in each pipe ";L$;"            =  ";H1
620 PRINT"-_-_-_-_-_-_-_-_-_-_-_-_-_-_-_-_-_-_"
630 PRINT
640 PRINT"do you wish to solve another problem(y/n) ";
650 INPUT P$
660 IF P$ = "y" THEN GOTO 50
670 STOP
680 R = V(I)*D(I)/N1
690 H = LOG(E(I)/(3.7*D(I))+5.74/R^.9)
700 P = 1.325/(H*H)
710 RETURN
720 G = 9.810001
730 L$ = "(m)"
740 K$ = "(sq.m/sec)"
750 Q$ = "(cu.m/sec)"
760 RETURN
770 G = 32.2
780 L$ = "(ft)"
790 K$ = "(sq.ft/sec)"
800 Q$ = "(cu.ft/sec)"
810 RETURN
820 END
```

FIGURE B.5 Parallel pipe program.

```
10 PRINT"This program solves branching pipe problems"
20 PRINT
30 DIM D(9),L(9),E(9),Q(9),H(9),R(9),S(9),T(9)
40 PRINT"title of the problem";
50 INPUT T$
60 PRINT "units(si/fps)";
70 INPUT U$
80 IF U$ = "si" THEN GOSUB 840
90 IF U$ = "fps" THEN GOSUB 890
100 PRINT"number of pipes ";
110 INPUT N
120 FOR I = 1 TO N
130 PRINT"diameter of pipe ";I;L$;
140 INPUT D(I)
150 PRINT"length of pipe ";I;L$;
160 INPUT L(I)
```

```
170 PRINT"epsilon for pipe ";I;L$;
180 INPUT E(I)
190 NEXT I
200 FOR I = 1 TO N
210 PRINT"elevation of reservoir ";I;L$;
220 INPUT H(I)
230 NEXT I
240 PRINT"k. viscosity of the fluid ";K$;
250 INPUT N1
260 FOR M = 1 TO 2
270 T(M) = H(M)-(H(M)-H(M+1))/2
280 GOSUB 620
290 S(M) = 0
300 FOR I = 1 TO N
310 S(M) =S(M)+Q(I)
320 NEXT I
330 NEXT M
340 B1 = (T(2)-T(1))/(S(2)-S(1))
350 A1 = T(1) - B1*S(1)
360 M = 1
370 T(1) = A1
380 GOSUB 620
390 S(M) = 0
400 FOR I = 1 TO N
410 S(M) = S(M) + Q(I)
420 NEXT I
430 IF ABS(S(M)) <= .01 GOTO 470
440 IF S(M) < 0 THEN A1 = A1 - .1
450 IF S(M) > 0 THEN A1 = A1 + .1
460 GOTO 370
470 PRINT
480 PRINT"-_-_-_-_-_-_-_-_-_-_-_-_-_-_-_-_-_-_-_-_"
490 PRINT T$
500 FOR I = 1 TO N
510 PRINT" Q for branch ";I,Q(I);Q$
520 NEXT I
530 PRINT" total head at junction          ";A1;L$
540 PRINT
550 PRINT"NOTE: negative Q means water flowing"
560 PRINT"       away from the junction"
570 PRINT"-_-_-_-_-_-_-_-_-_-_-_-_-_-_-_-_-_-_-_-_"
580 PRINT" do you wish to solve another problem (y/n) ";
590 INPUT P$
600 IF P$ = "y" THEN GO 23
610 STOP
620 FOR I = 1 TO N
630 R(I) = H(I) - T(M)
640 S1 = 1
650 IF R(I) < 0 GOTO 670
660 GOTO 690
670 R(I) = -R(I)
680 S1 = -1
690 A = SQR(2*G*D(I)*R(I)/L(I))
700 B = D(I)/N1
710 F = .02
720 V = A/SQR(F)
730 GOSUB 800
740 IF ABS(P-F)<=.0001 GOTO 770
750 F = P
760 GOTO 720
770 Q(I) = 3.1416*V*D(I)^2/4*S1
780 NEXT I
790 RETURN
800 R1 = V*B
810 C = LOG(E(I)/(3.7*D(I)) + 5.74/R1^.9)
820 P = 1.328/(C*C)
830 RETURN
840 G = 9.810001
850 L$ = "(m)"
860 K$ = "(sq.m/sec)"
```

```
870 Q$ = "(cu.m/sec)"
880 RETURN
890 G = 32.2
900 L$ = "(ft)"
910 K$ = "(sq.ft/sec)"
920 Q$ = "(cu.ft/sec)"
930 RETURN
940 END
```

FIGURE B.6 Branching pipe program.

```
10 DATA 5,2
20 DATA 50,0.25,130,10
30 DATA 100,0.5,130,90
40 DATA 75,0.5,130,30
50 DATA 100,0.5,130,30
60 DATA 50,0.25,130,50
70 DATA 3,1,3,-2
80 DATA 3,4,-5,-3
1000 PRINT"this program solves for a pipe network problem"
1010 PRINT
1020 PRINT"title of the problem";
1030 INPUT T$
1040 PRINT"system of units (si/fps)";
1050 INPUT U$
1060 IF U$ = "si" THEN GOSUB 1770
1070 IF U$ = "fps" THEN GOSUB 1800
1080 PRINT"number of iterations allowed";
1090 INPUT K
1100 PRINT"discharge tolerance";
1110 INPUT D1
1120 READ N1,N2
1130 DIM L(90),D(90),C(90),Q(90),R(90),A(90,10),N(10)
1140 FOR I = 1 TO N1
1150 READ L(I),D(I),C(I),Q(I)
1160 NEXT I
1170 REM
1180 REM ***read number of pipes in each loop***
1190 REM
1200 FOR I = 1 TO N2
1210 READ N(I)
1220 FOR J = 1 TO N(I)
1230 READ A(I,J)
1240 NEXT J
1250 NEXT I
1260 REM
1270 REM***compute r for each pipe***
1280 REM
1290 FOR I = 1 TO N1
1300 R(I) = C1*L(I)/C(I)^1.852/D(I)^4.8704
1310 NEXT I
1320 REM
1330 REM ***compute correction using hardy-cross method***
1340 REM
1350 K1 = 0
1360 K2 = 0
1370 FOR I = 1 TO N2
1380 N3 = 0
1390 N4 = 0
1400 FOR J = 1 TO N(I)
1410 A2 = 1!
1420 IF A(I,J) <0 THEN A2 = -1
1430 N5 = ABS(A(I,J))
1440 N3 = N3+A2*R(N5)*Q(N5)*ABS(Q(N5))^.852
1450 N4 = N4+R(N5)*1.852*ABS(Q(N5))^.852
1460 NEXT J
1470 N6 = -N3/N4
1480 IF ABS(N6) > K1 THEN K1 = ABS(N6)
```

```
1490 FOR J = 1 TO N(I)
1500 A2 = 1!
1510 IF A(I,J) < 0 THEN A2 = -1!
1520 N5 = ABS(A(I,J))
1530 Q(N5) = A2*Q(N5)+N6
1540 Q(N5) = A2*Q(N5)
1550 NEXT J
1560 NEXT I
1570 K2 = K2+1
1580 IF K1 <= D1 THEN 1620
1590 IF K2 = K THEN 1620
1600 K1 = 0
1610 GOTO 1370
1620 PRINT
1630 PRINT"****  ";T$;"  ****"
1640 PRINT
1650 PRINT TAB(10);"no. of pipe";TAB(30);"discharge"
1660 PRINT TAB(29);Q$
1670 FOR I = 1 TO N1
1680 PRINT TAB(15);I;TAB(30);Q(I)
1690 NEXT I
1700 PRINT"largest delq = ";K1;"no. of iterations = ";K2
1710 PRINT
1720 PRINT"NOTE:"
1730 PRINT"Negative discharge means that the direction"
1740 PRINT"of flow is opposite to that assumed initially."
1750 PRINT
1760 STOP
1770 C1 = 10.675
1780 Q$ = "(cu.m/sec)"
1790 RETURN
1800 C1 = 4.727
1810 Q$ = "(cu.ft/sec)"
1820 RETURN
1830 END
```

FIGURE B.7 Pipe network program.

```
10 PRINT "open channel program"
20 PRINT"title";
30 INPUT T$
40 PRINT"system of unit(si/fps)";
50 INPUT U$
60 IF U$ = "fps" THEN GOSUB 1610
70 IF U$ = "si" THEN GOSUB 1680
80 PRINT"use the following code to describe the"
90 PRINT"channel type"
100 PRINT" 1 for triangular"
110 PRINT" 2 for trapezoidal"
120 PRINT" 3 for rectangular"
130 PRINT" 4 for circular"
140 PRINT
150 F1 = 2!/3!
160 N1 = 1
170 PRINT"channel type";
180 INPUT H
190 IF H = 4 THEN GOTO 380
200 IF H = 1 THEN D$ = "triangular"
210 IF H = 2 THEN D$ = "trapezoidal"
220 IF H = 3 THEN D$ = "rectangular'
230 PRINT"bottom width (enter 0 for type 1) ";L$;
240 INPUT B
250 PRINT"left side slope (--:1) (enter 0 for type 3)";
260 INPUT Z1
270 PRINT"right side slope (--:1) (enter 0 for type 3)";
280 INPUT Z2
290 PRINT"bed slope";
300 INPUT S
```

```
310 PRINT"mannings n";
320 INPUT N
330 PRINT "normal depth ";L$;
340 INPUT Y
350 PRINT"discharge ";Q$;
360 INPUT Q
370 GOTO 490
380 D$ = "circular"
390 PRINT"diameter ";L$;
400 INPUT D
410 PRINT"bed slope";
420 INPUT S
430 PRINT"mannings n ";
440 INPUT N
450 PRINT"normal depth ";L$;
460 INPUT Y
470 PRINT"discharge ";Q$;
480 INPUT Q
490 IF Y = 0 GOTO 670
500 IF H = 4 GOTO 590
510 GOSUB 980
520 GOSUB 1260
530 N1 = 2
540 GOSUB 1440
550 P = B+Y*(SQR(1+Z1*Z1) + SQR(1+Z2*Z2))
560 S1 = (V1*N/(C*(A/P)^F1))^2
570 GOSUB 1260
580 GOTO 1750
590 GOSUB 1100
600 P = 2*R*T1
610 S1 = (V1*N/(C*(A/P)^F1))^2
620 GOSUB 1260
630 N1 = 2
640 GOSUB 1440
650 GOSUB 1260
660 GOTO 1750
670 IF H < 4 THEN GOTO 900
680 Y = .95*D
690 GOSUB 1100
700 IF (Q-Q1) < 0 THEN GOTO 750
710 PRINT"normal open channel flow is not possible for the"
720 PRINT"given conditions. discharge is too large. pipe will"
730 PRINT"flow full."
740 GOTO 770
750 GOSUB 1440
760 GOSUB 1260
770 N1 = 2
780 Y = .95*D
790 GOSUB 1100
800 IF (Q-Q1) < 0 THEN GOTO  850
810 PRINT"critical open channel flow is not possible for the"
820 PRINT"given conditions. discharge is too large. pipe will"
830 PRINT"flow full."
840 GOSUB 1750
850 GOSUB 1440
860 P = 2*R*T1
870 S1 = (V1*N/(C*(A/P)^F1))^2
880 GOSUB 1260
890 GOTO 1750
900 GOSUB 1440
910 GOSUB 1260
920 N1 = 2
930 GOSUB 1440
940 P = B + Y*(SQR(1+Z1*Z1) + SQR(1+Z2*Z2))
950 S1 = (V1*N/(C*(A/P)^F1))^2
960 GOSUB 1260
970 GOTO 1750
980 A = Y*(B+Y/2!*(Z1+Z2))
990 IF N1 = 2 THEN GOTO 1060
1000 P = B+Y*(SQR(1+Z1*Z1)+SQR(1+Z2*Z2))
```

```
1010 V1 = C/N*(A/P)^F1*SQR(S)
1020 Q1 = A*V1
1030 T = B+Y*(Z1+Z2)
1040 F2 = V1/SQR(G*A/T)
1050 IF N1 = 1 THEN GOTO 1090
1060 T = B + Y*(Z1+Z2)
1070 V1 = SQR(G*A/T)
1080 Q1 = A*V1
1090 RETURN
1100 R = D/2
1110 IF Y > D THEN Y = .95*D
1120 T1 = ATN(SQR(R^2-(R-Y)*(R-Y))/(R-Y))
1130 IF T1 < 0 THEN T1 = 3.141593 + T1
1140 A = R*R*(T1-SIN(2*T1)/2)
1150 IF N1 = 2 THEN GOTO 1220
1160 P = 2*R*T1
1170 V1 = C/N*(A/P)^F1*SQR(S)
1180 Q1 = A*V1
1190 T = 2*SQR(Y*(D-Y))
1200 F2 = V1/SQR(G*A/T)
1210 IF N1 = 1 THEN GOTO 1250
1220 T = 2*SQR(Y*(D-Y))
1230 V1 = SQR(G*A/T)
1240 Q1 = A*V1
1250 RETURN
1260 IF N1 = 1 THEN N$ = "normal"
1270 IF N1 = 2 THEN N$ = "critical"
1280 IF N1 = 2 THEN GOTO 1330
1290 PRINT
1300 PRINT
1310 PRINT T$
1320 PRINT D$;" channel problem "
1330 PRINT"***** ";N$;" flow calculations ";"*****"
1340 PRINT"rate of flow ";Q$;TAB(30);"=";TAB(33);Q1
1350 PRINT N$;" velocity ";V$;TAB(30);"=";TAB(33);V1
1360 PRINT N$;" depth ";L$;TAB(30);"=";TAB(33);Y
1370 PRINT N$;" top width ";L$;TAB(30);"=";TAB(33);T
1380 PRINT N$;" area of flow ";A$;TAB(30);"=";TAB(33);A
1390 IF N1 = 2 THEN GOTO 1420
1400 PRINT"froud number ";TAB(30);"=";TAB(33);F2
1410 GOTO 1430
1420 PRINT"critical slope ";TAB(30);"=";TAB(33);S1
1430 RETURN
1440 Y =1
1450 Y1 = 0
1460 IF Q = 0 THEN Q = Q1
1470 IF H = 4 THEN GOSUB 1100
1480 IF H < 4 THEN GOSUB 980
1490 IF (Q-Q1) > 0 THEN GOTO 1510
1500 GOTO 1530
1510 Y = 2*Y
1520 GOTO 1470
1530 IF ABS(Q-Q1) <= .001 GOTO 1600
1540 IF (Q-Q1) < 0 THEN Y2 = Y
1550 IF (Q-Q1) > 0 THEN Y1 = Y
1560 Y = (Y1+Y2)/2
1570 IF H = 4 THEN GOSUB 1100
1580 IF H < 4 THEN GOSUB 980
1590 GOTO 1530
1600 RETURN
1610 C = 1.486
1620 G = 32.2
1630 L$ = "(ft)"
1640 Q$ = "(cu.ft/sec)"
1650 V$ = "(ft/sec)"
1660 A$ = "(sq.ft)"
1670 RETURN
1680 C = 1
1690 G = 9.810001
1700 L$ = "(m)"
```

```
1710 Q$ = "(cu.m/sec)"
1720 V$ = "(m/sec)"
1730 A$ = "(sq.m)"
1740 RETURN
1750 PRINT
1760 PRINT"do you wish to solve another problem (y/n)";
1770 INPUT P$
1780 IF P$ = "y" THEN GOTO 20
1790 STOP
1800 END
```

FIGURE B.8 Open channel program.

```
10 PRINT"This program computes for water surface"
20 PRINT"profile in rectangular,trapezoidal or"
30 PRINT"triangular channel"
40 DEF FNA(Y1) = Y1*(B+Y1/2*(Z1+Z2))
50 DEF FNB(Y1) = B+Y1*(SQR(1+Z1*Z1)+SQR(1+Z2*Z2))
60 DEF FNC(Y1) = 1-Q*Q*(B+Y1*(Z1+Z2))/(G*FNA(Y1)^3)
70 DEF FND(Y1) = 1-Q*Q*((N/C1)^2/SO)/(FNA(Y1)^3.333/FNB(Y1)^1.333)
80 DEF FNE(Y1) = FNC(Y1)/(FND(Y1)*SO)
90 DEF FNF(Y1) = G1*(Y1*Y1*(B*.5+Y1/6*(Z1+Z2))+Q*Q/(G*FNA(Y1)))
100 DEF FNG(Y1) = Y1+Q*Q/(2*G*FNA(Y1)^2)
110 PRINT"title";
120 INPUT T$
130 PRINT"system of units (si/fps) ";
140 INPUT U$
150 IF U$ = "si" THEN GOSUB 1390
160 IF U$ = "fps" THEN GOSUB 1460
170 PRINT"length of reach ";L$;
180 INPUT X
190 PRINT"channel width ";L$;
200 INPUT B
210 PRINT"left side slope(---:1) ";
220 INPUT Z1
230 PRINT"right side slope(---:1) ";
240 INPUT Z2
250 PRINT"mannings n ";
260 INPUT N
270 PRINT"channel slope";
280 INPUT SO
290 PRINT"discharge ";Q$;
300 INPUT Q
310 PRINT"depth of flow at control section ";L$;
320 INPUT Y
330 PRINT"number of intervals ";
340 INPUT K
350 PRINT"              ************"
360 PRINT
370 PRINT T$
380 U = 30
390 D = 0
400 Y2 = 15
410 FOR I = 1 TO 15
420 IF FNC(Y2)<0 THEN GOTO 450
430 IF FNC(Y2) = 0 THEN GOTO 500
440 IF FNC(Y2)>0 THEN GOTO 470
450 D = Y2
460 GOTO 480
470 U = Y2
480 Y2 = (U+D)/2
490 NEXT I
500 IF Y = 0 THEN Y = Y2
510 IF SO<=0 THEN GOTO 670
520 U = 40
530 D = 0
540 Y3 = 20
550 FOR I = 1 TO 15
```

```
560 IF FND(Y3)<0 THEN GOTO 590
570 IF FND(Y3) = 0 THEN GOTO 640
580 IF FND(Y3)>0 THEN GOTO 610
590 D = Y3
600 GOTO 620
610 U = Y3
620 Y3 = (U+D)/2
630 NEXT I
640 PRINT"normal depth =";Y3;L$
650 PRINT"critical depth =";Y2;L$
660 GOTO 690
670 Y3 = 3*Y2
680 PRINT"critical depth =";Y2;L$
690 IF Y3<Y2 THEN GOTO 910
700 REM******
710 REM ***mild,adverse or horizontal channel, yn>yc **
720 REM ******
730 IF Y<Y2 THEN GOTO 840
740 REM ******
750 REM **subcritical flow, y>y2 **
760 REM ******
770 S = -1
780 Y4 = (Y-Y3)*.998/K
790 PRINT"control is downstream, depth =";Y;L$
800 GOTO 1050
810 REM ******
820 REM **supercritical flow, y<y2 **
830 REM ******
840 S = 1
850 Y4 = (Y2-Y)/K
860 PRINT"control is upstream, depth =";Y;L$
870 GOTO 1050
880 REM ******
890 REM **steep channel, yn<yc **
900 REM ******
910 IF Y<=Y2 THEN GOTO 1010
920 REM ******
930 REM ** subcritical flow, y>yc **
940 REM ******
950 S = -1
960 Y4 = (Y-Y2)/K
970 PRINT"control is downstream, depth =";Y;L$
980 GOTO 1050
990 REM ******
1000 REM **supercritical flow, y<yc **
1010 S = 1
1020 K = K*2
1030 Y4 = (Y3-Y)*.998/K
1040 PRINT"control is upstream, depth =";Y;L$
1050 S1 = 0
1060 Y5 = Y
1070 E = FNG(Y5)
1080 F = FNF(Y5)
1090 PRINT
1100 PRINT"distance           depth         sp. energy         sp. force"
1110 PRINT TAB(3);L$;TAB(20);L$;TAB(35);L$;TAB(51);F$
1120 PRINT TAB(1);S1;TAB(17);Y5;TAB(32);E;TAB(47);F
1130 REM ******
1140 REM ** water surface profile calculations **
1150 REM ******
1160 FOR I = 1 TO K STEP 2
1170 Y6 = Y + S*Y4*(I+1)
1180 X1 = Y4*(FNE(Y5)+FNE(Y6)+4*FNE(Y+S*I*Y4))/3
1190 S1 = S1+X1
1200 IF S1>X THEN GOTO 1300
1210 Y5 = Y6
1220 E = FNG(Y5)
1230 F = FNF(Y5)
1240 IF I = K-1 THEN GOTO 1260
1250 GOTO 1270
```

```
1260 IF S1<0 THEN S1 = X
1270 PRINT TAB(1);S1;TAB(17);Y5;TAB(32);E;TAB(47);F
1280 NEXT I
1290 GOTO 1340
1300 Y5 = Y6-S*2*Y4*(S1-X)/X1
1310 E = FNG(Y5)
1320 F = FNF(Y5)
1330 PRINT TAB(1);X;TAB(17);Y5;TAB(32);E;TAB(47);F
1340 PRINT"**************"
1350 PRINT"do you wish to continue (y/n) ";
1360 INPUT C$
1370 IF C$ = "y" THEN GOTO 110
1380 STOP
1390 G = 9.810001
1400 G1 = 9810
1410 C1 = 1!
1420 L$ = "(m)"
1430 Q$ = "(cu.m/sec)"
1440 F$ = "(N)"
1450 RETURN
1460 G = 32.2
1470 G1 = 62.4
1480 C1 = 1.486
1490 L$ = "(ft)"
1500 Q$ = "cu.ft/sec)"
1510 F$ = "(lbs)"
1520 RETURN
1530 END
```

FIGURE B.9 Water surface profile program.

REFERENCES

1. Addison, H., *Centrifugal and Other Rotodynamic Pumps*, Chapman and Hall, London, 1966.
2. Allen, J., *Scale Models in Hydraulic Engineering*, Longmans, London, 1952.
3. ASME, *Flow Meters*, 6th ed., ASME, New York, 1971.
4. Bernoulli, D., *Hydrodynamics*, Trans. Thomas Carmody and Helmut Kobus, Dover, New York, 1968.
5. Birkhoff, G. and Zarantonello, E. H., *Jets, Wakes and Cavities*, Academic Press, New York, 1957.
6. Blausius, H., *Das Ahnlichkeitsgesetz bei Reibungsvorgangen in Flussigkeiten*, Ver. Dtsch. Ing. Forschungsh., Vol. 131, 1913.
7. Brown, R. C., *The Fundamental Concepts Concerning Surface Tension and Capillarity*, Proc. Phys. Soc., Vol. 59, pp. 429–448, 1947.
8. Brown, R. C., *Mechanics and Properties of Matter*, Longmans, Green, London, 1961.
9. Brush, S. G., *Theories of Liquid Viscosity*, Chem. Rev., Vol. 62, pp. 513–548, 1962.
10. Buckingham, E., *Model Experiments and the Form of Emperical Equations*, Trans, ASME, Vol. 37, pp. 263–296, 1915.
11. Burdon, R. S., *Surface Tension and the Spreading of Liquids*, 2nd ed., Camb. Univ. Press, Cambridge, 1949.
12. Cameron, A. *et al.*, *Principles of Lubrication*, Longmans, London, 1979.
13. Carter, R. W., Einstein, H. A., Hinds, J., Powell, R. W. and Silberman, E., *Friction Factors in Open Channels*, J. Hydrau. Div., ASCE, Vol. 89, No. HY2, pp. 97–143, 1963.
14. Cheers, F., *Elements of Compressible Flow*, Wiley, New York, 1963.
15. Chow, V. T., *Open Channel Hydraulics*, McGraw-Hill, New York, 1959.
16. Colebrook, C. F., and White, C. M., *The Reduction of Carrying Capacity of Pipes with Age*, J. Inst. Civ. Engrs., London, 1937.
17. Colebrook, C. F., *Turbulent Flow in Pipes with Particular Reference to the Transition Region Between the Smooth and Rough Pipe Laws*, J. Inst. Civ. Engrs., London, Vol. 11, pp. 133–156, 1939.
18. Cross, H., *Analysis of Flow in Networks of Conduits or Conductors*, Bull. Illinois Univ. Engng. exp. stn. 286, 1936.
19. Church, A. H., *Centrifugal Pumps and Blowers*, Wiley, New York, 1944.
20. Daniels, F., and Alberty, R. A., *Physical Chemistry*, 5th ed., Wiley, New York, 1979.
21. Duncan, W. J., *The Principles of the Control and Stability of Aircraft*, Camb. Univ. Press, Cambridge, 1952.
22. Duncan, W. J., Thom, A. S., and Young, J. D., *An Elementary Treatise on the Mechanics of Fluids*, Arnold, London, 1970.
23. Durand, W. F., (ed.), *Aerodynamic Theory*, Vol. 1, Dover, New York, 1963.
24. Foster, H. A., *Construction of the flow net for hydraulic design*, Trans. ASCE, Vol. 110, pp. 1237–1251, 1945.
25. Goldstein, S., (ed.), *Modern Developments in Fluid Dynamics*, Dover, New York, 1965.

26. Griffith, W. C. and Bleakney, W., *Shock Waves in Gases*, Am. J. Phys., Vol. 22, pp. 597–612, 1954.

27. Hayes, W. D., *Gas-Dynamic Discontinuties*, Princeton University Press, 1960.

28. Hayward, A. T. J., *Flowmeters: A Basic Guide and Source-Book for Users*, Wiley, New York, 1979.

29. Henderson, F. M., *Open-channel Flow*, MacMillan, New York, 1966.

30. Hun Saker, J. C. and Rightmire, B. G., *Engineering Applications of Fluid Mechanics*, McGraw-Hill, New York, 1962.

31. Hydraulic Institute, *Engineering Data Book*, 1st ed. Hydraulic Inst., Cleveland, Ohio, 1979.

32. Ispen, D. C., *Units, Dimensions, and Dimensionless Numbers*, McGraw-Hill, New York, 1960.

33. John, J. E. A., *Gas Dynamics*, Allyn and Bacon, Boston, 1969.

34. Karman, T. von *Turbulence and Skin Friction*, J. Aeronaut. Sci., Vol. 1, No. 1, pp. 1, 1934.

35. Karman, T. von, *Aerodynamics, Selected Topics in the Light of Their Historical Development*, Cornell Univ. Press, Ithaca, New York, 1954.

35. Karman, T. von, *Aerodynamics, Selected Topics in the Light of Their Historical Development*, Cornell Univ. Press, Ithaca, New York, 1954.

36. Keith, H. D., *Simplified Theory of Ship Waves*, Am. J. Phys., Vol. 25, pp. 466–474, 1957.

37. Kline, S. J., *Similitude and Approximation Theory*, McGraw-Hill, New York, 1965.

38. Knapp, R. T., Daily, J. W. and Hammitt, F. G., *Cavitation*, McGraw-Hill, New York, 1970.

39. Lamb, H., *Hydrodynamics*, Camb. Univ. Press, Cambridge, 1932.

40. Langhaar, H. L., *Steady Flow in the Transition Length of a Straight Tube*, J. Appl. Mech., Vol. 9, pp. 55–58, 1942.

41. Langhaar, H. L., *Dimensional Analysis and Theory of Models*, Wiley, New York, 1951.

42. Lenz, A. T., *Viscosity and Surface Tension Effects of V Notch Weis Coefficients*, Trans. ASCE Vol. 108, pp. 759–802, 1943.

43. Linford, A., *Flow Measurement and Meters*, Spon, London, 1961.

44. Metzner, A. B., *Flow of Non-Newtonian Fluids*, Sec. 7, in Handbook of Fluid Dynamics, edited by V. L. Streeter, McGraw-Hill, New York, 1961.

45. Miller, R. W., *Flow Measurement Engineering Handbook*, McGraw-Hill, New York, 1983.

46. Moody, L. F., *Friction Factors for Pipe Flow*, Trans. ASME, Vol. 66, pp. 671-684.

47. Moody, L. F. and Zowski, T., *Hydraulic Machinery*, Section 26 in Handbook of Applied Hydraulics: edited by Davis and Sorensen, McGraw-Hill, New York, 1969.

48. Moss, S. A., Smith, C. W. and Foote, W. R., *Energy Transfer between a Fluid and a rotor for pump and turbine machinery*, Trans. ASME Vol. 64, pp. 567–597, 1942.
49. Nikuradse, J., *Gesetzmassigkeiten der turbulenten Stromung in glatten Rohren*, Ver. Dtsch. Ing. Forschungsh., Vol. 356, 1932.
50. Nikuradse, J., *Stromungsgesetze in Rauhen Rohren*, Ver. Dtsch. Ing. Forschungsh., Vol. 361, 1933.
51. Pinkus, O. and Sternlicht, B., *Theory of Hydrodynamic Lubrication*, McGraw-Hill, New York, 1961.
52. Prandtl, L., *Uber Flussigkeitsbewegung bei Sehr Kleiner Reibung*, Verh. III Int. Math-Kongr., Heidelb, 1904.
53. Prandtl, L., *Uber den Reibungswiderstand Stromender Luft*, Result. Aerodyn. Test Inst. Goett., III Lieferung, 1927.
54. Prandtl, L., and Schlichting, H., *Das Widerstandsgesetz Rauher Platten*, Werft, Reederli, Hafen, pp. 1, 1934.
55. Raudkivi, A. J., and Callander, R. A., *Advanced Fluid Mechanics*, Arnold, London, 1975.
56. Reid, R. C., Prausnitz, J. M., and Sherwood, T. K., *The Properties of Gases and Liquids*, 3rd ed., McGraw-Hill, New York, 1977.
57. Reynolds, O., *An Experimental Investigation of the Circumstances Which Determine Whether the Motion of Water Shall Be Direct or Sinuous, and of the Laws of Resistance in Parallel Channels*, Trans. R. Soc. Lond., Vol 174, 1983.
58. Rheingans, W. J., *Selecting Materials to Avoid Cavitation Damage*, Mater. Des. Eng., pp. 102–106, 1958.
59. Robertson, James, *Hydrodynamics in Theory and Application*, Prentice Hall, N. J., 1965.
60. Rosemary, D. R., *Fluid Flow Measurement: A Bibliography*, BHRA Fluid Engineering, Cranfield, Bedford, England, 1972.
61. Rosenhead, L. (editor), *Laminar Boundary Layers*, Clarendon Press, Oxford, 1963.
62. Roshko, A., *Experiments on the flow past a Circulation Cylinder at very high Reynolds numbers*, J. Fluid Mech., Vol. 10. pp. 345–356, 1961.
63. Rotty, R. M., *Introduction to Gas Dynamics*, Wiley, New York, 1962.
64. Rouse, H., and Ince, S., *History of Hydraulics*, Iowa Inst. of Hydr. Research, Iowa State Univ., 1957.
65. Rouse, H., *Fluid Mechanics for Hydraulic Engineers*, pp. 212–216, McGraw-Hill, New York, 1938.
66. Schlichting, H., *Boundary Layer Theory*, McGraw-Hill, New York, 1978.
67. Shames, I. H., *Mechanics of Fluids*, McGraw-Hill, New York, 1962.
68. Shamir, U., and Howard, C. D. D., *Water Distribution Systems Analysis*, J. Hydraul. Div., ASCE, Vol. 94, No. HY1, pp. 219–234, 1968.
69. Shapiro, A. H., *The Dynamics and Thermodynamics of Compressible Fluid Flow*, Ronald Press, New York, 1953.

70. Shepard, D. G., *Principles of Turbomachinery*, MacMillan, New York, 1961.
71. Smith, A. M. O. and Clutter, D. W., *The Smallest Height of Roughness Capable of Affecting Boundary-layer Transition*, J. Aero Space Sci., Vol. 26, pp. 229–245, 1959.
72. Stepanoff, A. J., *Centrifugal and Axial Flow Pumps*, Wiley, New York, 1957.
73. Stoner, M. A., *A New Way to Design Natural Gas System*, Pipe Line Ind., Vol. 32, No. 2, pp. 38–42, 1970.
74. Streeter, V. L., *The Kinetic Energy and Momentum Correction Factors For Pipes and Open Channels of Great Widths*, Civ. Eng. N.Y., Vol. 12, No. 4, pp. 212–213, 1942.
75. Sutton, G. W., *A. Photoelastic Study of Strain Waves Caused by Cavitation*, J. Appl. Mech., Vol. 24, No. 3, pp. 340–348, 1957.
76. Swamee, P. K. and Jain, A. K., *Explicit Equations for Pipe-flow Problems*, J. Hydr. Div., Proc. ASCE, pp. 657–664, May, 1976.
77. Tennekes, H. and Lumley, J. L., *A First Course in Turbulence*, The MIT Press, Cambridge, Mass., 1972.
78. Tipei, N., *Theory of Lubrication*, Stanford Univ. Press, California, 1962.
79. Walker, J. E., Whan, G. A. and Rothfus, R. R., *Fluid Friction in Non-circular Ducts*, J. Am. Inst. of Ch. Engrs., No. 3, pp. 484–489, 1957.
80. Washburn, E. W., (ed.), *International Critical Tables of Numerical Data*, McGraw-Hill, New York, 1929.
81. Weast, R. C. (ed.), *Handbook of Chemistry and Physics*, 59th ed., Chemical Rubber Company, 1978–79.
82. Weisbach, J., *Die Experimental-Hydraulic*, pp. 133, Englehardt, Freiburg, 1855.
83. Wilkinson, W. L., *Non-Newtonian Fluids*, Pergamon Press, New York, 1960.
84. Wislicenus, G. F., *Fluid Mechanics of Turbomachinery*, McGraw-Hill, New York, 1947.
85. Wislicenus, G. F., *Fluid Mechanics of Turbomachinery*, Dover, New York, 1965.
86. Wood, D. J., and Rayes, A. G., *Reliability of Algorithms for Pipe Network Analysis*, J. Hydraul. Div., ASCE, Vol. 107, No. HY10, pp. 1145–1161, 1981.

FILMS ON FLUID MECHANICS

1. Abernathy, F. H., *Fundamentals of Boundary Layers*, 24 min., Encyclopedia Britannica Films, Chicago, Illinois.
2. Bryson, A. E., *Waves in Fluids*, 33 min., EBF, Chicago, Illinois.
3. Eisenberg, P., *Cavitation*, 31 min., EBF, Chicago, Illinois.
4. Fultz, D., *Rotating Flows*, 29 min., EBF, Chicago, Illinois.

5. Hazen, D. C., *Boundary Layer Control*, 25 min., EBF, Chicago, Illinois.
6. Kline, S. J., *Flow Visualization*, 31 min., EBF, Chicago, Illinois.
7. Lighthill, M. J. and Ffowcs-Williams, J. E., *Aerodynamic Generation of Sound*, 44 min., EBF, Chicago, Illinois.
8. Lumley, J. L., *Eulerian and Lagrangian Description in Fluid Mechanics*, 27 min., EBF, Chicago, Illinois.
9. Rouse, H., *Introduction to the Study of Fluid Motion*, 24 min., University of Iowa, Iowa City, Iowa.
10. Rouse, H., *Fundamental Principles of Flow*, 23 min., University of Iowa, Iowa City, Iowa.
11. Rouse, H., *Fluid Motion in a Gravitational Field*, 23 min., University of Iowa, Iowa City, Iowa.
12. Rouse, H., *Characteristics of Laminar Turbulent Flow*, 26 min., University of Iowa, Iowa City, Iowa.
13. Rouse, H. *Form Drag, Lift and Propulsion*, 24 min., University of Iowa, Iowa City, Iowa.
14. Rouse, H., *Effects of Fluid Compressibility*, 17 min., University of Iowa, Iowa City, Iowa.
15. Saint Anthony Falls Hydraulics Lab., *Some Phenomenon of Open-Channel Flow*, 33 min., University of Minnesota, Minneapolis, Minn.
16. Saint Anthony Falls Hydraulics Lab., *Fluid Mechanics—The Boundary Layer*, 30 min., University of Minnesota, Minneapolis, Minn.
17. Shapiro, A. H., *Pressure Fields and Fluid Acceleration*, 27 min., EBF, Chicago, Illinois.
18. Shapiro, A. H., *The Fluid Dynamics of Drag*, 4 parts, 118 min., EBF, Chicago, Illinois.
19. Shapiro, A. H., *Vorticity*, 44 min., EBF, Chicago, Illinois.
20. Shell Film Library, *Approaching the Speed of Sound*, 20 min., Shell, Indianapolis, Ind.
21. Shell Film Library, *Beyond the Speed of Sound*, 20 min., Shell, Indianapolis, Ind.
22. Shell Film Library, *How an Airplane Flies*, 50 min., Shell, Indianapolis, Ind.
23. Stewart, R. W., *Turbulence*, 29 min., EBF, Chicago, Illinois.
24. Taylor, Sir G. I., *Low-Reynolds-Number Flows*, 33 min., EBF, Chicago, Illinois.
25. Trefethen, L. M., *Surface Tension in Fluid Mechanics*, 29 min., EBF, Chicago, Illinois.

INDEX